ABC Transporters in Microorganisms

Research, Innovation and Value as Targets against Drug Resistance

Edited by

Alicia Ponte-Sucre

Laboratorio de Fisiología Molecular
Instituto de Medicina Experimental
Facultad de Medicina
Universidad Central de Venezuela
Caracas
Venezuela

Caister Academic Press

Caister Academic Press
Norfolk, UK

www.caister.com

British Library Cataloguing-in-Publication Data
A catalogue record for this book is available from the British Library

ISBN: 978-1-904455-49-3

Printed and bound in Great Britain

Contents

Contributors

Suresh V. Ambudkar
Laboratory of Cell Biology
Center for Cancer Research
National Cancer Institute, NIH
Bethesda, MD
USA

ambudkar@mail.nih.gov

Patrick J. Bakkes
Department of Molecular Microbiology
Groningen Biomolecular Sciences
and
Biotechnology Institute and Zernike Institute for Advanced Materials
University of Groningen
Haren
The Netherlands

p.j.bakkes@rug.nl

Emilia Diaz
Laboratorio de Fisiología Molecular
Instituto de Medicina Experimental
Facultad de Medicina
Universidad Central de Venezuela
Caracas
Venezuela

emilid68@gmail.com

James M. Dorrian
School of Biomedical Sciences
Queen's Medical Centre
University of Nottingham
Nottingham
UK

mbxjd1@nottingham.ac.uk

Arnold J.M. Driessen
Department of Molecular Microbiology
Groningen Biomolecular Sciences
and
Biotechnology Institute and Zernike
Institute for Advanced Materials
University of Groningen
Haren
The Netherlands

a.j.m.driessen@biol.rug.nl

Daniel A. P. Gutmann
Department of Pharmacology
University of Cambridge
Cambridge
UK

dg358@cam.ac.uk

I. Barry Holland
Institut de Génétique et Microbiologie
Université de Paris
Orsay
France

barry.holland@igmors.u-psud.fr

Thorsten Jumpertz
Institute of Biochemistry
Heinrich Heine University
Düsseldorf
Germany

thorsten.jumpertz@uni-duesseldorf.de

Ian D. Kerr
School of Biomedical Sciences
Queen's Medical Centre
University of Nottingham
Nottingham
UK

Ian.Kerr@nottingham.ac.uk

Wil N. Konings
Department of Molecular Microbiology
Groningen Biomolecular Sciences
and
Biotechnology Institute
University of Groningen
Haren
The Netherlands

w.n.konings@biol.rug.nl

Herman Lage
Charité Campus Mitte
Institute of Pathology
Berlin
Germany

hermann.lage@charite.de

Danielle Légaré
Centre de Recherche en Infectiologie et
Division de Microbiologie
Faculté de Médecine
Université Laval
Ste-Foy, QC
Canada

danielle.legare@crchul.ulaval.ca

Philippe Leprohon
Centre de Recherche en Infectiologie et
Division de Microbiologie
Faculté de Médecine
Université Laval
Ste-Foy, QC
Canada

philippe.leprohon@crchul.ulaval.ca

Eve Orlandi-Pradines
Unité de Recherche en Biologie et Epidémiologie
Parasitaires – UMR6236
Institut de Recherche Biomédicale des Armées
Institut de Médecine Tropicale du Service de Santé
des Armées
Marseille Armées
France

eve.pradines@yahoo.fr

Marc Ouellette
Centre de Recherche en Infectiologie et
Division de Microbiologie
Faculté de Médecine
Université Laval
Ste-Foy, QC
Canada

marc.ouellette@crchul.ulaval.ca

Maritza Padrón Nieves
Laboratorio de Fisiología Molecular
Instituto de Medicina Experimental
Facultad de Medicina
Universidad Central de Venezuela
Caracas
Venezuela

mpadron43@gmail.com

Sneh Lata Panwar
Yeast Molecular Genetics Laboratories
School of Life Sciences
Jawaharlal Nehru University
New Delhi
India

sneh10@hotmail.com

Véronique Parquet
Unité de Recherche en Biologie et Epidémiologie
Parasitaires – UMR6236
Institut de Recherche Biomédicale des Armées
Institut de Médecine Tropicale du Service de Santé
des Armées
Marseille Armées
France

verogofio@hotmail.com

Bruno Pradines
Unité de Recherche en Biologie et Epidémiologie Parasitaires – UMR6236
Institut de Recherche Biomédicale des Armées
Institut de Médecine Tropicale du Service de Santé des Armées
Marseille Armées
France

bruno.pradines@free.fr

Rajendra Prasad
Membrane Biology Laboratory
School of Life Sciences
Jawaharlal Nehru University
New Delhi
India

rp47jnu@gmail.com

Alicia Ponte-Sucre
Laboratorio de Fisiología Molecular
Instituto de Medicina Experimental
Facultad de Medicina
Universidad Central de Venezuela
Caracas
Venezuela

aiponte@gmail.com

Hina Sanwal
Membrane Biology Laboratory
School of Life Sciences
Jawaharlal Nehru University
New Delhi,
India

hina.sanwal@gmail.com

Zuben E. Sauna
Laboratory of Cell Biology
Center for Cancer Research
National Cancer Institute, NIH
Bethesda, MD
USA

saunaz@mail.nih.gov

Lutz Schmitt
Institute of Biochemistry
Heinrich Heine University
Düsseldorf
Germany

lutz.schmitt@uni-duesseldorf.de

Bart van den Berg van Saparoea
Department of Molecular Microbiology
Groningen Biomolecular Sciences
and
Biotechnology Institute and Zernike
Institute for Advanced Materials
University of Groningen
Haren
The Netherlands

bergvans@science.uva.nl

Hendrik W. van Veen
Department of Pharmacology
University of Cambridge
Cambridge
UK

hwv20@cam.ac.uk

Foreword

About 40 years ago, when I started my scientific career, there was general consensus about ATP being the primary energy source for transport of solutes and macromolecules across cytoplasmic membranes. The finding of Kundig and Roseman at the end of the sixties, that in several microorganisms the energy for transport of some sugars across the cell membrane was supplied by phospho-enol pyruvate, concomitant with phosphorylation of the sugar, was considered a reasonable exception to the previously mentioned rule.

A major turn around was initiated by the acceptance of Mitchell's[1] chemiosmotic hypothesis (originally described in 1961) and the demonstration by Kaback and co-workers in the beginning of the seventies that the uphill transport of lactose in *Escherichia coli* was driven by a proton motive force. From there on so many different transporters in so many different organisms were shown to be proton motive force (or sodium-motive force) driven that direct energization of transport by means of ATP itself was hardly considered a reasonable option. The discovery of ABC transporters and the role of ATP as energy source around 1975 turned the views around as did the discovery of ATP-driven protein translocation systems.

These paradigmatic changes about the energy of transport that occurred in approximately forty years tells us about the flexibility of transport systems and illustrates the amazing capacity of organisms to recruit all possible mechanisms to get a job done. Additionally, it warns us as scientists to be open and not dogmatic about the various tools that organisms can mobilize to achieve their goal.

In terms of energy cost ATP-driven transporters (primary transporters) are usually more expensive than proton motive force driven transporters (secondary transporters). This is nicely reflected in the differences found in the substrates translocated by secondary and primary transporters. Substrates for secondary transporters are usually common anabolic or catabolic substrates or end products, while ABC transporters mainly transport solutes that are rare, and/or in high demand and present in low intracellular concentrations, or solutes that are highly membrane-permeable and can only be kept out of the cell by extremely high turnover rates of transporters.

The wide variety of solutes transported by ABC transporters, as well as the amazing variations found in their molecular structures and subunit compositions has fascinated scientists. Also, intriguing questions on how ABC transporters work, especially in drug resistance in cancer cells and in pathological microorganisms, are bound to keep enthusiastic scientists busy for many years to come. The state of art of these and many other aspects of ABC transporters will be covered in this book by experts in the field. It is an exciting book especially for those scientists interested in biomembranes and membrane transport.

Wil N. Konings
Stellenbosch, South Africa

1 Peter Dennis Mitchell (September 29, 1920–April 10, 1992) was a British biochemist who was awarded the 1978 Nobel Prize for Chemistry for his discovery of the chemiosmotic mechanism of ATP synthesis.

Preface

ABC transporters are fascinating molecular systems that catalyse the vectorial transport of a great variety of substrates across biological membranes. They constitute a large superfamily of primary active transport systems that are present in all kingdoms of life, and play a diversity of physiological roles. A prominent characteristic of these systems is that they share a highly conserved domain, the ATP binding cassette (ABC), which binds and hydrolyses ATP. The amino acid sequence of this cassette displays three major conserved motifs: the Walker A and Walker B motifs commonly found in P-loop containing ATPases or GTPases and a specific signature motif known as the linker peptide, the ABC signature motif, or simply the C loop.

The year 2006 marked the 20th anniversary of the identification of the ABC protein superfamily, and the 30th anniversary of the molecular identification of P-glycoprotein (MDR1, ABCB1). The gene of P-glycoprotein was sequenced in 1986 and it turned out to be the first example of the 48 individual transporters that belong to the ABC protein family in humans.

The cellular physiology of prokaryotes (including cellular processes and their regulation) critically relies on ABC systems. Among prokaryotes, ABC proteins segregate into 29 families belonging to three main functional branches of the ABC system (importers, exporters and 'other'). They facilitate the import of essential nutrients and the extrusion of toxins, and participate in DNA repair and translation or regulation of gene expression. On the other hand, in eukaryotes ABC proteins function only as efflux pumps playing multiple and important physiological roles.

As ABC transporters extrude an unusually large set of chemically diverse compounds, they are implicated in multidrug resistance (MDR) and pleiotropic drug resistance (PDR), and may hinder the clinical treatment of microbial infections and human cancers. However, there is accumulating evidence that MDR transporters also have natural physiological roles.

It is the aim of the present volume to give an overview of our current knowledge on ABC proteins present in microorganisms. The book may be divided in three sections. In the first section (Chapters 1–4) we present information regarding structure, evolution and molecular function of ABC transporters. The second section (Chapters 5–8) deals with the role of ABC transporters in the physiology of microorganisms like bacteria, parasites and fungi. In the final section (Chapters 9–10) we discuss classical and new approaches designed to fight against MDR.

Chapter 1 deals with the organization and topology of ABC transporters. This chapter also focuses on the three-dimensional architecture, domain–domain interactions and putative signaling pathways that guarantee efficient substrate recognition and translocation in an ATP-dependent manner. In the debate about the structure of ABC transporters and proteins, Chapter 4 discusses a mechanism related to resistance against macrolide, lincosamide and streptogramin antibiotics (MLS antibiotics) that involves an unusual class of ABC proteins, the antibiotic resistance element

(ARE). These ABC proteins do not have membrane-spanning segments within the polypeptide, are not linked to operons coding for membrane spanning regions, and constitute just another example of the diversity of ABC proteins and their potential modes of action. The primary focus of Chapter 2 is the evolution and structure–function relationship of microbial and mammalian ABC transporters responsible for MDR and the status of substrate specificity, as well as mechanistic aspects of mammalian ABC drug transporters are discussed. Finally, Chapter 3 deals with the mechanism(s) by which ATP binding/hydrolysis is coupled to transport and with the knowledge that quantitative structure activity analyses have provided about ABC transporters and their interaction with substrates, modulators and inhibitors.

Within the second section, Chapter 5 summarizes recent advances in the characterization of ABC-type transporters involved in bacterial multidrug resistance with implications for their physiological functions, and discusses the possible pitfalls that may occur in the functional analysis of ABC-type MDR-like transporters in heterologous systems. Malaria is by far the world's most important parasitic disease and the responsible parasites, the Plasmodia, consequently gather a lot of research. The same is true for the evolutionarily related protozoa *Leishmania* spp., Trypanosoma cruzi and *Trypanosoma brucei*. Unfortunately, the extensive use of drugs to fight these parasitic diseases led to the emergence of drug resistance, also mediated by ABC proteins. Our current understanding of the functions of ABC transporters in the biology of these medically important parasites, including drug resistance, is discussed in Chapters 6 and 7. Additionally, progress made in recent drug developments to fight parasite mediated drug disease is briefly summarized. In this same line of argument, Chapter 8 discusses the contribution of MDR and PDR proteins to drug resistance in yeast.

Innovative experimental therapeutic strategies for overcoming MDR are thus urgently needed. For that reason, the final section of the book includes a description of the progress made in recent years in the identification, design, availability and applicability of compounds that may work as ABC transporter blockers (Chapter 9). Finally, Chapter 10 analyses the potential application of RNA interference (RNAi) technology to microorganisms. Interestingly, these chapters additionally deal with: (i) the roles that ABC transporter blockers may play, not only by directly blocking ABC transporters involved in multidrug resistance, but also by affecting the function of ABC transporters of the host that may influence the pharmacokinetics and pharmacodynamics of therapeutic drugs used against infectious diseases; and (ii) how the RNAi technology (restricted to eukaryotic cells) may be helpful to block ABC transporter-mediated drug resistance in parasites. Notably this last chapter leaves an open window to the successful use of the RNAi technology for overcoming MDR in prokaryotic cells.

I would like to express my gratitude to all chapter authors for their dedication, to Horizon Scientific Press for asking me to plan and put together this interesting book, and to Hugh Griffin, member of the editorial board of Horizon Scientific Press, for his efforts to satisfy all the special requests the authors had, and for always being ready to answer the questions I had during the preparation of the book.

Alicia Ponte-Sucre

ABC Transporters – a Smart Example of Molecular Machineries

1

Thorsten Jumpertz, I. Barry Holland and Lutz Schmitt

Abstract

ABC transporters are fascinating molecular machines that use the energy of ATP to catalyse the transport of a huge variety of substrates across biological membranes in a vectorial fashion. Here, we will summarize our current knowledge of the domain organization and topology of this class of primary transporters. All ABC transporters analysed so far are composed of two nucleotide-binding domains (NBD) and two transmembrane domains (TMD) that can be arranged in any possible combination. However, additional transmembrane segments or extended NBDs, raise the possibility that these extensions act as platforms to interact with additional proteins with functional or regulatory consequences. In the light of the recent crystal structures of isolated NBDs in different functional states and the structures of intact ABC transporters, we will also focus on the three-dimensional architecture, domain–domain interactions and putative signaling pathways in these membrane proteins that guarantee efficient substrate recognition and translocation in an ATP-dependent manner.

Introduction

A short tale of ABC transporters

ABC transporters are one of the largest membrane protein families discovered so far. In *E. coli* ABC transport systems constitute 5% of the coding capacity of the organism (Linton and Higgins, 1998). The number of ABC transporters for different *E. coli* strains varies between 52 and 83 (Moussatova *et al.*, 2008) while in humans 48 genes for ABC proteins have been identified (Dean *et al.*, 2001), which are now assigned to seven subfamilies. Initially, proteins with an ATP binding cassette (ABC-) motif were identified and subsequently were shown to be part of a transport system composed of an integral membrane component and the ABC domain. This was the beginning of the discovery of a new superfamily of transport proteins. The first ABC transporters identified were the histidine permease and the maltose importer (Ames and Lever, 1970; Hazelbauer, 1975) driven by the ATPase motors, HisP and MalK respectively. Subsequently, ABC proteins were identified, however, with no associated transmembrane components and no transport function but that instead participate in DNA repair mechanisms (Husain *et al.*, 1986), antibiotic resistance (Jacquet *et al.*, 2008), or regulation of gene expression (Vazquez de Aldana *et al.*, 1995). ABC proteins are now divided into three major classes corresponding to their overall quaternary structure organization. Class 1 contains ABC transporters whose transmembrane components and ABC domains are fused in a single polypeptide chain. Class 2 is comprised of ABC proteins lacking an integral membrane component or transport function. Class 3 identifies ABC transport systems where the ABC and membrane components are encoded on separate polypeptide chains and where an additional component essential for import processes can be present. As may be expected it is hypothesized that all ABC proteins have evolved from a last universal common anchestor (LUCA) but this will not be discussed further here (Saurin *et al.*, 1999; Dassa and Bouige, 2001; Davidson *et al.*, 2008).

Most ABC proteins are essentially membrane proteins and function as transporters. The cellular membrane bilayer is impermeable towards charged molecules and substances that exceed a certain size. Proteins in the membrane are essential to mediate exchanges across this barrier allowing cells to import nutrients or expel toxic compounds. ABC transporters are essential players in these tasks, being involved in the transport of a huge variety of different molecules. The functions range from import of small ions to the extrusion of macromolecules that can be much larger than the transporter itself as in the case of the RTX polypeptide toxins such as HlyA (Holland *et al.*, 2005). An interesting class of transport substrates comprises mostly hydrophobic drugs (e.g. Hoechst, rhodamine, verapamil, and many more; see Sharom, 2008). For this reason, ABC transporters have been ascribed an important role in the development of multi drug resistance. To avoid possible confusion between hydrolysed substrates for an ABC protein like ATP as opposed to a *transport* substrate, we shall use the term 'allocrite' for the latter substrate.

In this chapter, knowledge available until the end of 2008, in terms of structure and function of ABC transporters, will be summarized, and recent models directed towards a more detailed molecular understanding are discussed. Several structures of complete ABC transporters have now been solved and many more isolated ABC motor domains and substrate binding proteins have been structurally described. Among the structures of complete transporters are importers as well as exporters captured in inward or outward facing conformations. These structures illustrate conversion of energy generation into mechanical movement involving communication between the motor domains and the transmembrane component.

Without doubt the most prominent examples of the ABC transporter family are P-glycoprotein (Pgp, MDR1, Multidrug resistance protein 1, ABCB1) and CFTR (Cystic fibrosis transmembrane conductance regulator, ACBC7). P-glycoprotein was first identified in cells showing resistance to drugs usually administered during cancer therapy (Kartner *et al.*, 1983) and is now recognized as a major problem in humans undergoing chemotherapy, conferring multidrug resistance, as expression levels are increased. P-glycoprotein is endogenously expressed in epithelial cells lining the colon or kidney proximal tubules and protects cells from toxic compounds (Leslie *et al.*, 2005). A knock-out of the gene Pgp disrupts the blood–brain barrier in mice and leads to neurotoxic effects (Doran *et al.*, 2005). After the first sequence of the Pgp clone appeared, Victor Ling and colleagues immediately discovered that the closest similarity was to the HlyB gene; this similarity allowed them to propose that Pgp was no doubt a drug pump. Interestingly, in addition to strong conservation in the NBDs (nucleotide binding domains) the two proteins have small shared motifs in the distal region of the TMDs (transmembrane domains) which, while still remaining a puzzle, may indicate some conservation of an energy coupling mechanism to transport in exporters.

Mutations in the CFTR gene cause cystic fibrosis with the most common mutation F508Δ leading to misfolding of the protein. This results in subsequent degradation of the transporter through the quality control system of the cell, while the wild type protein is transported to the cell surface. The F508Δ mutation is the most widespread cause of cystic fibrosis, the genetically most inherited disease among Caucasians (i.e. people from Europe, western Asia, North Africa and parts of the Indian subcontinent). The severe course of the disease, which is characterized by viscous mucus literally gluing together the bronchia of the lungs and the pancreas, often leads to death of the patients when they reach their mid-thirties, most frequently caused by bacterial infection of the lungs (Cystic Fibrosis Foundation, 2006; http://www.cff.org/).

At this point it should have become clear that a detailed knowledge of ABC transporters is essential since they function in important physiological and patho-physiological processes. To study these molecular systems different research groups have concentrated on ABC transporters found in prokaryotes. This preference stems from the fact that genes in bacteria are usually easier to manipulate and the protein products easier to purify. Proteins can be overproduced using commercially available and well characterized systems. The yield of protein in bacteria is often much higher compared to mouse or human cell lines. One can search for the best suitable bacterial homologue in terms of sequence identity compared to the eukaryotic protein or even stability of the transporter in various *in vitro* assays. Moreover, bacterial cell culture is much cheaper, which is quite important when proteins have to be produced in milligram quantities for structural studies. The strategy to use bacterial systems as models is not only appealing in terms of simplification, but is fundamentally supported by studies demonstrating that the evolutionary divergence of ABC proteins occurred prior to the separation of archea, bacteria, and eukarya. Additionally, these proteins have remained highly conserved in the nucleotide binding domain (NBD) (Saier *et al.*, 1998; Fuellen *et al.*, 2005). This justifies bacterial homologues of human ABC proteins as excellent model systems to evaluate and describe the underlying mechanisms common to all members of the ABC transporter family.

In recent years, some crystal structures of intact bacterial ABC transporters have been solved and many more isolated motor domains that fuel the import or export of substances have been described in great detail. Recently, a bioinformatics approach has allowed the listing of all the functional ABC transporters identified in *E. coli* and summarize the crystal structures of every ABC component (full transporter, isolated motor domain (NBD), substrate binding protein (SBP), or complexes of transporter and SBP) that have been deposited in the Brookhaven Protein Data Bank (http://www.rcsb.org) (Moussatova *et al.*, 2008). Over 90 structures of ABC transporter components have been obtained and represent the basis for structural and functional interpretation of the mechanisms of allocrite transport.

ABC transporters in molecular detail

In order to gain insight into how ABC transporters work it is important to understand the basic architecture of these proteins. In general, ABC transporters consist of a membrane spanning component (transmembrane domain, TMD) and a nucleotide binding or motor domain exposed to the cytoplasm. A fully assembled and functional transporter commonly consists of two TMDs and two NBDs, either as a homo- or heterodimer depending on the TMD and NDB subunits being identical e.g. $BtuC_2D_2$ (Locher *et al.*, 2002) or different e.g. $MalFGK_2$ (Oldham *et al.*, 2007). For exporters, usually a total of twelve transmembrane domains are detected, whereas the number for importers differs between ten and twenty TMDs per fully assembled transporter (Hollenstein *et al.*, 2007a). The motor domains bind ATP, which provides chemical energy upon hydrolysis, producing mechanical energy for transport. In some cases the NBDs have a regulatory domain at the C-terminal (Chen *et al.*, 2003a; Verdon *et al.*, 2003; Scheffel *et al.*, 2005; Gerber *et al.*, 2008; Kadaba *et al.*, 2008) whose proposed role is explained in more detail later. While ABC transporters involved in export normally require only the NBDs and TMDs for function, importers need a further component – the so-called substrate binding protein (SBP). The SBP binds the extracellular allocrite and delivers it to the transport domain.

ABC transporters involved in import are only present in prokaryotes and every component (TMD, TMD, NBD, NBD) is often encoded by separate genes with the protein then assembling into a functional transporter. Additionally, either both TMDs or both NBDs can be fused in a single polypeptide chain (Buckel *et al.*, 1986; Bohm *et al.*, 1996; Biemans-Oldehinkel *et al.*, 2006). The substrate binding protein can either be found diffusing freely in the periplasm for Gram-negative bacteria or attached to the membrane via lipid or protein anchors or covalently linked to the TMD for Gram-positive species (van der Heide and Poolman, 2002; Davidson and Chen, 2004; Borths *et al.*, 2005;). In the case of ABC importers each individual subunit (TMD, NBD, SBP) may be encoded as a separate polypeptide chain and the transport complex assembles from up to five modules.

For some ABC exporters both transmembrane domains and both nucleotide-binding domains are fused into a single polypeptide chain, designated as a full-size transporter. On the other hand, one TMD and one NDB can be found as separate polypeptide chains and these exporters are consequently called half-size transporters, with two copies of the NBD and TMD assembling as a homo- or heterodimer to form a functional transporter. Heterodimers can arise either with respect to the transmembrane components (MalFG; (Froshauer and Beckwith, 1984; Dassa and Hofnung, 1985;) or the nucleotide-binding domains (LmrCD; Lubelski *et al.*, 2006).

For structural analysis of ABC transporters initial attempts concentrated on the small, approximately 30 kDa NBDs since these relatively water soluble components of the transporter were easier to over-express and purify in reasonable amounts. Several structures of NDBs have now been obtained, both as monomers with and without nucleotide bound and also as dimers that reflect the relevant physiological state necessary for ATP hydrolysis. The first structure of the HisP ABC-ATPase was published in 1998 (Hung *et al.*, 1998) and revealed the now familiar, largely β-stranded lobe, constituting the catalytic domain, plus the helical domain containing the signature motif which was presumed to have a regulatory role.

Despite the fact that ABC transporters as a family have a broad range of transport substrate specificity, vectorial transport and even a variety of different architectures (number of TMDs, degenerate NBDs, additional regulatory domains), certain highly conserved amino acid motifs can be found in the NBDs. However, the transmembrane part is not highly conserved although recent crystal structures have revealed some structural similarities. The overall diversity of the amino acid sequence has led to the assumption that the TMDs of the transporter mediate allocrite selectivity. However, one strictly conserved motif in the TMDs of ABC importers in bacteria is the EAA motif located in a cytoplasmic loop implicated in an interaction with the NBDs. As will be discussed below, this region interacting with the NBD seems to be involved in communication between the transporting and energy generating domains. Contacts between the EAA loop and the NBDs in MalK were first shown in mutational studies and then in cross-linking experiments that demonstrated the close proximity of the helical domain of the NBD and the intracellular loops of MalF and MalG, each containing the EAA-motif (Hunke *et al.*, 2000; Mourez *et al.*, 1997).

In contrast to importers, ABC exporters lack the conserved EAA motif, although recently a sequence possibly undertaking the task of ensuring proper communication between TMDs and motor domains was identified (Dawson and Locher, 2006). Thus, in the Sav1866 crystal structure the region of TMD–NBD interaction is characterized by the TEVGERG sequence (designated as the x-loop) with a similar sequence in other ABC exporters, as well as by sequence alignments. This sequence is not located in the intracellular loops of the transmembrane part but lies within the structurally diverse region (SDR) of the NBD helical domain (see Fig. 1.8B). The SDR was first identified from an analysis of the crystal structure of the nucleotide-free state of the HlyB-NBD. A structural comparison with the crystal structures available at that time revealed that roughly 30 amino acids of the helical domain located just upstream of the C loop showed the highest structural variation within the analysed NBDs. Importantly, the SDR can be found in ABC exporters as well as importers and it was proposed that this variable structural motif was vital for the specificity of different ABC proteins, allowing the otherwise conserved NBD to interact uniquely with the cognate TMD.

Unlike the diversity shown by the TMDs, the NBDs display an extensive conservation of sequences. This is to be expected since the hydrolysis of ATP that fuels transport is a mechanism found in all ABC transporters throughout all kingdoms of life and irrespective of the allocrite specificity or transport direction. The conserved sequences are the Walker A and Walker B motifs, characteristic for the large family of P-loop NTPases (<u>n</u>ucleotide <u>t</u>ri<u>p</u>hosphatases) (Vetter and Wittinghofer, 1999), the C-loop/signature motif, D-loop, Pro-loop, H-loop and the <u>s</u>tructurally <u>d</u>iverse <u>r</u>egion (SDR) containing the x-loop, a motif restricted to exporters. The C-loop is localized to the helical domain of the NBD (Ames and Lecar, 1992), while the Walker A motif is located to the catalytic domain. The Walker A is involved in ATP binding and this has the consensus amino acid sequence GXXGXGKS/T where X stands for any amino acid (Walker *et al.*, 1982). The Walker B motif is composed of an amino acid stretch that resembles ΦΦΦΦD where Φ is any hydrophobic residue. The Walker B element is the typical nucleotide binding fold of P-loop ATPases and forms part of the binding site for nucleotides

(Vetter and Wittinghofer, 1999). Together, the Walker A and B motifs resemble part of the RecA-like (Story and Steitz, 1992) or F_1-like (Abrahams *et al.*, 1994) subdomain of the NBD which contains mostly β-sheet structures. Additionally, the D-loop (SALD) and H-loop, approximately 25 residues downstream of the Walker B motif, are localized in this catalytic subdomain.

The hallmark of ABC transporters is the signature motif or C-loop (Schmitt and Tampé, 2002). Its consensus sequence is highly conserved and can be found in any NBD with only minor variations (e.g. Q → E in BtuD). A degenerate C-loop sequence results in a loss of ATP-hydrolysis. Further motifs found close to the helical domain are the Q-loop, and the Pro-loop; in fact, these essentially connect the catalytic and helical domains.

This plethora of conserved and important motifs may be confusing at first sight, but the sequence along the polypeptide chain and the resulting three-dimensional arrangement are not too complex (Fig. 1.1 and 1.2A). Starting from the N-terminus of the NBD along the motor domain, the Walker A motif in the catalytic domain is encountered first. The next motif is the Q-loop serving as a hinge between the catalytic and helical domains. The helical domain contains the structurally diverse region (SDR) immediately followed by the signature motif or C-loop. About 15 amino acids beyond the C-loop, the Pro-loop, the second hinge for the movement of the helical domain appears and delineates simultaneously the way back to the catalytic domain. The remaining motifs are the Walker B sequence adjacent to the D-loop and the H-loop, the closest of all the consensus motifs to the C-terminus (see Chapter 3 for further details).

All stretches of conserved amino acids indicated above have specific tasks to ensure proper functioning of the motor domain and hence the transport process. We shall focus on all these structural motifs and explain how the characteristic amino acid side chains and the orientation of the residues and the nucleotide catalyse chemical reactions in a concerted manner. Since malfunctioning of several ABC transporters in humans is associated with severe diseases, a thorough understanding of such molecular details of the action of the ABC domain is a prerequisite for successful design of effective drugs in the future. Thus, our focus will be on residues interacting with the nucleotide, those that interact with the transmembrane part, and those that mediate NBD-NBD contacts in the functional dimer. A comprehensive schematic drawing of a nucleotide-binding domain is shown in Fig. 1.2 and the respective motifs are colour-coded.

Fig. 1.2A shows the overall structure of the nucleotide binding domain for HisP from the ABC histidine importer of *S.* Typhimurium. The structure of the monomeric NBD displays a more or less L-shaped form and can be divided into two subdomains that have certain unique functions. First the domain consisting of α-helices, therefore the 'helical domain', while the second subdomain (catalytic domain, RecA-like domain, F_1-like domain) contains a central antiparallel β-sheet flanked by α-helices. This arrangement was first described by Ames and coworkers in the crystal structure of the ABC motor domain from the histidine permease of *S.* Typhimurium ten years ago (Hung *et al.*, 1998). Important structural elements in the helical domain are the C-loop (pale green), the designated signature motif for ABC transporters and the SDR (black) (Schmitt *et al.*, 2003) which contains the x-loop (Dawson and Locher, 2006).

Under physiological conditions the NBDs of an ABC transporter have to dimerize to bring all residues necessary for hydrolysis of the substrate into close proximity. Thus, each monomer contributes to the binding pocket of ATP in a head-to-tail arrangement with the Walker A motif of one monomer and the signature motif of the other monomer sandwiching the ATP (Chen *et al.*, 2003a; Smith *et al.*, 2002; Zaitseva *et al.*, 2005a).

The structurally diverse region interacts with cytoplasmic loops extending from the transmembrane components of the ABC transporter and is therefore involved in conveying information about the actual status of the two major components of the ABC transporter, the NBD and TMD (Schmitt *et al.*, 2003). Thus, it is reasonable to suppose that binding of nucleotide, hydrolysis of ATP or binding of the allocrite (transport substrate) must induce intramolecular structural changes that in turn promote appropriate changes in the TMDs, finally leading to import or export of the allocrite via a membrane pathway.

```
MsbA_S.typhimurium      1     EKDEGKRVIDRATGDLEFRNVTFTYPGREVPALRNINLKIPAGKTVALVGRSGSGKSTIASLITRFY..DIDEGHILMDGHD.........LREYT
MsbA_E.coli             1     EKDEGKRVIERATGDVEFRNVTFTYPGRDVPALRNINLKIPAGKTVALVGRSGSGKSTIASLITRFY..DIDEGEILMDGHD.........LREYT
Sav1866_S.aureus        1  DYDIKNGVGAQPIEIKQGRIDIDHVSFQYNDNEAPILKDINLSIEKGETVAFVGMSGGGKSTLINLIPRFY..DVTSGQILIDGHN.........IKDFL
LmrA_L.lactis           1               MLSARHVDFAYDDSEQ.ILRDISFEAQPNSIIAFAGPSGGGKSTIFSLLERFY..QPTAGEITIDGQP.........IDNIS
HlyB_E.coli             1               DITFRNIRFRYKPDSPVILDNINLSIKQGEVIGIVGRSGSGKSTLTKLIQRFY..IPENGQVLIDGHD.........LALAD
MetN_E.coli             1             MIKLSNITKVFHQGTRTIQALNNVSLHVPAGQIYGVIGASGAGKSTLIRCVNLLE..RPTEGSVLVDGQ.........ELTTL
HisP_S.typhimurium      1             MMSENKLHVIDLHKRYGGHEVLKGVSLQARAGDVISIIGSSGSGKSTFLRCINFLE..KPSEGAIIVNGQNINLVRDKDGQLKVA
MalK_P.horikoshii       1         GNNIEVIKMVEVKLENLTKRFG.NFTAVNKLNLTIKDGEFLVLLGPSGCGKTTTLRMIAGLE..EPTEGRIYFGDRD.........VTYLP
1VCI_P.horikoshii       1        MGNNIEVIKMVEVKLENLTKRFG.NFTAVNKLNLTIKDGEFLVLLGPSGCGKTTTLRMIAGLE..EPTEGRIYFGDRD.........VTYLP
MalK_E.coli             1              MASVQLQNVTKAWG.EVVVSKDINLDIHEGEFVVFVGPSGCGKSTLLRMIAGLE..TITSGDLFIGEKR.........MNDTP
ModC_A.fulgidus         1               MFLKVRAEKRLG.NFRLN...VDFEMGRDYCVLLGPTGAGKSVFLELIAGIV..KPDRGEVRLNGAD.........ITPLP
ModC_M.acetivorans      1               MIEIESLSRKWK.NFSLDNLSLKVESG.EYFVILGPTGAGKTLFLELIAGFH..VPDSGRILLDGKD.........VTDLS
CysA_A.acidocaldarius   1       MRGHHHHHHGSMTIEFVGVEKIYPGGARSVRGVSFQIREGEMVGLLGPSGSGKTTILRLIAGLE..RPTKGDVWIGGKR.........VTDLP
MJ1267_E.coli           1             MIKLKNVTKTYKMGEEIIYALKNVNLNIKEGEFVSIMGPSGSGKSTMLNIIGCLD..KPTEGEVYIDNIK.........TNDLD
MJ0796_M.janaschii      1             MIKLKNVTKTYKMGEEIIYALKNVNLNIKEGEFVSIMGPSGSGKSTMLNIIGCLD..KPTEGEVYIDNIK.........TNDLD
GlcV_S.solfataricus     1            MVRIIVKNVSKVFKKGK..VVALDNVNINIENGERFGILGPSGAGKTTFMRIIAGLD..VPSTGELYFDDRL.........VASNG
SufC_E.coli             1                 MLSIKDLHVSVEDKAILRGLSLDVHPGEVHAIMGPNGSGKSTLSATLAGREDYEVTGGTVEFKGKD.........LLALS
SufC_T.thermophilus     1              MSQLEIRDLWASIDGETILKGVNLVVPKGEVHALMGPNGAGKSTLGKILAGDPEYTVERGEILLDGEN.........ILELS
TM0544_T.maritima       1       MGSDKIHHHHHHMGAVVVKDLRKRIGKKEILKGISFEIEEGEIFGLIGPNGAGKTTTLRIISTLIK..PSSGIVTVFGKN.........VVEEP
MJ1267_M.janaschii      1            MRDTMEILRTENIVKYFGEFKALDGVSISVNKGDVTLIIGPNGSGKSTLINVITGFL..KADEGRVYFENKD.........ITNKE
BtuD_E.coli             1                    MSIVMQLQDVAESTRLGPLSGEVRAGEILHLVGPNGAGKSTLLARMAGMT...SGKGSIQFAGQP.............
HI1470_H.influenzae     1            MNKALSVENLGFYYQAENFLFQQLNFDLNKGDILAVLGQNGCGKSTLLDLLLGIH..RPIQGKIEVYQSIG.............
                                                                   Walker A

MsbA_S.typhimurium     86 LAS....LRNQVALVSQNVHLFN...DTVANNIAYARTEEYSREQIEEAARMAYAMDFINKMDNGLDT.IIGENGVL.LSGGQRQRIAIARALLRDS...
MsbA_E.coli            86 LAS....LRNQVALVSQNVHLFN...DTVANNIAYARTEQYSREQIEEAARMAYAMDFINKMDNGLDT.VIGENGVL.LSGGQRQRIAIARALLRDS...
Sav1866_S.aureus       90 TGS....LRNQIGLVQQDNILFS...DTVKENILLGR.PTATDEEVVEAAKMANAHDFIMNLPQGYDT.EVGERGVK.LSGGQKQRLSIARIFLNNP...
LmrA_L.lactis          71 LEN....WRSQIGFVSQDSAIMA...GTIRENLTYGLEGDYTDEDLWQVLDLAFARSFVENMPDQLNT.EVGERGVK.ISGGQRQRLAIARAFLRNP...
HlyB_E.coli            72 PNW....LRRQVGVVLQDNVLLN...RSIIDNISLAN.PGMSVEKVIYAAKLAGAHDFISELREGYNT.IVGEQGAG.LSGGQRQRIAIARALVNNP...
MetN_E.coli            73 SESELTKARRQIGMIFQHFNLLS...SRTVFG.NVALPLELDN...TPKDEVKRRVTELLSLVGLGDK.HD.SYPSN.LSGGQKQRVAIARALASNP...
HisP_S.typhimurium     84 DKNQLRLLRTRLTMVFQHFNLWS...HMTVLENVMEAPIQVLG...LSKHDARERALKYLAKVGIDER.AQGKYPVH.LSGGQQQRVSIARALAMEP...
MalK_P.horikoshii      80 P......KDRNISMVFQSYAVWP...HMTVYE.NIAFPLKIKK...FPKDEIDKRVRWAAELLQIEE..LLNRYPAQ.LSGGQRQRVAVARAIVVEP...
1VCI_P.horikoshii      81 P......KDRNISMVFQSYAVWP...HMTVYE.NIAFPLKIKK...FPKDEIDKRVRWAAELLQIEE..LLNRYPAQ.LSGGQRQRVAVARAIVVEP...
MalK_E.coli            72 P......AERGVGMVFQSYALYP...HLSVAE.NMSFGLKLAG...AKKEVINQRVNQVAEVLQLAH..LLDRKPKA.LSGGQRQRVAIGRTLVAEP...
ModC_A.fulgidus        67 P......ERRGIGFVPQDYALFP...HLSVYR.NIAYGLRNVE...RVER..DRRVREMAEKLGIAH..LLDRKPAR.LSGGERQRVALARALVIQP...
ModC_M.acetivorans     69 P......EKHDIAFVYQNYSLFP...HMNVKK.NLEFGMR.MK...KIKD..PKRVLDTARDLKIEH..LLDRNPLT.LSGGEQQRVALARALVTNP...
CysA_A.acidocaldarius  84 P......QKRNVGLVFQNYALFQ...HMTVYD.NVSFGLREKR...VPKDEMDARVRELLRFMRLES..YANRFPHE.LSGGQQQRVALARALAPRP...
MJ1267_E.coli          74 DDELTKIRRDKIGFVFQQFNLIP...LLTALE.NVELPLIFKYRGAMSGEERRKRALECLKMAELEER.FANHKPNQ.LSGGQQQRVAIARALANNP...
MJ0796_M.janaschii     74 DDELTKIRRDKIGFVFQQFNLIP...LLTALE.NVELPLIFKYRGAMSGEERRKRALECLKMAELEER.FANHKPNQ.LSGGQQQRVAIARALANNP...
GlcV_S.solfataricus    74 KLIVPPEDR.KIGMVFQTWALYP...NLTAFE.NIAFPLTNMK...MSKEEIRKRVEEVAKILDIHH..VLNHFPRE.LSGGQQQRVALARALVKDP...
SufC_E.coli            72 PED...RAGEGIFMAFQYPVEIPGVSNQFFLQTALNAVRSYRGQETLDRFDFQDLMEEKIALLKMPED.LLTRSVNVGFSGGEKKRNDILQMAVLEP...
SufC_T.thermophilus    74 PDE...RARKGLFLAFQYPVEVPGVTIANFLRLALQAKLGR....EVGVAEFWTKVKKALELLDWDES.YLSRYLNEGFSGGEKKRNEILQLLVLEP...
TM0544_T.maritima      84 HEV.....RKLISYLPEEAGAYR.......NMQGIEYLRFVAGFYASSSSEIEEMVERATEIAGLGEK.IKDRVSTY..SKGMVRKLLIARALMVNP...
MJ1267_M.janaschii     76 PAELYHYGIVRTFQTPQPLKEMTVLENLLIGEINPGESPLNSLFYKKWIPKEEEMVEKAFKILEFLKLSHLYDRKAGELSGGQMKLVEIGRALMTNP...
BtuD_E.coli            63 ......LEAWSATKLALHRAYLS.....QQQTPPFATPVWHYLTLHQHDKTRTELLNDVAGALALDDK..LGRSTNQ.LSGGEWQRVRLAAVVLQITPQA
HI1470_H.influenzae    70 ..........FVPQFFSSPFAYS......VLDIVLMGRSTHIN...TFAKPKSHDYQVAMQALDYLNLTHLAKREFTSLSGGQRQLILIARAIASEC...
                                 Q-loop                               Structural diverse region   C-loop        Pro-loop
```

```
MsbA_S.typhimurium     174 ....PILILDEATSALDTESERAIQAALDELQK..NRTSLVIAHRLS.TIEQADEIVVVEDGIIVERGTHSELLAQ.HGVYAQLHKMQFGQ
MsbA_E.coli            174 ....PILILDEATSALDTESERAIQAALDELQK..NRTSLVIAHRLS.TIEKADEIVVVEDGVIVERGTHNDLLEH.RGVYAQLHKMQFGQ
Sav1866_S.aureus       177 ....PILILDEATSALDLESESIIQEALDVLSK..DRTTLIVAHRLS.TITHADKIVVIENGHIVETGTHRELIAK.QGAYEHLYSIQNL
LmrA_L.lactis          159 ....KILMLDEATASLDSESESMVQKALDSLMK..GRTTLVIAHRLS.TIVDADKIYFIEKGQITGSGKHNELVAT.HPLYAKYVSEQLTVGQ
HlyB_E.coli            159 ....KILIFDEATSALDYESEHVIMRNMHKICK..GRTVIIIAHRLS.TVKNADRIIVMEKGKIVEQGKHKELLSEPESLYSYLYQLQSD
MetN_E.coli            160 ....KVLLCDEATSALDPATTRSILELLKDINRRLGLTILLITHEMDVVKRICDCVAVISNGELIEQDTVSEVFSHPKTPLAQKFIQSTLHLDIPEDYQE
HisP_S.typhimurium     173 ....DVLLFDEPTSALDPELVGEVLRIMQQLAE.EGKTMVVVTHEMGFARHVSSHVIFLHQGKIEEEGDPEQVFGNPQSPRLQQFLKGSLKKLEH
MalK_P.horikoshii      161 ....DVLLMDEPLSNLDAKLRVAMRAEIKKLQQKLKVTTIYVTHDQVEAMTMGDRIAVMNRGQLLQIGSPTEVYLRPNSVFVATFIGAPEMNILEVSVGD
1VCI_P.horikoshii      162 ....DVLLMDEPLSNLDAKLRVAMRAEIKKLQQKLKVTTIYVTHDQVEAMTMGDRIAVMNRGQLLQIGSPTEVYLRPNSVFVATFIGAPEMNILEVSVGD
MalK_E.coli            153 ....SVFLLDQPLSNLDAALRVQMRIEISRLHKRLGRTMIYVTHDQVEAMTLADKIVVLDAGRVAQVGKPLELYHYPADRFVAGFIGSPKMNFLPVKVTA
ModC_A.fulgidus        146 ....RLLLLDEPLSAVDLKTKGVLMEELRFVQREFDVPILHVTHDLIEAAMLADEVAVMLNGRIVEKGKLKELFS.AKNGEVAEFL..........SAR
ModC_M.acetivorans     147 ....KILLLDEPLSALDPRTQENAREMLSVLHKKNKLTVLHITHDQTEARIMADRIAVVMDGKLIQVGKPEEIFEKPVEGRVASFVGFENVLKGRVISAE
CysA_A.acidocaldarius  165 ....QVLLFDEPFAAIDTQIRRELRTFVRQVHDEMGVTSVFVTHDQEEALEVADRVLVLHEGNVEQFGTPEEVYEKPGTLFVASFIGESNVWTRAVQNGR
MJ1267_E.coli          165 ....PIILADEPTGALDSKTGEKIMQLLKKLNEEDGKTVVVVTHD..............................INVARFG.............
MJ0796_M.janaschii     165 ....PIILADEPTGALDSKTGEKIMQLLKKLNEEDGKTVVVVTHD..............................INVARFG.............
GlcV_S.solfataricus    160 ....SLLLLDEPFSNLDARMRDSARALVKEVQSRLGVTLLVVSHDPADIFAIADRVGVLVKGKLVQVGKPEDLYDNPVSIQVASLIGEINELEGKVTNEG
SufC_E.coli            165 ....ELCILDESDSGLDIDALKVVADGVNSLRDGKRSFIIVT.HYQRILDYIKPDYVHVLYQGRIVKSGDFTLVKQLEEQGYGWLTEQQ
SufC_T.thermophilus    163 ....TYAVLDETDSGLDIDALKVVARGVNAMRGPNFGALVIT.HYQRILNYIQPDKVHVMMDGRVVATGGPELALELEAKGYEWLKEKVKEGA
TM0544_T.maritima      166 ....RLAILDEPTSGLDVLNAREVRKILKQASQE.GLTILVSSHNMLEVEFLCDRIALIHNGTIVETGTVEELKERYKAQNIEEVFEEVVKCSENF
MJ1267_M.janaschii     173 ....KMIVMDQPIAGVAPGLAHDIFNHVLELKAK.GITFLIIEHRLDIVLNYIDHLYVMFNGQIIAEGRGEEEIKNVLSDPKVVEIYIGE
BtuD_E.coli            149 NPAGQLLLLDEPMNSLDVAQQSALDKILSALCQQ.GLAIVMSSHDLNHTLRHAHRAWLLKGGKMLASGRREEVLTPPNLAQAYGMNFRRLDIEGHRMLIS
HI1470_H.inluenzae     148 ....KLILLDEPTSALDLANQDIVLSLLIDLAQSQNMTVVFTTHQPNQVVAIANKTLLLNKQNFKFGETRNILTSENLTALFHLPMFEQQAQYKESFFTH
                               Walker B D-loop                     H-loop
```

Figure 1.1 Amino acid sequence alignment of the nucleotide binding domains of prokaryotic ABC transporters whose structures have been solved. The conserved motifs are colour-coded: Walker A (red), Q-loop (magenta), structural diverse region (SDR; black), C-loop (pale green), Pro-loop (orange), Walker B (blue), D-loop (cyan), and H-loop (green). The function of the distinct residues is explained in detail in the text. A colour version of this figure is located in the plate section at the back of the book.

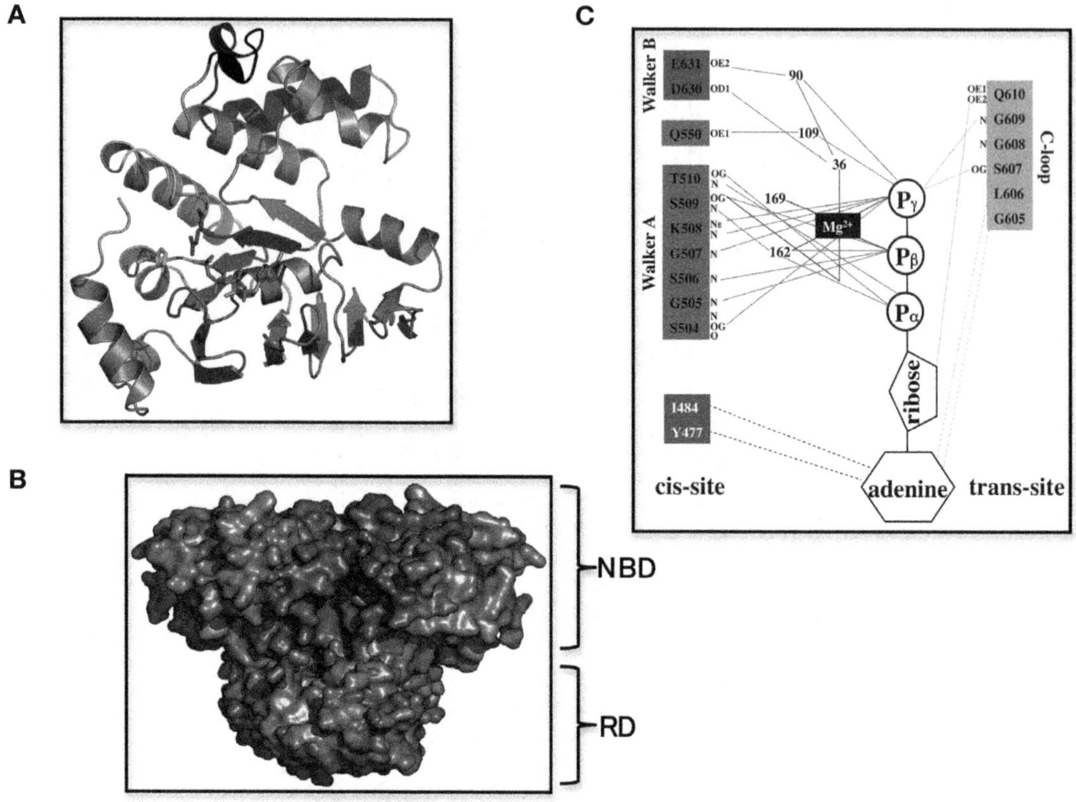

Figure 1.2 A Crystal structure of HisP, the nucleotide binding domain of the histidine permease from *S.* Typhimurium with bound ATP (pdb entry 1B0U). The helical domain contains the C-loop (pale green) and the structurally diverse region (SDR; black) and is linked via the Q-loop (magenta) and the Pro-loop (orange) to the catalytic domain. The catalytic domain contains the Walker A (red) and Walker B (blue) motifs and the D-loop (cyan) as well as the H-loop (green). The glutamate at the end of Walker B and the histidine of the H-loop are depicted in stick-representation as they are crucial in ATP hydrolysis. **B** Structural superposition of the nucleotide free (grey) and ATP-bound (magenta) structure of the nucleotide binding domain MalK (pdb entries: 1Q1E, 1Q12). The NBDs come into close contact upon ATP binding whereas only minor changes can be detected in the regulatory domains (RD). **C** Detailed network of ATP coordination in the binding pocket of the HlyB-NBD. Solid lines indicate hydrogen bonds and dashed lines show van der Waals interactions. Colour-coding is equivalent to Fig. 1.2A. The black numbers indicate water molecules that are part of the extensive network of interactions. Letters next to the amino acid labels indicate involvement via main-chain or side-chain interactions. A colour version of this figure is located in the plate section at the back of the book.

We shall discuss the binding of the nucleotide in more detail since this molecule constitutes the energy source for transport and therefore a closer look is of great importance to understand how this molecular motor functions (Fig. 1.2C). In all P-loop NTPases (Vetter and Wittinghofer, 1999) the Walker A motif wraps around the phosphate moiety of the nucleotide. Acidic residues located at the end of the Walker B motif further coordinate the γ-phosphate moiety. Exceptionally important is the glutamate residue, which stabilizes and positions the γ-phosphate group, together with the H-loop histidine, in a position where the P-O-bond can be attacked. These two residues, glutamate and histidine, likely constitute a catalytic dyad essential for the final hydrolysis of ATP in all ABC-NBDs (Zaitseva *et al.*, 2005a). In addition, a glutamine residue from the Q-loop is also implicated in the hydrolysis of ATP. This, together with the glutamate of the Walker B, forms hydrogen bonds to a water molecule that could fulfill the role of the attacking water to break the P-O-bond between the γ- and β-phosphate of ATP. However, in addition to contacts between the phosphate residues of the nucleotide and the protein, the adenine moiety of ATP interacts with a tyrosine residue in the catalytic domain via π–π-stacking.

Interestingly, due to this rather nonspecific interaction, the ABC motor domains generally can bind different nucleotides (at least *in vitro*) even though with less affinity and therefore with reduced activity. The Pro-loop (orange) and Q-loop (magenta) shown in Fig. 1.1, as indicated above, serve as hinges between the helical and the RecA-like domain. If the structures of nucleotide free and ATP-bound NBDs are compared, a relative movement of the helical domain towards the catalytic domain is clearly visible. This conformational change is referred to as a 'rigid body movement' or 'nucleotide-mediated induced fit' (Fig. 1.2B) (Karpowich *et al.*, 2001).

NBD dimers and the catalytic cycle

To further investigate how the mechanism of ATP hydrolysis is accomplished, it was essential to obtain structures of NBD dimers to investigate protein–protein interactions at the dimer interface and consequent structural changes upon formation of the complex. Several groups succeeded in crystallizing NBD dimers but not every assembly of the desired complex met the criteria expected for a physiological arrangement of the NBDs. From the information obtained from monomeric structures and biochemical studies it was clear that particular motifs should be in close contact in the dimers. Their absence in certain dimeric crystal structures could therefore be judged as non-physiological artifacts. For example, ortho-vanadate added to MalK (the NBD of the maltose importer) during the hydrolysis reaction becomes trapped in the position formerly engaged by the γ-phosphate of ATP and catalyses a photocleavage reaction. Analysis of the cleavage sites revealed that both the Walker A and the signature (LSGGQ) motif contribute residues to the ATP binding site (Fetsch and Davidson, 2002). Another hint indicating the importance of the C-loop in MalK came from mutagenesis studies showing that C-loop mutants had considerably reduced ATPase-activity (Schmees *et al.*, 1999). Surprisingly the monomeric structure of HisP showed that the Walker A motif and the C-loop were too far apart to interact with each other and a different dimer geometry was proposed (Hung *et al.*, 1998). Finally, the structures of the ABC proteins MJ0796 from *Methanococus jannaschii* (Smith *et al.*, 2002), MalK from *E. coli* (Chen *et al.*, 2003a), and the HlyB-NBD from *E. coli* (Zaitseva *et al.*, 2005a) were recognized to fulfill all the criteria necessary to represent a native conformation of the motor domain dimer of ABC transporters.

In order to represent an informative dimer ATP should be present in the structure but this is difficult to achieve with a protein specifically designed to hydrolyse this nucleotide. Thus, minor modifications to the protein were necessary to obtain suitable crystals. This problem was circumvented by mutating amino acids essential for hydrolysis of ATP. For MJ0796 the glutamate residue at the end of the Walker B motif was changed to a glutamine, while for the HlyB-NBD the histidine of the H-loop was mutated to an alanine. In the case of the MalK structure, Mg^{2+} was omitted from the crystallization protocol. For MJ0796 and HlyB, these mutations resulted in proteins where the ATPase activity was completely abolished (Davidson and Sharma, 1997; Nikaido and Ames, 1999; Moody *et al.*, 2002; Zaitseva *et al.*, 2005b). However, solving a crystal structure of a protein is just the beginning – the interpretation of the three dimensional arrangement of the amino acid residues and hypotheses with respect to molecular function then arise, and the accuracy of these hypotheses is dependent upon having good biochemical data on the properties of the enzyme. With this in hand an informative picture of how these molecular machines carry out their function can be obtained.

Now we may consider what insights have been revealed from the structures of the dimeric complex of the ABC motor domains. First of all it is clearly visible that the Walker A motif from one monomer and the C-loop of the second monomer complement each other to form the binding pocket for ATP (see Fig. 1.1B). This organization of the two subunits is called a 'head-to-tail' arrangement and interestingly this was anticipated earlier via a molecular modelling approach (Jones and George, 1999), before the first correct structure of an NBD dimer was published. The Walker A motif forms extensive backbone and side chain interactions with all three phosphate moieties of the ATP. A conserved lysine residue of the

Walker A interacts with the β- and γ-phosphates consequently fixing them in a defined orientation. The Walker B motif interacts mainly with the γ-phosphate of the nucleotide, as does the C-loop of the opposing monomer. Thus, ATP is sandwiched between the two NBD monomers and acts as molecular glue. Since the three phosphate moieties of ATP carry a distinct negative charge, which might result in electrostatic repulsion, a Mg^{2+} ion is coordinated by the β- and γ-phosphates of the nucleotide.

This Mg^{2+} cofactor is essential for hydrolysis and hence forms an extensive network of interactions (a detailed analysis of the interactions of ATP and Mg^{2+} with the HlyB-NBD is depicted in Fig. 1.2C). A residue from the Walker A motif interacts with water molecules and the β- and γ-phosphate residues constitute the octahedral coordination sphere of the Mg^{2+} ion – in total, six oxygen atoms from the surrounding residues have the ideal distances and angles to provide this octahedral coordination (Harding, 2000). Two further water molecules connect the γ-phosphate with the Walker B motif and the Q-loop. Analysis of the interaction between glutamine 550 from the Q-loop and aspartate 630 from the Walker B in the HlyB-NBD (Zaitseva *et al.*, 2005a) detects only van der Waals interactions in the nucleotide free form of the protein, while binding of ATP promotes hydrogen bond formation between those two residues, indicating a structural reorganization. Indeed, this change can be linked to an induced fit mechanism first described for MJ1267 that is assumed now to be a general feature of the motor domains of ABC transporters (Karpowich *et al.*, 2001). The Q- and Pro-loops function as linkers mediating an inward rotation of the helical domain of roughly 20° when ATP binds (Fig. 1.2B) (Lu *et al.*, 2005; Zaitseva *et al.*, 2005a). Thus, the effect of the presence of the γ-phosphate at the Walker B site is transferred to the Q-loop glutamine causing the rigid body movement of the helical domain. Through this mechanism the NBD 'senses' and responds to the binding of ATP. An additional structural change that also occurs during the conversion from the nucleotide free to the ATP-bound state has been observed in the NBD of the ABC transporter Haemolysin B. This is the conversion of the last three amino acids of the Walker A motif from a 3_{10} helix towards a regular α-helix upon binding of ATP. The 3_{10} helix/α-helix conversion is expected to be a mechanism to control the ATPase activity of this motor domain *in vivo*. Other examples of additional structural perturbations are SufC from *T. thermophilus* where a similar transition at the Walker B motif occurs (Watanabe *et al.*, 2005) and the effect of the regulatory inserts and extensions in the NBD of CFTR (Lewis *et al.*, 2004).

The most prominent inter-protein interaction in the ABC domain is obviously the completion of the nucleotide binding site when ATP is sandwiched between the Walker A and the C-loop. A second motif highly characteristic for ABC proteins in addition to the C-loop is the D-loop, which is almost exclusively involved in interactions between the two monomers. Close proximity between the D-loop and the highly conserved histidine residue involved in ATP hydrolysis lead to the assumption that information about the functional state of one ATP binding site is transferred to the other monomer. The histidine may therefore not only act catalytically but also as a phosphate sensor transmitting information about progress in the chemical reaction.

In order to ensure that all the inter-protein interactions described above are relevant to a hydrolysis competent NBD dimer, an analysis of the surface accessible solvent area (ASA) was performed. The results demonstrated that the buried protein surface at the dimer interface is in the range for protein–protein complexes and therefore not an artifact of crystal contacts (Bahadur and Zacharias, 2008). Additionally, it has also been shown that a specific microenvironment at the nucleotide binding sites is established upon ATP binding and dimerization. Local pK_a values of amino acid site chains (mostly in the Walker A and Walker B region as well as in the D- and H-loop) and the intrinsic pK_a of the ATP molecule are presumably essential to provide the basis for efficient catalysis (Hanekop *et al.*, 2006). A final remark concerning ATP binding should be made. Thus, for at least the NBDs, positive cooperativity of ATP hydrolysis has been demonstrated. This is consistent with dimerization being dependent on nucleotide binding to both NBDs although only one composite site competent for ATP hydrolysis is found in certain systems (Davidson *et al.*, 1996; Liu *et al.*, 1997; Steinfels *et al.*, 2004; Zaitseva *et al.*, 2005b). This aspect will be discussed later in this chapter.

Further studies have revealed the structures of monomeric and dimeric NBDs crystallized with and without nucleotide (ADP/ATP) or with non-hydrolysable substrates (AMP-PNP). However, the description of how residues are arranged in these crystal structures are not sufficient to understand the mechanism of catalytic activity since they represent only a snap-shot of the catalytic cycle of the NBDs. They must be complemented by relevant biochemical experiments. Thus, structural studies plus insights from biochemical studies have yielded a model of the catalytic cycle of the motor domains of ABC transporters. This model basically explains the molecular steps involved in generating and utilizing the energy provided by ATP binding/hydrolysis necessary to promote the essential structural changes required for transport (Fig. 1.3) (Lu *et al.*, 2005; Zaitseva *et al.*, 2006).

What might be the precise nature of the chemical mechanism of ATP hydrolysis? It has been shown that the NBDs are capable of cleaving the γ-phosphate of ATP and that the motor domains display positive cooperativity during this process (Davidson *et al.*, 1996; Nikaido *et al.*, 1997; Benabdelhak *et al.*, 2005; Zaitseva *et al.*, 2005b). Moreover, a chemical reaction rather than diffusion controlled processes is the rate-limiting step in hydrolysis. In fact, two general mechanisms can be proposed to explain the hydrolysis of ATP: 1. general base catalysis (Fersht, 1997), and 2. substrate assisted catalysis or SAC (Dall'Acqua and Carter, 2000). The first candidate for the general base catalysis in this case is the glutamate at the end of the Walker B motif. This was proposed to be responsible for proton abstraction or polarization of a catalytic water mediating contact between the γ-phosphate and the Walker B glutamate (Orelle *et al.*, 2003). In this model, this process should be rate-limiting in terms of enzyme kinetics. In studies with the HlyB-NBD, designed to distinguish between a general base and SAC mechanism, hydrolysis rates in D_2O (rather than H_2O) and pH profiles of activity of the WT and an E→Q mutant, were determined. This led to the conclusion that substrate assisted catalysis is the predominant mechanism in hydrolysis of ATP (Zaitseva *et al.*, 2005a). Importantly in this case, the pK_a values of the microenvironment at the ATP binding site support the idea of SAC. The C-loop, which completes the composite ATP binding site, drastically influences the pK_a values of both amino acid residues and ATP (Hanekop *et al.*, 2006).

In the course of the above described and similar experiments from other groups (Shyamala *et al.*, 1991; Davidson and Sharma, 1997; Moody *et al.*, 2002; Verdon *et al.*, 2003), it became clear nevertheless that both the glutamate at the end of the Walker B and the H-loop histidine play crucial roles in catalysis and likely form a 'catalytic dyad'. The histidine residue forms hydrogen bonds with the glutamate side chain and the γ-phosphate. These interactions position and stabilize the key residues, including the glutamate, in the optimum orientation for cleaving the P-O bond of the γ-phosphate. However, whether the histidine fixes the glutamate in a position competent for hydrolysis, or *vice versa,* remains an open question. Certainly, it is clear that deletion of either one of these residues can greatly decrease or abolish ATPase activity due to the resulting higher intrinsic flexibility of either the glutamate or the histidine preventing a correct alignment for catalysis of ATP (Smith *et al.*, 2002; Zaitseva *et al.*, 2006).

Comparison of ABC-NBD dimer structures reveals that either ADP or ATP is bound in the respective substrate-binding site, as expected from the mechanism of nucleotide-induced dimerization observed in biochemical experiments. Unfortunately, despite detailed insights into the network of interactions between nucleotide and NBD described above, no conclusion can be drawn whether one or two ATP molecules are hydrolysed per catalytic cycle of the ABC protein. Biochemical analyses similarly provide no definite conclusion with respect to the stoichiometry of ATP consumption (Senior *et al.*, 1995; Sauna and Ambudkar, 2001; Patzlaff *et al.*, 2003;). This difficulty is even more evident when huge allocrites are to be transported across the membrane, like the 110 kDa toxin HlyA (Holland *et al.*, 2003), where additionally the proton motive force is apparently required during transport (Koronakis *et al.*, 1991).

Our journey through the catalytic cycle of an ABC-NBD starts with the apo (i.e. nucleotide

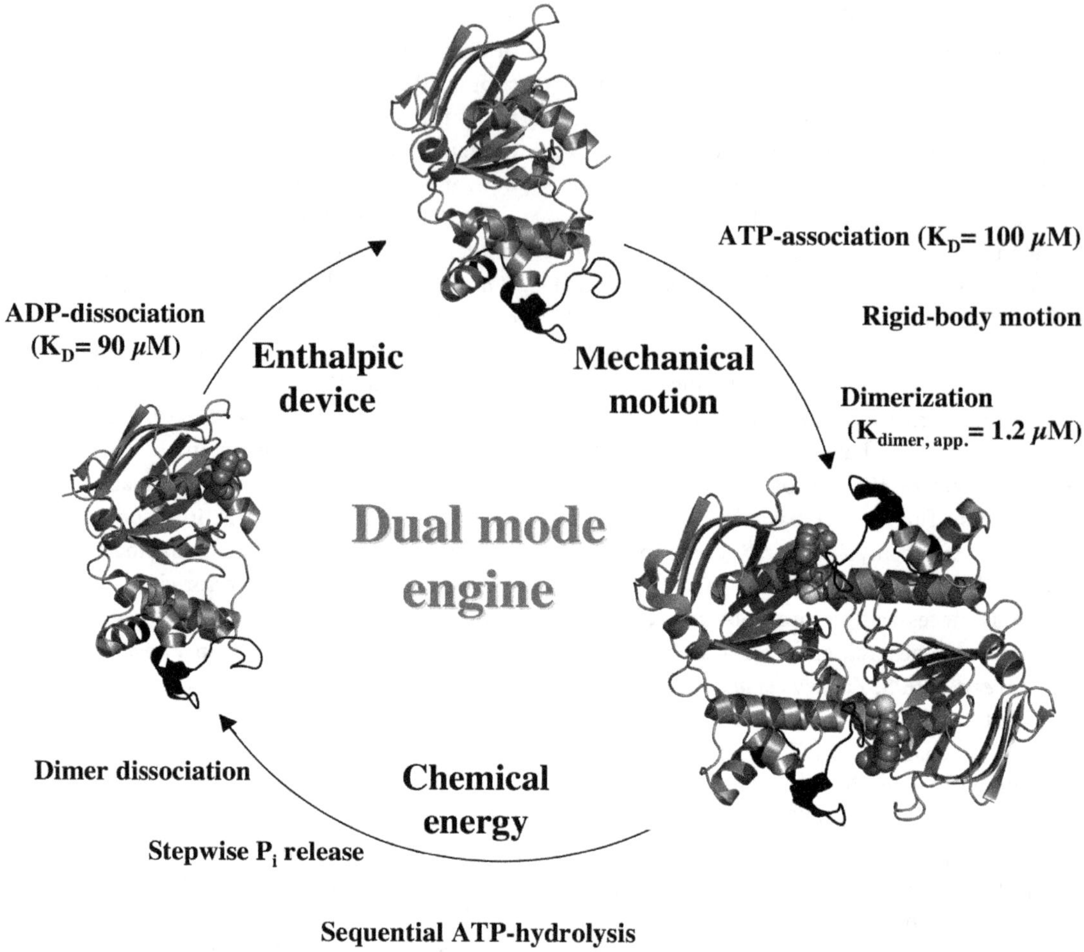

Figure 1.3 Schematic representation of the catalytic cycle of the HlyB-NBD. Colour-coding is identical to Fig. 1.2A. The cycle starts with the nucleotide free state of the monomer (pdb entry: 1MT0). Binding of ATP to both monomers induces dimer formation (ATP in brown and Mg^{2+} in yellow; pdb entry: 1XEF). After hydrolysis of ATP the dimer dissociates and ADP is subsequently released (pdb entry: 2FF7). A detailed description of each step can be found in the corresponding section of the text. A colour version of this figure is located in the plate section at the back of the book.

free) form of the motor domain (Fig. 1.3). When ATP is present and binds in its cognate binding pocket, dimerization and an inward rotation of the helical domain occur. The Q-loop and Pro-loop regions are involved in this movement that affects both subdomains of the NBD. When ATP is hydrolysed and ADP remains in the binding pocket, the Walker B and D-loop region, together with helix 6, in addition to the Q- and Pro-loop, are involved in structural realignment and dimer dissociation. However, an ADP-bound dimer is predicted to be unstable due to steric clashes of the opposing D-loops. Alternatively, in the structure of the MJ0796 dimer it was proposed that a negative charge distribution, resulting from ATP hydrolysis, accumulates with an electrostatic potential sufficiently high to lead to disruption of the NBD dimer (Smith *et al.*, 2002). This proposal, however, is not compatible with the electrostatics observed in the HlyB-NBD dimer structure (Zaitseva *et al.*, 2006).

Following dimer disruption and in order to reset the NBD to the nucleotide free state, the Walker B and D-loop region, as well as helix 6, appear to act as mechanical hinges to return to the initial state. As pointed out before, the

Q- and Pro-loops connect the catalytic domain to the helical subdomain, which in turn interacts with the transmembrane domain. On the other hand the Walker B and D-loop regions are also involved in contacts between the two monomers. Together, all these key motifs provide possible communication pathways in cis and in trans during every step of the catalytic cycle. This applies to interactions both between the motor domains and between the NBDs and the transmembrane domains. An interesting feature of helix 6 in the HlyB-NBD is its structural displacement in the ADP-bound state that is not present in the nucleotide free or ATP-bound state. Since the affinity of ADP for the NBDs is similar to that of ATP, spontaneous dissociation of the dinucleotide is unlikely. For this reason we suggest that some chemical energy from ATP hydrolysis might be stored in helix 6 acting as a potential mechanical spring that ultimately promotes the release of ADP and restores the resting state ready for the next ATP-hydrolysis/transport cycle (Lu *et al.*, 2005; Zaitseva *et al.*, 2006).

Previous extensive analysis of the NBD dimer structures of MJ0769, MalK and HlyB (Smith *et al.*, 2002; Chen *et al.*, 2003a; Zaitseva *et al.*, 2006), interestingly, revealed cavities inside the NBD dimer that led to a surprising discovery. This was the presence of tunnel-like formations extending from the surface to the inside of the NBDs leading directly to the ATP binding pocket. Thus, the adenine moiety is completely solvent accessible in these structures whereas the ribose and the phosphate moieties are shielded from the solvent. Moreover, two amino acid residues (D551 and R611 for the HlyB-NBD) form a salt bridge about 6 Å above the γ-phosphate, creating a dead end for the tunnel preventing it reaching the terminal phosphate residue. This scenario is true also for the MJ0796, MalK, and HlyB-NBD dimers crystallized in the presence of ATP. However, the analysis is more interesting when we consider the additional effect of Mg^{2+}. So far, there is only one NBD dimer structure available with ATP/Mg^{2+} bound. Consequently, the following discussion is based on the analysis of the HlyB-NBD. Thus, in particular in this ATP/Mg^{2+} bound state the asymmetry of interactions between the monomers in terms of NBD-ATP and NBD-NBD contacts is significantly increased compared to the ATP-bound structure. This is also reflected in the cavity analysis in the presence of Mg^{2+} where the salt bridge between D551 and R611 is broken in one of the two monomers.

In the dimer crystal structure of the HlyB-NBD with ATP/Mg^{2+} present, all the prerequisites for ATP hydrolysis are in place – with one major exception, the histidine essential for catalysis in this structure is mutated to an alanine. The snap-shot captured in these circumstances may reflect the state just before elimination of the terminal phosphate of ATP and the tunnel in one monomer is continuous all the way to the phosphate group. This architecture therefore allows for diffusion of free phosphate out of the NBD leaving ADP in the binding pocket. Interestingly, mutation of these amino acids engaged in this potential gating mechanism in the HlyB-NBD (D551A and R611A) abolishes the positive cooperativity observed for this and other ABC-NDBs (Davidson *et al.*, 1996; Nikaido *et al.*, 1997; Zaitseva *et al.*, 2005b). Based on all these data we can propose a 'sequential mode' of ATP hydrolysis in accordance with earlier biochemical studies of other ABC transporters (Smith *et al.*, 2002; Chen *et al.*, 2003a; Janas *et al.*, 2003). A prerequisite of the model is that the dimer is stable with one ATP and one ADP bound. Dissociation in this case should occur at the latest after hydrolysis of the second ATP, consistent with the second tunnel until then being still occluded by a salt bridge. The principle of an asymmetric phosphate exit tunnel in the NBD dimer was demonstrated for the prokarytic ABC transporter HlyB from *E. coli* (Zaitseva *et al.*, 2006) and now more recently with the mammalian transporter associated with antigen processing (TAP) from *Rattus norvegicus* (Procko and Gaudet, 2008).

Conformational changes in the NBDs

As a complementary approach to the analysis of the mechanism of action of the ABC-NBDs, electron paramagnetic resonance (EPR) experiments have been performed to elucidate possible conformational changes occurring during the catalytic and transport cycles (Grote *et al.*, 2008). Complementary cross-linking studies with linkers of different length support the measured distance changes between the NBDs during the transport cycle determined by EPR. Similar experiments were conducted with the lipid flippase MsbA and the results indicated a significant closing motion of the NBDs upon hydrolysis of ATP (Borbat *et al.*, 2007). For the $MalFGK_2$ complex, cysteine replacements inside the Q-loop sequence were generated and EPR spectra recorded for the apo-, ATP-bound-, vanadate-trapped- and post-hydrolysis states of the complete maltose importer. The results showed that upon transition from the apo- to the ATP-bound/vanadate-trapped state a decrease in distance between the NBDs of roughly 1 nm was observed. Thus, both NBDs appear to come into closer contact upon nucleotide binding (Fig. 1.2B). In the crystal structures of apo- compared to ATP-bound MalK a closing motion of 4 Å between the Walker A and the C-loop of opposing monomers was observed (Chen *et al.*, 2003a; Lu *et al.*, 2005). In contrast, spin labels placed on the surface of MalK through cysteine mutants or the use of wild type cysteines in the regulatory domain of MalK (for a definition of regulatory domain see below), showed no movement when ATP was bound at the dimer interface in the absence of the transport domain. In the post-hydrolysis state when ADP is bound and the ABC transporter is reset, an increase in distance between the two NBDs was detected in the EPR spectra.

Strikingly, the motor domains of the maltose importer have an additional C-terminal domain of roughly 150 amino acids, the so called regulatory domain. Such C-terminal elongations can also be found in other ABC transporters. It has been shown that these extensions play a role in the activity of the transporter when they bind substances that reversibly inhibit the transporter (Gerber *et al.*, 2008) or interact with transcription factors and hence regulate the expression of the transport system at the genetic level (Joly *et al.*, 2004; Richet *et al.*, 2005). In summary, these results encourage the idea of a tweezer-like motion of the NBDs during the transport cycle as proposed earlier by Chen *et al.* (Chen *et al.*, 2003a). In this model, the NBDs are connected through a specific regulatory domain at the C-terminus of MalK in the cytoplasm, to a pivotal-point at the membrane interface. It is proposed that binding of ATP induces movements of the NBDs towards each other without affecting the regulatory domain in terms of a major structural reorientation. With the regulatory domain fixed and the NBDs moving the picture of a tweezer-like motion emerges.

Substrate binding proteins (SBP)/periplasmic binding proteins (PBP)

Surprisingly, ABC importers are only found in prokaryotes and archaea (Davidson and Chen, 2004; see also Chapters 2 and 3 for further details). Nevertheless, they are vitally important, delivering nutrients and trace elements to the cells. The auxiliary binding proteins transfer the allocrite to the cognate transporter and therefore ensure substrate specificity and transport directionality (Boos and Lucht, 1996). In the case of Gram-positive bacteria possessing only a single cell membrane, the substrate binding proteins (SBPs) are tethered to the membrane via a lipid anchor (Boos and Lucht, 1996; Gilson *et al.*, 1988; Kempf *et al.*, 1997) or are covalently attached to the transmembrane domain of the ABC transporter in some cases (van der Heide and Poolman, 2000). In Gram-negative bacteria, ABC-allocrites must first cross the outer membrane and this is facilitated by specific porins, proteins present in the outer membrane. The binding proteins necessary for the next step are located in the periplasm and are therefore sometimes referred to as periplasmic binding proteins (PBPs). To ensure effective delivery of substrates to the transporter in the inner membrane the concentration of binding proteins is in great excess compared to the amount of ABC transporters. For the maltose import system there are approximately 30,000 SBPs per 1000 molecules of the ABC transporter (Treptow and Shuman, 1985).

The mode of action of binding proteins is rather 'simple'. First, the molecule can be divided into two subdomains that are linked via one, two, or three segments. In general, the subdomains contain a five-stranded antiparallel β-sheet that is surrounded by α-helices. The substrate/allocrite binding site is located between both subdomains (also referred to as lobes). Upon binding of the substrate molecule both lobes move towards each other in a fashion described as a 'Venus fly trap' mechanism (Sack *et al.*, 1989; Miles *et al.*, 2002). Without substrate the binding proteins exist largely in the open conformation with both subdomains spread apart (Fig. 1.4A and B) (Quiocho and Ledvina, 1996).

In the following part of the chapter, we shall describe in detail the well studied BtuF and MalE (MBP, maltose binding protein) as typical examples of SBPs. BtuF is the binding protein of the cognate ABC transporter BtuCD from *E. coli* and resides in the periplasm. Both subdomains of BtuF consist of a central five-stranded β-sheet surrounded by α-helices connected by a single α-helix and thus belongs to the so-called group III SBPs (Lawrence *et al.*, 1998; Lee *et al.*, 1999; Clarke *et al.*, 2000;). This class of SBPs apparently does not undergo major structural changes upon ligand binding due to restrictions of the helices spanning the back of the protein. In fact, in the case of BtuF no opening of the lobes was observed in the apo-form (Karpowich *et al.*, 2003) compared to the ligand-bound form of the SBP (Borths *et al.*, 2002). However, a significant change in the organization of BtuF occurs when it is co-crystallized with its cognate ABC transporter (BtuCD). In this complex, the lobes of BtuF are spread approximately 4 Å further apart compared to the crystal structure of BtuF alone (Hvorup *et al.*, 2007). The substrate binding site for vitamin B_{12} is located in a deep cleft between both lobes of BtuF (Borths *et al.*, 2002) and each subdomain contributes three aromatic residues that interact with vitamin B_{12} in the binding pocket. Despite the fact that even in the presence of BtuCD only a small structural change is induced in BtuF, this seems to be sufficient to displace the ligand. Periplasmic loops from the transmembrane domains of BtuCD that invade

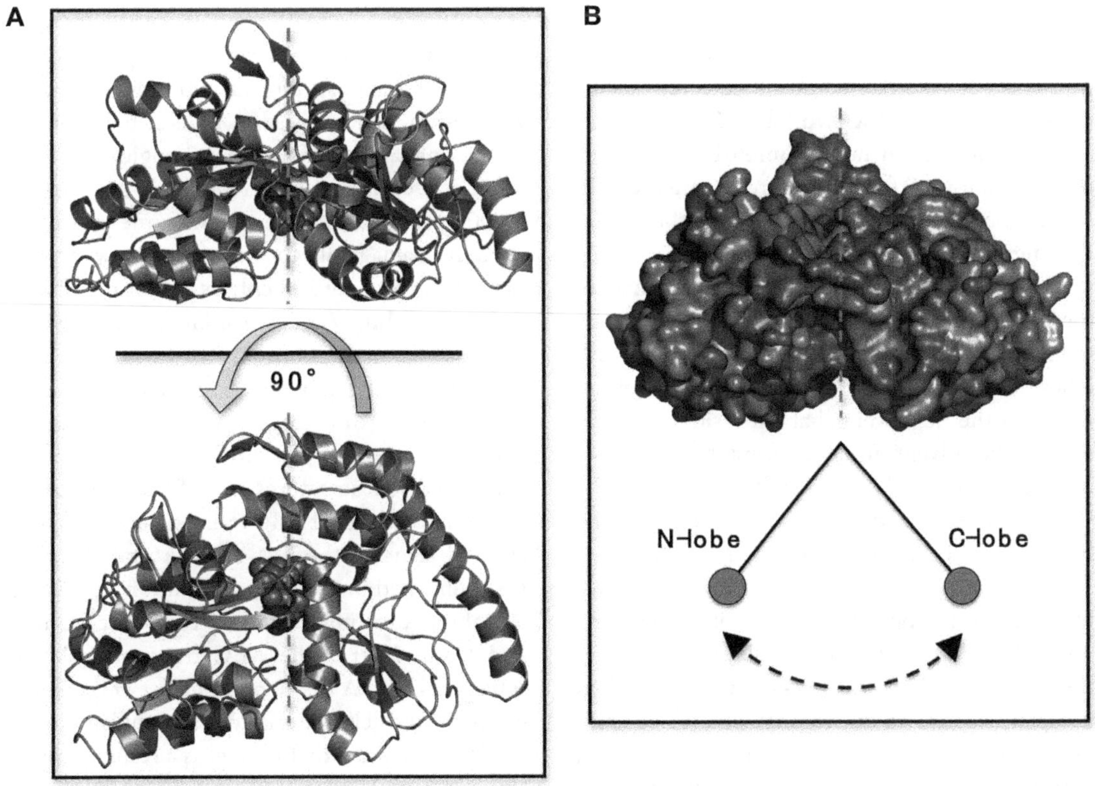

Figure 1.4 **A** In the upper part the side view of the periplasmic binding protein BtuF with bound vitamin B_{12} (magenta) is shown (pdb entry 1N2Z). The five-stranded antiparallel β-sheets are depicted in orange and blue, respectively. N- and C-lobe are connected by loop regions; here shown in green. In the lower part BtuF is rotated about 90° to look at the protein from above. **B** Structural superposition of the ligand-free (grey) and ligand-bound (blue) structure of the maltose binding protein. Upon binding of maltose a domain displacement leading to closure of the lobes of roughly 30° is observed. A colour version of this figure is located in the plate section at the back of the book.

the substrate binding site also participate in transferring B_{12} to the transmembrane translocation pore. For the Zn^{2+} SBP TroA a similar minor rearrangement is sufficient to displace the ion into the tranlocation pore (Lee *et al.*, 1999; Lee *et al.*, 2002). Interestingly, positive cooperativity and enhanced rates of transport can be observed when two or even more copies of a SBP are fused to the TMD of an ABC transporter (Biemans-Oldehinkel and Poolman, 2003).

The crystal structures of MBP in an open and a closed state were dependent on the presence of maltose (Quiocho *et al.*, 1997; Sharff *et al.*, 1992). In fact, it has long been assumed that an equilibrium between those two states exists in solution even in the absence of a suitable ligand. However, it took 15 years before this hypothesis was supported with the help of nuclear magnetic resonance paramagnetic relaxation enhancement (NMR-PRE) experiments and finally the results were surprising (Tang *et al.*, 2007). Indeed, a minor species of MBP in a closed-like conformation without substrate could be detected due to the convergence of the two lobes. However, this species, which is in rapid equilibrium with the open (ligand-free) form, is different from the holo (ligand-bound) state. Notably, since the ligand binding site is made up of negatively charged residues and consequently the convergence of these residues towards each other is energetically unfavorable. Therefore, the orientation of both lobes is distorted with respect to each other, exposing the negatively charged residues of one lobe to the solvent. In fact, a crystal structure of MBP exists where the ligand is bound only to the C-terminal lobe of the apo conformation of the binding protein (Duan and Quiocho, 2002).

The binding of maltose by MBP is necessary to bridge both lobes and induce domain closure. Thus, the analysis of mutants demonstrated that the stability of MBP decreases upon closure when no ligand is present, due to unfavorable exposure of distinct amino acid side chains (Millet *et al.*, 2003). In contrast, a different mechanism was observed in the glucose/galactose binding protein where a water-network mediates protein–ligand interactions and therefore the data from MBP cannot easily be transferred to the glucose/galactose SBP (Flocco and Mowbray, 1994; Quiocho and Ledvina, 1996).

An important property of the binding proteins is thought to be their participation in triggering the translocation process upon interacting with the TMDs of the transporter. Consequently, the structural characterization of transporters in complex with their SBPs has been a major goal in understanding ABC import systems. For further information on the mechanism of action and properties of SBPs several reviews are available (Boos and Lucht, 1996; Felder *et al.*, 1999; Quiocho and Ledvina, 1996).

Structure and functional analysis of ABC transporters

ABC importers

Crystallization of proteins is always a challenge – taking a complex molecule out of its dedicated environment (in the case of membrane proteins with the help of large amounts of detergents), concentrating it to high concentrations and precipitation under conditions leading to crystals, followed by exposure to an ultra-high-energetic x-ray beam, can easily lead, at least occasionally, to the appearance of non-physiological states of the desired protein. Nonetheless, roughly 53,000 protein structures are now present in the Protein Data Bank to date (http://www.rcsb.org) although only 168 entries are for membrane proteins (http://blanco.biomol.uci.edu/Membrane_Proteins_xtal.html).

After at the last 15 years of accumulating biochemical knowledge of ABC transport systems, and more recently the elucidation of several crystal structures of NBDs, the atomic structure of an intact ABC transporter was eagerly awaited. Finally, the first such crystal structure was published in 2001 at a resolution of 4.5 Å in the absence of nucleotide and allocrite (Chang and Roth, 2001). The ABC transporter MsbA transports lipid A to the outer leaflet of the *E. coli* envelope and cells die if lipid A accumulates in the inner leaflet of the membrane, as a result of mutations or complete loss of MsbA. Disappointingly, the crystal structure of MsbA did not agree with biochemical data or MD simulations (Campbell *et al.*, 2003; Haubertin *et al.*, 2006; Stenham *et al.*, 2003) and was retracted in 2006 due to problems in data analysis. The structure was re-published in 2007 at a resolution of 3.7 Å with AMP-PNP bound to the NBDs (Ward *et al.*, 2007). The next crystal structure of an ABC transporter was published one year later (Locher *et al.*, 2002). This was an importer for vitamin B_{12} from *E. coli*, BtuCD (Fig. 1.5A). The structure was solved at 3.2 Å resolution and was completely different from MsbA. The BtuCD structure consists of two copies of BtuC, the transmembrane domain and two copies of the motor domain, BtuD. The transmembrane subunit (BtuC) consists of ten transmembrane helices making up a total of 20 helices in the functional dimer.

What might be the function of accessory transmembrane helices, compared to the more usual number of 12? This structure of Btu was without bound nucleotide in the NBDs and also without allocrite but cyclovanadate, known to inhibit ABC-ATPases, is bound to the motor domains. The TMDs of this BtuCD structure are open towards the periplasm and closed towards the cytoplasm of the cell and this was interpreted to be an equivalent to the ATP-bound state. As discussed above, when ATP is bound to the NBDs, dimer formation is induced and consequently the interface of dimerization is buried from the solvent (Smith *et al.*, 2002; Chen *et al.*, 2003a; Zaitseva *et al.*, 2005a). In fact, the solvent accessible surface area of the BtuD interface was about 740 $Å^2$ which pointed towards crystal induced contacts rather than a physiological protein–protein interface (Bahadur and Zacharias, 2008). Nonetheless, this apparently more authentic structure of an ABC transporter revealed some important and interesting aspects. Viewed from the periplasmic side, each TMD has three exposed and conserved arginines that can interact with two glutamates, one on the very left and one on the right hand side of the binding protein BtuF (Borths *et al.*, 2002; Locher *et al.*, 2002). The substrate translocation pathway appears to

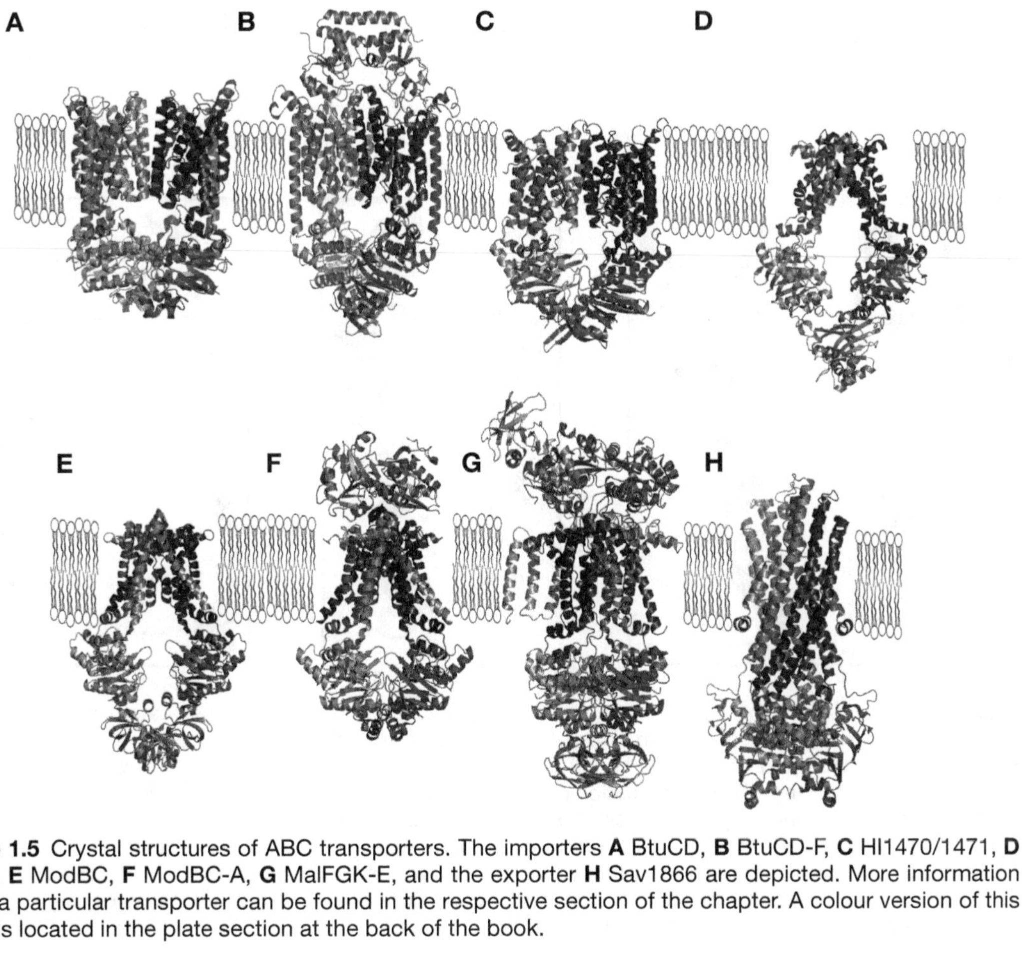

Figure 1.5 Crystal structures of ABC transporters. The importers **A** BtuCD, **B** BtuCD-F, **C** HI1470/1471, **D** MetNI, **E** ModBC, **F** ModBC-A, **G** MalFGK-E, and the exporter **H** Sav1866 are depicted. More information about a particular transporter can be found in the respective section of the chapter. A colour version of this figure is located in the plate section at the back of the book.

be formed by helices five and ten from opposing BtuC subunits, respectively (Fig. 1.6A). TM5 of one subunit packs against helix 10 of the second TMD. The resulting outward-facing cavity is large enough to incorporate the substrate vitamin B_{12} (Fig. 1.7A and B).

A closer look at the transmembrane part of BtuCD reveals an eye-catching structural element between TM helix 6 and TM helix 7. About 16 amino acids form an L-shaped helix–turn–helix motif (the 'L-loop' motif) which interacts extensively with the structurally diverse region (Schmitt *et al.*, 2003) in the helical domain of the NBDs (Fig. 1.8C). Examining the sequence of the L-loop reveals that this structural element contains the importer characteristic EAA-motif described earlier for bacterial importers (Mourez *et al.*, 1997). Thus, structural changes described for the helical domain of NBDs during the catalytic cycle may be transduced to the membrane embedded part of the transporter via interaction with the L-loop. Equally, a reverse rearrangement of distinct TMD segments upon allocrite binding or its subsequent release could be envisaged to be communicated to the motor domains.

It took five years before the next structure of an ABC importer was solved (Pinkett *et al.*, 2007). This was a putative metal-chelate-type ABC transporter HI1470(NBD)/1471(TMD) from *Haemophilus influenzae* (Fig. 1.5C) that displays high sequence identity with BtuCD, and both proteins are members of the same binding protein dependent transporter family. However, the ligand of HI1470/1471 has not been identified so far. The resolution of 2.4 Å for the HI1470/1471 structure is a substantial improvement in comparison to BtuCD. Moreover, HI1470/1471 composed of two copies of HI1470 and HI1471, respectively, is crystallized in an inward facing conformation. Interestingly, cavity analysis of the outward facing conformation of BtuCD and the inward facing state of HI1470/1471, both resulted in a hypothetical allocrite translocation pore formed by the transmembrane domains of approximately 10 Å at the widest point. In contrast to BtuCD, the transmembrane domains of HI1470/1471 are composed of ten helices and

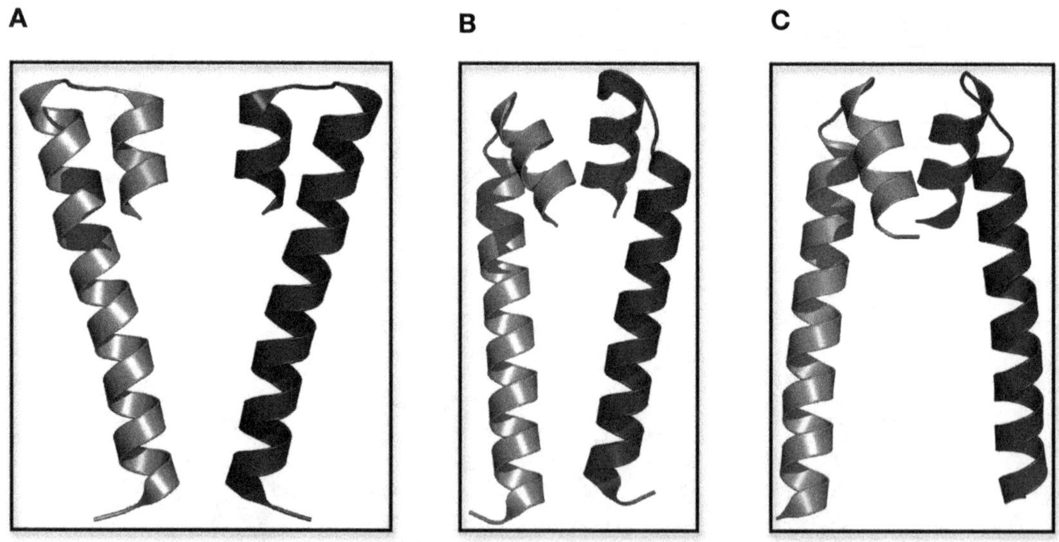

Figure 1.6 Helix 5/5a of the ABC importers has a pivotal role in forming the translocation pore **A** BtuCD, **B** BtuCD-F, and **C** HI1470/1471. The structure of BtuCD **A** was crystallized in an outward facing conformation with the translocation pore accessible from the periplasm. The structure of HI1470/1471 **C** reveals an inward facing conformation where the translocation pathway is only open to the cytoplasm. Figures **A** and **C,** where only helices 5 and 5a are depicted, show the involvement of this structural element in a postulated gating mechanism. In **B** the arrangement of the helices from the BtuCD-F structure is shown. Interaction of the substrate binding protein appears to provoke a conformation of the TMDs where the translocation pore is accessible neither from the periplasm nor the cytosol. The light and dark shaded helices belong to different transmembrane domains of the transporter.

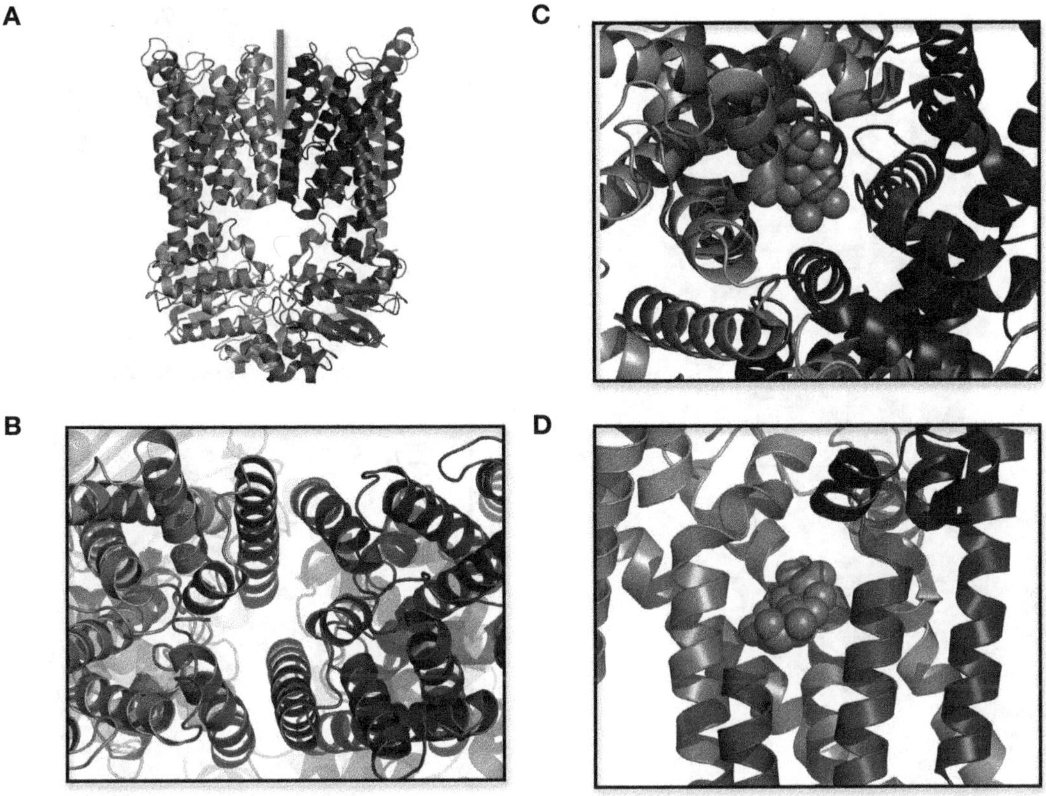

Figure 1.7 A Quaternary structure of the vitamin B_{12} importer BtuCD. **B** BtuCD is viewed from above through the postulated substrate translocation pore. Helices from each TMD subunit lining the translocation pore are coloured red and blue, respectively. **C** View from above into the translocation pore of the $MalEFGK_2$ transporter with bound maltose (magenta). **D** Side view of the maltose binding site in the transmembrane domains of the $MalEFGK_2$ importer complex. For clarity some helices have been omitted. A colour version of this figure is located in the plate section at the back of the book.

TM5 and TM10 mediate bulk contact between both TMD subunits. In HI1470/1471, TM-helix 5 deviates about 20° from its position in the BtuCD structure (Fig. 1.6C). This change implies a reorganization of the whole transmembrane domain architecture resulting in the inward facing conformation of HI1470/1471. Further differences include the relative orientation of the NBDs which are further apart in HI1470/1471 compared to the BtuCD structure. This arrangement likely presents therefore the post hydrolysis state of the NBDs after cleaving off the γ-phosphate of ATP (Chen *et al.*, 2003a; Lu *et al.*, 2005; Pinkett *et al.*, 2007).

The interpretation of the crystallized states of these proteins inevitably has to be highly descriptive. This is particularly so in the case of HI1470/1471, which is presumed to be a metal-chelate transporter. This assumption is based only on sequence homology and unfortunately, there is no biochemical data and no allocrite identified so far. Our ability to identify specific structural elements participating in a transition from an inward to an outward facing conformation is therefore limited in this case. Nevertheless, structural changes can be deduced and these mainly involve rigid body movements arising from the different arrangement of the NBDs in BtuCD and HI1470/1471. In fact, we cannot exclude some contribution to the deduced rigid body motions that result from lateral pressure of detergent, compensating for the absence of the native membranous environment. Alternatively, a contribution to apparent rigid body motions could even be caused by crystal lattice contacts.

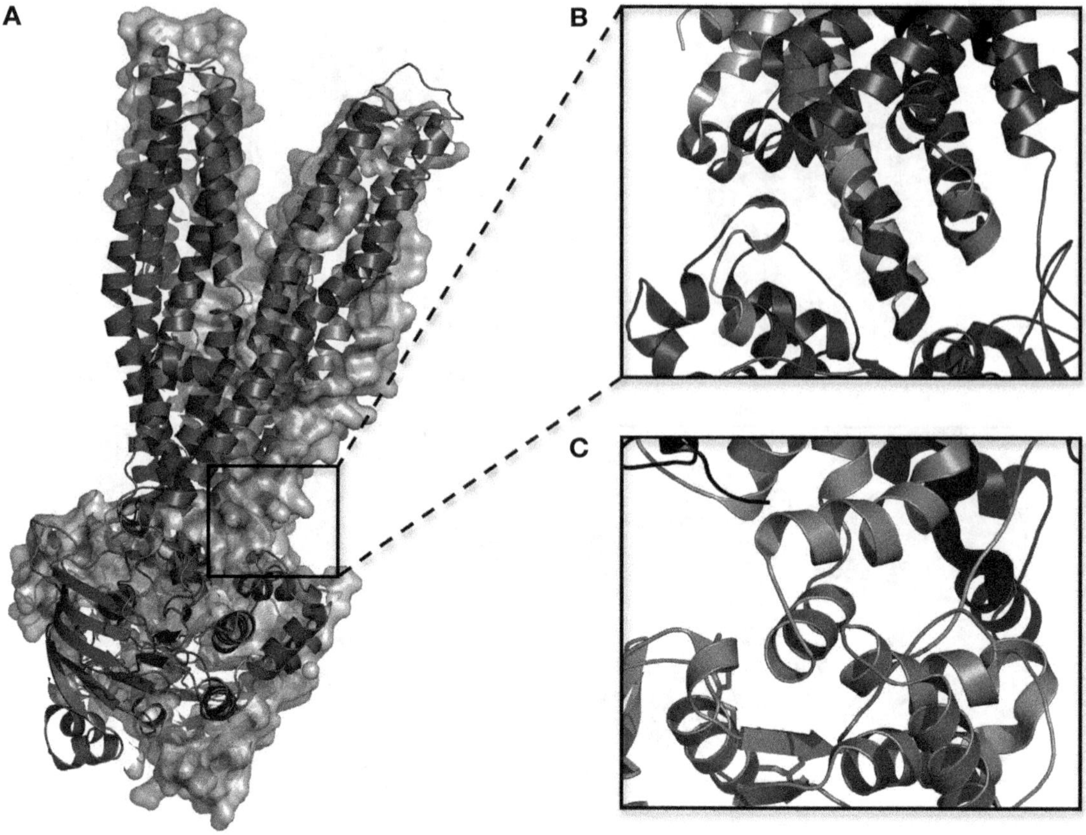

Figure 1.8 **A** Cartoon representation of the ABC exporter Sav1866. **B** The inset is a magnification of the postulated interaction site between TMDs and NBDs. In blue and slate intracellular loops from both membrane subunits are shown. Residues in red represent the TEVGERG sequence found to interact extensively with the intracellular loops. The C-loop is depicted in pale green. **C** Interaction site between TMD and NBD of the $MalEFGK_2$ complex. The transmembrane part is shown in grey and the NBD in brown. In the L-loop from the transmembrane domain the EAA motif characteristic for ABC importers (red) is highlighted. The C-loop is depicted in pale green and the D-loop in cyan. A colour version of this figure is located in the plate section at the back of the book.

To sum up: although the structures of two distinct, albeit closely similar ABC transporters, are available in two states probably resembling the extreme scenarios of the transport process with an inward and outward facing configuration, these findings are not sufficient to reveal the mechanistic basis of the transport process. In 2007, one year after the structure of HI1470/1471 was published, BtuCD in complex with its cognate binding protein BtuF was solved (Fig. 1.5B) (Hvorup *et al.*, 2007). Additionally as shown in Fig. 1.5E, another putative ABC transporter $ModB_2C_2$ from *Archaeoglobus fulgidus*, responsible for molybdate import, in complex with the binding protein ModA expanded the list of ABC transporter crystal structures (Hollenstein *et al.*, 2007b; Gerber *et al.*, 2008). Last but nor least, Fig. 1.5H shows the complex of MalFGK, together with the maltose binding protein in the presence of substrate and ATP/Mg^{2+} that has recently been solved (Oldham *et al.*, 2007).

BtuCDF was resolved to 2.6 Å and displays major differences when compared to the structure of the vitamin B_{12} importer without the complexed binding protein. The most intriguing observation is the difference in conformation between the two transmembrane subunits (BtuC) leading to closure of the translocation pathway both from the cytoplasm and the periplasmic space. Vitamin B_{12} is no longer present in the TMD region or in the binding pocket of the SBP BtuF. This has adopted a conformation with both lobes spread apart, indicating a post ATP hydrolysis state and consistent with release of Vitamin B_{12} into the cytoplasm. In summary, these findings are consistent with the new BtuCDF structure being a posttranslocation intermediate.

The two lobes of BtuF are not identical and therefore different loops of the BtuC subunits interact with the binding protein generating asymmetry in the interaction. Interestingly, the loop of TM helix 5, seen to participate intimately in TMD–TMD interaction, inserts into the binding pocket for vitamin B_{12} and may force displacement of the allocrite. TM helices 3 to 5a in BtuCDF interact differently with the C- and N-lobe of the binding protein. On one side, the orientation of the helices of one BtuC subunit is similar to the arrangement present in the BtuCD structure in the absence of binding protein, resembling an outward facing conformation. In the second BtuC subunit, the helix conformation resembles the one present in HI1470/1471 with its inward facing orientation. Consequently, this asymmetry could present a state that 'allows' the transporter to switch to the outward or inward-facing conformation. To support this rather unexpected finding of asymmetry, obviously induced by TM helix 5, spin labels were introduced in the region preceding and following this central structural element. Comparison of EPR spectra from BtuCD and BtuCDF revealed differences in mobile and immobile components that can be explained by rearrangements of structural elements observed in the crystal structure (Fig. 1.6B) (Hvorup *et al.*, 2007).

Also in 2007, the structure of the putative molybdate importer ModBC from *Archaeoglobus fulgidus*, in complex with the binding protein ModA, was solved at 3.1 Å resolution. The protein has 12 transmembrane helices (Fig. 1.5F) with helices 3 and 5 contributing to the gating region, which closes the translocation pathway towards the extracellular space (Hollenstein *et al.*, 2007b). Molybdate is bound to ModA in an oxyanion binding site providing an octahedral coordination sphere. ModA in turn interacts with extracellular loops of the TMD region of ModB. This interaction involves charged amino acid residues as demonstrated for BtuCD (Hvorup *et al.*, 2007). The NBDs are separated by a gap between the Walker A and C-loop which reflects the nucleotide free state. The focus of this structure is on the interaction of the ModBC units with the binding protein. In the absence of ATP, an inward facing conformation of the TMDs is visible. The binding protein displays a closed conformation, hence trapping the substrate inside. The cytosolic L-loops containing the EAA motif are also present in the ModBC structure and interact with the helical domain of the NBDs. In line with the generally accepted model, the structure is consistent with the binding of ATP inducing the motor domains to converge to dimerize. This would produce a rigid body motion of the helical domain relative to the catalytic domain. Through an interaction between the helical domain and the L-loops, the required structural rearrangement should flip the transmembrane part to an outward facing conformation. At the same time this latter movement will pull apart both lobes of the binding protein, inducing the open conformation of ModA and consequent displacement of molybdenum into the translocation cavity. Thus, despite obvious differences in the helix arrangement of ModBC compared to BtuCD and HI1470/1471, these structural studies have led to a plausible mechanism of allocrite import. As a note of caution, these ideas are still based on nucleotide free structures, SBPs analysed in different conformations with and without bound substrate, and *presumed* physiological inward and outward facing orientations.

Further experiments with a related molybdate/tungstate importer from *Methanosarcina acetivorans* recently revealed a striking example of how the structure of an ABC transporter can be modified through evolution to facilitate more efficient regulation of its activity. In contrast to the importer from *Archaeoglobus fulgidus*, the NBDs of this second molybdate transporter, as shown in Fig. 1.5E, contain a C-terminal extension that functions to inhibit the transporter when a certain concentration of substrate inside the cell is exceeded (Gerber *et al.*, 2008). Like the binding protein, ModA from *Archaeoglobus fulgidus*, an oxyanion binding site is present in the regulatory cytoplasmic extension of the NBDs where tungstate binds. The inhibitory effect results from interaction of the regulatory domains forcing the NBDs apart, resulting in an inward facing conformation of the transmembrane entity as seen in the crystal structure. The importer is thus locked in a configuration where ATP hydrolysis is precluded because binding of nucleotide is unable to bring the motor domains into close contact. Cross-linking studies with cysteines introduced in TM helix 4 indicated an inward directed movement of the transmembrane domain when ATP was omitted. This was consistent with a conformational change leading towards an outward facing conformation as observed for the BtuCD or MalFGK crystal structures (Locher *et al.*, 2002; Oldham *et al.*, 2007).

A similar elegant story of regulation comes from the methionine importer MetNI from *E. coli*, solved recently at a resolution of 3.7 Å. This transporter has the smallest set of TM helices of all ABC transporters described structurally so far (Fig. 1.5D), with 5 helices in each TMD. In this case, the regulatory extension of the NBDs, stringently controls methionine import via inhibition responding to the intracellular concentrations of methionine (Kadaba *et al.*, 2008; Kadner, 1975). Since inhibition is accomplished from the cytoplasmic side of the membrane, this phenomenon is called 'trans-inhibition'. Interestingly, the presence of an odd number of TM helices causes the C-terminus of the TMDs to reside in the cytoplasm whereas the N-terminus is in the extracellular space – an alignment not described for any other ABC transporter to this point. With methionine bound to the regulatory domain of the NBD, the NBDs are forced apart and as a result the rate of ATP hydrolysis and therefore substrate import is markedly decreased. An explanation for this fine-tuning of import function might reflect a need to conserve cellular energy. Since transport processes which work against concentration gradients need energy to fuel translocation of molecules, the effect of trans-inhibition immediately saves cellular energy by inactivating the transporter when the need for certain substances is satisfied.

In the structure of MetNI a large distance between the NBDs is observed. Such a configuration was first described for the lipid flippase MsbA whose crystal structures were subsequently withdrawn. Nonetheless, these structures suggest the possibility of substantial movement to separate the NBDs under some conditions. This has now actually been verified using EPR measurements to determine the distance between the motor domains in the nucleotide free and ATP bound states for MsbA in a reconstituted system (Borbat *et al.*, 2007). The relative motion of the NBDs was determined to be roughly 30 Å. In contrast to MetNI and ModBC from *Methanosarcina acetivorans*, where the substrate itself mediates inhibition of the transporter, a slightly different regulatory function can be assigned to the regulatory domains of MalK. A transcriptional activator (MalT) binds to MalK and thereby inactivates the ATPase activity (Joly *et al.*, 2004; Richet *et al.*, 2005).

An interesting approach to functional analysis that we have not discussed so far is overlaying of the structures of different transporters (Pinkett *et al.*, 2007). A good starting point is BtuCD with its outward facing conformation and HI1470/1471 which is open to the cytoplasm, reflecting the inward facing arrangement of the TMDs. The result of overlaying these structures is that possible translational movements along certain symmetry axes can be envisaged, in this case, through a reversible interconversion of an outward-facing into an inward-facing orientation. This projected movement can be described as a screw-like rotation, which aligns the NBDs of both transporters and propagates a conformational change to the transmembrane domains. Thus, structural changes in the motor domains, resulting from nucleotide binding or hydrolysis, as well as rearrangements of the transmembrane helices upon binding of the allocrite can apparently be transmitted in either direction, via simple geometrical movements.

The maltose import system of *E. coli* ($MalFGK_2$) has been subjected to extensive structural, biophysical and biochemical analysis and is probably the best studied prokaryotic ABC transporter (Davidson and Chen, 2004). MalF and MalG form a heterodimer in the membrane and MalK is the cytoplasmic motor domain. A striking feature of the heterodimeric maltose transporter is the obvious difference in the two transmembrane domains. MalF consists of eight transmembrane helices whereas MalG has only six. Additionally, MalF contains an extracytoplasmic domain P2 adopting the well-known Ig-like fold that interacts with the N-terminal domain of the maltose binding protein (MBP or MalE).

MalE delivers maltose (or maltodextrins) to the membrane-embedded transporter, following docking onto the MalFG complex before release of maltose into the transport channel. MalE exists in open and closed conformations that are in equilibrium in solution with or without bound substrate. In the open conformation, the affinity towards the substrate is substantially decreased. Thus, the open conformation is the one adopted while MalE is bound onto the cognate transporter. There are two possible explanations for this observation: (i) maltose binds to a low affinity site on the TMDs of the transporter or (ii) that maltose simply diffuses directly into the cytoplasm. When MalE binds to MalFG, it stimulates ATP hydrolysis and can result in vanadate trapping. Vanadate is an analogue of inorganic phosphate and mimics the transition state of the γ-phosphate moiety of ATP during hydrolysis. In a reconstituted MalE-MalFGK_2 system trapped with vanadate, no maltose can be detected. This is a clear indication that conformational changes occur in a concerted manner and maltose in such experiments is released only when the trapped transition state is formed (Chen *et al.*, 2001). This hypothesis is supported, as shown in Fig. 1.5G, through the crystallization of the transition state of the MalE–MalFGK_2 complex (Oldham *et al.*, 2007).

The resulting working model for the MBP–MalFGK_2 complex depends on certain assumptions and observations. The maltose binding protein has the ligand bound and exists in the closed ligated form. Binding of MalE to the MalFG transmembrane domains induces a conformational change in the binding protein producing the open form in which the affinity for maltose is substantially reduced. At the same time, an extracellular loop from MalG inserts into the cleft where the maltose is bound and aids in the displacement of the allocrite. As a result, maltose is then able to occupy a binding site in MalF. MalG does not participate in the maltose-binding site and in the structure, only one substrate-binding site in the transmembrane region is visible. Since the ATP concentration inside a cell is much higher than the K_M value of the NBDs towards the nucleotide, it is presumed that ATP occupies the motor domains even in the absence of maltose in the periplasm. Therefore, it is proposed that to prevent hydrolysis the nucleotide binding domains are held far apart. In this model (Oldham *et al.*, 2007), the signal to hydrolyse ATP results from the binding of the MalE/maltose complex to MalFG, causing a conformational change in the transmembrane helices. Consequently the motor domains come into close contact to form a functional dimer able to catalyse the elimination of the γ-phosphate. It is also proposed that the hydrolysis of ATP is required to reset the transporter and complete the transport cycle.

It should be mentioned that the crystal structure of the MalEFGK_2 complex was successfully obtained by creating an NBD mutant, in which a glutamate at position 159 was replaced by a glutamine (E159Q). This mutation completely prevents hydrolysis of ATP and traps the ABC transporter in a conformation where both motor domains, although in close contact with ATP bound, remain as a stable dimer.

There are two other crystal structures of ABC importers described with bound SBP, BtuCDF and ModB_2C_2A. In contrast to MalEFGK_2, these structures are devoid of the allocrite, for which maltose is found within the transporter pathway. The absence of B_{12} is surprising since the affinity of vitamin B_{12} is reasonably high, in the range of 15 nM (Cadieux *et al.*, 2002). ATP was also omitted from the crystallization conditions for BtuCDF and ModABC since it interfered with crystal lattice formation. The maltose import system constitutes therefore the most complete view of an ABC transporter to date.

ABC exporters

ABC exporters are of a special interest because of their participation in some cases in the induction of multidrug resistance (MDR) in humans undergoing chemotherapy (Kartner *et al.*, 1983). The most fascinating feature of these mammalian exporters (and their bacterial equivalents) is the wide variety of substrates that can be transported. The only apparent common characteristic of these compounds is their hydrophobicity – size or steric arrangements are of minimal relevance (Sharom, 2008; see also Chapters 4 and 9). There are of course examples of exporters with well characterized, apparently single allocrites, as in the case of the Haemolysin B transporter from *E.*

coli. The allocrite in this case is a 110 kDa toxin secreted by enterohaemorrhagic *E. coli* strains via the Type I secretion system. Interestingly, however, by fusing the C-terminal secretion signal of this toxin to other proteins, many other proteins can be secreted (Blight and Holland, 1994; Fernandez *et al.*, 2000; Fraile *et al.*, 2004).

Much of our knowledge of the action of membrane proteins such as P-gp depends on *in vitro* experiments and although these are carried out under controlled conditions, the physiological significance *in vivo* of the many chemical compounds tested with the endogenous ABC exporters remains to be established. An interesting feature of ABC exporters such as Pgp (MDR-1) is that transport can also be affected by substances that bind to certain regulatory sites that then modulate transport of the true allocrite. However, the molecular details of allocrite and modulator binding sites are not yet clearly established (Sharom, 2008).

The only structure of an ABC exporter derived so far (see Fig. 1.5H) is the putative MDR (multidrug resistance) protein Sav1866 from *S. aureus* (Dawson and Locher, 2006). Sav1866 is homodimeric protein (a half size transporter) with 12 transmembrane helices and the structure was derived from crystals scattering up to 3 Å. A remarkable feature of Sav1866 is the fact that the transmembrane helices of the TMDs are twisted and embrace each other. This arrangement leads to extensive interactions between the TMD and NBD including, surprisingly, crossover interactions between the TMD of one monomer and the NBD of the opposing monomer. Crystals were formed in the presence of ATP but ADP is present in the structure and the NBDs remain in close contact with the Walker A motif and the C-loop interacting in a hydrolysis competent orientation. Another structure of Sav1866 with a non-hydrolysable ATP analogue (Dawson and Locher, 2007) revealed the same outward facing orientation of the transmembrane domain as in the structure with ADP. The L-loop of the TMDs in importers discussed earlier that interacts with the NBDs apparently to transmit reciprocal structural changes between the cytoplasmic and the transmembrane domains, is not, however, present in the Sav1866 structure. Instead, two intracellular loops protruding from one TMD and connecting helices 2 and 3 and helices 4 and 5, both interact with the cis and trans NBDs (Fig. 1.8A and B). Notably, despite the absence of the L-loop, the site of interaction in the NBD, still coincides with the structurally diverse region preceding the C-loop sequence.

A closer look at the putative translocation pathway of Sav1866, reveals mostly hydrophilic amino acid residues lining a cavity in accordance with the expected properties of a low affinity-binding site, necessary for the extrusion of hydrophobic drugs from the cell. When the crystal structure of Sav1866 was published, no biochemical evidence for transport activity or any information about potential substrates was available. However, Sav1866 displayed increased ATPase activity in the presence of verapamil, tetraphenylphosphonium, Hoechst 33342 or vinblastine – all compounds previously used in functional assays with other multidrug transporters. Importantly, a recent study of Sav1866 provided direct evidence for a drug transport activity. Thus, transport of Hoechst 33342 by purified and reconstituted Sav1866 in proteoliposomes was demonstrated (Velamakanni *et al.*, 2008).

Estimations from the crystal structures of the importer BtuCD and the exporter Sav1866 indicated that considerable amounts of the transmembrane part of Sav1866, in contrast to BtuCD, protrude into the cytoplasm (Lomize *et al.*, 2006a; Lomize *et al.*, 2006b). This provides a hint that interaction with drug molecules for export is established via the TMDs of the transporter since a large interface of possible interactions is made available. Transport studies *in vitro* in fact show that drug molecules for MDR ABC transporters can be picked up from either the cytoplasm or the membrane (Bolhuis *et al.*, 1996; Sharom *et al.*, 2005). However, classical models of the action of MDR transporters envisage direct access of drugs from the inner leaflet of the bilayer into the transporter.

With structural data at hand and a system that can be described by biochemical and biophysical approaches, the foundations for the characterization of at least one bacterial MDR ABC transporter have been laid. Future developments can be expected to provide a fundamental understanding of the Sav1866 molecular machine that will be usefully extrapolated to the mechanism of action of the mammalian ABC drug transporters.

How ABC transporters transport substrates

Although we have outlined above how ABC transporters like MalEFGK might function, some problems with the models still remain. How can one imagine all the requirements of the process of substrate transport via an ABC system? To address this question in relation to importers, the initial state of an ABC transporter remains to be established. In principle, two contrary scenarios are possible when no ATP is bound and the transporter is in its resting state: (i) the translocation pore is closed to the extracellular space and open to the cytoplasm (Oldham *et al.*, 2007) or (ii) the translocation pathway is open to the exterior and closed towards the inside of the cell (Locher *et al.*, 2002).

Two distinct mechanisms of ABC transporter function that depend on the action of the motor domain are presently in vogue. These are the *processive clamp model* (Smith *et al.*, 2002;Chen *et al.*, 2003a; Janas *et al.*, 2003; van der Does and Tampe, 2004; Zaitseva *et al.*, 2005a) and the *alternating site model* (Senior *et al.*, 1995; Urbatsch *et al.*, 1995; Sharma and Davidson, 2000; Tombline *et al.*, 2004a; Tombline *et al.*, 2004b). In the processive clamp model, two ATP molecules are present in the two composite binding sites, involving interactions at the NBD dimer interface following ATP induced dimerization. The nucleotides are hydrolysed sequentially before dissociation of the NBD dimer. This idea is in agreement with the crystal structures of NBD dimers where two ATP molecules are present at the dimer interface in their respective binding sites. Biochemical studies *in vitro* have revealed the presence of either two ATP (Smith *et al.*, 2002; Chen *et al.*, 2003a; Zaitseva *et al.*, 2005a) or two ADP molecules (Janas *et al.*, 2003; Dawson and Locher, 2006), or an ATP/ADP mixture bound to the NBD dimer (Chen *et al.*, 2001; Janas *et al.*, 2003). The existence of a state where two ATP molecules are bound is plausible if we assume that ATP bound monomers form dimers. The ATP/ADP state is in agreement with analyses revealing a phosphate exit tunnel for inorganic phosphate in the assembled dimer (Zaitseva *et al.*, 2006; Procko and Gaudet, 2008;). The presence of two ADP molecules in the NBD dimer has been detected but was only possible if a BeF_X trapped and hence stabilized system was generated (Janas *et al.*, 2003). Thus, a two-ADP state seems to be rather unstable under physiological conditions.

In the alternating site model, the starting point is also a dimer with two ATP molecules bound. Hydrolysis of one ATP opens up the dimer at the corresponding site and exchange of ADP for ATP is achieved. Following closing of the dimer the hitherto unaffected ATP is hydrolysed and the alternating cycle starts again following the second APD/ATP exchange. Support for the alternating site model comes from experiments where vanadate trapping was used to capture the hydrolysed ATP in the binding pocket since vanadate has the ability to mimic the transition state of the γ-phosphate on the point of elimination (Goodno, 1982; Shimizu, 1981; Smith and Rayment, 1996). The vanadate trapping experiments demonstrated that only one nucleotide molecule per NBD dimer could be detected, pointing towards an intermediate state where exchange of nucleotides is possible. This is good evidence that a mechanism for concerted hydrolysis dependent on alternating opening of the NBD at the two sites might exist (Vergani *et al.*, 2005). Of course, both models were established for different proteins and this leaves open the possibility that for different transporters different mechanisms may be required.

An intensely discussed question in the ABC field is also the number of allocrite binding sites in an ABC transporter and where these are located. For importers, it can clearly be stated that the binding site is inside the translocation pore formed by the transmembrane domains (Oldham *et al.*, 2007). For exporters that receive their substrate from the membrane bilayer the site of interaction is also likely to be the transmembrane domain. However, it remains an open question for other exporters that transport compounds from the cytoplasm, whether the allocrite interacts directly with the transmembrane domain, with cytoplasmic loops of the TMDs or whether the first contact is made with the transporter via NBD interactions. And how, if the translocation pore is not entered directly, can one imagine the targeting of the allocrite to the exit tunnel? First, the number of binding sites needs to be addressed. Since it has been shown that importers do not necessarily have only one substrate binding protein but two or even four per transporter, this question becomes even more crucial (Biemans-Oldehinkel and Poolman, 2003; Schuurman-Wolters and Poolman, 2005). In the past, three major models have evolved that attempt to describe the concerted transport of substrate coupled to ATP hydrolysis. There was no consensus concerning whether there are one or two binding sites present in the TMDs, whether high and low-affinity sites can be interconverted and whether the energy provided by ATP is used to shuttle the allocrite from one binding site to the other or to induce structural rearrangements in the transmembrane helices.

The approach presented in the two-cylinder engine model, which incorporates alternating site hydrolysis of ATP (van Veen *et al.*, 2000), describes two allocrite binding sites that have either high or low affinity for a drug molecule in the bacterial multidrug ABC exporter LmrA. When ATP is bound at the NBD, a high affinity drug binding site becomes accessible which is subsequently transformed into a low affinity binding site by hydrolysis of ATP. At this stage in this model, one NBD with bound ADP and another with ATP will be present as well as one high and one low affinity binding site. Difficulties remain in identifying and characterizing the binding sites and the additional important specific sites where substances can bind that modulate allocrite transport. Another difficulty is the fact that multidrug resistance transporters are able to accept a wide variety of structurally diverse compounds as transport substrates. Therefore, a single binding site for all allocrites can almost be ruled out. For P-glycoprotein, for example, at least two substrate-binding sites (H- and R-site) were identified that are also accessible to inhibitors (Lugo and Sharom, 2005; Shapiro and Ling, 1998; Sharom, 2008).

An alternative model for MDR ABC mediated transport is the interconversion of a single binding site from a high to a low affinity site, also coupled to the alternating catalytic site model for ATP hydrolysis described above (Senior *et al.*, 1995; Urbatsch *et al.*, 2003). Yet another possibility to explain the experimental results is the presence of two distinct binding sites with high or low affinity, respectively, where the substrate is transferred physically from one to the other binding site with the energy provided by ATP hydrolysis. The second ATP will in this model be used to reset the transport cycle (Sauna and Ambudkar, 2000).

To clarify the developing confusion concerning which model best describes the molecular processes of binding, recognition, and consequent structural rearrangements, the 'ATP switch model' (Higgins and Linton, 2004) has tried to summarize recent advances in ABC transporter function. This presumes that the NBDs of the transporter are initially in an open, nucleotide free-state and the high affinity substrate binding site is probably located close to the inner membrane. Binding of transport substrate induces binding of ATP to the NBDs which then results in NDB dimer formation. This dimerization of the motor domains induces a conformational change in the transmembrane domains of the ABC transporter leading to substrate release from a low affinity binding site. ATP hydrolysis and ADP + P_i release triggers NBD dimer decomposition and reconstitution of the initial state. This scheme should be applicable to both exporters and importers. However, predicted outcomes concerning sequential or alternate hydrolysis or altered affinity of ATP, or the exact nature of the binding sites, are not considered with respect to the general applicability of the model. To stress the point, we need to briefly reconsider the information available from the crystal structure of the MalEFGK$_2$ complex, with ATP and allocrite present. The SBP is able to bind to the translocator in the presence of ATP and maltose is found in the translocation pathway where a single allocrite binding site is visible. A possible scenario, in line with the ATP switch model for the entire transportation process could be the following: the allocrite interacts with the ABC transporter in the presence of ATP this initiates dimerization of the motor domains which induces structural changes in the transmembrane domain. This then leads to the situation reflected in the crystal structure. The next step would probably be the hydrolysis of the nucleotide. The energy provided could be used to finalize transport of the allocrite while the binding site may undergo structural changes to assure efficient displacement of the drug or allocrite. Dimer dissociation and ADP release would then re-establish an 'initial state' competent for a new cycle of transport. This is by no means the final story but offers for the first time an insight into the transport mechanism at a molecular level and provides a perspective based on all theories developed so far.

Conservation versus degeneration in ABC transporters

Conservation of certain sequence motifs, particularly in the NBDs, has always been considered an important feature of ABC transporters. These motifs are used to identify ABC proteins and play a crucial role in predicting and modelling approaches to infer functions of putative ABC transporters (Fig. 1.9). We need to emphasize, however, that the most highly conserved motifs can be found in only one composite catalytic ATP binding site, in a wide variety of eukaryotic ABC transporters, including families of the ABC C-subfamily (CFTR, MRP, SUR). In CFTR, the second site is degenerate and unable to hydrolyse the nucleotide (Gadsby *et al.*, 2006). Another prominent example is the transporter associated with antigen processing (TAP). The TAP1-NBD possesses the canonical signature sequence of the C-loop (LSGGQ), whereas the TAP2-NBD has a degenerate C-loop sequence LAAGQ (Chen *et al.*, 2004). The only prokaryotic ABC transporter displaying such asymmetry with respect to its NBD motifs, is LmrCD from *Lactococcus lactis* (Lubelski *et al.*, 2006).

Following the discussion above, reducing the number of sites for ATP hydrolysis has major

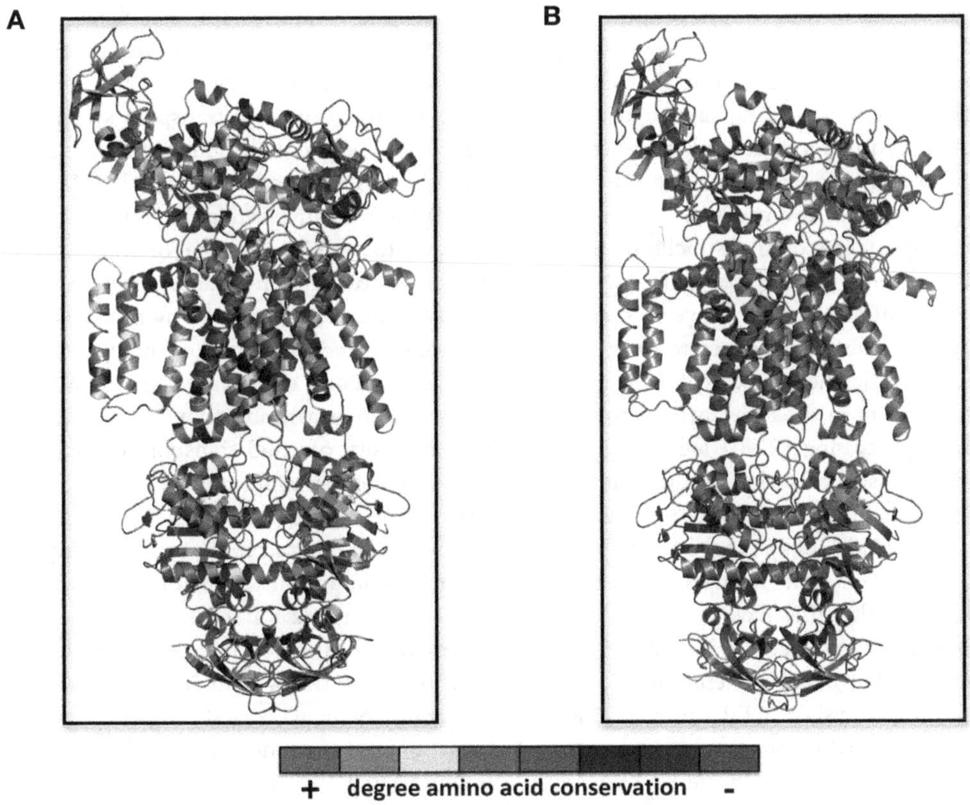

Figure 1.9 **A** Analysis of conserved amino acids in the MalEFGK$_2$ complex. The amino acid sequence of this importer was compared with 50 other ABC transporters and conserved amino acids are highlighted with the ConSurf Server (http://consurf.tau.ac.il/). **B** Only the highest conserved amino acids are shown in red. As explained in the text, the highest conservation is found in the nucleotide binding domains. A colour version of this figure is located in the plate section at the back of the book.

implications for the nature of the catalytic cycle of the motor domains and in particular for the coupling of ATP binding and/or hydrolysis to transport. When only one active site is available, only one ATP molecule can be hydrolysed, the energy provided has to be sufficient to fuel translocation. With respect to CFTR, since this protein functions as a chloride channel, this might be considered a special case (Vergani *et al.*, 2005). Nevertheless, a single active hydrolytic site still raises the important question of the stoichiometry of the reaction: is one ATP sufficient to fulfill transport requirements?

Up to this point we have considered that the power stroke in the transport process is solely dependent on the energy provided by the hydrolysis of ATP. Taking a closer look at the catalytic cycle illustrated from structural studies of the isolated HlyB-NBD, the Gibbs free energy derived experimentally from dissociation constants is similar for ATP hydrolysis and NBD dimerization (Zaitseva *et al.*, 2006). Consequently, either of these steps might be sufficient to generate the power stroke. To phrase this differently, dimerization itself may be capable of complementing any deficiencies in nucleotide hydrolysis which might be crucial in an ABC transporter with only one canonical ATP binding site. Nevertheless, speculation about the importance of the stoichiometry of ATP consumption, begs the question why single active sites in ABC transporters have successfully evolved and whether the degenerate site retains some function. A detailed analysis of the degenerated motifs present in the transporter associated with antigen processing (TAP) led to a model that provides essential tasks in the transport process for the ATPase site lacking key catalytic residues (Ernst *et al.*, 2006; Procko *et al.*, 2006). As discussed above, starting from an open conformation, where both NBDs have separated, ADP-ATP exchange is stimulated by allocrite binding (Chen *et al.*, 2003b). Subsequent dimerization of the NBDs, hydrolysis of ATP, and consequent structural changes in the transmembrane domains drive export of the allocrite. In the case of TAP, the idea that dimerization is controlled by peptide binding offers a nice solution explaining the efficient use of ATP with prevention of consumption in the absence of the allocrite. For CFTR, for example, it was suggested that the degeneration in one of the two ATP composite sites results in an active, ATP hydrolysing site, and one site that has a regulatory function. However, degenerate sites and their functional consequence in ABC transporters remain a matter of debate and we still await a clear experimental answer to this puzzle.

Conclusions and future trends

Finally, we must return to the starting point and clarify why we propose that ABC transporters constitute such a smart example. First, it is quite impressive that ABC proteins and their associates account for roughly 5% of the total coding capacity of the *E. coli* genome, clearly evidence of their important role in this and most microorganisms. It is equally significant that the highly conserved ABC motor is universally distributed throughout all kingdoms of life playing vitally important roles. Import of nutrients, export of toxic compounds or building the essential constituents of biological membranes are examples contributing to the fundamental requisites for life. Since the first definition of an ABC transporter system more than 25 years ago (Higgins *et al.*, 1982) many more of these amazing transport systems have been identified and the understanding of their function has advanced enormously. Many ABC transporters are involved in the immunity mechanism, detoxification, uptake of trace elements, mediation of drug resistance, osmoprotection, bile salt homeostasis and many more have been revealed. Moreover, studies of the proteins cover many disciplines, employing the whole range of techniques now available to modern science. The very diversity of ABC transporters in terms of sequence conservation or degeneration, allocrite import or export, modular assembly, and of course function from microorganisms to higher organisms simply adds to their fascination.

The main task for the future is to obtain a more detailed and thorough understanding of the ABC proteins. This will no doubt be aided by the availability of many more structures. Importantly, these should include their dynamic properties indicated for example, from NMR studies combined with physical and biochemical studies *in vitro* and especially *in vivo*. In this way the true physiological roles of these fundamentally important proteins should be revealed. Certainly, many questions still remain to be answered: the precise sequence of different steps involving binding of nucleotide, allocrite, the dimerization of NBDs and finally the 'reversible' reorientation of the transmembrane helices completing the transport cycle are at the atomic level still not solved. Emerging techniques in the field of computational biology with the power to simulate processes together with previously intractable techniques such as single molecule flourescence spectroscopy or NMR are becoming more and more common and will provide many important new insights. These approaches will lead to answers to questions concerning protein dynamics and will enhance substantially our ideas gained from the static pictures of crystal structures.

ABC transporters are associated with severe diseases such as, for example, cystic fibrosis, multidrug resistance towards anti-tumour drugs, cholestasis (liver disease), adrenoleukodystrophy (damage of the nervous system), Dubin-Johnson-syndrome (liver disease), or the consequences of multi-drug resistance during cancer chemotherapy. Thus, in addition to our intense curiosity to see one of the largest families of membrane proteins characterized and understood fully at the structural and functional level, their medical importance is surely by itself sufficient to keep the interest in ABC transporters alive.

Acknowledgements

We apologize to all our colleagues, whose work could not be adequately referenced due to space limitations. We are indebted to all former and current members of our laboratories. Especially, we would like to acknowledge the financial support of the EU (FP-7, EDICT) and the DFG (Emmy Noether program, SFBs 575 and 628 and grants Schm1279 4–1, 5–3, 7–1) to L.S., and CNRS and the University of Paris-Sud to I.B.H.

References

Abrahams, J.P., Leslie, A.G., Lutter, R., and Walker, J.E. (1994). Structure at 2.8 A resolution of F1-ATPase from bovine heart mitochondria. Nature *370*, 621–628.

Ames, G.F., and Lecar, H. (1992). ATP-dependent bacterial transporters and cystic fibrosis: analogy between channels and transporters. FASEB J 6, 2660–2666.

Ames, G.F., and Lever, J. (1970). Components of histidine transport: histidine-binding proteins and hisP protein. Proc. Natl. Acad. Sci. U.S.A. *66*, 1096–1103.

Bahadur, R.P., and Zacharias, M. (2008). The interface of protein–protein complexes: analysis of contacts and prediction of interactions. Cell Mol. Life Sci. *65*, 1059–1072.

Benabdelhak, H., Schmitt, L., Horn, C., Jumel, K., Blight, M.A., and Holland, I.B. (2005). Positive cooperative activity and dimerization of the isolated ABC-ATPase domain of HlyB from E.coli. Biochem. J. *368*, 1–7.

Biemans-Oldehinkel, E., Doeven, M.K., and Poolman, B. (2006). ABC transporter architecture and regulatory roles of accessory domains. FEBS Lett. *580*, 1023–1035.

Biemans-Oldehinkel, E., and Poolman, B. (2003). On the role of the two extracytoplasmic substrate-binding domains in the ABC transporter OpuA. EMBO J. *22*, 5983–5993.

Blight, M.A., and Holland, I.B. (1994). Heterologous protein secretion and the versatile *Eschericichia coli* haemolysin translocator. TIBTECH *12*, 450–455.

Bohm, B., Boschert, H., and Koster, W. (1996). Conserved amino acids in the N- and C-terminal domains of integral membrane transporter FhuB define sites important for intra- and intermolecular interactions. Mol. Microbiol. *20*, 223–232.

Bolhuis, H., van Veen, H.W., Molenaar, D., Poolman, B., Driessen, A.J., and Konings, W.N. (1996). Multidrug resistance in *Lactococcus lactis*: evidence for ATP-dependent drug extrusion from the inner leaflet of the cytoplasmic membrane. EMBO J. *15*, 4239–4245.

Boos, W., and Lucht, J.M. (1996). Periplasmic binding protein-dependent ABC transporters. In *Escherichia coli* and *Salmonella* Cellular and Molecular Biology, F.C. Neidhard, Curtiss R., III, Ingraham, J. L., Lin, E. C. C., Low, K. B., Magasanik, B., Reznikoff, W. S., Riley, M., Schaechter, M. & Umbarger, H. E., eds (Washington, DC, American Society for Microbiology Press), pp. 1175–1209.

Borbat, P.P., Surendhran, K., Bortolus, M., Zou, P., Freed, J.H., and McHaourab, H.S. (2007). Conformational motion of the ABC transporter MsbA induced by ATP hydrolysis. PLoS Biol. *5*, e271.

Borths, E.L., Locher, K.P., Lee, A.T., and Rees, D.C. (2002). The structure of *Escherichia coli* BtuF and binding to its cognate ATP binding cassette transporter. Proc. Natl. Acad. Sci. U.S.A. *99*, 16642–16647.

Borths, E.L., Poolman, B., Hvorup, R.N., Locher, K.P., and Rees, D.C. (2005). In vitro functional characterization of BtuCD-F, the *Escherichia coli* ABC transporter for vitamin B12 uptake. Biochemistry *44*, 16301–16309.

Buckel, S.D., Bell, A.W., Rao, J.K., and Hermodson, M.A. (1986). An analysis of the structure of the product of the rbsA gene of *Escherichia coli* K12. J. Biol. Chem. *261*, 7659–7662.

Cadieux, N., Bradbeer, C., Reeger-Schneider, E., Koster, W., Mohanty, A.K., Wiener, M.C., and Kadner, R.J. (2002). Identification of the periplasmic cobalamin-binding protein BtuF of *Escherichia coli*. J. Bacteriol. *184*, 706–717.

Campbell, J.D., Biggin, P.C., Baaden, M., and Sansom, M.S. (2003). Extending the structure of an ABC transporter to atomic resolution: modeling and simulation studies of MsbA. Biochemistry 42, 3666–3673.

Chang, G., and Roth, C.B. (2001). Structure of MsbA from *E. coli*: a homolog of the multidrug resistance ATP binding cassette (ABC) transporters. Science *293*, 1793–1800.

Chen, M., Abele, R., and Tampe, R. (2003b). Peptides induce ATP hydrolysis at both subunits of the transporter associated with antigen processing. J. Biol. Chem. *278*, 29686–29692.

Chen, M., Abele, R., and Tampe, R. (2004). Functional non-equivalence of ATP-binding cassette signature motifs in the transporter associated with antigen processing (TAP). J. Biol. Chem. *279*, 46073–46081.

Chen, J., Lu, G., Lin, J., Davidson, A.L., and Quiocho, F.A. (2003a). A tweezers-like motion of the ATP-binding cassette dimer in an ABC transport cycle. Mol. Cell *12*, 651–661.

Chen, J., Sharma, S., Quiocho, F.A., and Davidson, A.L. (2001). Trapping the transition state of an ATP-binding cassette transporter: evidence for a concerted mechanism of maltose transport. Proc. Natl. Acad. Sci. U.S.A. *98*, 1525–1530.

Clarke, T.E., Ku, S.Y., Dougan, D.R., Vogel, H.J., and Tari, L.W. (2000). The structure of the ferric siderophore binding protein FhuD complexed with gallichrome. Nat. Struct Biol. *7*, 287–291.

Dall'Acqua, W., and Carter, P. (2000). Substrate-assisted catalysis: molecular basis and biological significance. Protein Sci. *9*, 1–9.

Dassa, E., and Bouige, P. (2001). The ABC of ABCS: a phylogenetic and functional classification of ABC systems in living organisms. Res. Microbiol. *152*, 211–229.

Dassa, E., and Hofnung, M. (1985). Sequence of gene malG in *E. coli* K12: homologies between integral membrane components from binding protein-dependent transport systems. EMBO J. *4*, 2287–2293.

Davidson, A.L., and Chen, J. (2004). ATP-binding cassette transporters in bacteria. Annu Rev. Biochem. *73*, 241–268.

Davidson, A.L., Dassa, E., Orelle, C., and Chen, J. (2008). Structure, function, and evolution of bacterial ATP-binding cassette systems. Microbiol. Mol. Biol. Rev. *72*, 317–364, table of contents.

Davidson, A.L., Laghaeian, S.S., and Mannering, D.E. (1996). The maltose transport system of *Escherichia coli* displays positive cooperativity in ATP hydrolysis. J. Biol. Chem. *271*, 4858–4863.

Davidson, A.L., and Sharma, S. (1997). Mutation of a single MalK subunit severely impairs maltose transport activity in *Escherichia coli*. J. Bacteriol. *179*, 5458–5464.

Dawson, R.J., and Locher, K.P. (2006). Structure of a bacterial multidrug ABC transporter. Nature *443*, 180–185.

Dawson, R.J., and Locher, K.P. (2007). Structure of the multidrug ABC transporter Sav1866 from *Staphylococcus aureus* in complex with AMP-PNP. FEBS Lett. *581*, 935–938.

Dean, M., Rzhetsky, A., and Allikmets, R. (2001). The human ATP-binding cassette (ABC) transporter superfamily. Genome Res. *11*, 1156–1166.

Doran, A., Obach, R.S., Smith, B.J., Hosea, N.A., Becker, S., Callegari, E., Chen, C., Chen, X., Choo, E., Cianfrogna, J., Cox, L.M., Gibbs, J.P., Gibbs, M.A., Hatch, H., Hop, C.E., Kasman, I.N., Laperle, J., Liu, J., Liu, X., Logman, M., Maclin, D., Nedza, F.M., Nelson, F., Olson, E., Rahematpura, S., Raunig, D., Rogers, S., Schmidt, K., Spracklin, D.K., Szewc, M., Troutman, M., Tseng, E., Tu, M., Van Deusen J.W., Venkatakrishnan, K., Walens, G., Wang, E.Q., Wong, D., Yasgar, A.S., and Zhang, C. (2005). The impact of P-glycoprotein on the disposition of drugs targeted for indications of the central nervous system: evaluation using the MDR1A/1B knockout mouse model. Drug Metab Dispos 33, 165–174.

Duan, X., and Quiocho, F.A. (2002). Structural evidence for a dominant role of nonpolar interactions in the binding of a transport/chemosensory receptor to its highly polar ligands. Biochemistry *41*, 706–712.

Ernst, R., Koch, J., Horn, C., Tampe, R., and Schmitt, L. (2006). Engineering ATPase activity in the isolated ABC cassette of human TAP1. J. Biol. Chem. *281*, 27471–27480.

Felder, C.B., Graul, R.C., Lee, A.Y., Merkle, H.P., and Sadee, W. (1999). The Venus flytrap of periplasmic binding proteins: an ancient protein module present in multiple drug receptors. AAPS PharmSci. *1*, E2.

Fernandez, L.A., Sola, I., Enjuanes, L., and de Lorenzo, V. (2000). Specific secretion of active single-chain Fv antibodies into the supernatants of *Escherichia coli* cultuRes. by use of the hemolysin system. Appl Environ Microbiol. *66*, 5024–5029.

Fersht, A. (1997). Enzyme Structure and Mechanism. W H Freeman and Company *second edition*.

Fetsch, E.E., and Davidson, A.L. (2002). Vanadate-catalyzed photocleavage of the signature motif of an ATP-binding cassette (ABC) transporter. Proc. Natl. Acad. Sci. U.S.A. *99*, 9685–9690.

Flocco, M.M., and Mowbray, S.L. (1994). The 1.9 A x-ray structure of a closed unliganded form of the periplasmic glucose/galactose receptor from *Salmonella typhimurium*. J. Biol. Chem. *269*, 8931–8936.

Fraile, S., Munoz, A., de Lorenzo, V., and Fernandez, L.A. (2004). Secretion of proteins with dimerization capacity by the haemolysin type I transport system of *Escherichia coli*. Mol. Microbiol. 53, 1109–1121.

Froshauer, S., and Beckwith, J. (1984). The nucleotide sequence of the gene for malF protein, an inner membrane component of the maltose transport system of *Escherichia coli*. Repeated DNA sequences are found in the malE-malF intercistronic region. J. Biol. Chem. *259*, 10896–10903.

Fuellen, G., Spitzer, M., Cullen, P., and Lorkowski, S. (2005). Correspondence of function and phylogeny of ABC proteins based on an automated analysis of 20 model protein data sets. Proteins *61*, 888–899.

Gadsby, D.C., Vergani, P., and Csanady, L. (2006). The ABC protein turned chloride channel whose failure causes cystic fibrosis. Nature *440*, 477–483.

Gerber, S., Comellas-Bigler, M., Goetz, B.A., and Locher, K.P. (2008). Structural basis of trans-inhibition in a molybdate/tungstate ABC transporter. Science *321*, 246–250.

Gilson, E., Alloing, G., Schmidt, T., Claverys, J.P., Dudler, R., and Hofnung, M. (1988). Evidence for high affinity binding-protein dependent transport systems in gram-positive bacteria and in Mycoplasma. EMBO J. *7*, 3971–3974.

Goodno, C.C. (1982). Myosin active-site trapping with vanadate ion. Methods EnzyMol. *85 Pt B*, 116–123.

Grote, M., Bordignon, E., Polyhach, Y., Jeschke, G., Steinhoff, H.J., and Schneider, E. (2008). A comparative EPR study of the nucleotide-binding domains' catalytic cycle in the assembled maltose ABC-importer. Biophys. J. *95*, 2924- 2938

Hanekop, N., Zaitseva, J., Jenewein, S., Holland, I.B., and Schmitt, L. (2006). Molecular insights into the mechanism of ATP-hydrolysis by the NBD of the ABC-transporter HlyB. FEBS Lett. *580*, 1036–1041.

Harding, M.M. (2000). The geometry of metal–ligand interactions relevant to proteins. II. Angles at the metal atom, additional weak metal–donor interactions. Acta Crystallogr D Biol. Crystallogr *56 (Pt 7)*, 857–867.

Haubertin, D.Y., Madaoui, H., Sanson, A., Guerois, R., and Orlowski, S. (2006). Molecular dynamics simulations of *E. coli* MsbA transmembrane domain: formation of a semipore structure. Biophys. J. *91*, 2517–2531.

Hazelbauer, G.L. (1975). Maltose chemoreceptor of *Escherichia coli*. J. Bacteriol. *122*, 206–214.

Higgins, C.F., Haag, P.D., Nikaido, K., Ardeshir, F., Garcia, G., and Ames, G.F. (1982). Complete nucleotide sequence and identification of membrane components of the histidine transport operon of *S. typhimurium*. Nature *298*, 723–727.

Higgins, C.F., and Linton, K.J. (2004). The ATP switch model for ABC transporters. Nat. Struct Mol. Biol. *11*, 918–926.

Holland, I.B., Benabdelhak, H., Young, J., De Lima Pimenta, A., Schmitt, L., and Blight, M.A. (2003). Bacterial ABC transporters involved in protein translocation. In ABC proteins: From bacteria to man, I.B. Holland, S.P. Cole, K. Kuchler, and C. Higgins, eds. (London, Academic Press), pp. 209–241.

Holland, I.B., Schmitt, L., and Young, J. (2005). Type 1 protein secretion in bacteria, the ABC-transporter dependent pathway (review). Mol. Membr Biol. *22*, 29–39.

Hollenstein, K., Dawson, R.J., and Locher, K.P. (2007a). Structure and mechanism of ABC transporter proteins. Curr Opin Struct Biol. *17*, 412–418.

Hollenstein, K., Frei, D.C., and Locher, K.P. (2007b). Structure of an ABC transporter in complex with its binding protein. Nature *446*, 213–216.

Hung, L.-W., Wang, I.X., Nikaido, K., Liu, P.-Q., Ferro-Luzzi Ames, G., and Kim, S.-H. (1998). Crystal structure of the ATP-binding domain of an ABC transporter. Nature *396*, 703–707.

Hunke, S., Mourez, M., Jehanno, M., Dassa, E., and Schneider, E. (2000). ATP modulates subunit–subunit interactions in an ATP-binding cassette transporter (MalFGK2) determined by site-directed chemical cross- linking. J. Biol. Chem. *275*, 15526–15534.

Husain, I., Van Houten, B., Thomas, D.C., and Sancar, A. (1986). Sequences of *Escherichia coli* uvrA gene and protein reveal two potential ATP binding sites. J. Biol. Chem. *261*, 4895–4901.

Hvorup, R.N., Goetz, B.A., Niederer, M., Hollenstein, K., Perozo, E., and Locher, K.P. (2007). Asymmetry in the structure of the ABC transporter-binding protein complex BtuCD-BtuF. Science *317*, 1387–1390.

Jacquet, E., Girard, J.M., Ramaen, O., Pamlard, O., Levaique, H., Betton, J.M., Dassa, E., and Chesneau, O. (2008). ATP hydrolysis and pristinamycin IIA inhibition of the *Staphylococcus aureus* Vga(A), a dual ABC protein involved in streptogramin A resistance. J. Biol. Chem. *283*, 25332- 25338.

Janas, E., Hofacker, M., Chen, M., Gompf, S., van der Does, C., and Tampe, R. (2003). The ATP hydrolysis cycle of the nucleotide-binding domain of the mitochondrial ATP-binding cassette transporter Mdl1p. J. Biol. Chem. *278*, 26862–26869.

Joly, N., Bohm, A., Boos, W., and Richet, E. (2004). MalK, the ATP-binding cassette component of the *Escherichia coli* maltodextrin transporter, inhibits the transcriptional activator malt by antagonizing inducer binding. J. Biol. Chem. *279*, 33123–33130.

Jones, P.M., and George, A.M. (1999). Subunit interactions in ABC transporters: towards a functional architecture. FEMS Microbiol. Lett. *179*, 187–202.

Kadaba, N.S., Kaiser, J.T., Johnson, E., Lee, A., and Rees, D.C. (2008). The high-affinity *E. coli* methionine ABC transporter: structure and allosteric regulation. Science *321*, 250–253.

Kadner, R.J. (1975). Regulation of methionine transport activity in *Escherichia coli*. J. Bacteriol. *122*, 110–119.

Karpowich, N., Martsinkevich, O., Millen, L., Yuan, Y.R., Dai, P.L., MacVey, K., Thomas, P.J., and Hunt, J.F. (2001). Crystal structuRes. of the MJ1267 ATP binding cassette reveal an induced-fit effect at the ATPase active site of an ABC transporter. Structure *9*, 571–586.

Karpowich, N.K., Huang, H.H., Smith, P.C., and Hunt, J.F. (2003). Crystal structuRes. of the BtuF periplasmic-binding protein for vitamin B12 suggest a functionally important reduction in protein mobility upon ligand binding. J. Biol. Chem. *278*, 8429–8434.

Kartner, N., Riordan, J.R., and Ling, V. (1983). Cell surface P-glycoprotein associated with multidrug resistance in mammalian cell lines. Science *221*, 1285–1288.

Kempf, B., Gade, J., and Bremer, E. (1997). Lipoprotein from the osmoregulated ABC transport system OpuA of Bacillus subtilis: purification of the glycine betaine binding protein and characterization of a functional lipidless mutant. J. Bacteriol. *179*, 6213–6220.

Koronakis, V., Hughes, C., and Koronakis, E. (1991). Energetically distinct early and late stages of HlyB/HlyD-dependent secretion across both *Escherichia coli* membranes. EMBO J. *10*, 3263–3272.

Lawrence, M.C., Pilling, P.A., Epa, V.C., Berry, A.M., Ogunniyi, A.D., and Paton, J.C. (1998). The crystal structure of pneumococcal surface antigen PsaA reveals a metal-binding site and a novel structure for a putative ABC-type binding protein. Structure *6*, 1553–1561.

Lee, Y.H., Deka, R.K., Norgard, M.V., Radolf, J.D., and Hasemann, C.A. (1999). Treponema pallidum TroA is a periplasmic zinc-binding protein with a helical backbone. Nat. Struct Biol. *6*, 628–633.

Lee, Y.H., Dorwart, M.R., Hazlett, K.R., Deka, R.K., Norgard, M.V., Radolf, J.D., and Hasemann,

C.A. (2002). The crystal structure of Zn(II)-free Treponema pallidum TroA, a periplasmic metal-binding protein, reveals a closed conformation. J. Bacteriol. *184*, 2300–2304.

Leslie, E.M., Deeley, R.G., and Cole, S.P. (2005). Multidrug resistance proteins: role of P-glycoprotein, MRP1, MRP2, and BCRP (ABCG2) in tissue defense. Toxicol Appl Pharmacol *204*, 216–237.

Lewis, H.A., Buchanan, S.G., Burley, S.K., Conners, K., Dickey, M., Dorwart, M., Fowler, R., Gao, X., Guggino, W.B., Hendrickson, W.A., *et al.* (2004). Structure of nucleotide-binding domain 1 of the cystic fibrosis transmembrane conductance regulator. EMBO J. *23*, 282–293.

Linton, K.J., and Higgins, C.F. (1998). The *Escherichia coli* ATP-binding cassette (ABC) proteins. Mol. Microbiol. *28*, 5–13.

Liu, C.E., Liu, P.Q., and Ames, G.F.L. (1997). Characterization of the adenosine-triphosphatase activity of the periplasmic histidine permease, a traffic atpase (abc transporter). J. Biol. Chem. *272*, 21883–21891.

Locher, K.P., Lee, A.T., and Rees, D.C. (2002). The *E. coli* BtuCD structure: a framework for ABC transporter architecture and mechanism. Science *296*, 1091–1098.

Lomize, M.A., Lomize, A.L., Pogozheva, I.D., and Mosberg, H.I. (2006b). OPM: orientations of proteins in membranes database. Bioinformatics 22, 623–625.

Lomize, A.L., Pogozheva, I.D., Lomize, M.A., and Mosberg, H.I. (2006a). Positioning of proteins in membranes: a computational approach. Protein Sci. *15*, 1318–1333.

Lu, G., Westbrooks, J.M., Davidson, A.L., and Chen, J. (2005). ATP hydrolysis is required to reset the ATP-binding cassette dimer into the resting-state conformation. Proc. Natl. Acad. Sci. U.S.A. *102*, 17969–17974.

Lubelski, J., van Merkerk, R., Konings, W.N., and Driessen, A.J. (2006). Nucleotide-binding sites of the heterodimeric LmrCD ABC-multidrug transporter of *Lactococcus lactis* are asymmetric. Biochemistry *45*, 648–656.

Lugo, M.R., and Sharom, F.J. (2005). Interaction of LDS-751 with P-glycoprotein and mapping of the location of the R drug binding site. Biochemistry *44*, 643–655.

Miles, G., Bayley, H., and Cheley, S. (2002). Properties of Bacillus cereus hemolysin II: a heptameric transmembrane pore. Protein Sci. *11*, 1813–1824.

Millet, O., Hudson, R.P., and Kay, L.E. (2003). The energetic cost of domain reorientation in maltose-binding protein as studied by NMR and fluorescence spectroscopy. Proc. Natl. Acad. Sci. U.S.A. *100*, 12700–12705.

Moody, J.E., Millen, L., Binns, D., Hunt, J.F., and Thomas, P.J. (2002). Cooperative, ATP-dependent association of the nucleotide binding cassettes during the catalytic cycle of ATP-binding cassette transporters. J. Biol. Chem. *277*, 21111–21114.

Mourez, M., Hofnung, M., and Dassa, E. (1997). Subunit interactions in ABC transporters: a conserved sequence in hydrophobic membrane proteins of periplasmic permeases defines an important site of interaction with the ATPase subunits. EMBO J. *16*, 3066–3077.

Moussatova, A., Kandt, C., O'Mara, M.L., and Tieleman, D.P. (2008). ATP-binding cassette transporters in *Escherichia coli*. Biochim Biophys. Acta. *1778*, 1757- 1771.

Nikaido, K., and Ames, G.F. (1999). One intact ATP-binding subunit is sufficient to support ATP hydrolysis and translocation in an ABC transporter, the histidine permease. J. Biol. Chem. *274*, 26727–26735.

Nikaido, K., Liu, P.Q., and Ames, G.F. (1997). Purification and characterization of HisP, the ATP-binding subunit of a traffic ATPase (ABC transporter), the histidine permease of *Salmonella typhimurium*. Solubility, dimerization, and ATPase activity. J. Biol. Chem. *272*, 27745–27752.

Oldham, M.L., Khare, D., Quiocho, F.A., Davidson, A.L., and Chen, J. (2007). Crystal structure of a catalytic intermediate of the maltose transporter. Nature *450*, 515–521.

Orelle, C., Dalmas, O., Gros, P., Di Pietro, A., and Jault, J.M. (2003). The conserved glutamate residue adjacent to the Walker-B motif is the catalytic base for ATP hydrolysis in the ATP-binding cassette transporter BmrA. J. Biol. Chem. *278*, 47002–47008.

Patzlaff, J.S., van der Heide, T., and Poolman, B. (2003). The ATP/substrate stoichiometry of the ATP-binding cassette (ABC) transporter OpuA. J. Biol. Chem. *278*, 29546–29551.

Pinkett, H.W., Lee, A.T., Lum, P., Locher, K.P., and Rees, D.C. (2007). An inward-facing conformation of a putative metal-chelate-type ABC transporter. Science *315*, 373–377.

Procko, E., Ferrin-O'Connell, I., Ng, S.L., and Gaudet, R. (2006). Distinct structural and functional properties of the ATPase sites in an asymmetric ABC transporter. Mol. Cell 24, 51–62.

Procko, E., and Gaudet, R. (2008). Functionally important interactions between the nucleotide-binding domains of an antigenic peptide transporter. Biochemistry 47, 5699–5708.

Quiocho, F.A., and Ledvina, P.S. (1996). Atomic structure and specificity of bacterial periplasmic receptors for active transport and chemotaxis: variation of common themes. Mol. Microbiol. *20*, 17–25.

Quiocho, F.A., Spurlino, J.C., and Rodseth, L.E. (1997). Extensive featuRes. of tight oligosaccharide binding revealed in high-resolution structuRes. of the maltodextrin transport/chemosensory receptor. Structure 5, 997–1015.

Richet, E., Joly, N., and Danot, O. (2005). Two domains of MalT, the activator of the *Escherichia coli* maltose regulon, bear determinants essential for anti-activation by MalK. J. Mol. Biol. 347, 1–10.

Sack, J.S., Saper, M.A., and Quiocho, F.A. (1989). Periplasmic binding protein structure and function. J. Mol. Biol. *206*, 171–191.

Saier, M.H., Jr., Paulsen, I.T., Sliwinski, M.K., Pao, S.S., Skurray, R.A., and Nikaido, H. (1998). Evolutionary origins of multidrug and drug-specific efflux pumps in bacteria. FASEB J *12*, 265–274.

Sauna, Z.E., and Ambudkar, S.V. (2000). Evidence for a requirement for ATP hydrolysis at two distinct steps during a single turnover of the catalytic cycle of human P-glycoprotein. Proc. Natl. Acad. Sci. U.S.A. *97*, 2515–2520.

Sauna, Z.E., and Ambudkar, S.V. (2001). Characterization of the catalytic cycle of ATP hydrolysis by human P-glycoprotein. The two ATP hydrolysis events in a single catalytic cycle are kinetically similar but affect different functional outcomes. J. Biol. Chem. *276*, 11653–11661.

Saurin, W., Hofnung, M., and Dassa, E. (1999). Getting in or out: early segregation between importers and exporters in the evolution of ATP-binding cassette (ABC) transporters. J. Mol. Evol *48*, 22–41.

Scheffel, F., Demmer, U., Warkentin, E., Hulsmann, A., Schneider, E., and Ermler, U. (2005). Structure of the ATPase subunit CysA of the putative sulfate ATP-binding cassette (ABC) transporter from Alicyclobacillus acidocaldarius. FEBS Lett. *579*, 2953–2958.

Schmees, G., Stein, A., Hunke, S., Landmesser, H., and Schneider, E. (1999). Functional consequences of mutations in the conserved 'signature sequence' of the ATP-binding-cassette protein MalK. Eur. J. Biochem. *266*, 420–430.

Schmitt, L., Benabdelhak, H., Blight, M.A., Holland, I.B., and Stubbs, M.T. (2003). Crystal structure of the nucleotide binding domain of the ABC-transporter haemolysin B: Identification of a variable region within ABC helical domains. J. Mol. Biol. *330*, 333–342.

Schmitt, L., and Tampé, R. (2002). Structure and mechanism of ABC transporters. Cur Opin Struc Biol. *12*, 754–760.

Schuurman-Wolters, G.K., and Poolman, B. (2005). Substrate specificity and ionic regulation of GlnPQ from *Lactococcus lactis*. An ATP-binding cassette transporter with four extracytoplasmic substrate-binding domains. J. Biol. Chem. *280*, 23785–23790.

Senior, A.E., al-Shawi, M.K., and Urbatsch, I.L. (1995). The catalytic cycle of P-glycoprotein. FEBS Lett. *377*, 285–289.

Shapiro, A.B., and Ling, V. (1998). The mechanism of ATP-dependent multidrug transport by P-glycoprotein. Acta Physiol. Scand Suppl. *643*, 227–234.

Sharff, A.J., Rodseth, L.E., Spurlino, J.C., and Quiocho, F.A. (1992). Crystallographic evidence of a large ligand-induced hinge-twist motion between the two domains of the maltodextrin binding protein involved in active transport and chemotaxis. Biochemistry *31*, 10657–10663.

Sharma, S., and Davidson, A.L. (2000). Vanadate-induced trapping of nucleotides by purified maltose transport complex requires ATP hydrolysis. J. Bacteriol. *182*, 6570–6576.

Sharom, F.J. (2008). ABC multidrug transporters: structure, function and role in chemoresistance. Pharmacogenomics *9*, 105–127.

Sharom, F.J., Lugo, M.R., and Eckford, P.D. (2005). New insights into the drug binding, transport and lipid flippase activities of the p-glycoprotein multidrug transporter. J. Bioenerg Biomembr *37*, 481–487.

Shimizu, T. (1981). Steady-state kinetic study of vanadate-induced inhibition of ciliary dynein adenosinetriphosphatase activity from *Tetrahymena*. Biochemistry *20*, 4347–4354.

Shyamala, V., Baichwal, V., Beall, E., and Ames, G.F. (1991). Structure–function analysis of the histidine permease and comparison with cystic fibrosis mutations. J. Biol. Chem. *266*, 18714–18719.

Smith, C.A., and Rayment, I. (1996). X-ray structure of the magnesium(II).ADP.vanadate complex of the Dictyostelium discoideum myosin motor domain to 1.9 A resolution. Biochemistry *35*, 5404–5417.

Smith, P.C., Karpowich, N., Millen, L., Moody, J.E., Rosen, J., Thomas, P.J., and Hunt, J.F. (2002). ATP binding to the motor domain from an ABC transporter drives formation of a nucleotide sandwich dimer. Mol. Cell *10*, 139–149.

Steinfels, E., Orelle, C., Fantino, J.R., Dalmas, O., Rigaud, J.L., Denizot, F., Di Pietro, A., and Jault, J.M. (2004). Characterization of YvcC (BmrA), a multidrug ABC transporter constitutively expressed in Bacillus subtilis. Biochemistry *43*, 7491–7502.

Stenham, D.R., Campbell, J.D., Sansom, M.S., Higgins, C.F., Kerr, I.D., and Linton, K.J. (2003). An atomic detail model for the human ATP binding cassette transporter P-glycoprotein derived from disulfide cross-linking and homology modeling. FASEB J *17*, 2287–2289.

Story, R.M., and Steitz, T.A. (1992). Structure of the recA protein–ADP complex. Nature *355*, 374–376.

Tang, C., Schwieters, C.D., and Clore, G.M. (2007). Open-to-closed transition in apo maltose-binding protein observed by paramagnetic NMR. Nature *449*, 1078–1082.

Tombline, G., Bartholomew, L., Gimi, K., Tyndall, G.A., and Senior, A.E. (2004a). Synergy between conserved ABC signature Ser residues in P-glycoprotein catalysis. J. Biol. Chem. *279*, 5363–5373.

Tombline, G., Bartholomew, L.A., Urbatsch, I.L., and Senior, A.E. (2004b). Combined mutation of catalytic glutamate residues in the two nucleotide binding domains of P-glycoprotein generates a conformation that binds ATP and ADP tightly. J. Biol. Chem. *279*, 31212–31220.

Treptow, N.A., and Shuman, H.A. (1985). Genetic evidence for substrate and periplasmic-binding-protein recognition by the MalF and MalG proteins, cytoplasmic membrane components of the *Escherichia coli* maltose transport system. J. Bacteriol. *163*, 654–660.

Urbatsch, I.L., Sankaran, B., Weber, J., and Senior, A.E. (1995). P-glycoprotein is stably inhibited by vanadate-induced trapping of nucleotide at a single catalytic site. J. Biol. Chem. *270*, 19383–19390.

Urbatsch, I.L., Tyndall, G.A., Tombline, G., and Senior, A.E. (2003). P-glycoprotein catalytic mechanism: studies of the ADP-vanadate inhibited state. J. Biol. Chem. *278*, 23171–23179.

van der Does, C., and Tampe, R. (2004). How do ABC transporters drive transport? Biol. Chem. *385*, 927–933.

van der Heide, T., and Poolman, B. (2000). Osmoregulated ABC-transport system of *Lactococcus lactis* senses water stress via changes in the physical state of the membrane. Proc. Natl. Acad. Sci. U.S.A. *97*, 7102–7106.

van der Heide, T., and Poolman, B. (2002). ABC transporters: one, two or four extracytoplasmic substrate-binding sites? EMBO Rep. 3, 938–943.

van Veen, H.W., Margolles, A., Muller, M., Higgins, C.F., and Konings, W.N. (2000). The homodimeric ATP-binding cassette transporter LmrA mediates multidrug transport by an alternating two-site (two-cylinder engine) mechanism. EMBO J. *19*, 2503–2514.

Vazquez de Aldana, C.R., Marton, M.J., and Hinnebusch, A.G. (1995). GCN20, a novel ATP binding cassette protein, and GCN1 reside in a complex that mediates activation of the eIF-2 alpha kinase GCN2 in amino acid-starved cells. EMBO J. *14*, 3184–3199.

Velamakanni, S., Yao, Y., Gutmann, D.A., and van Veen, H.W. (2008). Multidrug Transport by the ABC Transporter Sav1866 from *Staphylococcus aureus*. Biochemistry *47*, 9300- 9308.

Verdon, G., Albers, S.V., van Oosterwijk, N., Dijkstra, B.W., Driessen, A.J., and Thunnissen, A.M. (2003). Formation of the productive ATP-Mg^{2+}-bound dimer of GlcV, an ABC-ATPase from Sulfolobus solfataricus. J. Mol. Biol. *334*, 255–267.

Vergani, P., Lockless, S.W., Nairn, A.C., and Gadsby, D.C. (2005). CFTR channel opening by ATP-driven tight dimerization of its nucleotide-binding domains. Nature *433*, 876–880.

Vetter, I.R., and Wittinghofer, A. (1999). Nucleoside triphosphate-binding proteins: different scaffolds to achieve phosphoryl transfer. Q Rev. Biophys. 32, 1–56.

Walker, J.E., Saraste, M., Runswick, M.J., and Gay, N.J. (1982). Distantly related sequences in the alpha- and beta-subunits of ATP synthase, myosin, kinases and other ATP-requiring enzymes and a common nucleotide binding fold. EMBO J. *1*, 945–951.

Ward, A., Reyes, C.L., Yu, J., Roth, C.B., and Chang, G. (2007). Flexibility in the ABC transporter MsbA: Alternating access with a twist. Proc. Natl. Acad. Sci. U.S.A. *104*, 19005–19010.

Watanabe, S., Kita, A., and Miki, K. (2005). Crystal structure of atypical cytoplasmic ABC-ATPase SufC from Thermus thermophilus HB8. J. Mol. Biol. 353, 1043–1054.

Zaitseva, J., Jenewein, S., Jumpertz, T., Holland, I.B., and Schmitt, L. (2005a). H662 is the linchpin of ATP hydrolysis in the nucleotide-binding domain of the ABC transporter HlyB. EMBO J. *24*, 1901–1910.

Zaitseva, J., Jenewein, S., Wiedenmann, A., Benabdelhak, H., Holland, I.B., and Schmitt, L. (2005b). Functional characterization and ATP-induced dimerization of the isolated ABC-domain of the haemolysin B transporter. Biochemistry *44*, 9680–9690.

Zaitseva, J., Oswald, C., Jumpertz, T., Jenewein, S., Wiedenmann, A., Holland, I.B., and Schmitt, L. (2006). A structural analysis of asymmetry required for catalytic activity of an ABC-ATPase domain dimer. EMBO J. *25*, 3432–3443.

Evolution and Function of the Multi-drug Resistance-linked ABC Transporters in Bacteria and Cancer Cells

2

Zuben E. Sauna and Suresh V. Ambudkar

Abstract

The ATP-binding cassette (ABC) genes constitute one of the largest super-families in living organisms. The majority of ABC proteins encode membrane proteins that utilize the energy of ATP binding and hydrolysis to transport a diverse set of compounds. ABC systems are critical to the cellular physiology of prokaryotes, facilitating the import of nutrients, the extrusion of toxins and for DNA repair. In eukaryotes, ABC proteins function primarily as efflux pumps (or as import pumps in intracellular organelles) playing important roles in protecting organisms from xenobiotics, antigen presentation, cholesterol and lipid transport. As ABC transporters extrude an unusually large set of chemically diverse compounds, they impede the treatment of microbial infections and human cancers and are implicated in multidrug resistance (MDR). Among prokaryotes, ABC proteins segregate into 29 families belonging to three main classes approximating the functional divisions of the ABC system (importers, exporters and 'other'). However, the vertebrate ABC proteins do not have the functional or organizational diversity of the prokaryotic proteins. The primary focus of this review is to discuss the evolution and structure–function relationship of microbial and mammalian ABC transporters responsible for MDR and the current status of substrate specificity and mechanistic aspects of mammalian ABC drug transporters.

The evolution of membrane proteins: complex systems from simple units

Primitive cells had to survive in a perilous environment where a fluctuating osmotic milieu was a principal threat that necessitated the evolution of membrane proteins to regulate volume and prevent cell lysis (Wilson and Lin, 1980; Wehner *et al.*, 2004). It is therefore not surprising that membrane proteins comprise some of the most ancient of protein families (Pohorille *et al.*, 2005). In addition, membrane proteins have evolved to take on more complex functions, are present in all organisms and mediate a multitude of roles, many of which are vital to life. Thus, for example, membrane proteins are the primary mediators of energy production and transduction, control the influx of nutrients and ions into cells as well as the removal of waste products, protect cells from xenobiotics, and the receptors that transmit signals from the environment, mediate intracellular communication, and control cell growth and volume regulation (Pohorille *et al.*, 2005). It is therefore not at all surprising that in the genomes sequenced thus far, about 30% of the gene products are known or predicted to be membrane proteins (Wallin and von Heijne, 1998). Moreover, the fact that approximately 60% of drugs approved for use in humans by the Food and Drug Administration target membrane proteins including enzymes, receptors, channels and transport proteins

(Lundstrom, 2006), underscores the importance of these proteins even in complex organisms.

Most membrane proteins in higher organisms are large and complex and are structurally and functionally diverse. This has given rise to two fundamental questions from an evolutionary perspective: (i) How did the basic structural unit(s) of membrane proteins emerge? (ii) How did these units evolve into systems with higher-order structures and function? It is intriguing, that despite their overall complexity most membrane proteins are composed of relatively simple conserved structural motifs. The most common structural element in membrane proteins is the α-helix. Extensive computer simulations have been performed by Wilson and coworkers (Pohorille *et al.*, 2005) that demonstrate how simple peptides could fold at a water–membrane interface, be inserted into the membrane and assemble themselves. These studies show that amino acid sequences where the hydrophobic and hydrophilic residues are at opposite faces (amphipathic peptides) readily fold at the water–membrane interface. Interestingly the composition of the peptide (vis-à-vis identity of the amino acids) appears to be secondary to a match between the peptides and the environment. These authors therefore argue that as such peptides had to fulfill 'very modest sequence constraints' they must have been relatively abundant even in a protocellular environment (Pohorille *et al.*, 2005). Examples of simple structures that can exhibit complex functional attributes can be found in viral proteins. One example is a 25-amino-acid fragment of the M2 protein from the influenza virus (Lamb *et al.*, 1985). This is considerably smaller than an average membrane protein, yet its tetramer forms a channel that transports protons. It thus appears that a small sub-set of protein domains associated with membrane proteins are conserved and evolve to form more complex structures in response to the changing needs of organisms.

The second question is how did the membrane proteins evolve in size and complexity? The consensus, based on the work of several groups studying the molecular evolution of membrane proteins, is that gene-duplication and gene-fusion have been the principal driving forces for the evolution of these proteins (Shimizu *et al.*, 2004). This view is supported by the structures of membrane proteins, a significant number of which exhibit a two-fold symmetry (von Heijne, 2006). Thus, proteins are composed of transmembrane helices organized into two domains, which often show a high degree of sequence identity. The individual domains may be composed of odd or even numbers of transmembrane helices and may be oriented either parallel or antiparallel to each other (Rapp *et al.*, 2007). In addition to the gene-duplication and gene-fusion events, mutations can alter the topology of the transmembrane helices whose orientation is determined by the 'positive inside' rule (Vonheijne, 1986). Furthermore, gene deletion and gain also drive the evolution of membrane proteins (Vert, 2002). Cordero and coworkers have determined the complete history of selected families of membrane proteins in terms of gene gain, deletion, duplication and fusion (Cordero *et al.*, 2008). A remarkable finding was that in addition to these events, co-evolution of genes plays an important role in directing the evolution of membrane proteins. What this means is that those genes that interact functionally evolve together. Taken together, the current view appears to be that although membrane proteins display a remarkable diversity, they have evolved from simple units (that may have occurred even in probiotic systems) via relatively straightforward processes that require a small number of steps. This diversity of form and function is nowhere more evident than in the role of transport proteins, which act as cellular gate-keepers for an almost infinite variety of substances.

The importance of being a transporter

In the last decade, an unprecedented quantity of sequence data has been generated. In addition to the complete human genome and those of other eukaryotes, over a thousand bacterial genomes have been sequenced (see the Genomes Online database, http://www.genomesonline.org/). These genomes have been exploited to identify new enzymes, as many conserved hypothetical genes have orthologues which are present in a large proportion of the genomes (Davidson *et al.*, 2008). It is much more difficult to identify transporters based on sequence homology alone. This is illustrated by the fact that approximately 75%

of proteins identified as transmembrane were classified as being of unknown function (Gelfand and Rodionov, 2008). This is troubling because transmembrane transport proteins constitute a large class of proteins that are critical to cellular physiology and provide important therapeutic targets. One very large and clinically important class of membrane proteins is the ATP-binding cassette (ABC) superfamily, containing more than 2000 members (Gottesman and Ambudkar, 2001).

The class of transport proteins now recognized as ABC proteins were first characterized in Gram-negative bacteria and designated binding protein-dependent or periplasmic transport systems. This nomenclature was found to be inaccurate, as exporters or the equivalent proteins in eukaryotes do not require a soluble binding protein to present the substrate to the transporter. Several other attempts to name this family were also misleading. For example, the name 'ATP-dependent transport systems' could not be used, as other ATP-dependent transport systems are known which are otherwise unrelated to this superfamily. These transporters were also called the 'Traffic ATPases' (Ames *et al.*, 1990). Finally, in 1990, based on the fact that these proteins exhibit a unique conserved ATP-binding cassette, Higgins and coworkers (Higgins *et al.*, 1990), suggested that this superfamily of transport systems be designated as 'ABC (ATP-binding cassette) transporters.' The presence of a 12-amino acid signature region upstream of the Walker B motif in NBDs is a diagnostic feature of the ABC proteins (Higgins, 1992).

The ABC transporters use energy from ATP hydrolysis to pump an unusually diverse set of compounds across biological membranes. Importers involved in the uptake of nutrients in Gram negative bacteria that used ATP hydrolysis rather than proton motive force were the first ABC proteins to be studied (Weiner and Heppel, 1971). It was soon discovered, however, that bacteria also use ABC transporters to remove various compounds from the cytoplasm and for protein secretion (Holland *et al.*, 2005) and in eukaryotes, all ABC transporters present at the cell surface are involved in efflux (Dean *et al.*, 2001). Some ABC proteins are involved in the export of very specific compounds such as cholesterol, peptides, lipids, etc. while others appear to have a more general protective role via the excretion of xenobiotics from the cytoplasm (Gottesman and Ambudkar, 2001). However, when such proteins extrude antimicrobials from disease-causing bacteria (Lubelski *et al.*, 2007) or chemotherapeutic agents from cancer cells, there are severe negative consequences (Gottesman *et al.*, 2002). The efflux of a broad range of structurally and functionally unrelated compounds from pathogenic bacteria, fungi and protozoa as well as cancerous human cells renders these cells resistant to therapeutic interventions.

The efflux of therapeutic compounds by several ABC proteins has implicated them in multidrug resistance (MDR). It must, however, be emphasized that not all ABC proteins are associated with MDR. In humans, for example, ABC proteins play a variety of roles (Table 2.1) and mutations in these are associated with human diseases such as cystic fibrosis, persistent hypoglycaemia of infancy, Stargardt disease, age-related macular degeneration and Tangier disease (Gottesman and Ambudkar, 2001). Although the majority of ABC proteins are involved in transport of molecular cargo across cell membranes, there are examples of ABC proteins that are not involved in transport of any kind and some authors have argued that the term ABC systems be used in lieu of the term ABC transporters (Davidson *et al.*, 2008).

Despite the diversity of function, tissue and organ localization and occurrence in almost all life forms, ABC proteins show a common architecture. All functional ABC proteins are made up of four domains, two transmembrane domains (TMDs) and two cytoplasmic nucleotide binding domains (NBDs) or ATP sites (see Chapter 1 by Jumpertz, Holland and Schmitt and Chapter 3 by Gutmann and van Veen for additional information). The transport substrate binds and is translocated via the TMDs, while ATP is hydrolysed by the NBDs (Sauna and Ambudkar, 2007). In most importers (which are only found in prokaryotes) these four domains are generally encoded by separate genes. On the other hand, the TMD and the corresponding NBD of exporters is encoded by a single gene (Dassa and Bouige, 2001). Furthermore, the functional ABC protein which consists of two TMDs and two NBDs may

Table 2.1 ABC sub-families in vertebrates

Sub-family/ gene	Function	Disease	Conserved genes[a]
ABCA			7/17 (41%)
ABCA1	Cholesterol efflux	Tangier disease and Familial HDL deficiency	
ABCA2	Drug resistance		
ABCA3	Phosphatidylcholine efflux	Surfactant deficiency	
ABCA4	Retinol, phospholipid derivatives	Stargardt disease, Age-related macular degeneration	
ABCA5			
ABCA6			
ABCA7			
ABCA8			
ABCA9			
ABCA10			
ABCA11			
ABCA12		Lamellar ichthyosis	
ABCA13			
ABCB			9/12 (75%)
ABCB1	Multidrug resistance	Cancer	
ABCB2	Peptide transport	Immune deficiency	
ABCB3	Peptide transport	Immune deficiency	
ABCB4	Phospatidylcholine transport	Progressive familial intrahepatic cholestasis	
ABCB5	Drug resistance	Cancer, melanoma	
ABCB6	Iron, porphyrin transport		
ABCB7	Fe/S transport	Sideroblastic anaemia and ataxia	
ABCB8			
ABCB9			
ABCB10			
ABCB11	Bile salt transport	Progressive familial intrahepatic cholestasis	
ABCC			9/13 (69%)
ABCC1	Multidrug resistance	Cancer	
ABCC2	Organic anion transport	Dubin-Johnson syndrome	
ABCC3	Drug resistance		
ABCC4	Nucleoside transport		
ABCC5	Nucleoside transport		

Sub-family/ gene	Function	Disease	Conserved genes[a]
ABCC6		Pseudoxanthoma elasticum	
ABCC7	Chloride channel	Cystic Fibrosis	
ABCC8	Sulfonylurea receptor-potassium channel regulator	Familial persistent hypoglycaemia of infancy	
ABCC9	Sulfonylurea receptor-potassium channel regulator	Dilated cardiomyopathy with ventricular tachycardia	
ABCC10			
ABCC11			
ABCC12			
ABCD			4/4 (100%)
ABCD1	VLCFA transport regulation	Adrenoleukodystrophy	
ABCD2			
ABCD3			
ABCD4			
ABCE			1/1 (100%)
ABCE1	Elongation factor complex		
ABCF			3/3 (100%)
ABCF1			
ABCF2			
ABCF3			
ABCG			5/6 (83%)
ABCG1	Cholesterol transport		
ABCG2	Multidrug resistance	Cancer	
ABCG4	Cholesterol transport		
ABCG5	Sterol transport	Sitosterolaemia	
ABCG8	Sterol transport	Sitosterolaemia	

a Number of genes conserved in vertebrates/ total number of genes (percentage in parenthesis).

either be expressed as a single polypeptide chain or be a homo- or hetero-dimer (Fig. 2.1).

The ABC proteins: role in drug resistance

The ABC transporters associated with MDR in human cancers have been extensively studied over the last 20 years. Approximately 50% of all cancers are metastatic at first diagnosis and systemic chemotherapy is the treatment of choice (and often the only option). Successful clinical outcomes following chemotherapy are still quite rare and MDR is a significant factor in this failure (Gottesman *et al.*, 2002). A number of mechanisms have been suggested as the cause for MDR; of these, the reduced accumulation of chemotherapeutic agents has received the most attention (Sauna *et al.*, 2007). Cancer cells become resistant to a single drug by mechanisms involving alterations in the molecular target(s). These

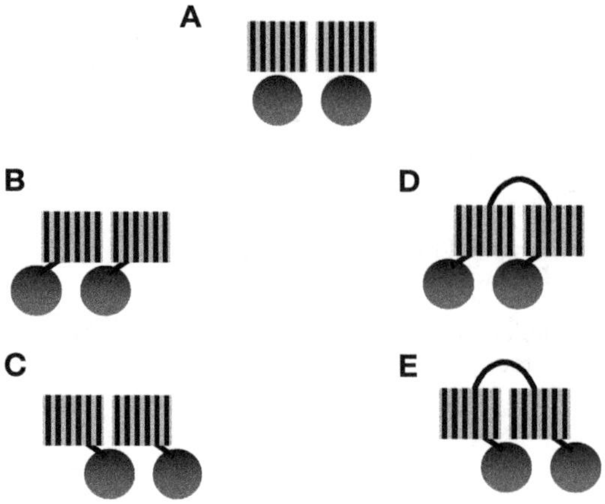

Figure 2.1 Domain organization of prokaryote and eukaryote ABC drug transporters. The transmembrane domain is depicted as a rectangle with six transmembrane helices (solid lines) and the NBDs or ATP sites are shown as circles. Panels **A** to **E** depict various combinations of gene products constituting functional ABC transporters. See text for additional details.

resistant cells are, however, often also resistant to other structurally and mechanistically unrelated drugs, i.e., these cells exhibit a multidrug resistant phenotype. Amphipathic or hydrophobic drugs (e.g. the natural products vinblastine, vincristine, doxorubicin, daunorubicin, actinomycin D, etoposide, colchicine, and paclitaxel) diffuse across the plasma membrane without any specific drug carriers. The efflux of these drugs by ABC transporters results in reduced cellular concentrations of the cytotoxic agent and the failure of chemotherapy (Ambudkar *et al.*, 1999). The first of these transporters to be identified and characterized was ABCB1 (P-glycoprotein; P-gp). Subsequently, two other ABC proteins, ABCC1 (MRP1) and ABCG2 (MXR or BCRP) have also been implicated in MDR associated with human cancers (for a first-person historical perspective see Gottesman and Ling, 2006).

ABCB1 transports a broad spectrum of hydrophobic drugs which are either neutral or positively charged (Ambudkar *et al.*, 1999). This transporter, under normal circumstances, does not transport anionic substrates. ABCC1 (MRP1), which was isolated from an MDR lung cancer cell line that did not express ABCB1 (Cole *et al.*, 1992) transports anionic drugs that are frequently conjugated with glutathione or are co-transported with glutathione (Haimeur *et al.*, 2004). Thus, while ABCB1 transports neutral and cationic hydrophobic compounds, ABCC1 can transport anionic compounds. Unlike ABCB1 and ABCC1, ABCG2 is a half-transporter that functions as a dimer (Ejendal and Hrycyna, 2002). The repertoire of drugs transported by ABCG2 is very similar to that of ABCB1. However, mitoxantrone, which is a poor substrate for ABCB1 (and ABCC1), is transported very well by ABCG2 (Miyake *et al.*, 1999). Moreover, ABCB1 is expressed in a limited number of tissues with barrier function, including epithelia of the liver, kidney, small and large intestine and capillary endothelial cells in the brain, ovary, and testis, while ABCC1 is expressed in most tissues (Gottesman *et al.*, 2002). Although *in vitro* as well as *in vivo* studies suggest a role for ABCB1 and ABCC1 in conferring MDR, studies have not provided conclusive evidence correlating expression of these proteins and the clinical manifestation of MDR.

In addition to the role that ABC transporters play in circumventing the effective chemotherapy of human cancers, ABC transporter-mediated MDR is a serious concern in the treatment of fungal infections. This is a particular problem in immune-compromised patients (such as those infected with HIV) who develop systemic infections. There is a significant incidence of antifungal resistance in such patients (Kontoyiannis and Lewis, 2002). For example, many patients develop systemic candidiasis, caused by the fungal pathogen *Candida albicans*. Antifungal azoles are substrates for fungal ABC transporters such as Candida drug resistance proteins (Cdr1p and Cdr2p), which extrude the drugs fluconazole, ketoconazole, itraconazole and voriconazole resulting in a MDR fungal pathogen (see Chapter 8 by Panwar and Prasad).

The role played by efflux pumps in bacterial drug resistance is quite complex and many more pumps are involved. Although the clinical importance of efflux pumps compared to specific enzymatic systems in antibiotic resistance has not been always clear, efflux systems are now thought to play an important role in bacterial drug resistance (Lubelski *et al.*, 2007). Furthermore, bacterial transporters play three distinct roles. First, many bacterial pumps have been identified that can efflux structurally and functionally unrelated compounds (often antibiotics) giving rise to a phenomenon similar to mammalian MDR (Lubelski *et al.*, 2007). Secondly, bacterial pumps have a central role in extruding compounds produced by the host that may be deleterious to the bacterium (Piddock, 2006). Finally, some bacteria that produce antibiotics (e.g., the *Actinomycetes*) use efflux pumps to transport the antibiotic out of the cell (Mendez and Salas, 2001). Thus, these pumps provide 'immunity' against the antibiotic. It is important to note that not all bacterial efflux pumps are ABC proteins. In fact, the predominant energy source for bacterial MDR transporter is the proton motive force, not ATP hydrolysis (Saier, 2000a, b). Nonetheless, MDR mediated by ABC transporters has been described in bacteria (see Lubelski *et al.*, 2007, for a recent review).

Organization of the ABC cassette

As stated above, a typical ABC protein consists of two transmembrane domains (TMDs) and two nucleotide-binding domains (NBDs). The TMDs show little sequence homology, whereas the NBDs exhibit very high sequence homology. As genes are classified as ABC transporters based on the sequence identity of the ATP-binding domain(s), we will briefly describe the organization of the NBDs. For a detailed description on the structure and function of NBDs refer to Chapter 1 by Jumpertz, Holland and Schmitt. Fig. 2.2A shows that the ATP binding cassettes of ABC proteins consist of several conserved motifs. The Walker A and Walker B motifs are involved in nucleotide binding. These domains are not unique to ABC proteins, but are also found in other ATPases such as the P-loop ATPase family (Sauna *et al.*, 2007). Critical residues in these domains include a conserved D residue in the Walker B motif that plays a role in the coordination of Mg^{2+} (Hrycyna *et al.*, 1999) and an adjacent E residue which is often referred to as the catalytic carboxylate base (Urbatsch *et al.*, 2000). The signature region, also called the C region, lies between the Walker A & B motifs, consists of the conserved sequence LSGGQ and is a hallmark of ABC transporters (Higgins, 1992).

In addition to these domains, there are several motifs that are defined by a single conserved residue. Analyses of high-resolution structures of NBDs have shown that the adenine ring of ATP interacts via π–π stacking, cation–π interaction or hydrogen bonding with an aromatic residue. In ABC proteins this is most often a conserved tyrosine approximately 25 amino acids upstream of the Walker A. This residue defines the A-loop (Aromatic residue interacting with the Adenine ring of ATP; see Kim *et al.*, 2006). Other conserved residues include those that define the Q loop, the H loop (also called the switch region) and another diagnostic marker of ABC transporters, the conserved D-loop (Sauna and Ambudkar, 2007). All these domains (and their relative placement) show a remarkable level of conservation among ABC proteins. In addition, recent biochemical and structural studies clearly show that the overall architecture of the ATP sites is also conserved. Biochemical, mutational and structural studies have shown that the two NBDs come together in a head-to-tail fashion, and form an ATP-binding domain using residues from both ATP sites. The essential feature of the nucleotide-binding pocket is that the ATP is sandwiched between the Walker A, Walker B, Q-loop, and H-loop of one NBD and the D-loop and signature sequence of the apposing NBD (Smith *et al.*, 2002; Zaitseva *et al.* 2005). Fig. 2.2B shows the organization of the conserved domains of two NBDs of mammalian P-gp to form functional ATP hydrolysis sites.

Evolution of prokaryotic ABC systems

Several groups have attempted to catalogue and classify the bacterial ABC proteins. It is clear from biochemical studies that bacterial ABC proteins perform three different functions. Two of these involve transport, viz. the export and import of molecular cargo. The third class of ABC proteins comprises those associated with cellular processes such as DNA repair, translation and gene expression (Dassa and Bouige, 2001). The fact that ABC proteins have highly conserved NBDs allows the application of computational methods to study the evolution of these proteins. Early studies compared ABC proteins from a single family or those with functional similarity (Kuan

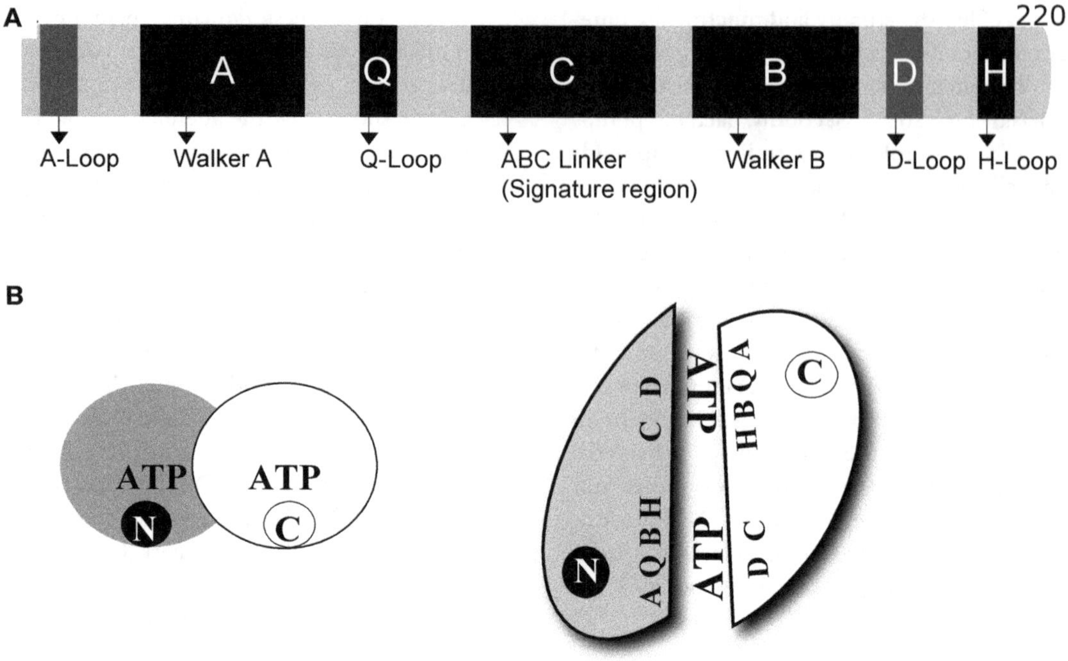

Figure 2.2 Schematic representation of the conserved motifs in the ATP-binding cassette (panel **A** and the interaction of two NBDs in a full-length ABC transporter such as ABCB1 to form a functional ATP site (panel **B**. In panel **A**, a typical ABC domain of about 200 amino acids (denoted at the top) with various conserved motifs is shown. The amino terminal boundary appears to be marked by the recently described A-loop (Kim *et al.*, 2006). The biochemical data has demonstrated that for P-gp, both NBDs are essential for ATP hydrolysis and drug transport (two NBDs denoted as **N**, the N-terminal and **C**, the carboxy-terminal). The ATP is sandwiched between two domains and the functional site is formed by the residues in the A-loop, the Walker **A** and **B**, the Q-loop and the H-loop of the N-terminal NBD and the signature region and the D-loop of the C-terminal NBD. Although two ATP molecules are shown to be bound, only one ATP molecule is occluded per P-gp molecule during E·S or E·P transition-state-like intermediate steps during ATP hydrolysis (Sauna *et al.*, 2006).

et al., 1995; Quentin *et al.*, 1999). The genomes of hundreds of organisms have been added to databases in recent years, permitting more extensive studies.

Dassa and Bouige designed the first global study devoted exclusively to ABC proteins (Dassa and Bouige, 2001). Over 600 ABC proteins were analysed to generate a phylogenetic tree which shows that these proteins appear to segregate very early into three distinct groups (see Fig. 2.1 in Dassa and Bouige, 2001). An interesting feature of this analysis was that the ABC proteins segregate based on (i) function and (ii) domain organization. Most ABC proteins in Class 1 are exporters in which the NBDs are fused with the TMDs. Class 2 represents the 'non-transporters', i.e., ABC proteins associated with DNA repair, translation, etc. These proteins do not possess TMDs. Finally, Class 3 ABC proteins are mostly importers. These proteins possess TMDs as well as NBDs but these moieties are expressed as separate polypeptide chains. Moreover, most class 3 ABCs are associated with a high-affinity binding protein that presents the transport-substrate to the ABC transporter at the extracellular face (Dassa and Bouige, 2001). This clustering of ABC proteins was based on multiple alignments of the NBDs of the proteins and it is impossible to perform similar studies with the TMDs which lack homology (this is also true of the binding proteins that occur in class 3 ABC importers). The different motifs that constitute the ATP binding cassette (Fig. 2.2) are highly conserved among all ABC proteins. Thus, the ABCs segregate based on differences in the region between the Walker A and B motifs. Notably, although the ABC proteins are segregated based on alignment of the NBDs, the fact that there is a strong correlation with the direction of transport and the domain organization suggests that the NBDs and TMDs of ABC systems have coevolved (Cordero *et al.*, 2008).

Milton Saier and coworkers have attempted a comprehensive classification of the prokaryotic ABC-type transporters. They have classified the transporters into 19 families that are uptake systems (TC 3.A.1.1 to TC 3.A.1.19) and 19 families that are primarily efflux systems (TC 3.A.1.101 to TC 3.A.1.119). Even with the prokaryotes alone, the efflux and uptake systems cluster separately on a phylogenetic tree (Saier, 2003). A noteworthy aspect of the clustering of the 19 ABC families involved in uptake of nutrients is that many of these families involve the transport of distinctive compounds. Thus, two are primarily concerned with polar and non-polar amino acid transport; a third is largely concerned with peptide transport; a fourth is primarily concerned with polyamine and opine uptake; a fifth is concerned with quaternary amine uptake; and a sixth functions to take up taurine. Similarly, among the prokaryotic ABC pumps that are exporters, three are associated with drug efflux, five mediate peptide export and two are involved in protein transport (Holland *et al.*, 2005). This clustering of families on the basis of types of substrate transported is extremely intriguing.

Evolution of vertebrate ABC systems

A total of 56 genes have been identified in all mammalian species and Dean and Annilo have determined that 38 of these 56 (or 68%) genes are present in all vertebrate genomes, suggesting that the ABC proteins are highly conserved (Dean and Annilo, 2005). As with prokaryotic ABC systems (see the previous discussion), the highly conserved NBDs are used for phylogenetic analysis of the vertebrate ABC proteins and the proteins are classified into 8 sub-families. Based on this segregation of the vertebrate ABC proteins, a systematic scheme for the nomenclature of ABC proteins has been advanced. The ABC superfamily has been divided into ABCA to ABCH sub-families, with numerals appended to indicate individual proteins.

The ABCA sub-family has 17 genes, but only 7 are conserved in all vertebrates, the lowest rate among the eight sub-families (Dean and Annilo, 2005). Functionally, the mammalian ABCA1 gene is associated with the transport of cholesterol (Oram and Lawn, 2001), while the mammalian ABCA4 gene is localized to photoreceptor cells and facilitates the transport of retinoid–lipid complexes out of these cells (Allikmets *et al.*, 1997a; Allikmets *et al.*, 1997b).

The ABCB family has 12 genes of which 9 are conserved among the vertebrates. Moreover, except for the fact that ABCB1 exists in two forms in rodent genomes, all ABCB genes are found to occur in all mammals. Thus, these genes have important and conserved functions. Some of the ABC transporters that are the most important clinically belong to this sub-family. These include ABCB1, overexpression of which is associated with the MDR phenotype in cancer cells, ABCB4, which transports phosphatidylcholine into bile (Pauli-Magnus *et al.*, 2004), and ABCB2 and ABCB3, which encode half-transporters that form a heterodimer (ABCB2/ABCB3 or Tap1/Tap2) which is a peptide transporter involved in antigen presentation (Abele and Tampe, 2004).

Similar to the ABCB sub-family, the ABCC sub-family is also well conserved among vertebrates (Dean and Annilo, 2005). This is a large sub-family of 13 genes that exhibit considerable functional diversity. The group of ABC transporters previously characterized as the MRPs all belong to this family (Haimeur *et al.*, 2004). These transporters efflux compounds that are either conjugated with glutathione or co-transported with glutathione (Haimeur *et al.*, 2004). ABCC7 (Cystic Fibrosis Transmembrane Regulator, CFTR) is an unusual ABC protein in that it is a channel rather than a transporter (Kerem *et al.*, 1989). This protein is an important therapeutic target, as mutations in this protein cause cystic fibrosis.

The sub-families ABCD, ABCE and ABCF are all small sub-families having only 4, 1 and 3 members respectively. However, all three families show 100% conservation among vertebrates. The ABCD genes are all half-transporters localized to the peroxisomes and are implicated in very long chain fatty acid metabolism (Smith *et al.*, 1999). This sub-family of genes is not only conserved among vertebrates but also among insects and worms (Dean and Annilo, 2005). The ABCE sub-family has only one gene which nonetheless shows an extraordinary level of conservation. This gene has been identified in all eukaryotes studied as well as plants and archaebacteria. It is not clear what the function of this gene is. However, the ABCE1-like gene RL11 in yeast has been shown to be part of the pre-initiation complex during protein synthesis (Dong *et al.*, 2004). Such a central role in cellular physiology would explain the fact that this gene is so well conserved. The three ABCF genes are also well conserved but very little is known about their function. All proteins belonging to the ABCG sub-family are half-transporters (Ejendal and Hrycyna, 2002). The human ABCG2, which is implicated in the MDR of cancer cells, is the most extensively characterized member of this sub-group. This transporter is also expressed at high levels in stem cells studied so far, indicating that it plays an important role in stem cell biology (Hirschmann-Jax *et al.*, 2004). ABCG1 and G4 are both induced by cholesterol (Cserepes *et al.*, 2004). While the former is found in macrophages the latter is localized to the brain. An interesting aspect of these genes is the fact that alternatively spliced transcripts have different tissue distributions. In addition, the heterodimer of ABCG5 and ABCG8 plays an important role in uptake of plant sterols and mutations in these genes cause the disease sitosterolaemia (Hubacek *et al.*, 2001; Kusuhara and Sugiyama, 2007). Finally, the ABCH family is composed of half transporters. This sub-family is probably the least understood in terms of structure and function.

There are 16–32 ABC genes in all single-cell eukaryotes and 47–68 ABC genes in all multi-cellular animal species (Dean and Annilo, 2005). It has therefore been suggested that the ABC genes conserved across different phyla play a role in functions that are fundamental to all cells. The ABC genes have also evolved to control functions required in multi-cellular organisms and these genes only occur in higher organisms. One major role of ABC proteins (particularly those that belong the sub-families ABCB and ABCC) is the efflux of toxic compounds (Lubelski *et al.*, 2007). It is, thus, not surprising that plants, which frequently use toxins to combat pathogens and competing plants, possess over 100 ABC proteins (Martinoia *et al.*, 2002). ABC proteins have attracted considerable attention because a significant number of human ABC genes (17/46) are linked to clinical disorders. However, the fact that this gene family is one of the few with large numbers of genes in all phyla makes it an attractive candidate to study the evolution of gene families.

The modular rearrangement of TMDs and NBDs in ABC systems

Fig. 2.1 shows that a functional ABC protein is composed of four modules: two TMDs, each composed of 6 transmembrane helices and two NBDs. These four modules organize into a functional protein in several ways. (i) The NBDs and TMDs are encoded by separate genes and the four polypeptides organize themselves into a tetramer (Fig. 2.1A). (ii) The gene encodes an NBD-TMD pair, with the NBD at the N-terminal of the polypeptide and the TMD at the C-terminal (Fig. 2.1B). (iii) The gene encodes an NBD-TMD pair; however the TMD is at the N-terminal and the NBD at the C-terminal (Fig. 2.1C). (iv) The gene codes for the entire ABC protein with each NBD preceding the TMD (Fig. 2.1D). (v) The gene codes for the entire ABC protein but each NBD follows the TMD (Fig. 2.1E). The existence of a diverse arrangement of basic modular units in ABC systems has led Dassa and coworkers to propose a hypothetical scenario for the evolution of ABC systems (Dassa and Bouige, 2001; Davidson *et al.*, 2008). They envision a putative ancestral organism that carried the genes for the TMD and NBD modules. Prokaryotes maintained such a system and also developed fusion proteins with the TMD-NBD orientation. Similar TMD-NBD fusion proteins in eukaryotes are often localized to organelle membranes, such as the TAP (ABCB2 and ABCB3) proteins in the endoplasmic reticulum (Abele and Tampe, 2004). It is therefore suggested that eukaryotes acquired such TMD-NBD systems through symbiotic bacteria that are often considered to have given rise to organelles. The $(\text{TMD-NBD})_2$ and $(\text{NBD-TMD})_2$ systems that gave rise to proteins such as ABCB1 (P-gp) and yeast Pdr5p are considered to be the result of several independent duplication–fusion events.

Another interesting aspect of ABC proteins is the fact that in the vast majority, the functional proteins are composed of twelve transmembrane helices and each TMD module consists of six transmembrane helices. This has led Saier (Saier, 2003) to suggest that there are two possible explanations. One is structural: 12 is the optimal number of transmembrane helices to comprise a helical bundle for the formation of a translocation pathway. However, other numbers of transmembrane helices (8, 10, 11, 13, 14, for example) can and do form helical bundles that function efficiently. The other explanation is evolutionary and based on probability. If TMDs have evolved via duplication events, there are more 'routes' to six than to five or seven. Thus, for example, 1 transmembrane helix × 6, two transmembrane helices × 3 or three transmembrane helices × 2 all yield the standard number of helices in a TMD. It is therefore argued that the greater probability of generating a TMD with six helices accounts for the preponderance of such a unit.

Overlapping substrate specificity of MDR-linked mammalian ABC drug transporters

We have discussed above that an intriguing aspect of the prokaryotic importers was that a significant number of these families were involved in the uptake of distinctive compounds. However, the mammalian efflux proteins do not cluster based on substrate specificity. The three mammalian ABC proteins associated with MDR are ABCB1, ABCC1 and ABCG2 (Gottesman *et al.*, 2002; Ambudkar *et al.*, 2003). All three proteins belong to distinct sub-families. ABCB1 can efflux a large number of structurally diverse amphipathic drugs that do not appear to exhibit common structural characteristics other than that they are generally hydrophobic and between 300–1300 Da (Ambudkar *et al.*, 1999).

A confounding issue for many years however was the existence of drug resistant cell lines that did not express ABCB1. This led to the discovery of multidrug-resistance-associated protein 1 (MRP1, or ABCC1) (Cole *et al.*, 1992). Additional studies showed that ABCC1 recognizes neutral and anionic hydrophobic natural products, and transports glutathione and other conjugates of these drugs (Haimeur *et al.*, 2004). ABCC1 also in some cases (e.g. vincristine) co-transports unconjugated glutathione (Borst and Elferink, 2002). Despite the fact that ABCB1 and ABCC1 belong to different sub-families and ABCC1 transports conjugated drugs, there is a large overlap in the drugs transported by these two transporters (Fig. 2.3). Some anticancer drugs, most notably mitoxantrone, being a substrate for both these transporters, mitoxantrone resistant cells were found to express a more distant member of the ABC transporter family, the mitoxantrone-resistance protein (MXR or ABCG2). Here again, we find that while there are a few compounds that are transported exclusively by ABCG2, there are many more that are transported by either ABCG2 and ABCB1 or ABCG2 and ABCC1.

Fig. 2.3 depicts a Venn diagram that illustrates the range of drug-substrates transported by ABCB1, ABCC1 and ABCG2. There is a core group of compounds that include the clinically important daunorubicin and doxorubicin that are substrates for all three transporters. In addition, there are significant over-laps in the compounds transported by any two of these three ABC proteins. The absence of core structural elements that define these substrates, lack of segregation of these proteins on the basis of substrates transported and the fact that the TM helices show no sequence homology are all consistent with the view that the drug-binding site in these transporters is a large polymorphous domain (Schumacher and Brennan, 2002). It has also been suggested that while such a drug-binding domain can accommodate a wide range of compounds, interactions between structural elements of different substrates and amino acid side chains of the drug-binding pocket permit more specific interactions (Schumacher *et al.*, 2004). This is borne out by the fact that several single amino-acid substitutions significantly alter the substrate specificity of P-gp (Ambudkar *et al.*, 1999). Similarly, the first two ABCG2 genes that were cloned encoded proteins that were gain of function mutants with either a threonine or glycine substituted for arginine at amino acid 482, giving them much broader substrate specificity (Gottesman *et al.*, 2002). However, it needs to be pointed out that these gain of function mutations have been observed only in tissue culture cells, not in patient tumour samples.

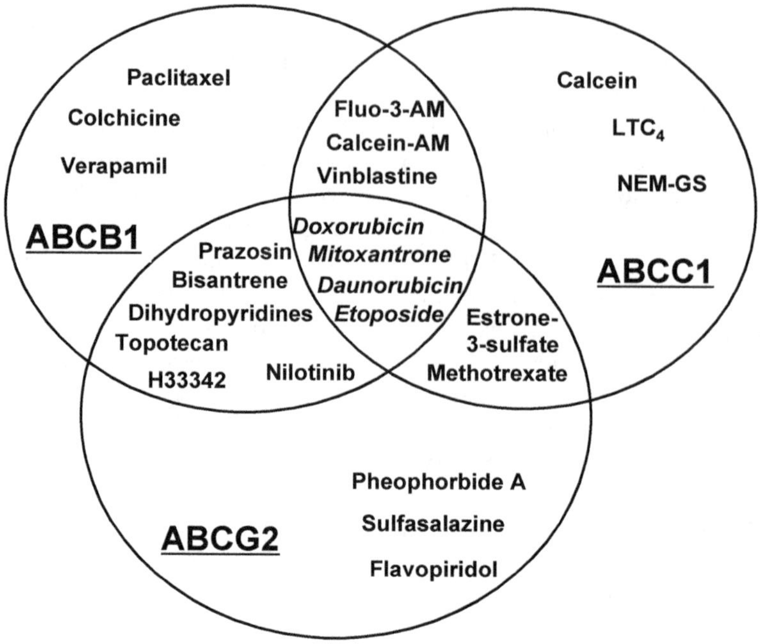

Figure 2.3 Venn diagram showing the overlapping substrate specificity of the three major mammalian ABC drug transporters, ABCB1, ABCC1 and ABCG2. The anticancer agents transported by all three transporters are shown in italics. For clarity, only selected substrates are shown. H33342, Hoechst33342; LTC_4, leukotriene C_4; and NEM-GS, N-ethylmaleimide–glutathione conjugate.

Transport mechanism of the mammalian MDR-linked ABCB1

It is generally recognized that transport involves the binding of ATP and/or drug, conformational changes at the NBDs due to binding and/or hydrolysis, the coupling of these conformational changes to cause a high-affinity to low-affinity switch at the transport–substrate site and finally the resetting of the pump (Sauna and Ambudkar, 2007). The mechanism of the P-gp-mediated transport pathway has been extensively studied and serves as a model for understanding mammalian efflux pumps. Numerous studies have exploited the P_i analogue, ortho-vanadate (Vi), to trap the post-hydrolysis reaction intermediate, P-gp•MgADP•Vi (for review see Sauna *et al.*, 2001) which represents the Enzyme-Product (E•P) state of the reaction. On the other hand, double mutants of the conserved glutamate in the Walker B domain in both NBDs are hydrolysis deficient and cannot generate ADP and occlude ATP in a pre-hydrolysis intermediate, P-gp (E556Q/E1201Q)•MgATP (Sauna and Ambudkar, 2007). This mutant is thus used to arrest the P-gp-mediated ATPase reaction in the pre-hydrolysis reaction intermediate which represents the Enzyme–Substrate (E•S) state.

As stated above, the two NBDs of ABC transporters come together in a head-to-tail fashion to form an ATP-binding domain using residues from both domains (see Fig. 2.2B). This so-called ATP sandwich is thought to be formed by the ATP-driven dimerization of the two NBDs. The P-gp(E556Q/E1201Q)•MgATP (E•S reaction intermediate) represents this dimer (Sauna *et al.*, 2006). Both the Senior group and we have demonstrated that unlike the ATP-driven dimers obtained with isolated NBDs, which exhibit a stoichiometry of 2 mol ATP per dimer (see for example Smith *et al.*, 2002), P-gp shows a maximal stoichiometry of 1 mol ATP occluded per mol P-gp (Tombline *et al.*, 2005; Sauna *et al.*, 2006). We have also provided evidence that ATP occlusion in mutant P-gp involves conformational changes. It has therefore been hypothesized that nucleotide binding initially leads to a symmetric dimer with ATPs bound at both NBDs, which progresses to an asymmetric occluded state where one of the two ATPs is tightly bound in a non-exchangeable form (see Fig. 2.1 in Tombline and Senior, 2005). Our studies with a P-gp mutant (E556Q/E1201Q) also allow us to address the question of what role, if any, the occluded nucleotide conformation plays in the transport pathway.

We have demonstrated that occlusion of [α-^{32}P]ATP and that the formation of the occluded dimer is accompanied by a temperature-dependent reduced binding of the drug-substrate analogue [^{125}I]IAAP (Sauna *et al.*, 2006). The former suggests that occlusion of ATP is accompanied by conformational changes at the NBDs and the latter that these conformational changes are transmitted to the TMDs. The formation of the occluded nucleotide conformation of the mutant human P-gp, E556Q/E1201Q, has an activation energy of ~60 kJ/mol and the activation energies for nucleotide trapping and decrease in drug binding are equivalent (Ambudkar *et al.*, 2006). This suggests that the occlusion of ATP in ATP sites and the high affinity → low affinity switch at the drug-binding site located in the transmembrane domains are also coupled. The recently solved structure of the full-length ABC protein Sav1866, a multidrug transporter from *S. aureus*, suggests that the coupling between the NBDs and TMDs occurs via a region of the NBD between the Walker A domain and the signature sequence of the ATP biding cassette (Dawson and Locher, 2006). Biochemical and genetic evidence from the yeast multidrug transporter Pdr5 supports this view (Sauna *et al.*, 2008). Moreover, binding of nucleotide in the absence of occlusion does not result in a high affinity to low affinity switch at the drug-binding site (Sauna and Ambudkar, 2001), suggesting that an early formation of the E•S reaction intermediate during the catalytic cycle of P-gp mediated ATP hydrolysis provides the power-stroke for drug transport. Events associated with the resetting of the pump, however, still remain controversial (Sauna and Ambudkar, 2000). The resolution of the structure of a mammalian drug efflux pump such as P-gp will greatly facilitate further experimental work to elucidate the mechanism of drug transport by these transporters.

Conclusions and future trends

ABC proteins are ubiquitous and have been found to occur in all genomes sequenced to date (Davidson *et al.*, 2008). The complete sequences of over 500 genomes are available. Among each of the groups (archea, bacteria and eukaryotes) there is a very strong correlation between the total number of genes in an organism and the number of ABC transporters (Konstantinidis and Tiedje, 2004). All these facts suggest the importance of this class of proteins in cellular physiology. However, bacteria and archea have far greater numbers of ABC proteins than do the eukaryotes, which possibly reflects the more challenging environment of these organisms and the greater dependence on transport proteins for survival. Although multicellular organisms possess fewer numbers of ABC proteins, they exhibit a diversity of function unique to these organisms. As a consequence of the important physiological roles played by ABC transporters, they offer numerous potential therapeutic targets (Gottesman and Ambudkar, 2001). Thus, transport proteins in bacteria, fungi and cancer cells are associated with the MDR phenotype and permit these cells to evade chemotherapy (Davidson *et al.*, 2008). Similarly, mutations in 17 of the 46 human ABC transporters are linked to diseases that show Mendelian inheritance (Dean and Annilo, 2005). It is for these reasons that the genetics, biochemistry and structural biology of this family of membrane proteins have been intensely studied. Although there has been much progress in this field, there are many gaps in understanding how these transporters work and several of the chapters in this volume address these issues. The literature reviewed in this chapter strongly suggests that ABC transporters constitute a very ancient family of proteins and despite the fact that they perform diverse functions in higher organisms, ABC proteins probably evolved from a common primitive ancestor. We have also discussed that almost all ABC proteins utilize core structural elements to form a functional protein. Thus, it is very probable that ABC proteins utilize common mechanisms of action. The overlapping substrate specificity of major MDR-linked ABC transporters as well as the ability of these transporters to transport a broad range of drugs which are currently in clinical use is a factor affecting the bioavailability of orally administered drugs. This, the strategy of focusing on detailed structural, biophysical and biochemical studies should provide insights into the mode of action of drug ABC transporters. This should allow development of mechanism-based modulators to increase the efficiency of chemotherapy for cancer patients.

Acknowledgements

We thank Mr. George Leiman for editorial assistance. This research was supported by the Intramural Research Program of the National Institutes of Health, National Cancer Institute, Center for Cancer Research.

References

Abele, R., and Tampe, R. (2004). The ABCs of immunology: Structure and function of TAP, the transporter associated with antigen processing. Physiology *19*, 216–224.

Allikmets, R., Shroyer, N.F., Singh, N., Seddon, J.M., Lewis, R.A., Bernstein, P.S., Peiffer, A., Zabriskie, N.A., Li, Y.X., Hutchinson, A., Dean, M., Lupski, J.R., and Leppert, M. (1997a). Mutation of the stargardt disease gene (Abcr) in age-related macular degeneration. Science *277*, 1805–1807.

Allikmets, R., Singh, N., Sun, H., Shroyer, N.E., Hutchinson, A., Chidambaram, A., Gerrard, B., Baird, L., Stauffer, D., Peiffer, A., Rattner, A., Smallwood, P., Li, Y.X., Anderson, K.L., Lewis, R.A., Nathans, J., Leppert, M., Dean, M., and Lupski, J.R. (1997b). A photoreceptor cell-specific ATP-binding transporter gene (Abcr) is mutated in recessive Stargardt macular dystrophy. Nature Genet. *15*, 236–246.

Ambudkar, S.V., Dey, S., Hrycyna, C.A., Ramachandra, M., Pastan, I., and Gottesman, M.M. (1999). Biochemical, cellular, and pharmacological aspects of the multidrug transporter. Annu. Rev. Pharmacol. Toxicol. *39*, 361–398.

Ambudkar, S.V., Kim, I.W., and Sauna, Z.E. (2006). The power of the pump: Mechanisms of action of p-glycoprotein (ABCB1). Eur. J. Pharm. Sci. *27*, 392–400.

Ambudkar, S.V., C. Kimchi-Sarfaty, C., Sauna, Z.E., and Gottesman, M.M. (2003). Oncogene *22*, 7468–7485.

Ames, G.F., Mimura, C.S., and Shyamala, V. (1990). Bacterial periplasmic permeases belong to a family of transport proteins operating from *Escherichia coli* to human – traffic ATPases. Fems Microbiol. Rev. *75*, 429–446.

Borst, P., and Elferink, R.O. (2002). Mammalian abc transporters in health and disease. Annu. Rev. Biochem. *71*, 537–592.

Cole, S.P., Bhardwaj, G., Gerlach, J.H., Mackie, J.E., Grant, C.E., Almquist, K.C., Stewart, A.J., Kurz, E.U., Duncan, A.M., and Deeley, R.G. (1992). Overexpression of a transporter gene in a multidrug-resistant human lung cancer cell line. Science *258*, 1650–1654.

Cordero, O.X., Snel, B., and Hogeweg, P. (2008). Coevolution of gene families in prokaryotes. Genome Res. *18*, 462–468.

Cserepes, J., Szentpetery, Z., Seres, L., Ozvegy-Laczka, C., Langmann, T., Schmitz, G., Glavinas, H., Klein, I., Homolya, L., Varadi, A., Sarkadi, B., and Elkind, N.B. (2004). Functional expression and characterization of the human ABCG1 and ABCG4 proteins: Indications for heterodimerization. Biochem. Biophys. Res. Commun. *320*, 860–867.

Dassa, E., and Bouige, P. (2001). The ABC of ABCs: A phylogenetic and functional classification of ABC systems in living organisms. Res. Microbiol. *152*, 211–229.

Davidson, A.L., Dassa, E., Orelle, C., and Chen, J. (2008). Structure, function, and evolution of bacterial ATP-binding cassette systems. Microbiol. Mol. Biol. Rev.*72*, 317–364.

Dawson, R.J.P., and Locher, K.P. (2006). Structure of a bacterial multidrug ABC transporter. Nature *443*, 180–185.

Dean, M., and Annilo, T. (2005). Evolution of the atp-binding cassette (ABC) transporter superfamily in vertebrates. Annu. Rev. Genomics Hum. Genet. *6*, 123–142.

Dean, M., Hamon, Y., and Chimini, G. (2001). The human ATP-binding cassette (ABC) transporter superfamily. J. Lipid Res. *42*, 1007–1017.

Dong, J.S., Lai, R., Nielsen, K., Fekete, C.A., Qiu, H.F., and Hinnebusch, A.G. (2004). The essential ATP-binding cassette protein RLI 1 functions in translation by promoting preinitiation complex assembly. J. Biol. Chem. *279*, 42157–42168.

Ejendal, K.F.K., and Hrycyna, C.A. (2002). Multidrug resistance and cancer: The role of the human ABC transporter ABCG2. Curr. Protein Pept. Sci. 3, 503–511.

Gelfand, M.S., and Rodionov, D.A. (2008). Comparative genomics and functional annotation of bacterial transporters. Phys. Life Rev. *5*, 22–49.

Gottesman, M.M., and Ambudkar, S.V. (2001). Overview: ABC transporters and human disease. J. Bioenerg. Biomembr. 33, 453–458.

Gottesman, M.M., Fojo, T., and Bates, S.E. (2002). Multidrug resistance in cancer: Role of atp-dependent transporters. Nat. Rev. Cancer 2, 48–58.

Gottesman, M.M., and Ling, V. (2006). The molecular basis of multidrug resistance in cancer: The early years of p-glycoprotein research. Febs Lett. *580*, 998–1009.

Haimeur, A., Conseil, G., Deeley, R.G., and Cole, S.P.C. (2004). The MRP-related and BCRP/ABCG2 multidrug resistance proteins: Biology, substrate specificity and regulation. Curr. Drug Metab. *5*, 21–53.

Higgins, C.F. (1992). ABC transporters: From micro-organisms to man. Annu. Rev. Cell Biol. *8*, 67–113.

Higgins, C.F., Hyde, S.C., Mimmack, M.M., Gileadi, U., Gill, D.R., and Gallagher, M.P. (1990). Binding protein-dependent transport-systems. J. Bioenerg. Biomembr. *22*, 571–592.

Hirschmann-Jax, C., Foster, A.E., Wulf, G.G., Nuchtern, J.G., Jax, T.W., Gobel, U., Goodell, M.A., and Brenner, M.K. (2004). A distinct 'Side population' Of cells with high drug efflux capacity in human tumor cells. Proc. Natl. Acad. Sci. U.S.A. *101*, 14228–14233.

Holland, I.B., Schmitt, L., and Young, J. (2005). Type 1 protein secretion in bacteria, the ABC-transporter dependent pathway. Mol. Membr. Biol. *22*, 29–39.

Hrycyna, C.A., Ramachandra, M., Germann, U.A., Cheng, P.W., Pastan, I., and Gottesman, M.M. (1999). Both ATP sites of human P-glycoprotein are essential but not symmetric. Biochemistry *38*, 13887–13899.

Hubacek, J.A., Berge, K.E., Cohen, J.C., and Hobbs, H.H. (2001). Mutations in ATP-cassette binding proteins G5 (ABCG5) and G8 (ABCG8) causing sitosterolemia. Hum. Mutat. *18*, 359–360.

Kerem, B.S., Rommens, J.M., Buchanan, J.A., Markiewicz, D., Cox, T.K., Chakravarti, A., Buchwald, M., and Tsui, L.C. (1989). Identification of the cystic-fibrosis gene – genetic-analysis. Science *245*, 1073–1080.

Kim, I.W., Peng, X.H., Sauna, Z.E., FitzGerald, P.C., Xia, D., Muller, M., Nandigama, K., and Ambudkar, S.V. (2006). The conserved tyrosine residues 401 and 1044 in ATP sites of human P-glycoprotein are critical for ATP binding and hydrolysis: Evidence for a conserved subdomain, the A-loop in the ATP-binding cassette. Biochemistry *45*, 7605–7616.

Konstantinidis, K.T., and Tiedje, J.M. (2004). Trends between gene content and genome size in prokaryotic species with larger genomes. Proc. Natl. Acad. Sci. U.S.A. *101*, 3160–3165.

Kontoyiannis, D.P., and Lewis, R.E. (2002). Antifungal drug resistance of pathogenic fungi. Lancet *359*, 1135–1144.

Kuan, G., Dassa, E., Saurin, W., Hofnung, M., and Saier, M.H. (1995). Phylogenetic analyses of the ATP-binding constituents of bacterial extracytoplasmic receptor-dependent ABC-type nutrient-uptake permeases. Res. Microbiol. *146*, 271–278.

Kusuhara, H., and Sugiyama, Y. (2007). ATP-binding cassette, subfamily G (ABCG family). Pflugers Arch. *453*, 735–744.

Lamb, R.A., Zebedee, S.L., and Richardson, C.D. (1985). Influenza virus-m2 protein is an integral membrane-protein expressed on the infected-cell surface. Cell *40*, 627–633.

Lubelski, J., Konings, W.N., and Driessen, A.J.M. (2007). Distribution and physiology of abc-type transporters contributing to multidrug resistance in bacteria. Microbiol. Mol. Biol. Rev. *71*, 463–475.

Lundstrom, K. (2006). Structural genomics for membrane proteins. Cell Mol. Life Sci. *63*, 2597–2607.

Martinoia, E., Klein, M., Geisler, M., Bovet, L., Forestier, C., Kolukisaoglu, U., Muller-Rober, B., and Schulz, B. (2002). Multifunctionality of plant ABC transporters – more than just detoxifiers. Planta *214*, 345–355.

Mendez, C., and Salas, J.A. (2001). The role of ABC transporters in antibiotic-producing organisms: Drug secretion and resistance mechanisms. Res. Microbiol. *152*, 341–350.

Miyake, K., Mickley, L., Litman, T., Zhan, Z., Robey, R., Cristensen, B., Brangi, M., Greenberger, L., Dean, M., Fojo, T., and Bates, S.E. (1999). Molecular cloning of cdnas which are highly overexpressed in mitoxantrone-resistant cells: Demonstration of homology to ABC transport genes. Cancer Res. *59*, 8–13.

Oram, J.F., and Lawn, R.M. (2001). ABCA1: The gatekeeper for eliminating excess tissue cholesterol. J. Lipid Res. *42*, 1173–1179.

Pauli-Magnus, C., Lang, T., Meier, Y., Zodan-Marin, T., Jung, D., Breymann, C., Zimmermann, R., Kenngott, S., Beuers, U., Reichel, C., Kerb, R., Penger, A., Meier, P.J., and Kullak-Ublick, G.A. (2004). Sequence analysis of bile salt export pump (ABCB11) and multidrug resistance p-glycoprotein 3 (ABCB4, MDR3) in patients with intrahepatic cholestasis of pregnancy. Pharmacogenetics *14*, 91–102.

Piddock, L.J.V. (2006). Multidrug-resistance efflux pumps – not just for resistance. Nature Rev. Microbiol. *4*, 629–636.

Pohorille, A., Schweighofer, K., and Wilson, M.A. (2005). The origin and early evolution of membrane channels. Astrobiology *5*, 1–17.

Quentin, Y., Fichant, G., and Denizot, F. (1999). Inventory, assembly and analysis of bacillus subtilis abc transport systems. J. Mol. Biol. *287*, 467–484.

Rapp, M., Seppala, S., Granseth, E., and von Heijne, G. (2007). Emulating membrane protein evolution by rational design. Science *315*, 1282–1284.

Saier, M.H. (2000a). Families of transmembrane transporters selective for amino acids and their derivatives. Microbiology-Sgm *146*, 1775–1795.

Saier, M.H. (2000b). Vectorial metabolism and the evolution of transport systems. J. Bacteriol. *182*, 5029–5035.

Saier, M.H. (2003). Tracing pathways of transport protein evolution. Mol. Microbiol. *48*, 1145–1156.

Sauna, Z.E., and Ambudkar, S.V. (2000). Evidence for a requirement for atp hydrolysis at two distinct steps during a single turnover of the catalytic cycle of human p-glycoprotein. Proc. Natl. Acad. Sci. U.S. A. *97*, 2515–2520.

Sauna, Z.E., and Ambudkar, S.V. (2001). Characterization of the catalytic cycle of ATP hydrolysis by human P-glycoprotein: The two ATP hydrolysis events in a single catalytic cycle are kinetically similar but affect different functional outcomes. J. Biol. Chem. *276*, 11653–11661.

Sauna, Z.E., and Ambudkar, S.V. (2007). About a switch: How P-glycoprotein (ABCB1) harnesses the energy of ATP binding and hydrolysis to do mechanical work. Mol. Cancer Ther. *6*, 13–23.

Sauna, Z.E., Bohn, S.S., Rutledge, R., Dougherty, M.P., Cronin, S., May, L., Xia, D., Ambudkar, S.V., and Golin, J. (2008). Mutations define cross-talk between the N-terminal nucleotide-binding domain and transmembrane helix-2 of the yeast multidrug transporter Pdr5: Possible conservation of a signaling interface for coupling atp hydrolysis to drug transport. J. Biol. Chem. *283*, 35010–35022.

Sauna, Z.E., Kim, I.W., and Ambudkar, S.V. (2007). Genomics and the mechanism of P-glycoprotein (ABCB1). J. Bioenerg. Biomembr. *39*, 481–487.

Sauna, Z.E., Nandigama, K., and Ambudkar, S.V. (2006). Exploiting reaction intermediates of the atpase reaction to elucidate the mechanism of transport by P-glycoprotein (ABCB1). J. Biol. Chem. *281*, 26501–26511.

Sauna, Z.E., Smith, M.M., Muller, M., Kerr, K.M., and Ambudkar, S.V. (2001). The mechanism of action of multidrug-resistance-linked p-glycoprotein. J. Bioenerg. Biomembr. 33, 481–491.

Schumacher, M.A., and Brennan, R.G. (2002). Structural mechanisms of multidrug recognition and regulation by bacterial multidrug transcription factors. Mol. Microbiol. 45, 885–893.

Schumacher, M.A., Miller, M.C., and Brennan, R.G. (2004). Structural mechanism of the simultaneous binding of two drugs to a multidrug-binding protein. EMBO J. 23, 2923–2930.

Shimizu, T., Mitsuke, H., Noto, K., and Arai, M. (2004). Internal gene duplication in the evolution of prokaryotic transmembrane proteins. J. Mol. Biol. 339, 1–15.

Smith, K.D., Kemp, S., Braiterman, L.T., Lu, J.F., Wei, H.M., Geraghty, M., Stetten, G., Bergin, J.S., Pevsner, J., and Watkins, P.A. (1999). X-linked adrenoleukodystrophy: Genes, mutations, and phenotypes. Neurochem. Res. 24, 521–535.

Smith, P.C., Karpowich, N., Millen, L., Moody, J.E., Rosen, J., Thomas, P.J., and Hunt, J.F. (2002). ATP binding to the motor domain from an ABC transporter drives formation of a nucleotide sandwich dimer. Mol. Cell *10*, 139–149.

Tombline, G., Muharemagic, A., White, L.B., and Senior, A.E. (2005). Involvement of the 'Occluded nucleotide conformation' of p-glycoprotein in the catalytic pathway. Biochemistry *44*, 12879–12886.

Tombline, G., and Senior, A.E. (2005). The occluded nucleotide conformation of P-glycoprotein. J. Bioenerg. Biomembr. 37, 497–500.

Urbatsch, I.L., Julien, M., Carrier, I., Rousseau, M.E., Cayrol, R., and Gros, P. (2000). Mutational analysis of conserved carboxylate residues in the nucleotide binding sites of p-glycoprotein. Biochemistry *39*, 14138–14149.

Vert, J.-P. (2002). A tree kernel to analyse phylogenetic profiles. Bioinformatics *18*, S276–284.

von Heijne, G. (2006). Membrane-protein topology. Nat. Rev. Mol. Cell Biol. 7, 909–918.

Vonheijne, G. (1986). Mitochondrial targeting sequences may form amphiphilic helices. EMBO J. *5*, 1335–1342.

Wallin, E., and von Heijne, G. (1998). Genome-wide analysis of integral membrane proteins from eubacterial, archaean, and eukaryotic organisms. Protein Sci. *7*, 1029–1038.

Wehner, F., Olsen, H., Tinel, H., Kinne-Saffran, E., and Kinne, R.K.H. (2004). Cell volume regulation: Osmolytes, osmolyte transport, and signal transduction. Rev. Physiol. Bioch. P. *148*, 1–80.

Weiner, J.H., and Heppel, L.A. (1971). Binding protein for glutamine and its relation to active transport in *Escherichia coli*. J. Biol. Chem. *246*, 6933–6939.

Wilson, T.H., and Lin, E.C.C. (1980). Evolution of membrane bioenergetics. J. Supramol. Struct. *13*, 421–446.

Zaitseva, J., Jenewein, S., Jumpertz, T., Holland, B., and Schmitt, L (2005). H662 is the linchpin of ATP hydrolysis in the nucleotide-binding domain of the ABC transporter HlyB. EMBO J. *24*, 1901–1910.

Structure–Function Relationships in ABC Multidrug Transporters

3

Daniel A. P. Gutmann and Hendrik W. van Veen

Abstract

Investigations on ATP binding cassette (ABC) transporters have not only been an academic pursuit; these membrane transporters are central to many inherited disorders and are associated with multidrug resistance in cancer cells and pathogenic microorganisms. ABC multidrug transporters are capable of exporting a vast number of structurally unrelated compounds. Quantitative structure–activity relationship (QSAR) analyses of transported substrates, site-directed mutagenesis and use of thiol-reactive or photo-activatable drug analogues have provided useful information regarding the determinants of drug binding by ABC transporters. Recent studies on MsbA from *Escherichia coli* indicated that residues important for drug selectivity are located in the membrane domains of the transporter close to the leaflet–leaflet interface of the phospholipid bilayer. These and other insights in substrate binding, substrate translocation and energy-coupling by ABC extrusion systems will be reviewed.

Introduction

The cytosol of all living cells is enclosed by a membrane that prevents loss of ions, metabolites and enzymes, and that allows the selective accumulation of nutrients and extrusion of waste products. For this purpose, a wide range of specialized membrane channels and transporters has evolved. In terms of the underlying bioenergetics, these transport proteins belong to the categories of primary-active or secondary-active transporters (Saier, 2000a). Primary-active transporters are able to directly employ the chemical energy of ATP binding/hydrolysis to drive transport against a concentration gradient. Secondary-active transporters couple the metabolic energy stored in transmembrane electrochemical ion gradients to the transport of substrate across the membrane. The ABC protein superfamily and the secondary-active major facilitator superfamily (MFS) constitute two of the largest families of membrane transporters in organisms ranging from prokaryotes to human (Saier, 2000b; Higgins and Linton, 2004).

This chapter focuses on the ubiquitous group of ABC transporters (Higgins *et al.*, 1986; Holland and Blight, 1999). The importance of ABC transporters is highlighted by their function in virtually all cellular processes, conservation in all five kingdoms of life (Saurin *et al.*, 1999) and presence in both plasma membrane as well as in membranes surrounding intracellular organelles. The ABC superfamily also comprises ABC domain-containing soluble proteins, which encompass diverse roles ranging from transcription activation to protein folding (Dassa and Bouige, 2001). These soluble ABC proteins are discussed in more detail in Chapter 4. ABC proteins are evolutionary ancient and are thought to be present in the last universal common ancestor (LUCA) before life split into prokarya and archaea (Dassa and Bouige, 2001; see also Chapter 2).

All ABC transporters share the four-domain 'core' organization, comprising two membrane domains (MDs) in addition to two nucleotide-binding domains (NBDs). ABC transporters are classified either as exporters or importers (Biemans-Oldehinkel *et al.*, 2006; see also Chapter 1). Importers often utilize an accessory substrate binding protein, which is essential for their function (Shuman, 1982; Dassa and Bouige, 2001). Multidrug resistance in pathogenic bacteria and tumours can be associated with the increased expression of ABC transporters that actively extrude multiple drugs from the cell's interior (Higgins, 2007). In addition, mutations in some of the 48 human ABC transporter genes constitute the basis for a number of severe hereditary diseases (Dean *et al.*, 2001) including adrenoleukodystrophy (McGuinness *et al.*, 2003), cystic fibrosis (Riordan *et al.*, 1989) and several forms of intrahepatic cholestasis (Alvarez *et al.*, 2004; Carlton *et al.*, 2004).

Knowledge about the mechanisms by which ABC transporters interact with, and transport their substrates might reveal important information regarding their physiological role(s) and the question how one could enhance or inhibit the activity of these systems. In the following section, we will provide an outline of the structural and functional characteristics of ABC multidrug transporters.

Energy-coupling in ABC transporters

All ABC transporters possess two highly conserved NBDs at the cytoplasmic side of the membrane. They function as a dimeric ATPase to make energy available for the transport of substrates across the membrane by the MD. The NBDs are highly conserved throughout evolution (Higgins *et al.*, 1986), reflecting the importance of their common role in nucleotide binding and hydrolysis in diverse members of the ABC superfamily. More interestingly, amino acid sequence comparisons have revealed that NBDs cluster according to their role in importers and exporters (Schneider and Hunke, 1998). Although MD also clusters according to this role, these domains are more divergent, thus reflecting the plethora of substrates that are transported by these transporters, which ranges from small ions to large substrates such as peptides.

NBD structure/motor domain

NBDs function as either homo- or heterodimers, and interact in a dimer arrangement that is often referred to as 'head-to-tail' (Schmitt and Tampe, 2002) (see also Chapter 1). Both NBDs in the dimer contribute amino acid residues that help to coordinate the binding of two nucleotides at their interface (Fig. 3.1); this arrangement is observed in a variety of crystal structures of ABC NBDs including those in the DNA repair enzyme Rad50 (Hopfner *et al.*, 2000), DNA mismatch repair protein MutS (Lamers *et al.*, 2000), vitamin B12 ABC importer BtuCD (Locher *et al.*, 2002), prokaryotic ABC transporter NBD dimer MJ0796 (Yuan *et al.*, 2001), and multidrug ABC transporter Sav1866 (Dawson and Locher, 2006, 2007). Several lines of biochemical evidence support the physiological relevance of this architecture (Hunke *et al.*, 2000; Urbatsch *et al.*, 2000; Loo and Clarke, 2000a). The majority of the conserved motifs are found in the 'core' lobe of each NBD (Fig. 3.1 – yellow), which has a structural fold that resembles that of other ATPases such as the DNA recombinase RecA (Story *et al.*, 1992), and members of the P-type ATPase family. The Walker A motif (GXXGXGKS/T, where X is any residue; Fig. 3.1 – orange) contains several conserved glycine residues and one lysine which interacts with the β and γ phosphate moieties in ATP. The Walker B motif ($\phi\phi\phi\phi$D, where ϕ is any hydrophobic residue; Fig. 3.1 – dark green) is characterized by 4 hydrophobic residues followed by a glutamate that plays a role in ATP hydrolysis. The adenosine ring of ATP is stabilized by π–π interactions with the so-called 'stacking' aromatic amino acid, a tyrosine (Fig. 3.1 – Y, white circle) located ~ 20 residues upstream of the Walker A motif. The Q-loop glutamine (Fig. 3.1 – Q, light blue) contacts with the γ-phosphate indirectly via a water molecule and is thought to be involved in NBD-MD cross talk. The γ-phosphate is also coordinated by the H-loop histidine (Fig. 3.1 – H, grey), which is implicated in the hydrolysis of ATP and has been proposed to form a 'catalytic dyad' (Zaitseva *et al.*, 2005) together with an adjacent glutamate residue. The core lobe also carries the SALD motif or D-loop (Fig. 3.1 – SALD, pink) and the P-loop (Fig. 3.1 – P, red). The D-loop follows the Walker B motif, and is highly conserved but functionally poorly characterized. The P-loop precedes the Walker B motif and carries an almost invariant proline residue (Zaitseva *et al.*, 2006). In addition, the NBD contains a small helical lobe (Fig. 3.1 – slate), hosting the ABC signature

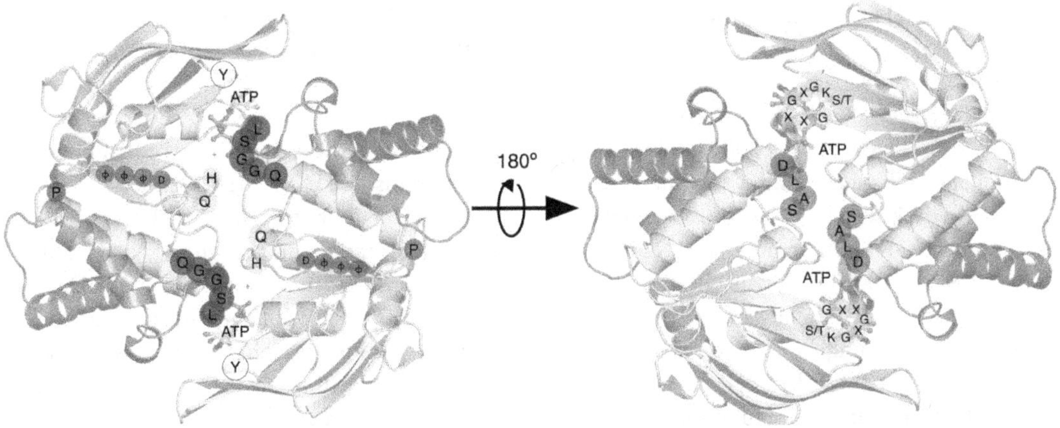

Figure 3.1 NBD dimer in complex with two ATP molecules. The core lobe is rendered in yellow and the small helical lobe in slate. The Walker A (GXXGXGKS/T) and Walker B (ϕϕϕϕD) motifs (orange and dark green, respectively) form the nucleotide-binding fold also observed in the P-loop ATPase family. Residues from the Walker B (dark green), the ABC signature motif (LSGGQ; brown), and the stacking aromatic (Y), form the bulk of the binding site. The H-loop (H grey) histidine and D-loop aspartate (also called SALD-motif; pink) make direct contact with the γ-phosphate. The Q-loop glutamine (Q; light blue) makes indirect contact with the ATP γ-phosphate via a water molecule. The almost invariable P-loop proline is rendered in red. Figures were generated with PyMol (DeLano, 2002). A colour version of this figure is located in the plate section at the back of the book.

motif (Fig. 3.1 – LSGGQ, brown), which is the hallmark of ABC proteins (Schneider and Hunke, 1998). This short motif coordinates the γ phosphate and catalytic Mg^{2+} in the ATP binding site of the opposite NBD. The ABC signature motif follows an amino acid stretch referred to as Structurally Diverse Region (SDR). The potential importance of this region in conformational coupling between MDs and NBDs in ABC exporters will be discussed later.

The crystal structures of Rad50 (Hopfner *et al.*, 2000), MutS (Lamers *et al.*, 2000), BtuCD (Locher *et al.*, 2002), MJ0796 (Yuan *et al.*, 2001), and Sav1866 (Dawson and Locher, 2006; Dawson and Locher, 2007) show NBDs dimers with perfect symmetry between the nucleotide binding sites. Biochemical studies on ABCB1 (Urbatsch *et al.*, 1995) and the binding protein (MBP)-dependent maltose uptake system MBP-MalFGK2 from *E. coli* (Sharma and Davidson, 2000) however, indicate that the two nucleotide-binding sites are non-equivalent during ATP turnover. Subsequent molecular dynamics simulations (Oloo and Tieleman, 2004; Ivetac *et al.*, 2007) suggest that at a time point during the turnover of ATP, asymmetry in the NBD–NBD interface exists. Indeed, for ABCB1 (Sauna and Ambudkar, 2001) at least three distinct ATP bound states of the NBD can be identified: (i) 'loose' ATP-bound, (ii) 'occluded' (asymmetrical) ATP-bound and (iii) the post-hydrolysis state where ADP-Pi is still bound.

This data is consistent with a model where ATP hydrolysis occurs in an alternating fashion between both ATP binding sites. The symmetry between the two NBDs in the dimer observed in the crystal structures of Rad50 (Hopfner *et al.*, 2000), MutS (Lamers *et al.*, 2000), BtuCD (Locher *et al.*, 2002), MJ0796 (Yuan *et al.*, 2001), Sav1866 (Dawson and Locher, 2006), and MBP-MalFGK_2 (Oldham *et al.*, 2007) could therefore be a result of the truncation or mutation of the NBD domains and/or use of artificial nucleotide analogues required to stabilize NBD dimers, whilst the dimerization undergone by wild-type NBDs supplied with Mg-ATP is at best transient (Benabdelhak *et al.*, 2005). In fact, until now, appreciable structural asymmetry (i.e. non-equivalence) between the two NBDs in the dimer has been detected only in the crystal structure of dimeric HlyB NBD of the type I secretion machinery of *E. coli* that is involved in the efflux of haemolysin (HlyA) across the cell envelope (Zaitseva *et al.*, 2006), as well as in the NBDs present in the BtuCD–F complex (BtuCD crystallized with a bound, soluble periplasmic binding protein BtuF) (Hvorup *et al.*, 2007). Interestingly, a phosphate exchange channel was identified in the case of HlyB, which might be associated with the asymmetry in the structure of the NBD dimer (Zaitseva *et al.*, 2006). HlyB is the only crystal structure that has been analysed in the presence of Mg^{2+} in the binding site; these results suggest that the presence of this divalent metal ion might be important in maintaining functional asymmetry (see also Chapter 1). The BtuCD–F complex was crystallized without the ATPase inhibitor vanadate (Hvorup *et al.*, 2007) in contrast to the previously published BtuCD structure (Locher *et al.*, 2002). It seems that BtuF imposes a degree of asymmetry on the BtuCD–F complex. Indeed, the asymmetry evident in the crystal structures of HlyB and the BtuCD–F complex might be the structural explanation for the 'occluded' state with tight binding of ATP observed in the catalytic cycle of ABCB1 (Tombline *et al.*, 2005).

At present, an increasing number of crystal structures of full-length ABC transporters, not just isolated NBDs, are becoming available; this new information allows the integration of biochemical data with structural data. For example, previous biochemical experiments on ABC transporters have suggested that the free energy of ATP hydrolysis can power the transmembrane transport of the substrate by inducing conformational changes in the MDs with a concomitant change in the accessibility of substrate binding pockets in these domains (see also Chapter 1). At present, the structures of Sav1866 (Dawson and Locher, 2007) and BtuCD (Locher *et al.*, 2002) suggest that their MDs are in an outward-facing orientation, while their NBDs are locked in a dimeric configuration with bound ATP analogue AMP-PNP and ATPase inhibitor cyclotetravanadate, respectively.

These emerging structures are shaping a new paradigm for the catalytic cycle of ABC transporters, which began with the reported crystal structure of the NBD dimer of Rad50 (Hopfner *et al.*, 2000). The structural configuration seen in these and additional crystal structures (as seen in Fig. 3.1) allowed formulating a transport cycle based on ATP driven association and dissociation of the NBDs (Smith *et al.*, 2002). As a result, the 'ATP switch model' (Higgins and Linton, 2004) constitutes one of the eminent models that has evolved from the original work on Rad50. This model considers ATP driven dimerization – and dissociation following ATP hydrolysis – may it occur alternating or in concert, as the driving force behind rearrangements in the MDs during substrate translocation. In the resting state, the ABC exporter is thought to expose a high-affinity binding site for the substrate at the inside surface of the membrane. Substrate binding to this high affinity site is thought to stimulate ATP binding to the NBDs. The free energy provided by the NBD dimerization is then thought to 'switch' the MDs to an outward-facing low-affinity configuration allowing substrate release. ATP hydrolysis and NBD dissociation then occurs as the final step, which resets the transporter to an inward facing conformation. The ATP switch model is largely consistent with the overall transport process and is structurally supported by the very recent importer structures of the putative iron transporter HI1470/1 (Pinkett *et al.*, 2007) and putative molybdate transporter ModBC (Gerber *et al.*, 2008) that show an inward facing cavity formed by the MDs (see also chapter 1).

The ABC switch model does not include a role for a 'nucleotide-occluded' state as has been observed in ABCB1, or with the asymmetry seen in the HlyB crystal structure. This prompted the development of another important model, which sees the formation of the 'nucleotide-occluded' state as the actual power-stroke of substrate transport (Tombline *et al.*, 2005; Ambudkar *et al.*, 2006). In fact, in the mitochondrial ABC transporter Mdl1, the formation of the 'nucleotide-occluded' state would be associated with a free energy change of ~70 kJ mol^{-1}, indicating that it might occur together with a large conformational change; this data is consistent with the 'nucleotide-occlusion model' (van der Does *et al.*, 2006). Further investigations are required to resolve the intricate molecular details at play.

Structural coupling between MDs and NBDs in importers and exporters

The MD–NBD interface in the importers BtuCD (Locher *et al.*, 2002), HI1470/1 (Pinkett *et al.*,

2007) and ModBC (Gerber *et al.*, 2008) seems to be confined to the loosely conserved EAA-loop in the MD (Mourez *et al.*, 1997) which interacts with a cleft in the NBD containing the Q-loop (NBD) in a *cis* (MD1:NBD1 and MD2:NBD2) configuration (se also Chapter 1). Since the Q-loop glutamine interacts with the γ-phosphate of ATP through a water molecule, the EAA–Q-loop interface is thought to transmit information on the nucleotide occupancy and/or identity from the NBD to the MDs. On the other hand, the MD–NBD interactions present in the structure of the ABC exporter Sav1866 are rather more elaborate (Dawson and Locher, 2006, Dawson and Locher, 2007). In this case, two interfaces can be identified and their general role in exporters is supported by their sequence conservation. One interface is in a *cis*-configuration (MD1:NBD1 and MD2:NBD2) due to the ICL-2–Q-loop interaction that is analogous to the EAA loop–Q-loop interaction seen in importers. The second interaction at the exporter MD–NBD interface is made in a *trans*-configuration (MD1–NBD2 and MD2–NBD1) to the opposite NBD. This second interaction is mediated by the recently described TEVGERV motif (Dawson and Locher, 2006), which is only seen in the SDR region in exporter NBDs.

Thus, the TEVGERV motif might be instrumental in determining the directionality of transport by ABC transporters. By contacting the NBDs both in a *cis*- and *trans*-configuration, the MDs are forced to twist along their dimer axis. The ICL-1–TEVGERV interface is located in close proximity to the signature motif, that coordinates the catalytic magnesium and γ-phosphate of one bound ATP nucleotide. This combination of *cis*- and *trans*-interactions might allow each half of the MD to respond to the nucleotide occupancy of both nucleotide binding sites of the NBD dimer.

Drug specificity of ABC multidrug exporters

Initial studies

Multidrug resistance was first described in Chinese hamster ovary (CHO) cell lines that were selected for their resistance to colchicine (Juliano and Ling, 1976; also described in detail in Chapter 1), and which were found to overexpress the *ABCB1* gene (Ueda *et al.*, 1987). In the following years many more ABC transporters were found to contribute to multidrug resistance in various cancers and antibiotic resistance in microorganisms. In man, ABCB1 is found in apical cells lining the intestine, brain capillaries, liver, kidney and the pregnant uterus. Due to its location and multidrug transport activity, ABCB1 expression is an important element for the bioavailability of therapeutic drugs; e.g. all currently prescribed HIV protease inhibitors are substrates for ABCB1 (Hegedus *et al.*, 2002; Kim *et al.*, 1998; Lee *et al.*, 1998). Due to the clinical implications of ABCB1 and other ABC transporters in multidrug resistance, a considerable effort has gone into the development of inhibitors (discussed in Chapter 9). Several ABCB1 inhibitors have undergone clinical trials (Ferry *et al.*, 1996; Fisher *et al.*, 1996; Fisher and Sikic, 1995). The ability of ABCB1 to bind and transport many structurally unrelated drugs has been a core focus of investigations.

The role of the lipid bilayer in drug specificity

The role of the surrounding lipid bilayer on the function of ABCB1 has long been established. It has been suggested that ABCB1 can bind drugs from the cytoplasmic membrane leaflet (Shapiro *et al.*, 1997; Shapiro and Ling, 1997b); indeed ABCB1 is often linked to the 'hydrophobic vacuum cleaner model' (Higgins and Gottesman, 1992). Due to their hydrophobicity, the substrates to be transported are estimated to be 300 – 2000 fold more concentrated in the membrane bilayer than in the aqueous phase (Romsicki and Sharom, 1997). One example for the interaction of ABCB1 with substrates from the membrane can potentially be found in general anaesthetics. These hydrophobic aromatic compounds are chemically similar to many ABCB1 substrates. Pre-exposure of ABCB1 to anaesthetics prevents the photoaffinity labelling of ABCB1 with the transport substrate [125I]-iodoarylazidoprazosin (Regev *et al.*, 1999). Molecular modelling studies with L-type calcium channels suggest that the interaction between the anaesthetic benziothiazepine and these channels parallels drug interactions with ABCB1-type ABC transporters (Tikhonov and Zhorov, 2008). Measurements of the binding affinity of drug binding pockets should therefore take into consideration that drug will partition into the membrane from which transport might occur. The quantitative analysis of drug binding from the membrane to the transporter is therefore a rather complex task.

Omote and Al-Shawi approached the issue of ABCB1-mediated transport of substrates from the membrane using a mixed molecular dynamics and molecular modelling approach (Omote and Al-Shawi, 2006). The behavior of typical ABCB1 substrates in a lipid bilayer was characterized following their route *via* a simulation lasting up to 20 ns. Particular attention was paid to the frequency of hydrogen bonding of these drugs to phosphatidylcholine head groups and water at the membrane-water interfacial region. This frequency of interaction was found to correlate with the experimentally determined ATPase stimulation of ABCB1 by these compounds. Based on this data the authors proposed that ABCB1 recognizes and binds substrates by a solvation exchange mechanism. As the substrate diffuses laterally into the binding pocket, hydrogen bonds formed with residues in the MD at the cytoplasmic membrane leaflet replace those made with water and the phosphatidyl head groups in the membrane. The hydrogen bonds made in the hydrophobic environment of the MD are stronger than those formed in the aqueous environment of the membrane water interface. In a further step, the hydrophobic portion of the drug is bound by hydrophobic side chains within the drug-binding pocket. Eventually, ATP binding to the NBD is thought to open an exit tunnel in ABCB1, and to reduce the affinity of the binding site for the substrate, thereby facilitating drug release.

Ligand-based approaches into drug recognition

In order to uncover the factors which exert a significant impact on the interaction between substrates/inhibitors and ABCB1, a large body of data has been generated on structure–activity relationship (SAR) analyses including molecular modelling of substrates, two-dimensional (2D) and three-dimensional (3D) parameter-frame-setting analysis, and quantitative structure activity relationship (QSAR) analysis among substrates/modulators. The earliest SAR analyses invoked the requirement of hydrophobicity (Gottesman and Pastan, 1993). This idea was based upon the fact that several structurally related drugs that differ in their ability to be transported by ABCB1 differ in their relative hydrophobicities, as measured by their water/octanol partitioning ratio (logP) (Zamora *et al.*, 1988). This observation was used as evidence for the proposition that ABCB1 and related transporters act as a flippase (Higgins and Gottesman 1992). Thus, a drug might be intercalated into the lipid bilayer before it interacts with a binding site on the efflux protein (Van Veen and Konings, 1997).

Didziapetris and colleagues analysed 220 compounds that are transported by ABCB1, and derived from this study the general rule that ABCB1 substrates are hydrophobic compounds with (i) ^{3} 8 N and O atoms, (ii) molecular weight > 400 Da and (iii) pka < 4 (Didziapetris *et al.*, 2003). On the other hand, Seelig defined ABCB1 substrates in terms of spatial arrangement of hydrogen bond acceptor groups (Seelig, 1998), information which is related to the number of N and O atoms described by Didziapetris. Ecker and co-workers investigated the role of the N atom in the activity of ABCB1 inhibitors. A clear relationship was observed between the hydrogen bond acceptor strengths and the EC50 values for ABCB1 inhibition, even for a compound that lacked the N atom (Ecker *et al.*, 1999), addressing attention to the important role of hydrogen-bonding interactions between substrate and the transport proteins.

Orlowski and co-workers determined affinities and mutual relationships from the changes in P-glycoprotein ATPase activity induced by a series of cyclic peptides and peptide-like compounds, used alone or in combination (Garrigues *et al.*, 2002). Modelling of the intramolecular distribution of the hydrophobic and polar surfaces of this series of molecules made it possible to superimpose some of these surface elements. On each of these pharmacophores, the ligands competed with each other. The typical ABCB1 substrates, verapamil, cyclosporin A, and actinomycin D, bound to pharmacophore 1, whereas vinblastine bound to pharmacophore 2. Thus, according to these authors, the multi-specific binding pocket of P-glycoprotein can be seen as sites, located near one another, that bind ligands according to the distribution of their hydrophobic and polar elements rather than their chemical motifs.

Drug–protein interactions in ABC multidrug transporters

Several techniques have been used to study drug–protein interactions in ABC multidrug transporters. Inside out membrane vesicles are a popular target which have been used successfully in studies on the drug-stimulated ATPase activity of these transport systems, but have been used less successfully in measurements of ATP-dependent transport of drugs. ABC multidrug transporters have also been studied as purified proteins in detergent solution and when functionally reconstituted in proteoliposomes. Drug recognition has been examined by specific techniques including equilibrium drug-binding, cysteine scanning mutagenesis in combination with thiol-reactive drug analogues, and cross-linking with photoreactive drug analogues followed by peptide mapping. Some important underlying themes have emerged although the molecular details of drug binding and transport remain undefined.

Drug transport assays in cells revealed that ABCB1 has at least two positively cooperative drug-binding sites, one selective for Rhodamine 123 (R-site) and anthracyclines, the other for Hoechst 33342 (H-site) and colchicine (Shapiro and Ling, 1997a). As a consequence, drug transport from either site will stimulate drug transport from the other (Shapiro and Ling, 1997a). This underlines the modest specificity of the binding sites. Ligand binding assays on LmrA, the best-studied bacterial homologue of ABCB1 (van Veen *et al.*, 1998), has equally revealed the presence of two allosterically linked drug binding sites in the protein (van Veen *et al.*, 2000).

Beyond functional studies, scanning mutagenesis and photoaffinity labelling have delivered information on the location of the drug binding sites. For example, Loo and Clarke built a library of 189 single cysteine ABCB1 mutants that were tested for their ability to cross link with thiol reactive substrates such as MTS-verapamil, MTS-rhodamine 6G or dibromobimane in the presence and absence of non-thiol containing verapamil or rhodamine 6G. This was possible because both cysteine-less ABCB1 and most of the single cysteine mutants were transport-active (Loo and Clarke, 1995, 1996). Verapamil and rhodamine 6G protected distinct areas in ABCB1 against covalent modification by the thiol reactive substrates, thereby indicating areas in the protein where these substrates can bind (Loo and Clarke, 1997, 1999a, 2000b, 2002). The resulting data implicated transmembrane helices (TMHs) 6, 9 and 12 in forming the R-site pocket. The H-site would be formed by residues located on TMHs 4, 6, 10, 11 and 12. Cysteines in helices 6 and 12 cross react to both verapamil- and rhodamine-MTS highlighting some overlap in the binding sites.

LmrA, which can complement ABCB1 in lung fibroblast cells, displays equal promiscuity towards drugs as its human counterpart (van Veen *et al.*, 1998). Truncated versions of ABCB1 and LmrA, lacking the NBDs still retain the ability to bind drugs (Loo and Clarke, 1999b; Venter *et al.*, 2003). Using LmrA as a model, Poelarends and co-workers created a library of 40 single cysteine LmrA mutants (Poelarends and Konings, 2002). Primarily residues in TMH6, corresponding to helices 6 and 12 in ABCB1 were chosen for mutagenesis. Eleven out of 15 aromatic residues in TMH 6 were accessible to the highly soluble compound fluorescein maleimide. It was concluded that the MD of LmrA would form an aqueous chamber that is exposed to the cytosol. This helix corresponds to TMH 6 and 12 of the heterodimeric ABCB1, which had been shown to interact with both verapamil and rhodamine 6G (Loo and Clarke, 1997, 1999a, 2000b). The model of a cytoplasmically exposed aqueous drug binding chamber in ABC multidrug transporters received further support by studies undertaken on ABCB1 (Loo *et al.*, 2004). The water soluble cysteine-reactive compound sodium (2-sulfonatoethyl)methanethiosulfonate (MTSES) was found to successfully compete with other transport substrates for site-specific labelling of single cysteine mutants, leading the authors to the conclusion that the drug binding pocket must be hydrated (Loo *et al.*, 2004). Photoaffinity labelling analyses further expanded the knowledge about the TMHs involved in drug binding, without the need to create cysteine mutants. The results were largely complementary to those obtained from cysteine scanning mutagenesis (Pleban *et al.*, 2005).

The increasing availability of structural data on full-length ABC transporters has also begun to guide functional studies. In a comprehensive literature study, Shilling and co-workers identified several change-in-specificity mutations of ABCB1, which affect the specificity of ABCB1 for specific drugs (Shilling *et al.*, 2006). Woebking *et al.* (2008) observed a clustering of these mutations in the grooves that run along the two 'wings' formed by the MDs in the corrected crystal structure of *Salmonella enterica* serovar Typhimurium MsbA (Ward *et al.*, 2007), which is a close functional and structural homologue of LmrA and ABCB1 (Reuter *et al.*, 2003). Therefore, the positions in MsbA of these change-in-specificity mutants of ABCB1 were used as a starting point for structure–functional analysis in MsbA. A double alanine mutant was constructed of the respective residues S289 and S290 in a cluster of change-in-specificity mutants in Helix 6 of *E. coli* MsbA near the leaflet–leaflet interface of the membrane (Woebking *et al.*, 2008). Although this mutant, called SASA MsbA, showed a similar basal ATPase activity as wildtype MsbA, the stimulation of the ATPase activity by taxol and ethidium, and the binding and transport of these substrates were strongly inhibited in the SASA mutant. In contrast, the interactions with erythromycin and Hoechst 33342 remained unaltered. This work highlights the importance of residues in TMH 6 in drug binding and transport by MsbA. Furthermore, the importance and validity of structure-guided functional studies is underlined by this work.

Crystal structures of drug binding pockets

A crystallographic structure of ABCB1 that could reveal insights into the nature and location of drug-binding surfaces has not yet been published. Fortunately, crystal structures of proteins with equally multispecific binding sites have been solved. Several crystal structures of QacR, the soluble repressor of the *Staphylococcus aureus* multidrug resistance transporter QacA, have been obtained in the presence of different cytotoxic drugs (Grkovic *et al.*, 2001; Schumacher and Brennan, 2003; Murray *et al.*, 2004; Brooks *et al.*, 2007; Peters *et al.*, 2008;). Some concepts derived from these drug-bound crystal structures may help to explain the observations made on drug interactions in ABCB1 and other multidrug transporters. The obtained data revealed a considerable plasticity in the binding site of QacR. In one example, the ternary QacR–ethidium–proflavine complex (Fig. 3.2 A) is stabilized by multiple aromatic stacking and hydrogen bonds (Schumacher *et al.*, 2004). Binding is facilitated by several glutamate residues, which act as hydrogen bond donors in agreement with the postulated ABCB1 binding site model (Seelig, 1998). In another QacR–drug complex, the positive charges on pentamidine and hexamidine were neutralised by the carboxylates of E57 and E120 (Brooks *et al.*, 2007). This type of charge neutralization by carboxylate side chains might have relevance for multidrug transporters, as it was observed that replacement of E26 in the *E. coli* multidrug resistance protein MdfA abolished transport of cationic drugs but not of neutral compounds such as chloramphenicol (Edgar and Bibi, 1999). The drug–protein interactions in the binary QacR–dequalinium complex (Fig. 3.2 B) (Schumacher *et al.*, 2001) are facilitated by the same residues as in the ternary QacR–ethidium–proflavine complex (Schumacher *et al.*, 2004). However, the superimposition of the residues involved in the binding shows considerable side chain rearrangement, which is particularly clear for Trp-61 and Tyr-103. In addition to this plasticity, Brooks and co-workers demonstrated that QacR is able to simultaneously bind at least two drugs, the divalent benzamidines DB75 and DB359 in alternative orientations with near equal affinity (Brooks *et al.*, 2007). These alternative modes of binding allow a certain degree of chemical redundancy in these binding pockets. A similar redundancy might be present in the bacterial ABCB1 homologue MsbA (Woebking *et al.*, 2008) where mutations lead to a change in specificity instead of outright loss of binding and transport activity.

Conclusions and future trends

We have begun to grasp the importance of ABC transporters in key biological processes. Despite concerted efforts, many of the functions carried out by these systems are still unknown or poorly understood at the molecular level. Our current understanding of the mechanisms by which ABC multidrug transporters recognize and transport drugs was gained by a synergy of biochemical and structural approaches, which will most likely also be the backbone of investigations in the near

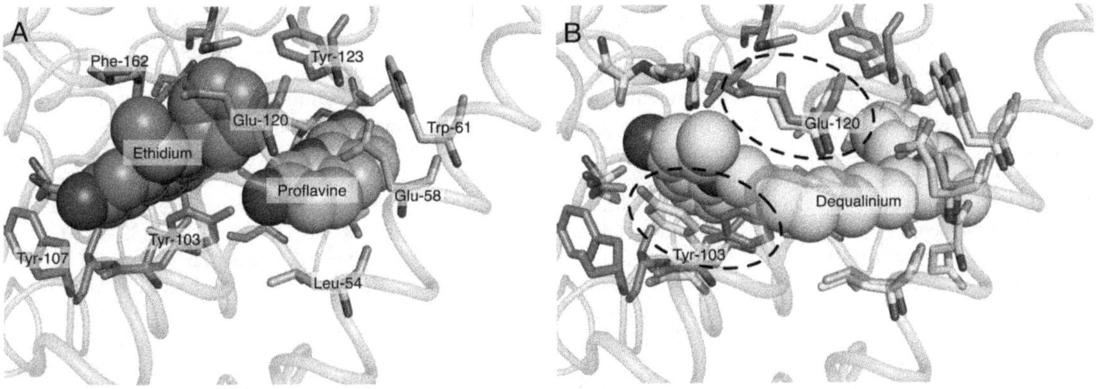

Figure 3.2 A Plasticity in the QacR multidrug binding site. Ethidium and proflavine are bound to QacR simultaneously (pdb code 1QVU). The drugs are stabilised by hydrophobic stacking and polar interactions. **B** Dequalinium occupies a site equivalent to the volume of ethidium and proflavine combined (pdb code 1JT6). Side chains are superimposed for improved clarity and visualization of the structural rearrangements to accommodate the dequalinium moiety. Broken circles indicate residues undergoing rearrangement to accommodate different drugs in the binding site. Figures were generated with PyMol (DeLano, 2002). A colour version of this figure is located in the plate section at the back of the book.

future. From a clinical point of view, a better understanding of the mechanisms by which ABC transporters operate and recognize their substrates might reveal new ways of drug-based intervention using compounds that can inhibit/enhance transport activity. From an academic point of view studies with such modulators will be interesting as these compounds might help to identify rate-limiting steps in the translocation mechanism of these transporters, inhibition/activation of which will affect the overall activity the most. As chemotherapy represents a major line of treatment of cancers and infections by pathogenic microorganisms, basic knowledge about the roles and molecular mechanisms of membrane transporters in multidrug resistance will continue to be important for advancements in the clinic.

Acknowledgements

This work was funded by the Biotechnology and Biological Sciences Research Council (BBRSC).

References

Alvarez, L., Jara, P., Sanchez-Sabate, E., Hierro, L., Larrauri, J., Diaz, M.C., Camarena, C., De la Vega, A., Frauca, E., Lopez-Collazo, E. and Lapunzina, P. (2004). Reduced hepatic expression of farnesoid X receptor in hereditary cholestasis associated to mutation in ATP8B1. Hum. Mol. Genet. *13*, 2451–2460.

Ambudkar, S.V., Kim, I.W., and Sauna, Z.E. (2006). The power of the pump: mechanisms of action of P-glycoprotein (ABCB1). Eur. J. Pharm. Sci. *27*, 392–400.

Benabdelhak, H., Schmitt, L., Horn, C., Jumel, K., Blight, M.A., and Holland, I.B. (2005). Positive co-operative activity and dimerization of the isolated ABC ATPase domain of HlyB from *Escherichia coli*. Biochem. J. *386*, 489–495.

Biemans-Oldehinkel, E., Doeven, M.K., and Poolman, B. (2006). ABC transporter architecture and regulatory roles of accessory domains. Febs Lett. *580*, 1023–1035.

Brooks, B.E., Piro, K.M., and Brennan, R.G. (2007). Multidrug-binding transcription factor QacR binds the bivalent aromatic diamidines DB75 and DB359 in multiple positions. J. Am. Chem. Soc. *129*, 8389–8395.

Carlton, V.E., Pawlikowska, L., and Bull, L.N. (2004). Molecular basis of intrahepatic cholestasis. Ann. Med. *36*, 606–617.

Dassa, E., and Bouige, P. (2001). The ABC of ABCS: a phylogenetic and functional classification of ABC systems in living organisms. Res. Microbiol. *152*, 211–229.

Dawson, R.J., and Locher, K.P. (2006). Structure of a bacterial multidrug ABC transporter. Nature *443*, 180–185.

Dawson, R.J., and Locher, K.P. (2007). Structure of the multidrug ABC transporter Sav1866 from *Staphylococcus aureus* in complex with AMP-PNP. FEBS Lett. *581*, 935–938.

Dean, M., Rzhetsky, A., and Allikmets, R. (2001). The human ATP-binding cassette (ABC) transporter superfamily. Genome Res. *11*, 1156–1166.

DeLano, W.L., (2002) The PyMol. Molecular Graphics System. DeLano Scientific, Palo Alto, CA, USA.

Didziapetris, R., Japertas, P., Avdeef, A., and Petrauskas, A. (2003). Classification analysis of P-glycoprotein substrate specificity. J. Drug Target *11*, 391–406.

Ecker, G., Huber, M., Schmid, D., and Chiba, P. (1999). The importance of a nitrogen atom in modulators of multidrug resistance. Mol. Pharmacol. *56*, 791–796.

Edgar, R., and Bibi, E. (1999). A single membrane-embedded negative charge is critical for recognizing positively charged drugs by the *Escherichia coli* multidrug resistance protein MdfA. EMBO J. *18*, 822–832.

Ferry, D.R., Traunecker, H., and Kerr, D.J. (1996). Clinical trials of P-glycoprotein reversal in solid tumours. Eur. J. Cancer *32A*, 1070–1081.

Fisher, G.A., Lum, B.L., Hausdorff, J., and Sikic, B.I. (1996). Pharmacological considerations in the modulation of multidrug resistance. Eur. J. Cancer *32A*, 1082–1088.

Fisher, G.A., and Sikic, B.I. (1995). Clinical studies with modulators of multidrug resistance. Hematol. Oncol. Clin. North. Am. *9*, 363–382.

Garrigues, A., Loiseau, N., Delaforge, M., Ferté, J., Garrigos, M., André, F., and Orlowski, S. (2002) Characterization of two pharmacophoRes. on the multidrug transporter P-glycoprotein. Mol. Pharmacol. *62*, 1288–1298

Gerber, S., Comellas-Bigler, M., Goetz, B.A., and Locher, K.P. (2008). Structural basis of trans-inhibition in a molybdate/tungstate ABC transporter. Science *321*, 246–250.

Gottesmann, M. and Pastan, I. (1993) Biochemistry of multidrug resistance mediated by the multidrug transporter. Annu. Rev. Biochem. *62*, 385–427.

Grkovic, S., Brown, M.H., Schumacher, M.A., Brennan, R.G., and Skurray, R.A. (2001). The staphylococcal QacR multidrug regulator binds a correctly spaced operator as a pair of dimers. J. Bacteriol. *183*, 7102–7109.

Hegedus, T., Orfi, L., Seprodi, A., Varadi, A., Sarkadi, B., and Keri, G. (2002). Interaction of tyrosine kinase inhibitors with the human multidrug transporter proteins, MDR1 and MRP1. Biochim. Biophys. Acta *1587*, 318–325.

Higgins, C.F. (2007). Multiple molecular mechanisms for multidrug resistance transporters. Nature *446*, 749–757.

Higgins, C.F., and Gottesman, M.M. (1992). Is the multidrug transporter a flippase? Trends Biochem. Sci. *17*, 18–21.

Higgins, C.F., Hiles, I.D., Salmond, G.P., Gill, D.R., Downie, J.A., Evans, I.J., Holland, I.B., Gray, L., Buckel, S.D., Bell, A.W., and Hermodson, M.A. (1986). A family of related ATP-binding subunits coupled to many distinct biological processes in bacteria. Nature *323*, 448–450.

Higgins, C.F., and Linton, K.J. (2004). The ATP switch model for ABC transporters. Nat. Struct. Mol. Biol. *11*, 918–926.

Holland, I.B., and Blight, M.A. (1999). ABC-ATPases, adaptable energy generators fuelling transmembrane movement of a variety of molecules in organisms from bacteria to humans. J. Mol. Biol. *293*, 381–399.

Hopfner, K.P., Karcher, A., Shin, D.S., Craig, L., Arthur, L.M., Carney, J.P., and Tainer, J.A. (2000). Structural biology of Rad50 ATPase: ATP-driven conformational control in DNA double-strand break repair and the ABC-ATPase superfamily. Cell *101*, 789–800.

Hunke, S., Mourez, M., Jehanno, M., Dassa, E., and Schneider, E. (2000). ATP modulates subunit–subunit interactions in an ATP-binding cassette transporter (MalFGK2) determined by site-directed chemical cross-linking. J. Biol. Chem. *275*, 15526–15534.

Hvorup, R.N., Goetz, B.A., Niederer, M., Hollenstein, K., Perozo, E., and Locher, K.P. (2007). Asymmetry in the structure of the ABC transporter-binding protein complex BtuCD-BtuF. Science *317*, 1387–1390.

Ivetac, A., Campbell, J.D., and Sansom, M.S. (2007). Dynamics and function in a bacterial ABC transporter: simulation studies of the BtuCDF system and its components. Biochemistry *46*, 2767–2778.

Juliano, R.L., and Ling, V. (1976). A surface glycoprotein modulating drug permeability in Chinese hamster ovary cell mutants. Biochim. Biophys. Acta *455*, 152–162.

Kim, R.B., Fromm, M.F., Wandel, C., Leake, B., Wood, A.J., Roden, D.M., and Wilkinson, G.R. (1998). The drug transporter P-glycoprotein limits oral absorption and brain entry of HIV-1 protease inhibitors. J. Clin. Invest. *101*, 289–294.

Lamers, M.H., Perrakis, A., Enzlin, J.H., Winterwerp, H.H., de Wind, N., and Sixma, T.K. (2000). The crystal structure of DNA mismatch repair protein MutS binding to a G x T mismatch. Nature *407*, 711–717.

Lee, C.G., Gottesman, M.M., Cardarelli, C.O., Ramachandra, M., Jeang, K.T., Ambudkar, S.V., Pastan, I., and Dey, S. (1998). HIV-1 protease inhibitors are substrates for the MDR1 multidrug transporter. Biochemistry *37*, 3594–3601.

Locher, K.P., Lee, A.T., and Rees, D.C. (2002). The *E. coli* BtuCD structure: a framework for ABC transporter architecture and mechanism. Science *296*, 1091–1098.

Loo, T.W., Bartlett, M.C., and Clarke, D.M. (2004). The drug-binding pocket of the human multidrug resistance P-glycoprotein is accessible to the aqueous medium. Biochemistry *43*, 12081–12089.

Loo, T.W., and Clarke, D.M. (1995). Membrane topology of a cysteine-less mutant of human P-glycoprotein. J. Biol. Chem. *270*, 843–848.

Loo, T.W., and Clarke, D.M. (1996). Inhibition of oxidative cross-linking between engineered cysteine residues at positions 332 in predicted transmembrane segments (TM) 6 and 975 in predicted TM12 of human P-glycoprotein by drug substrates. J. Biol. Chem. *271*, 27482–27487.

Loo, T.W., and Clarke, D.M. (1997). Identification of residues in the drug-binding site of human P-glycoprotein using a thiol-reactive substrate. J. Biol. Chem. *272*, 31945–31948.

Loo, T.W., and Clarke, D.M. (1999a). Identification of residues in the drug-binding domain of human P-glycoprotein. Analysis of transmembrane segment 11 by cysteine-scanning mutagenesis and inhibition by dibromobimane. J. Biol. Chem. *274*, 35388–35392.

Loo, T.W., and Clarke, D.M. (1999b) The transmembrane domains of the human multidrug resistance P-glycoprotein are sufficient to mediate drug binding and trafficking to the cell surface, J. Biol. Chem. *274*, 24759–24765.

Loo, T.W., and Clarke, D.M. (2000a). Drug-stimulated ATPase activity of human P-glycoprotein is blocked by disulfide cross-linking between the nucleotide-binding sites. J. Biol. Chem. *275*, 19435–19438.

Loo, T.W., and Clarke, D.M. (2000b). Identification of residues within the drug-binding domain of the human multidrug resistance P-glycoprotein by cysteine-scanning mutagenesis and reaction with dibromobimane. J. Biol. Chem. *275*, 39272–39278.

Loo, T.W., and Clarke, D.M. (2002). Location of the rhodamine-binding site in the human multidrug resistance P-glycoprotein. J. Biol. Chem. *277*, 44332–44338.

McGuinness, M.C., Lu, J.F., Zhang, H.P., Dong, G.A., Heinzer, A.K., Watkins, P.A., Powers, J., and Smith, K.D. (2003). Role of ALDP (ABCD1) and mitochondria in x-linked adrenoleukodystrophy. Mol. Cell Biol. *23*, 744–753.

Mourez, M., Hofnung, M., and Dassa, E. (1997). Subunit interactions in ABC transporters: a conserved sequence in hydrophobic membrane proteins of periplasmic permeases defines an important site of interaction with the ATPase subunits. EMBO J. *16*, 3066–3077.

Murray, D.S., Schumacher, M.A., and Brennan, R.G. (2004). Crystal structuRes. of QacR–diamidine complexes reveal additional multidrug-binding modes and a novel mechanism of drug charge neutralization. J. Biol. Chem. *279*, 14365–14371.

Oldham, M.L., Khare, D., Quiocho, F.A., Davidson, A.L., and Chen J. (2007) Crystal structure of a catalytic intermediate of the maltose transporter. Nature *450*, 515–521.

Oloo, E.O., and Tieleman, D.P. (2004). Conformational transitions induced by the binding of MgATP to the vitamin B12 ATP-binding cassette (ABC) transporter BtuCD. J. Biol. Chem. *279*, 45013–45019.

Omote, H., and Al-Shawi, M.K. (2006). Interaction of transported drugs with the lipid bilayer and P-glycoprotein through a solvation exchange mechanism. Biophys. J. *90*, 4046–4059.

Peters, K.M., Schuman, J.T., Skurray, R.A., Brown, M.H., Brennan, R.G., and Schumacher, M.A. (2008). QacR-cation recognition is mediated by a redundancy of residues capable of charge neutralization. Biochemistry *47*, 8122–8129.

Pinkett, H.W., Lee, A.T., Lum, P., Locher, K.P., and Rees, D.C. (2007). An inward-facing conformation of a putative metal-chelate-type ABC transporter. Science *315*, 373–377.

Pleban, K., Kopp, S., Csaszar, E., Peer, M., Hrebicek, T., Rizzi, A., Ecker, G.F., and Chiba, P. (2005). P-glycoprotein substrate binding domains are located at the transmembrane domain/transmembrane domain interfaces: a combined photoaffinity labeling–protein homology modeling approach. Mol. Pharmacol. *67*, 365–374.

Poelarends, G.J., and Konings, W.N. (2002). The transmembrane domains of the ABC multidrug transporter LmrA form a cytoplasmic exposed, aqueous chamber within the membrane. J. Biol. Chem. *277*, 42891–42898.

Regev, R., Assaraf, Y.G., and Eytan, G.D. (1999). Membrane fluidization by ether, other anesthetics, and certain agents abolishes P-glycoprotein ATPase activity and modulates efflux from multidrug-resistant cells. Eur. J. Biochem. *259*, 18–24.

Reuter, G., Janvilisri, T., Venter, H., Shahi, S., Balakrishnan, L., and van Veen, H.W. (2003). The ATP binding cassette multidrug transporter LmrA and lipid transporter MsbA have overlapping substrate specificities. J. Biol. Chem. *278*, 35193–35198.

Riordan, J.R., Rommens, J.M., Kerem, B.S., Alon, N., Rozmahel, R., Grzelczak, Z., Zielenski, J., Lok, S., Plavsic, N., Chou, J.L., Drumm, M.L., Iannuzzi, M.C., Collins, F.C., and Tsui, L.-C. (1989). Identification of the Cystic-Fibrosis Gene – Cloning and Characterization of Complementary-DNA. Science *245*, 1066–1072.

Romsicki, Y., and Sharom, F.J. (1997). Interaction of P-glycoprotein with defined phospholipid bilayers: a differential scanning calorimetric study. Biochemistry *36*, 9807–9815.

Saier, M.H. (2000a). A functional-phylogenetic classification system for transmembrane solute transporters. Microbiol. Mol. Biol. Rev. *64*, 354–411.

Saier, M.H. (2000b). Families of proteins forming transmembrane channels. J. Memb. Biol. *175*, 165–180.

Sauna, Z.E., and Ambudkar, S.V. (2001). Characterization of the catalytic cycle of ATP hydrolysis by human P-glycoprotein. The two ATP hydrolysis events in a single catalytic cycle are kinetically similar but affect different functional outcomes. J. Biol. Chem. *276*, 11653–11661.

Saurin, W., Hofnung, M., and Dassa, E. (1999). Getting in or out: early segregation between importers and exporters in the evolution of ATP-binding cassette (ABC) transporters. J. Mol. Evol. *48*, 22–41.

Schmitt, L., and Tampe, R. (2002). Structure and mechanism of ABC transporters. Curr Opin Struct Biol. *12*, 754–760.

Schneider, E., and Hunke, S. (1998). ATP-binding-cassette (ABC) transport systems: functional and structural aspects of the ATP-hydrolyzing subunits/domains. FEMS Microbiol. Rev. *22*, 1–20.

Schumacher, M.A., and Brennan, R.G. (2003). Deciphering the molecular basis of multidrug recognition: crystal structuRes. of the *Staphylococcus aureus* multidrug binding transcription regulator QacR. Res. Microbiol. *154*, 69–77.

Schumacher, M.A., Miller, M.C., and Brennan, R.G. (2004). Structural mechanism of the simultaneous binding of two drugs to a multidrug-binding protein. EMBO J. *23*, 2923–2930.

Schumacher, M.A., Miller, M.C., Grkovic, S., Brown, M.H., Skurray, R.A., and Brennan, R.G. (2001). Structural mechanisms of QacR induction and multidrug recognition. Science *294*, 2158–2163.

Seelig, A. (1998). A general pattern for substrate recognition by P-glycoprotein. Eur. J. Biochem. *251*, 252–261.

Shapiro, A.B., Corder, A.B., and Ling, V. (1997). P-glycoprotein-mediated Hoechst 33342 transport out of the lipid bilayer. Eur. J. Biochem. *250*, 115–121.

Shapiro, A.B., and Ling, V. (1997a). Positively cooperative sites for drug transport by P-glycoprotein with distinct drug specificities. Eur. J. Biochem. *250*, 130–137.

Shapiro, A.B., and Ling, V. (1997b). Extraction of Hoechst 33342 from the cytoplasmic leaflet of the plasma membrane by P-glycoprotein. Eur. J. Biochem. *250*, 122–129.

Sharma, S., and Davidson, A.L. (2000). Vanadate-induced trapping of nucleotides by purified maltose transport complex requires ATP hydrolysis. J. Bacteriol. *182*, 6570–6576.

Shilling, R.A., Venter, H., Velamakanni, S., Bapna, A., Woebking, B., Shahi, S., and van Veen, H.W. (2006). New light on multidrug binding by an ATP-binding-cassette transporter. Trends Pharmacol. Sci. *27*, 195–203.

Shuman, H.A. (1982). Active transport of maltose in *Escherichia coli* K12. Role of the periplasmic maltose-binding protein and evidence for a substrate recognition Site in the cytoplasmic membrane. J. Biol. Chem. *257*, 5455–5461.

Smith, P.C., Karpowich, N., Millen, L., Moody, J.E., Rosen, J., Thomas, P.J., and Hunt, J.F. (2002). ATP binding to the motor domain from an ABC transporter drives formation of a nucleotide sandwich dimer. Mol. Cell *10*, 139–149.

Story, R.M., Weber, I.T., and Steitz, T.A. (1992). The structure of the *E. coli* recA protein monomer and polymer. Nature *355*, 318–325.

Tikhonov, D.B., and Zhorov, B.S. (2008). Molecular modeling of benzothiazepine binding in the L-type calcium channel. J. Biol. Chem. *283*, 17594–17604.

Tombline, G., Muharemagic, A., White, L.B., and Senior, A.E. (2005) Involvement of the 'occluded nucleotide conformation' of P-glycoprotein in the catalytic pathway. Biochemistry *44*, 12879–86.

Ueda, K., Cardarelli, C., Gottesman, M.M., and Pastan, I. (1987). Expression of a full-length cDNA for the human 'MDR1' gene confers resistance to colchicine, doxorubicin, and vinblastine. Proc. Natl. Acad. Sci. U.S.A. *84*, 3004–3008.

Urbatsch, I.L., Julien, M., Carrier, I., Rousseau, M.E., Cayrol, R., and Gros, P. (2000). Mutational analysis of conserved carboxylate residues in the nucleotide binding sites of P-glycoprotein. Biochemistry *39*, 14138–14149.

Urbatsch, I.L., Sankaran, B., Weber, J., and Senior, A.E. (1995). P-glycoprotein is stably inhibited by vanadate-induced trapping of nucleotide at a single catalytic site. J. Biol. Chem. *270*, 19383–19390.

van der Does, C., Presenti, C., Schulze, K., Dinkelaker, S., and Tampe, R. (2006). Kinetics of the ATP hydrolysis cycle of the nucleotide-binding domain of Mdl1 studied by a novel site-specific labeling technique. J. Biol. Chem. *281*, 5694–5701.

van Veen, H.W. and Konings, W.N. (1997) Multidrug transporters from bacteria to man: similarities in structure and function. Semin. Cancer Biol. *8*, 183–191.

van Veen, H.W., Callaghan, R., Soceneantu, L., Sardini, A., Konings, W.N., and Higgins, C.F. (1998). A bacterial antibiotic-resistance gene that complements the human multidrug-resistance P-glycoprotein gene. Nature *391*, 291–295.

van Veen, H.W., Margolles, A., Muller, M., Higgins, C.F., and Konings, W.N. (2000). The homodimeric ATP-binding cassette transporter LmrA mediates multidrug transport by an alternating two-site (two-cylinder engine) mechanism. EMBO J. *19*, 2503–2514.

Venter, H., Shilling, R.A., Velamakanni, S., Balakrishnan, L., and van Veen, H.W. (2003). An ABC transporter with a secondary-active multidrug translocator domain. Nature *426*, 866–870.

Ward, A., Reyes, C.L., Yu, J., Roth, C.B., and Chang, G. (2007). Flexibility in the ABC transporter MsbA: Alternating access with a twist. Proc. Natl. Acad. Sci. U.S.A. *104*, 19005–19010.

Woebking, B., Velamakanni, S., Federici, L., Seeger, M.A., Murakami, S., and van Veen, H.W. (2008). Functional role of transmembrane helix 6 in drug binding and transport by the ABC transporter MsbA. Biochemistry *47*, 10904–10914.

Yuan, Y.R., Blecker, S., Martsinkevich, O., Millen, L., Thomas, P.J., and Hunt, J.F. (2001). The crystal structure of the MJ0796 ATP-binding cassette. Implications for the structural consequences of ATP hydrolysis in the active site of an ABC transporter. J. Biol. Chem. *276*, 32313–32321.

Zaitseva, J., Jenewein, S., Jumpertz, T., Holland, I.B., and Schmitt, L. (2005). H662 is the linchpin of ATP hydrolysis in the nucleotide-binding domain of the ABC transporter HlyB. EMBO J. *24*, 1901–1910.

Zaitseva, J., Oswald, C., Jumpertz, T., Jenewein, S., Wiedenmann, A., Holland, I.B., and Schmitt, L. (2006). A structural analysis of asymmetry required for catalytic activity of an ABC-ATPase domain dimer. EMBO J. 25, 3432–3443.

Zamora, J., Pearce, H.L., and Beck, W.T. (1988) Physical-chemical properties shared by compounds that modulate MDR in human leukemic cells. Mol. Pharmacol. 33, 454–462.

Can ABC Proteins Confer Drug Resistance in Microorganisms Without being Export Pumps?

4

James M. Dorrian and Ian D. Kerr

Abstract

Inhibition of bacterial protein translation is a key point of anti-microbial intervention therapy, with numerous antibiotics functioning as inhibitors of protein synthesis. The macrolide, lincosamide and streptogramin antibiotics (MLS antibiotics) all cause inhibition of protein synthesis by binding to the 50S ribosomal subunit. Resistance to this class of antibiotics is mediated by many different mechanisms, at least one of which involves an unusual class of ABC proteins. These antibiotic resistance element (ARE) type ABC proteins do not include membrane-spanning segments within the polypeptide and are not linked in operons to membrane spanning regions. A straightforward transport-based explanation for the function of these ABC proteins in antibiotic resistance is therefore difficult to support. We will illustrate the diversity of these ABC proteins, discuss their clinical relevance, and illustrate three potential modes of action.

Introduction

ATP binding cassette transporters and drug resistance

The phenomenon of drug resistance is observed in essentially every clinical setting. Drug export mediated by ATP binding cassette (ABC) proteins is at least partially responsible for the drug-resistant phenotype of tumour cells, parasites and bacteria. The diversity and mechanisms of these transporters are presented elsewhere in this volume (see chapters 1 – 3). In the current chapter we focus on a group of ABC proteins which may act to confer drug resistance without exporting the drugs in question. Specifically, we discuss the antibiotic resistance element (ARE) type ABC proteins, which are distinguished from other ABC transporters by the absence of membrane spanning regions either from the primary amino acid sequence, or from the operon in which the ARE gene is located. In spite of this, ARE proteins are able to confer resistance to numerous ribosomally active antibiotics. We begin with a discussion of the action of these antibiotics, and the mechanisms that can contribute to resistance. We then focus on ARE proteins considering their distribution in antibiotic producing organisms as well as antibiotic resistant isolates. The functional mechanism of these proteins is under investigation in our lab and in others', and we highlight recent developments in the latter half of the chapter.

The action of macrolide, lincosamide and streptogramin (MLS) antibiotics on the ribosome

Inhibition of protein synthesis is one of the key functional mechanisms of clinical antibiotics, originating with the discovery of streptomycin in the 1940s and erythromycin in the 1950s. As resistance to these macrolide compounds has emerged, so the development of a battery of similarly acting chemicals, including ketolides (ketone derivatives of macrolides), lincosamides and streptogramins (Fig. 4.1) has become a priority for the pharmaceutical industry (Retsema and Fu, 2001; Poehlsgaard and Douthwaite, 2002; Waugh and Long, 2002; Hermann, 2005; Bottger,

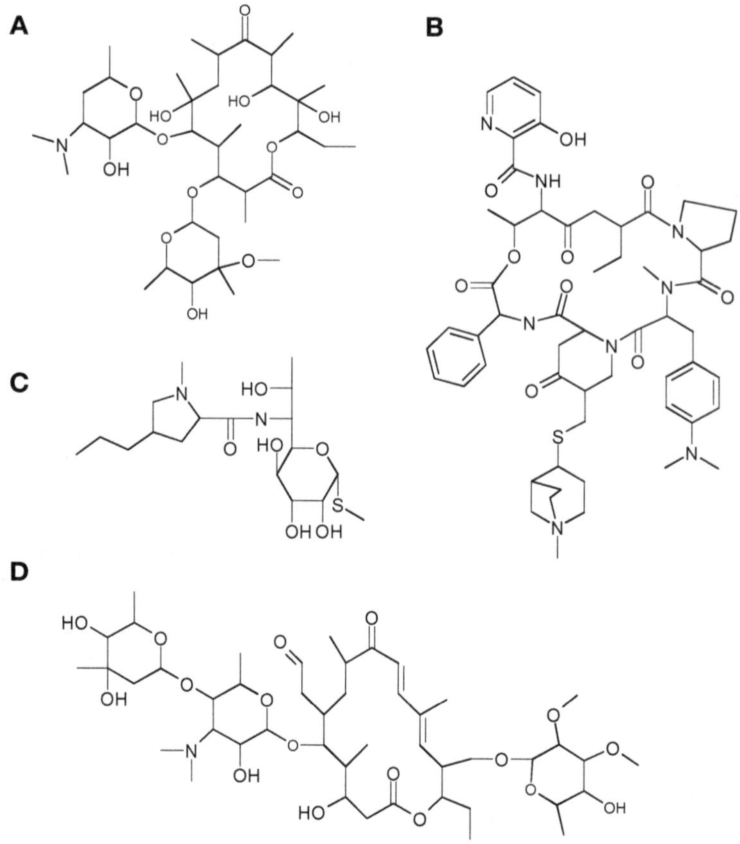

Figure 4.1 Chemical structures of antibiotics to which AREs confer resistance. Representative members of the 14 and 16-membered macrolides (erythromycin, **A**, and tylosin, **D** respectively), streptogramins **B** (quinupristin, **B**), and lincosamides (lincomycin, **C**) are shown.

2007). Although the functional basis for translation inhibition has been known for many years, the structural basis has only been unravelled with the advent of crystal structures for ribosomal subunits in complex with antibiotics in the last decade (Schlunzen *et al.*, 2001; Hansen *et al.*, 2002; Steitz, 2005; Tu *et al.*, 2005).

Structural examination of 50S ribosomal subunits from *Haloarcula marismortui* demonstrates that the binding sites for macrolides, lincosamides and streptogramins B (MLS_B antibiotics) are overlapping, and are present primarily on the 23S rRNA subunit (Hansen *et al.*, 2002; Steitz, 2005; Tu *et al.*, 2005). Similar structural data has been obtained for MLS_B antibiotics bound to the 50S subunit of the ribosome from *Deinococcus radiodurans* (Schlunzen *et al.*, 2001), although as discussed by Steitz there are differences both in the specific mode of interaction of antibiotic with the ribosome, and in the conformation of the antibiotic at its binding site (Steitz, 2005). Notably, in the *H. marismortui* crystal structures the conformations of the antibiotics bound to the ribosome are very similar to their small molecule crystal structures suggesting that there is little requirement for structural alterations within the antibiotic as it binds (Tu *et al.*, 2005). In contrast, antibiotics bound to the *D. radiodurans* ribosome adopt different conformations (Schlunzen *et al.*, 2001). Notwithstanding the differences in the structural data from these two different ribosomes (*H. merismortui* is an archaebacteria, *D. radiodurans* is a eubacteria), it is evident that MLS_B antibiotics act by preventing the egress of newly synthesized polypeptide through the polypeptide exit tunnel. This inhibition is largely steric in nature and is consistent with functional data demonstrating that although ribosomes already translating mRNA are resistant to MLS_B antibiotics (Andersson and Kurland, 1987), and that translation does initiate in the presence of antibiotics (Tenson *et al.*, 2003), synthesis of a nascent polypeptide beyond 2–7 amino acids is inhibited (depending on the antibiotic used; Tenson *et al.*, 2003). It is worth adding at this stage that an additional mechanism of translation inhibition, namely the inhibition of 50S particle assembly, has been attributed to macrolides and ketolides (Champney *et al.*, 1998; Douthwaite and Champney, 2001). The relative ability of macrolides and ketolides to cause translation inhibition parallels their ability to inhibit 50S particle assembly and supports the idea that at least part of the inhibitory effect of macrolides and ketolides is due to interference in translation initiation.

Mechanisms of resistance to MLS_B antibiotics

Resistance to MLS_B antibiotics is a multifaceted issue with transport being just one of the contributory mechanisms (Pechere, 2001; Leclercq, 2002; Wright, 2003). The most well documented mechanisms of resistance to MLS_B antibiotics are target site modification and increased drug efflux. The former is best understood in terms of the methylation of nucleotide A2058 (numbering is with respect to the *Escherichia coli*) of the 23S rRNA (Weisblum, 1985), although other sites of methylation do exist that confer resistance to tylosin when methylated (Liu and Douthwaite, 2002; Douthwaite *et al.*, 2004). The addition of methyl groups to these bases by ribosomal methyltransferase enzymes (erm) reduces the affinity of ribosomes for the antibiotic and enhances resistance to MLS_B antibiotics by orders of magnitude (Weisblum, 1995), and the erm genes are prevalent causes of clinical antibiotic resistance (see below). The export of antibiotics from microorganisms is also a key mediator of the drug resistance phenotype, and is discussed for MLS_B antibiotics below. Finally, chemical modification of antibiotics (Leclercq, 2002) and mutations of both rRNA and ribosomal proteins are also mechanisms by which resistance to MLS antibiotics is conferred (Malbruny *et al.*, 2002; Pihlajamaki *et al.*, 2002).

The transport/export of antibiotics out of bacteria contributes to the resistance phenotype. Broadly speaking, 5 classes of transporters capable of exporting diverse drugs (so called multidrug pumps; MDR, see Chapter 5) have been identified in bacteria. Of these, the resistance nodulation division (RND) family, the major facilitator (MFS) family, the multidrug and toxic-compound extrusion (MATE) family, and the small multidrug resistance (SMR) family have all been implicated in MLS_B antibiotic efflux. Due to space restrictions these mechanisms are not discussed further here and the interested reader should refer to the following excellent reviews (Kumar and Schweizer, 2005; Cagliero *et al.*, 2006; Piddock, 2006). The fifth major class of MDR transporter in bacteria is the ABC transporters, and these transporters form the primary subject of discussion from hereon.

For the purposes of this chapter, ABC transporters can be divided into two classes: those where a transmembrane domain (TMD) is readily identifiable (potentially within the same polypeptide as the nucleotide binding domain; NBD), and those where no TMD has been identified. The former are covered in Chapters 2 and 3 and will not be discussed further, although at least two (MacAB and MsbA) are involved in erythromycin resistance (Kobayashi *et al.*, 2001; Rouquette-Loughlin *et al.*, 2005; Ghanei *et al.*, 2007; Tikhonova *et al.*, 2007).

Identification of ABCs without membrane spanning domains as MLS resistance factors

An observation made in the late 1980s provides hints to describe a new class of proteins able to confer resistance to MS_B antibiotics (Ross *et al.*, 1989). The cloning and sequencing of an open-reading frame present on a plasmid from a clinical isolate of *Staphylococcus epidermidis* was shown to be sufficient to confer cross-resistance to both macrolides and streptogramins B (Ross *et al.*, 1989; Ross *et al.*, 1990). The open reading frame encoded a protein of 488 amino acids, termed MsrA, and contained two highly similar repeats of a classical ABC transporter NBD (Ross *et al.*, 1990), separated by a sequence known as a Q-linker (Wootton and Drummond, 1989). Moreover, the presence of the twin-NBD protein (NBD_2) was alone sufficient to cause the MS_B phenotype when transformed into antibiotic sensitive *St. aureus* (Ross *et al.*, 1995, 1996). Since this observation, several ABC proteins, apparently lacking TMDs yet able to confer resistance to MLS_B antibiotics have been identified. These proteins broadly fall in to one of two categories; they are either encoded by antibiotic producers (*Streptomyces* – abbreviated *S.*) or by potentially pathogenic bacteria (*Staphylococcus* – *St.*, *Enterococcus* and *Streptococcus* – *Stp.*). They include, among others, the proteins Lsa from *Enterococcus faecium* (Singh *et al.*, 2002; Singh and Murray, 2005), and Vga(A and B) from *St. aureus* (Allignet *et al.*, 1992; Allignet and El Solh, 1997).

Phylogeny and distribution of AREs

The genus *Streptomyces* contains over 500 members of which many have evolved the ability to produce secondary metabolites that function as antibiotics. In parallel with this function has been the evolution of resistance mechanisms to prevent the adverse affects on the antibiotic on the antibiotic producer itself (Mendez and Salas, 2001). Typically, the genes for ARE proteins and other resistance determinants are found within the biosynthetic gene cluster for that particular antibiotic. For example, the operon that encodes for the synthesis and export of spiramycin from *S. ambofaciens* contains over 30 genes, including the ARE SrmB (Karray *et al.*, 2007). Clearly these biosynthetic gene clusters are of significant medical interest and they have been investigated thoroughly. As a consequence many antibiotic resistance genes have been identified. In many instances the AREs present on these operons have been shown to confer resistance to the specific antibiotic when they are expressed in related species. For example, the ARE protein OleB from *S. antibioticus* has been shown to confer resistance to oleandomycin when expressed in *S. albus* (Rodriguez *et al.*, 1993). Several similar examples are listed in Table 4.1.

In contrast to the ARE proteins of the antibiotic producers, a related group of ARE proteins have evolved in pathogenic bacteria, presumably as a response to antibiotics found in the extracellular environment. Typically their genes are found on mobile genetic elements which allow their rapid spread through the population. The first ARE to be studied extensively was MsrA which, as discussed above, was isolated from a plasmid from *St. epidermidis* and has been shown to confer resistance to erythromycin and streptogramin-B (Ross *et al.*, 1995). Again, similar examples of ARE proteins involved in resistance in pathogenic bacteria are listed in Table 4.2.

Sequences of ARE proteins contain all the hallmarks of ABC transporter NBDs. Each NBD contains classical Walker-A and Walker-B motifs, and the glutamine-loop, His-loop, signature sequence and D-loop are also present in both NBDs (e.g. see Lawson *et al.*, 2008, for discussion of the role of these sequences), although the signature sequence tends to contain a glutamate residue rather than a glutamine following the two conserved glycine residues. Thus, each ARE protein has the two consensus ATP binding sites expected if the two NBDs interact in the classical sandwich dimer orientation (Chen *et al.*, 2003). The major difference between the two groups of ARE proteins is their overall lengths: the antibiotic producers tend to have AREs of 550 amino acids, whereas those from pathogenic bacteria are typically 60–80 amino acids shorter (Tables 4.1 and 4.2). The reason for this differ-

Table 4.1 ARE proteins of antibiotic producers

Protein	Organism	Length	Demonstrated role in resistance	Reference
OleB	*S. antibioticus*	569	In *S. albus*	Rodriguez *et al.*, 1993; Olano *et al.*, 1995
LmrC	*S. lincolnensis*	550	In *S. lividans*	Peschke *et al.*, 1995
Car(A)	*S. thermotolerans*	551	In *S. griseofuscus*	Epp *et al.*, 1987
tlrC	*S. fradiae*	548	In a sensitive strain of *S. fradiae*	Schoner *et al.*, 1992
srmB	*S. ambofaciens*	550	*S. griseofuscus*	Karray *et al.*, 2007

Table 4.2 ARE proteins from pathogenic bacteria

Protein	Organism	Length	Demonstrated role in resistance	Reference
msrA	*St. epidermidis*	488	In *St. aureus*	Ross *et al.*, 1989
Msr(C)	*E. faecium*	492	In *St. aureus*	Reynolds and Cove, 2005a
Msr(D)	*Stp. pneumoniae*	487	In sensitive *Stp. pneumoniae*	Daly *et al.*, 2004
Lsa	*E. faecalis*	498	In previously sensitive strains	Singh *et al.*, 2002
Vga(A)	*St. aureus*	522	In previously sensitive strains	Allignet *et al.*, 1992; Chesneau *et al.*, 2005

ence is not within the linker separating the two NBDs – indeed the identification of a Q-linker in MsrA is not borne out in other ARE sequences (Wootton and Drummond, 1989). Instead the difference occurs in a region of NBD1 between the conserved glutamine loop and the signature sequence. This region has been shown to be structurally diverse in other ABC proteins, and these structural differences are proposed to contribute to the interaction of the NBDs with other domains of the transporter (Oswald *et al.*, 2006).

Sequence analysis of ARE proteins from antibiotic producers and pathogenic bacteria has been carried out to classify existing AREs and to identify other putative AREs. Repeated cycles of PSI-BLAST (http://blast.ncbi.nlm.nih.gov/Blast.cgi; Altschul *et al.*, 1997) were performed with a selection of ARE proteins of known function as the search sequence (CarA, SrmB, Lsa, LmrC, TlC, MsrA, VgaA). The data of each search were combined resulting in a list of several hundred putative ARE proteins. An abridged list of sequences was prepared such that no sequence was more than 90% identical at the amino acid level to any other sequence (http://www.expasy.ch/tools/redundancy/). This list contained 130 proteins and open reading frames. A multiple sequence alignment was assembled using ClustalW (http://www.ebi.ac.uk/Tools/clustalw2/index.html; (Thompson *et al.*, 1994)) and a phylogenetic tree constructed from the alignment using the Phylip package of programs (http://mobyle.pasteur.fr/cgi-bin/MobylePortal/portal.py; Fig. 4.2).

In Fig. 4.2 we list the organism and protein name or the Entrez protein accession code. There are several major groups identified: two branch points include all of the *Streptomyces* AREs, and several other clusters are evident including proteins related to Lsa, to Msr, and to Vga. The phylogenetic tree identifies a large number of putative ARE proteins from *Streptomyces* species, as well as putative ARE proteins from other pathogens such as *Listeria* and *Clostridia.*

Clinical prevalence of AREs

The emergence of resistance mechanisms, including MsrA homologues, to multiple antibiotics has led to the use of multiplex PCR and DNA microarrays to determine the distribution of resistance genes in clinical isolates of antibiotic resistant gram positive organisms, primarily staphylococci. It is clear from the data obtained that the ribosome methylation mechanism is the most important determinant of the resistance phenotype in bacteria but that additional mechanisms are also important factors that contribute to this phenotype (Gatermann *et al.*, 2007). ARE genes are present within clinical isolates as plasmid-borne elements (Ross *et al.*, 1990), as chromosomal genes (Singh *et al.*, 2002) or as part of a large mobile genetic elements (e.g. *Tn1207.1*) containing several resistance genes (Santagati *et al.*, 2000; Ambrose *et al.*, 2005). A number of studies have examined the presence of MsrA as a contributing factor to both inducible and constitutive resistance phenotypes. The percentage of positive samples range from just 2% in a sample of 168 erythromycin resistant *St. aureus* samples from Beijing hospitals (Zhu *et al.*, 2007) to over 50% of *St. hemolyticus* samples in the Czech Republic. Within these reports the frequency with which MsrA is found as the sole resistance determinant varies from less than 1/10 (Nawaz *et al.*, 2000) to essentially totality (Eady *et al.*, 1993). This variation reflects to some extent how many of the dif-

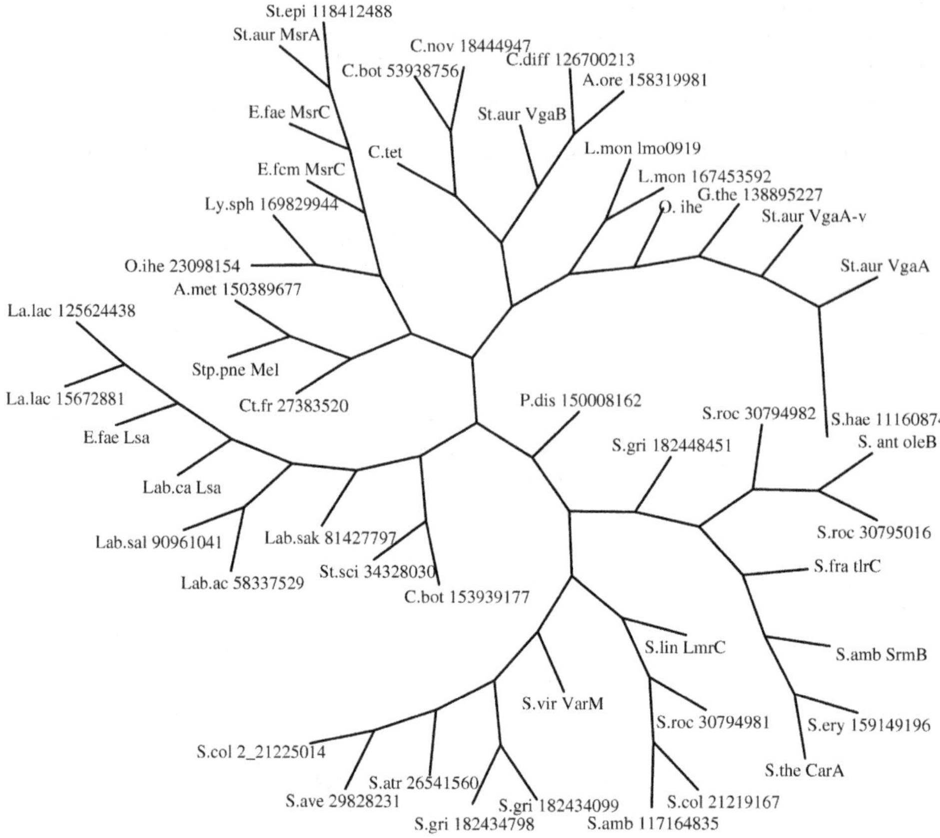

Figure 4.2 Phylogenetic analysis of the ARE proteins from antibiotic producers and pathogenic gram positive bacteria. The unrooted tree was constructed as described in the text using PSI_BLAST (http://blast.ncbi.nlm.nih.gov/Blast.cgi) and ClustalW (http://www.ebi.ac.uk/Tools/clustalw2/index.html) and was displayed using Phylip (http://mobyle.pasteur.fr/cgi-bin/MobylePortal/portal.py). The sequence of LmrC from *Streptomyces lincolnensis* is our own corrected version of this (Kerr, I. D. unpublished), rather than the database sequence. A small branch of putative ARE proteins mostly from the *Bacteroides* genus has been removed for clarity. The accession numbers for two proteins are not shown for clarity, namely O. ihe 23097904 and C. tet 28212014. Organism abbreviations are as follows: A. met – *Alkaliphilus metalliredigens*; A. ore – *Alkaliphilus oremlandii;* C. bot/dif/nov/tet – *Clostridium botulinum/difficile/novyi/ tetani;* Ct. fr – *Citrobacter freundii;* E. fae/fcm – *Enterococcus faecalis/faecium;* G. the – *Geobacillus thermodenitrificans;* L. mon – *Listeria monocytogenes;* Lab. sa/aci/sak/cas – *Lactobacillus salivarius/acidophilus/sakei/casei;* Ly. sp – *Lysinibacillus sphaericus;* La. lac – *Lactococcus lactis;* O. ihe – *Oceanobacillus iheyensis;* P. dis – *Parabacteroides distasonis;* S. amb/ant/atr/ave/col/eur/fra/gri/lin/roc/the/vir – *Streptomyces ambofaciens/antibioticus/atroolivaceus/avermitilis/coelicolour/eurythermus/fradiae/ griseus/lincolnensis/rochei/thermotolerans/virginae;* St. epi/hae/aur/sci – *Staphylococcus epidermidis/haemolyticus/aureus/sciuri;* St. pne – *Streptococcus pneumoniae.*

ferent resistance genes have been analysed in each study, but reproducibly, about 25% of resistant strains appear to use MsrA as the sole resistance determinant (Martineau *et al.*, 2000; Aktas *et al.*, 2007; Gatermann *et al.*, 2007; Zhu *et al.*, 2007). In a large study of over 800 strains (Schmitz *et al.*, 2000), MsrA was found in 13% of samples, with one-third of these samples showing a constitutive resistance to erythromycin and the remainder displaying a resistance that was inducible by low macrolide concentrations, consistent with the original observations of the MsrA-conferred resistance phenotype in disk diffusion assays (Ross *et al.*, 1990). Finally, it seems evident that MsrA is found both within methicillin resistant and methicillin sensitive multiple resistant *St. aureus* (Schmitz *et al.*, 2000; Novotna *et al.*, 2005; Aktas *et al.*, 2007).

The cloning of MsrA homologues has also prompted efforts to determine how widespread these are in *Staphylococcal* isolates. The ARE gene Vga (first identified in *S. aureus*; Allignet *et al.*, 1992) has been identified in highly variable percentages in studies of resistance factors in *St. epidermidis* – perhaps the variability stems from the relatively small numbers of M or MS_B resistant samples analysed (Loncle *et al.*, 1993; Allignet *et al.*, 1996; Bouchami *et al.*, 2007; Novotna *et al.*, 2007; Achour *et al.*, 2008). MsrC has been found to be in almost all *E. faecium* strains tested and is believed to be an inherent resistance factor in this organism, as the gene is absent in other *Enterococcal* strains (Portillo *et al.*, 2000; Singh *et al.*, 2001), although there is some disagreement as to whether the MsrC gene is conferring a selective advantage under antibiotic pressure (Werner *et al.*, 2001).

Resistance conferred by ARE proteins

A number of ARE proteins have been investigated for their potential to confer resistance by expressing them in Gram-positive, antibiotic sensitive organisms. Such experiments are usually performed in a closely related drug susceptible species. Table 4.3 summarizes the results for the most well studied AREs from pathogenic bacteria. A clear dichotomy exists between the conferral of high level of resistance (e.g. of MsrA to erythromycin) and the conferral of only low level resistance (e.g. MsrA to pristinamycin IA). This evidence of substrate specificity of the AREs is discussed below.

The results of such studies have been useful to examine two phenomena associated with AREs, which are also displayed by other resistance

Table 4.3 ARE proteins and the resistance conferred in heterologous expression studies

ARE	Expression system	Resistance phenotype	Reference:
MsrA	*St. aureus*/plasmid	ERY (250)[a], CLR (250), PIA (2), VSI (2), OSB (2), TEL (10), OL (100)	Matsuoka *et al.*, 2003; Reynolds and Cove, 2005a; Reynolds and Cove, 2005b
MsrA	*St. aureus*/chromosome	ERY (100), CLR (100), PIA (2), VSI (2), CLI (4),TEL (4)	Reynolds and Cove, 2005a; Reynolds and Cove, 2005b
MsrC	*St. aureus*/plasmid	ERY (100), PIA (2), VSI (2), OSB (2), TEL (100)	Reynolds and Cove, 2005a; Reynolds and Cove, 2005b
MsrC	*St. aureus*/chromosome	ERY (100), CLR (40), VSI (2), OSB (2), CLI (4), TEL (10)	Reynolds and Cove, 2005a; Reynolds and Cove, 2005b
MsrD	*St. aureus*/plasmid	ERY (10), TEL (4)	Reynolds and Cove, 2005b
MsrD	*Stp. pneumoniae*/plasmid	ERY (10), CLR (100), TEL (10)	Daly *et al.*, 2004
Vga(A)	*St. aureus*/plasmid *St. epidermidis*/plasmid	SGA (32), PT (2), LC (4), CLI (2), PIIA (16)	Chesneau *et al.*, 2005; Novotna and Janata, 2006
$Vga(A)_{LC}$	*St. aureus/plasmid*	LC (8), CLI (10), PIIA (4)	Novotna and Janata, 2006
Vga(A)v	*St. aureus*/plasmid *St. epidermidis*/plasmid	SGA (30), PT (2), LC (4), CLI (2)	Chesneau *et al.*, 2005
Vga(B)	*St. aureus*/plasmid *St. epidermidis*/plasmid	SGA (4), PT (8), LC (2), CLI (2)	Chesneau *et al.*, 2005
Lsa	*E. faecalis*/disruption mutant	CLI (200), Q-D (40), DAL (100)	Singh *et al.*, 2002
Lsa	*E. faecium*/plasmid	CLI ((100), Q-D (10), DAL (100)	Singh *et al.*, 2002

ERY, erythromycin; OL, oleandomycin; CLR, calithromycin; TEL, telithromycin; CLI, clindamycin; PIA, pristinamycin IA; OSB, osteogricin B; VSI, virginiamycin S1;SGA, streptomgrammin A; LC, lincomycin; PT, pristinamycin; Q-D, quinupristin-dalfopristin.

a Numbers in brackets refer to the fold increase in MIC comparing parental bacterial strains with those expressing the ARE.

elements in bacteria (Weisblum, 1985). The first phenomenon is the inducible resistance, in which resistance may be heightened by the co-incubation of the bacteria in the presence of sub-cytotoxic concentrations of an antibiotic. For example, the resistance conferred by MsrA and MsrC towards pristinamycin IA can be increased a further 2-fold in the presence of low concentrations of erythromycin (Reynolds and Cove, 2005b). The second phenomenon relates to the necessity of the short leader sequences that precede many AREs. This leader sequence has been shown to be dispensable for the antibiotic resistance phenotype (Ross *et al.*, 1996), and is thus distinguished from the short leader sequences upstream of methylase genes which control the translation of erm proteins (Karray *et al.*, 2007).

Biochemical analysis of ARE proteins

Characterization of ARE proteins *in vitro* is essential if we are to understand their function. A number of questions could be answered in such studies. For example, do ARE proteins hydrolyse ATP? Do they require two consensus ATP binding pockets? Does the ARE itself contribute to the specificity for the antibiotic? However, detailed biochemical characterizations of ARE proteins are somewhat sparse in the literature and, to date, only two of these proteins have been studied *in vitro*, namely OleB and Vga(A) (Aparicio *et al.*, 1996; Buche *et al.*, 1997; Jacquet *et al.*, 2008). Of the three published studies two are difficult to interpret because the complete ARE protein was not studied. Instead, the N-terminal half of the OleB protein was fused to maltose binding protein (to ensure solubility) and the resultant fusion protein was analysed for ATPase activity (by release of inorganic phosphate) and oleandomycin binding (by quenching of the intrinsic fluorescence of OleB) (Aparicio *et al.*, 1996; Buche *et al.*, 1997). The presence of the fusion protein and the absence of the second nucleotide-binding domain make unambiguous interpretation hazardous.

More recently, recombinant Vga(A) has been purified as a monomeric, full length protein following expression in *E. coli* (Jacquet *et al.*, 2008). The protein displayed a maximal ATPase activity of ca. 100 nmoles/min/mg protein, with a Km for ATP of 78 μM, both values of which are comparable with purified ABC transporters. The ATPase activity seems to be non-equivalent between the two NBDs as mutation of the conserved Walker-B glutamate in NBD1 was reasonably well tolerated, in contrast to the equivalent mutation in NBD2. Intriguingly, the ATPase activity of Vga(A) was inhibited by pristinamycin IIA, a compound to which Vga(A) confers resistance when expressed in *St. aureus* (Jacquet *et al.*, 2008, and Table 4.3), suggesting a direct interaction of the ARE protein and its substrate – a possibility that has been raised previously (Buche *et al.*, 1997; Kerr *et al.*, 2005). However, there are indications that additional factors are present in the *in vivo* environment; firstly, the inhibition of ATPase activity required a concentration of pristinamycin IIA 10-fold higher than the minimal inhibitory concentration (MIC), and secondly, the rank order of ATPase activities of purified wild type and mutant proteins did not correlate with the rank order of MICs when the same constructs were expressed in *St. aureus* (Jacquet *et al.*, 2008). Clearly similarly detailed studies of other intact ARE proteins are required.

Functional mechanisms of ARE proteins

At least three hypotheses exist for the function of AREs and, additionally, the plasticity of NBD proteins in general means that a combination of more than one mechanism may certainly exist. In Fig. 4.3 we schematically depict the MLS antibiotics preventing translation, either through direct occlusion of the peptide exit site, or by inhibition of ribosomal subunit assembly. The three resistance mechanisms that AREs may underlie are shown to the right hand side of the figure and discussed below.

Transport based

Early evidence that ARE proteins confer resistance to MLS antibiotics showed that when MsrA was expressed in *Staphylococcus aureus* (Ross *et al.*, 1990) these cells were made resistant to erythromycin. Additionally, it has been demonstrated that MsrA expression is associated with reduced accumulation of radio-labelled erythromycin (Ross *et al.*, 1990), although whether this is due to a transport phenomenon is unclear. Erythromycin is able to passively diffuse across the cell

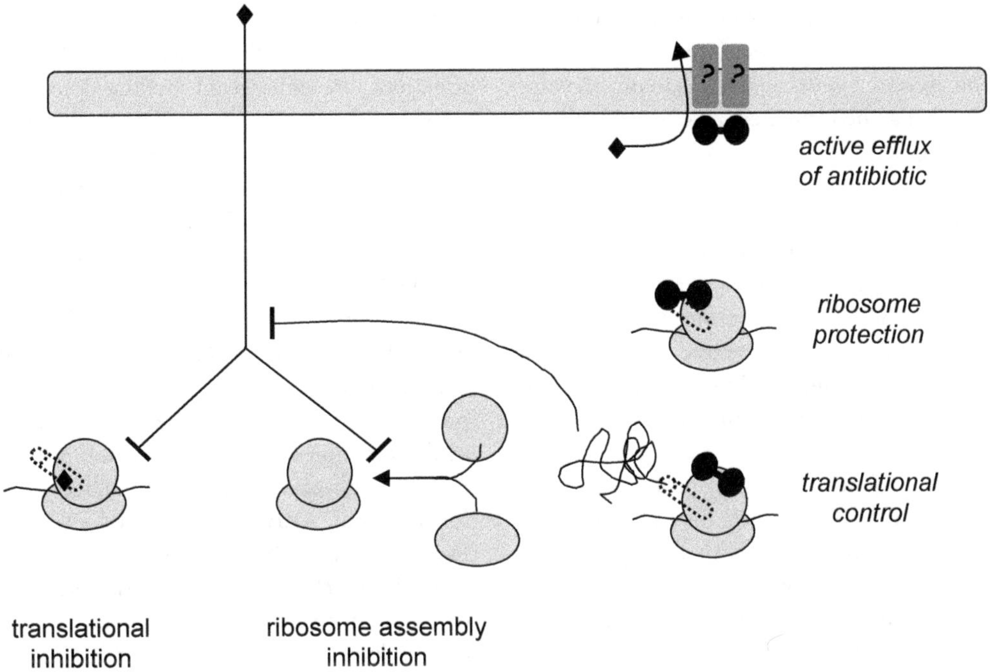

Figure 4.3 Models for ARE protein mechanism. The inhibition by MLS type antibiotics on translation and ribosome assembly (shown for simplicity as the 50S and 30S interaction) is shown on the left hand side of the figure. The antibiotic is represented as a filled diamond, and the polypeptide exit tunnel as a cylinder. On the right hand side the three discussed mechanisms are displayed schematically. In each case the ARE protein is represented as a filled dumbbell to denote the two NBDs. For the efflux mechanism the identity of the membrane spanning proteins (oblongs) is unknown. For the ribosome protection mechanism the ARE protein is shown schematically as preventing antibiotic access to its binding site. For the translational control mechanism the ARE is denoted as promoting the translation of a protein (scrambled line) which itself blocks the antibiotic action.

membrane with the ribosome acting as a 'sink' driving the influx process (Capobianco and Goldman, 1990; Goldman and Capobianco, 1990). If MsrA blocks the binding site at the ribosome less of the macrolide would diffuse into the cell, potentially resulting in an apparent efflux of the drug. Furthering this concern, adding streptogramin B to *St. aureus* that had previously been exposed to erythromycin also caused a decrease in cell associated erythromycin demonstrating this 'ribosomal sink effect' (Reynolds *et al.*, 2003).

If an ARE acts as a component of a transport system then it needs a transmembrane partner. The plasmid which encodes MsrA also encodes some putative transmembrane proteins (Ross *et al.*, 1995). The presence of these genes (designated *stp* and *smp*) with MsrA in erythromycin resistant staphylococci and their presence in the chromosomes of *St. aureus* and *St. hominis* and a plasmid of *St. epidermidis* gave credence to their involvement in MsrA-mediated antibiotic resistance. However these proteins have been shown not to confer macrolide resistance when independently expressed in *St. aureus* and *E. coli* (Ross *et al.*, 1995) and the necessity of them was disproved in later work (Ross *et al.*, 1996) where a *St. aureus* Δstp-smp mutant was generated. MICs for this mutant expressing MsrA were identical to wild type *St. aureus* expressing MsrA. No further potential transmembrane candidates have been identified for MsrA, however that is not to say that they don't exist.

Work on the Vga group and their mechanism of action produced some intriguing results which show the membrane localization of Vga(A) by western blot of membrane and cytosolic subcellular fractions (Chesneau *et al.*, 2005). Vga was found not to localize with the ribosomal L24 protein, instead being found in the same fractions as the β-subunit of the F_1-F_0-ATPase (Chesneau *et al.*, 2005). Clearly this result requires confirmation with other ARE proteins, and some caution must be exercised with the data, as in prokaryotes ribosomes may be intimately associated with the cell membrane.

Ribosome protection

The theory that ARE proteins function by somehow protecting the ribosome from the binding of MLS antibiotics stems from the sequence similarity with ribosomally active proteins and a parallel story with tetracycline resistance mediated by the Tet proteins. While there is no direct biochemical evidence to support this theory the recent evidence that ARE proteins interact with the antibiotics to which they confer resistance would be compatible with this (Jacquet *et al.*, 2008). Additionally, there is an intriguing and clear relationship between the antibiotic specificity of ARE proteins and the binding sites of MLS antibiotics (Reynolds *et al.*, 2003; Kerr *et al.*, 2005). For example, Msr proteins confer resistance to macrolides and streptogramin B, which have overlapping binding sites on the ribosome. Similarly, Lsa and Vga proteins confer resistance to lincosamides and streptogramin A which also have overlapping binding sites.

The potential for ARE proteins to be involved in ribosomal protection is also extrapolated from the phenomenon exemplified by the Tet proteins which confer resistance to tetracycline in *Campylobacter jejuni* (Spahn *et al.*, 2001; Connell *et al.*, 2003a; Connell *et al.*, 2003b). Tet(O) is a bacterial GTPase wich shares extensive sequence similarity with eukaryotic elongation factor proteins and functions by binding to the ribosome and promoting the release of tetracycline by inducing a conformational change in the ribosome structure (Connell *et al.*, 2003b). Biochemical and structural data from electron microscopy support a model whereby Tet(O) occupies the same binding site on the ribosome that tetracycline would interact with. This example of a bacterial GTPase conferring ribosomal protection provides the hypothesis that the ARE proteins confer resistance through competitive binding to the 50S ribosomal subunit. The existence of eukaryotic ABC proteins which are also ribosomally active is discussed below.

Translational control

Hypotheses regarding bacterial ARE protein functions are influenced by the known functions of eukaryotic non-transport-associated ABC proteins of the E and F subfamilies (Kerr, 2004). Although the ABCE family are involved in ribosome biogenesis (Karcher *et al.*, 2005; Kispal *et al.*, 2005), as well as other processes (Bisbal *et al.*, 1995; Zimmerman *et al.*, 2002), they are not discussed further here as they are characterized by highly conserved N-terminal iron-sulphur clusters which are absent from all bacterial ARE protein sequences (Kerr, 2004). Determination of the function of eukaryotic ABCF proteins offers hypotheses for translation-related functions for bacterial ARE proteins. ABCF1 (ABC50; Richard *et al.*, 1998) is an 800 residue phosphoprotein, which promotes translation initiation through involvement in the assembly of 40S, 60S, eukaryotic initiation factor 2 (eIF2), and the initiating methionine-loaded tRNA (Tyzack *et al.*, 2000). This interaction is dependent upon the 270 amino acid N-terminal region, which is distinct in sequence from the two NBDs which reside in the latter 530 amino acids (Paytubi *et al.*, 2008).

Similarly, ABCF3 proteins (such as yeast GCN20, GCN referring to the general amino acid control response to starvation) is also a ribosome-interacting translation initiation factor, required for the phosphorylation of eIF2 by GCN2 (Vazquez de Aldana *et al.*, 1995; Marton *et al.*, 1997). Again, GCN20 contains additional domains to its two C-terminal ABC domains, and indeed the N-terminal 120 residues is critical for function, in contrast to the two ABCs which can be deleted with only partial loss of the GCN response (Marton *et al.*, 1997).

Finally, yeast elongation factor 3 (EF3, ABCF4 (Kerr, 2004)) is a twin ABC protein involved in translation elongation (Skogerson and Engelhardt, 1977). Restricted to yeast (Belfield and Tuite, 1993), these proteins are believed to act as release factors, controlling the removal of deacylated tRNA from the ribosome, a process necessary for the binding of subsequent amino acyl t-RNA to the A-site (Uritani and Miyazaki, 1988; Kamath and Chakraburtty, 1989; Anand *et al.*, 2006). Yeast EF3 proteins have been purified and show a high affinity for ribosomes and, in parallel with numerous ABC transporters, have a low level basal ATPase activity as well as a stimulated activity. In this case 30-fold stimulation of ATPase activity is observed upon ribosome addition (Sarthy *et al.*, 1998). Bacterial ARE proteins are not direct homologues of yeast EF3; only the latter require this elongation factor, and EF3 proteins are 600 amino acids longer than the bacterial AREs, containing an N-terminal domain that is critical for ribosome interaction and function. However, the X-ray structure of yeast EF3 has been determined, and cryo-electron microscopy images of EF3 in complex with yeast ribosomes been obtained, enabling a direct visualization of this factor in contact with the ribosome (Andersen *et al.*, 2006). The structure demonstrates that interaction with the ribosome is dependent on both the N-terminal domain as well as with the second ABC domain – demonstrating that ABC domains are capable of ribosome interactions (Andersen *et al.*, 2006).

Thus, four of the five eukaryotic ABCE/ABCF sub-families have been shown to play a role in ribosomal events (Kerr, 2004). In Fig. 4.3 we therefore depict a mechanism where the ARE somehow influences the translation of mRNAs which themselves are responsible for the resistance phenotype. This rather speculative model awaits experimental validation. However we note that the *Agrobacterium tumefaciens* gene ChvD (which is not an ARE but belongs to the closely related REG (regulatory subfamily) of twin ABC proteins (Dassa and Bouige, 2001)) influences virulence by effecting the expression of the virulence response regulator (virG) (Liu *et al.*, 2001), potentially though a translational control mechanism.

Conclusions and future trends

In this short chapter we have highlighted the gaps in our knowledge of the mechanism of the ARE sub-family of ABC proteins in bacteria. Certainly we are moving forwards and experiments such as immuno-precipitations and subsequent proteomic approaches will enable us to determine what interaction partners ARE proteins have. Similarly, we may use other post-genomic approaches to determine whether the expression of an ARE has a wider influence on the organism's proteome. Finally, using such technologies, determining whether AREs function through a single, or multiple mechanisms is a realistic goal for researchers.

Acknowledgements

JMD is funded through a BBSRC DTA award. IDK thanks the British Society for Antimicrobial Chemotherapy for a Project Grant (GA582).

References

Achour, W., Bouchami, O., Galopin, S., Leclercq, R., and Ben Hassen, A. (2008). Analysis of pristinamycin-resistant Staphylococcus epidermidis isolates in the Tunisian Bone Marrow Transplant Center. Lett. App. Microbiol. *46*, 358–363.

Aktas, Z., Aridogan, A., Kayacan, C.B., and Aydin, D. (2007). Resistance to macrolide, lincosamide and streptogramin antibiotics in staphylococci isolated in Istanbul, Turkey. J. Microbiol. (Seoul, Korea) *45*, 286–290.

Allignet, J., Aubert, S., Morvan, A., and el Solh, N. (1996). Distribution of genes encoding resistance to streptogramin A and related compounds among staphylococci resistant to these antibiotics. Antimicrob. Agents Chemother. *40*, 2523–2528.

Allignet, J., and El Solh, N. (1997). Characterization of a new staphylococcal gene, vgaB, encoding a putative ABC transporter conferring resistance to streptogramin A and related compounds. Gene *202*, 133–138.

Allignet, J., Loncle, V., and el Sohl, N. (1992). Sequence of a staphylococcal plasmid gene, vga, encoding a putative ATP-binding protein involved in resistance to virginiamycin A-like antibiotics. Gene *117*, 45–51.

Altschul, S.F., Madden, T.L., Schaffer, A.A., Zhang, J., Zhang, Z., Miller, W., and Lipman, D.J. (1997). Gapped BLAST and PSI-BLAST: a new generation of protein database search programs. Nucleic Acids Res. *25*, 3389–3402.

Ambrose, K.D., Nisbet, R., and Stephens, D.S. (2005). Macrolide efflux in *Streptococcus pneumoniae* is mediated by a dual efflux pump (mel and mef) and is erythromycin inducible. Antimicrob. Agents Chemother. *49*, 4203–4209.

Anand, M., Balar, B., Ulloque, R., Gross, S.R., and Kinzy, T.G. (2006). Domain and nucleotide dependence of the interaction between *Saccharomyces cerevisiae* translation elongation factors 3 and 1A. J. Biol. Chem. *281*, 32318–32326.

Andersen, C.B., Becker, T., Blau, M., Anand, M., Halic, M., Balar, B., Mielke, T., Boesen, T., Pedersen, J.S., Spahn, C.M., Kinzy, T.G., Andersen, G.R., and Beckmann, R. (2006). Structure of eEF3 and the mechanism of transfer RNA release from the E-site. Nature *443*, 663–668.

Andersson, S., and Kurland, C.G. (1987). Elongating ribosomes *in vivo* are refractory to erythromycin. Biochimie *69*, 901–904.

Aparicio, G., Buche, A., Mendez, C., and Salas, J.A. (1996). Characterization of the ATPase activity of the N-terminal nucleotide binding domain of an ABC transporter involved in oleandomycin secretion by Streptomyces antibioticus. FEMS Microbiol. Lett. *141*, 157–162.

Belfield, G.P., and Tuite, M.F. (1993). Translation elongation factor 3: a fungus-specific translation factor? Mol. Microbiol. *9*, 411–418.

Bisbal, C., Martinand, C., Silhol, M., Lebleu, B., and Salehzada, T. (1995). Cloning and characterization of a RNAse L inhibitor. A new component of the interferon-regulated 2–5A pathway. J. Biol. Chem. *270*, 13308–13317.

Bottger, E.C. (2007). Antimicrobial agents targeting the ribosome: the issue of selectivity and toxicity – lessons to be learned. Cell. Mol. Life Sci. *64*, 791–795.

Bouchami, O., Achour, W., and Ben Hassen, A. (2007). Prevalence and mechanisms of macrolide resistance among Staphylococcus epidermidis isolates from neutropenic patients in Tunisia. Clin. Microbiol. Infect. *13*, 103–106.

Buche, A., Mendez, C., and Salas, J.A. (1997). Interaction between ATP, oleandomycin and the OleB ATP-binding cassette transporter of Streptomyces antibioticus involved in oleandomycin secretion. Biochem. J. *321 (Pt 1)*, 139–144.

Cagliero, C., Mouline, C., Cloeckaert, A., and Payot, S. (2006). Synergy between efflux pump CmeABC and modifications in ribosomal proteins L4 and L22 in conferring macrolide resistance in Campylobacter jejuni and Campylobacter coli. Antimicrob. Agents Chemother. *50*, 3893–3896.

Capobianco, J.O., and Goldman, R.C. (1990). Erythromycin and azithromycin transport into Haemophilus influenzae ATCC 19418 under conditions of depressed proton motive force (delta mu H). Antimicrob. Agents Chemother. *34*, 1787–1791.

Champney, W.S., Tober, C.L., and Burdine, R. (1998). A comparison of the inhibition of translation and 50S ribosomal subunit formation in *Staphylococcus aureus* cells by nine different macrolide antibiotics. Curr. Microbiol. *37*, 412–417.

Chen, J., Lu, G., Lin, J., Davidson, A.L., and Quiocho, F.A. (2003). A tweezers-like motion of the ATP-binding cassette dimer in an ABC transport cycle. Mol. Cell *12*, 651–661.

Chesneau, O., Ligeret, H., Hosan-Aghaie, N., Morvan, A., and Dassa, E. (2005). Molecular analysis of resistance to streptogramin A compounds conferred by the Vga proteins of staphylococci. Antimicrob. Agents Chemother. *49*, 973–980.

Connell, S.R., Tracz, D.M., Nierhaus, K.H., and Taylor, D.E. (2003a). Ribosomal protection proteins and their mechanism of tetracycline resistance. Antimicrob. Agents Chemother. *47*, 3675–3681.

Connell, S.R., Trieber, C.A., Dinos, G.P., Einfeldt, E., Taylor, D.E., and Nierhaus, K.H. (2003b). Mechanism of Tet(O)-mediated tetracycline resistance. EMBO J. *22*, 945–953.

Daly, M.M., Doktor, S., Flamm, R., and Shortridge, D. (2004). Characterization and prevalence of MefA, MefE, and the associated msr(D) gene in *Streptococcus pneumoniae* clinical isolates. J. Clin. Microbiol. *42*, 3570–3574.

Dassa, E., and Bouige, P. (2001). The ABC of ABCS: a phylogenetic and functional classification of ABC systems in living organisms. Res. Microbiol. *152*, 211–229.

Douthwaite, S., and Champney, W.S. (2001). StructuRes. of ketolides and macrolides determine their mode of interaction with the ribosomal target site. J. Antimicrob. Chemother. *48 Suppl. T1*, 1–8.

Douthwaite, S., Crain, P.F., Liu, M., and Poehlsgaard, J. (2004). The tylosin-resistance methyltransferase RlmA(II) (TlrB) modifies the N-1 position of 23S rRNA nucleotide G748. J. Mol. Biol. *337*, 1073–1077.

Eady, E.A., Ross, J.I., Tipper, J.L., Walters, C.E., Cove, J.H., and Noble, W.C. (1993). Distribution of genes encoding erythromycin ribosomal methylases and an erythromycin efflux pump in epidemiologically distinct groups of staphylococci. J. Antimicrob. Chemother. *31*, 211–217.

Epp, J.K., Burgett, S.G., and Schoner, B.E. (1987). Cloning and nucleotide sequence of a carbomycin-resistance gene from Streptomyces thermotolerans. Gene *53*, 73–83.

Gatermann, S.G., Koschinski, T., and Friedrich, S. (2007). Distribution and expression of macrolide resistance genes in coagulase-negative staphylococci. Clin. Microbiol. Infect. *13*, 777–781.

Ghanei, H., Abeyrathne, P.D., and Lam, J.S. (2007). Biochemical characterization of MsbA from Pseudomonas aeruginosa. J. Biol. Chem. *282*, 26939–26947.

Goldman, R.C., and Capobianco, J.O. (1990). Role of an energy-dependent efflux pump in plasmid pNE24-mediated resistance to 14- and 15-membered macrolides in Staphylococcus epidermidis. Antimicrob. Agents Chemother. *34*, 1973–1980.

Hansen, J.L., Ippolito, J.A., Ban, N., Nissen, P., Moore, P.B., and Steitz, T.A. (2002). The structuRes. of four macrolide antibiotics bound to the large ribosomal subunit. Mol. Cell *10*, 117–128.

Hermann, T. (2005). Drugs targeting the ribosome. Curr. Opin. Struct. Biol. *15*, 355–366.

Jacquet, E., Girard, J.M., Ramaen, O., Pamlard, O., Levaique, H., Betton, J.M., Dassa, E., and Chesneau, O. (2008). ATP hydrolysis and pristinamycin IIA inhibition of the *Staphylococcus aureus* Vga(A), a dual ABC protein involved in streptogramin A resistance. J. Biol. Chem. *283*, 25332–25339.

Kamath, A., and Chakraburtty, K. (1989). Role of yeast elongation factor 3 in the elongation cycle. J. Biol. Chem. *264*, 15423–15428.

Karcher, A., Buttner, K., Martens, B., Jansen, R.P., and Hopfner, K.P. (2005). X-ray structure of RLI, an essential twin cassette ABC ATPase involved in ribosome biogenesis and HIV capsid assembly. Structure (Camb) *13*, 649–659.

Karray, F., Darbon, E., Oestreicher, N., Dominguez, H., Tuphile, K., Gagnat, J., Blondelet-Rouault, M.H., Gerbaud, C., and Pernodet, J.L. (2007). Organization of the biosynthetic gene cluster for the macrolide antibiotic spiramycin in Streptomyces ambofaciens. Microbiology *153*, 4111–4122.

Kerr, I.D. (2004). Sequence analysis of twin ATP binding cassette proteins involved in translational control, antibiotic resistance and ribonuclease L inhibition. Biochem. Biophys. Res. Commun. *315*, 166–174.

Kerr, I.D., Reynolds, E.D., and Cove, J.H. (2005). ABC proteins and antibiotic drug resistance: is it all about transport? Biochem. Soc. Trans. *33*, 1000–1002.

Kispal, G., Sipos, K., Lange, H., Fekete, Z., Bedekovics, T., Janaky, T., Bassler, J., Aguilar Netz, D.J., Balk, J., Rotte, C., and Lill, R. (2005). Biogenesis of cytosolic ribosomes requires the essential iron-sulphur protein Rli1p and mitochondria. EMBO J. *24*, 589–598.

Kobayashi, N., Nishino, K., and Yamaguchi, A. (2001). Novel macrolide-specific ABC-type efflux transporter in *Escherichia coli*. J. Bacteriol. *183*, 5639–5644.

Kumar, A., and Schweizer, H.P. (2005). Bacterial resistance to antibiotics: active efflux and reduced uptake. Adv. Drug Deliv. Rev. *57*, 1486–1513.

Lawson, J., O'Mara M, L., and Kerr, I.D. (2008). Structure-based interpretation of the mutagenesis database for the nucleotide binding domains of P-glycoprotein. Biochim. Biophys. Acta *1778*, 376–391.

Leclercq, R. (2002). Mechanisms of resistance to macrolides and lincosamides: nature of the resistance elements and their clinical implications. Clin. Infect. Dis. *34*, 482–492.

Liu, M., and Douthwaite, S. (2002). Resistance to the macrolide antibiotic tylosin is conferred by single methylations at 23S rRNA nucleotides G748 and A2058 acting in synergy. Proc. Natl. Acad. Sci. U.S.A. *99*, 14658–14663.

Liu, Z., Jacobs, M., Schaff, D.A., McCullen, C.A., and Binns, A.N. (2001). ChvD, a chromosomally encoded ATP-binding cassette transporter-homologous protein involved in regulation of virulence gene expression in Agrobacterium tumefaciens. J. Bacteriol. *183*, 3310–3317.

Loncle, V., Casetta, A., Buu-Hoi, A., and el Solh, N. (1993). Analysis of pristinamycin-resistant Staphylococcus epidermidis isolates responsible for an outbreak in a Parisian hospital. Antimicrob. Agents Chemother. *37*, 2159–2165.

Malbruny, B., Nagai, K., Coquemont, M., Bozdogan, B., Andrasevic, A.T., Hupkova, H., Leclercq, R., and Appelbaum, P.C. (2002). Resistance to macrolides in clinical isolates of Streptococcus pyogenes due to ribosomal mutations. J. Antimicrob. Chemother. *49*, 935–939.

Martineau, F., Picard, F.J., Lansac, N., Menard, C., Roy, P.H., Ouellette, M., and Bergeron, M.G. (2000). Correlation between the resistance genotype determined by multiplex PCR assays and the antibiotic susceptibility patterns of *Staphylococcus aureus* and Staphylococcus epidermidis. Antimicrob. Agents Chemother. *44*, 231–238.

Marton, M.J., Vazquez de Aldana, C.R., Qiu, H., Chakraburtty, K., and Hinnebusch, A.G. (1997). Evidence that GCN1 and GCN20, translational regulators of GCN4, function on elongating ribosomes in activation of eIF2alpha kinase GCN2. Mol. Cell Biol. *17*, 4474–4489.

Matsuoka, M., Inoue, M., Endo, Y., and Nakajima, Y. (2003). Characteristic expression of three genes, msr(A), mph(C) and erm(Y), that confer resistance to macrolide antibiotics on *Staphylococcus aureus*. FEMS Microbiol. Lett. *220*, 287–293.

Mendez, C., and Salas, J.A. (2001). The role of ABC transporters in antibiotic-producing organisms: drug secretion and resistance mechanisms. Res. Microbiol. *152*, 341–350.

Nawaz, M.S., Khan, S.A., Khan, A.A., Khambaty, F.M., and Cerniglia, C.E. (2000). Comparative molecular analysis of erythromycin-resistance determinants in staphylococcal isolates of poultry and human origin. Molec. Cell Probes *14*, 311–319.

Novotna, G., Adamkova, V., Janata, J., Melter, O., and Spizek, J. (2005). Prevalence of resistance mechanisms against macrolides and lincosamides in methicillin-resistant coagulase-negative staphylococci in the Czech Republic and occurrence of an undefined mechanism of resistance to lincosamides. Antimicrob. Agents Chemother. *49*, 3586–3589.

Novotna, G., and Janata, J. (2006). A new evolutionary variant of the streptogramin A resistance protein, Vga(A)LC, from Staphylococcus haemolyticus with shifted substrate specificity towards lincosamides. Antimicrob. Agents Chemother. *50*, 4070–4076.

Novotna, G., Spizek, J., and Janata, J. (2007). In vitro activity of telithromycin and quinupristin/dalfopristin against methicillin-resistant coagulase-negative staphylococci with defined resistance genotypes. Folia Microbiol. *52*, 593–599.

Olano, C., Rodriguez, A.M., Mendez, C., and Salas, J.A. (1995). A second ABC transporter is involved in oleandomycin resistance and its secretion by Streptomyces antibioticus. Mol. Microbiol. *16*, 333–343.

Oswald, C., Holland, I.B., and Schmitt, L. (2006). The motor domains of ABC transporters. What can structuRes. tell us? Naunyn-Schmiedeberg's Arch. Pharmacol. *372*, 385–399.

Paytubi, S., Morrice, N.A., Boudeau, J., and Proud, C.G. (2008). The N-terminal region of ABC50 interacts with eukaryotic initiation factor eIF2 and is a target for regulatory phosphorylation by CK2. Biochem. J. *409*, 223–231.

Pechere, J.C. (2001). Macrolide resistance mechanisms in Gram-positive cocci. Int. J. Antimicrob. Agents *18 Suppl. 1*, S25–28.

Peschke, U., Schmidt, H., Zhang, H.Z., and Piepersberg, W. (1995). Molecular characterization of the lincomycin-production gene cluster of Streptomyces lincolnensis 78–11. Mol. Microbiol. *16*, 1137–1156.

Piddock, L.J. (2006). Multidrug-resistance efflux pumps – not just for resistance. Nature Rev. *4*, 629–636.

Pihlajamaki, M., Kataja, J., Seppala, H., Elliot, J., Leinonen, M., Huovinen, P., and Jalava, J. (2002). Ribosomal mutations in *Streptococcus pneumoniae* clinical isolates. Antimicrob. Agents Chemother. *46*, 654–658.

Poehlsgaard, J., and Douthwaite, S. (2002). The macrolide binding site on the bacterial ribosome. Curr. Drug Targets Infect. Disord. *2*, 67–78.

Portillo, A., Ruiz-Larrea, F., Zarazaga, M., Alonso, A., Martinez, J.L., and Torres, C. (2000). Macrolide resistance genes in *Enterococcus* spp. Antimicrob. Agents Chemother. *44*, 967–971.

Retsema, J., and Fu, W. (2001). Macrolides: structuRes. and microbial targets. Int. J. Antimicrob. Agents *18 Suppl. 1*, S3–10.

Reynolds, E., and Cove, J.H. (2005a). Enhanced resistance to erythromycin is conferred by the enterococcal msrC determinant in *Staphylococcus aureus*. J. Antimicrob. Chemother. *55*, 260–264.

Reynolds, E., Ross, J.I., and Cove, J.H. (2003). Msr(A) and related macrolide/streptogramin resistance determinants: incomplete transporters? Int. J. Antimicrob. Agents *22*, 228–236.

Reynolds, E.D., and Cove, J.H. (2005b). Resistance to telithromycin is conferred by msr(A), msrC and msr(D) in *Staphylococcus aureus*. J. Antimicrob. Chemother. *56*, 1179–1180.

Richard, M., Drouin, R., and Beaulieu, A.D. (1998). ABC50, a novel human ATP-binding cassette protein found in tumor necrosis factor-alpha-stimulated synoviocytes. Genomics *53*, 137–145.

Rodriguez, A.M., Olano, C., Vilches, C., Mendez, C., and Salas, J.A. (1993). Streptomyces antibioticus contains at least three oleandomycin-resistance determinants, one of which shows similarity with proteins of the ABC-transporter superfamily. Mol. Microbiol. *8*, 571–582.

Ross, J.I., Eady, E.A., Cove, J.H., and Baumberg, S. (1995). Identification of a chromosomally encoded ABC-transport system with which the staphylococcal erythromycin exporter MsrA may interact. Gene *153*, 93–98.

Ross, J.I., Eady, E.A., Cove, J.H., and Baumberg, S. (1996). Minimal functional system required for expression of erythromycin resistance by msrA in *Staphylococcus aureus* RN4220. Gene *183*, 143–148.

Ross, J.I., Eady, E.A., Cove, J.H., Cunliffe, W.J., Baumberg, S., and Wootton, J.C. (1990). Inducible erythromycin resistance in staphylococci is encoded by a member of the ATP-binding transport super-gene family. Mol. Microbiol. *4*, 1207–1214.

Ross, J.I., Farrell, A.M., Eady, E.A., Cove, J.H., and Cunliffe, W.J. (1989). Characterisation and molecular cloning of the novel macrolide-streptogramin B resistance determinant from Staphylococcus epidermidis. J. Antimicrob. Chemother. *24*, 851–862.

Rouquette-Loughlin, C.E., Balthazar, J.T., and Shafer, W.M. (2005). Characterization of the MacA-MacB efflux system in *Neisseria gonorrhoeae*. J. Antimicrob. Chemother. *56*, 856–860.

Santagati, M., Iannelli, F., Oggioni, M.R., Stefani, S., and Pozzi, G. (2000). Characterization of a genetic element carrying the macrolide efflux gene mef(A) in *Streptococcus pneumoniae*. Antimicrob. Agents Chemother. *44*, 2585–2587.

Sarthy, A.V., McGonigal, T., Capobianco, J.O., Schmidt, M., Green, S.R., Moehle, C.M., and Goldman, R.C. (1998). Identification and kinetic analysis of a functional homolog of elongation factor 3, YEF3 in *Saccharomyces cerevisiae*. Yeast *14*, 239–253.

Schlunzen, F., Zarivach, R., Harms, J., Bashan, A., Tocilj, A., Albrecht, R., Yonath, A., and Franceschi, F. (2001). Structural basis for the interaction of antibiotics with the peptidyl transferase centre in eubacteria. Nature *413*, 814–821.

Schmitz, F.J., Sadurski, R., Kray, A., Boos, M., Geisel, R., Kohrer, K., Verhoef, J., and Fluit, A.C. (2000). Prevalence of macrolide-resistance genes in *Staphylococcus aureus* and *Enterococcus faecium* isolates from 24 European university hospitals. J. Antimicrob. Chemother. *45*, 891–894.

Schoner, B., Geistlich, M., Rosteck, P., Jr., Rao, R.N., Seno, E., Reynolds, P., Cox, K., Burgett, S., and Hershberger, C. (1992). Sequence similarity between macrolide-resistance determinants and ATP-binding transport proteins. Gene *115*, 93–96.

Singh, K.V., Malathum, K., and Murray, B.E. (2001). Disruption of an *Enterococcus faecium* species-specific gene, a homologue of acquired macrolide resistance genes of staphylococci, is associated with an increase in macrolide susceptibility. Antimicrob. Agents Chemother. *45*, 263–266.

Singh, K.V., and Murray, B.E. (2005). Differences in the *Enterococcus faecalis* lsa locus that influence susceptibility to quinupristin-dalfopristin and clindamycin. Antimicrob. Agents Chemother. *49*, 32–39.

Singh, K.V., Weinstock, G.M., and Murray, B.E. (2002). An *Enterococcus faecalis* ABC homologue (Lsa) is required for the resistance of this species to clindamycin and quinupristin-dalfopristin. Antimicrob. Agents Chemother. *46*, 1845–1850.

Skogerson, L., and Engelhardt, D. (1977). Dissimilarity in protein chain elongation factor requirements between yeast and rat liver ribosomes. J. Biol. Chem. *252*, 1471–1475.

Spahn, C.M., Blaha, G., Agrawal, R.K., Penczek, P., Grassucci, R.A., Trieber, C.A., Connell, S.R., Taylor, D.E., Nierhaus, K.H., and Frank, J. (2001). Localization of the ribosomal protection protein Tet(O) on the ribosome and the mechanism of tetracycline resistance. Mol, Cell 7, 1037–1045.

Steitz, T.A. (2005). On the structural basis of peptide-bond formation and antibiotic resistance from atomic structuRes. of the large ribosomal subunit. FEBS Lett, *579*, 955–958.

Tenson, T., Lovmar, M., and Ehrenberg, M. (2003). The mechanism of action of macrolides, lincosamides and streptogramin B reveals the nascent peptide exit path in the ribosome. J, Mol, Biol, *330*, 1005–1014.

Thompson, J.D., Higgins, D.G., and Gibson, T.J. (1994). CLUSTAL W: improving the sensitivity of progressive multiple sequence alignment through sequence weighting, position-specific gap penalties and weight matrix choice. Nucleic Acids Res, *22*, 4673–4680.

Tikhonova, E.B., Devroy, V.K., Lau, S.Y., and Zgurskaya, H.I. (2007). Reconstitution of the *Escherichia coli* macrolide transporter: the periplasmic membrane fusion protein MacA stimulates the ATPase activity of MacB. Mol, Microbiol, *63*, 895–910.

Tu, D., Blaha, G., Moore, P.B., and Steitz, T.A. (2005). StructuRes. of MLSBK antibiotics bound to mutated large ribosomal subunits provide a structural explanation for resistance. Cell *121*, 257–270.

Tyzack, J.K., Wang, X., Belsham, G.J., and Proud, C.G. (2000). ABC50 interacts with eukaryotic initiation factor 2 and associates with the ribosome in an ATP-dependent manner. J, Biol, Chem, *275*, 34131–34139.

Uritani, M., and Miyazaki, M. (1988). Role of yeast peptide elongation factor 3 (EF-3) at the AA-tRNA binding step. J. Biochem. *104*, 118–126.

Vazquez de Aldana, C.R., Marton, M.J., and Hinnebusch, A.G. (1995). GCN20, a novel ATP binding cassette protein, and GCN1 reside in a complex that mediates activation of the eIF-2 alpha kinase GCN2 in amino acid-starved cells. EMBO J. *14*, 3184–3199.

Waugh, A.C., and Long, P.F. (2002). Prospects for generating new antibiotics. Sci. Prog *85*, 73–88.

Weisblum, B. (1985). Inducible resistance to macrolides, lincosamides and streptogramin type B antibiotics: the resistance phenotype, its biological diversity, and structural elements that regulate expression--a review. J. Antimicrob. Chemother. *16 Suppl. A*, 63–90.

Weisblum, B. (1995). Erythromycin resistance by ribosome modification. Antimicrob. Agents Chemother. *39*, 577–585.

Werner, G., Hildebrandt, B., and Witte, W. (2001). The newly described msrC gene is not equally distributed among all isolates of *Enterococcus faecium*. Antimicrob. Agents Chemother. *45*, 3672–3673.

Wootton, J.C., and Drummond, M.H. (1989). The Q-linker: a class of interdomain sequences found in bacterial multidomain regulatory proteins. Protein Eng. *2*, 535–543.

Wright, G.D. (2003). Mechanisms of resistance to antibiotics. Curr Opin Chem. Biol. *7*, 563–569.

Zhu, L.X., Zhang, Z.W., Wang, C., Yang, H.W., Jiang, D., Zhang, Q., Mitchelson, K., and Cheng, J. (2007). Use of a DNA microarray for simultaneous detection of antibiotic resistance genes among staphylococcal clinical isolates. J. Clin. Microbiol. *45*, 3514–3521.

Zimmerman, C., Klein, K.C., Kiser, P.K., Singh, A.R., Firestein, B.L., Riba, S.C., and Lingappa, J.R. (2002). Identification of a host protein essential for assembly of immature HIV-1 capsids. Nature *415*, 88–92.

ABC-type Multidrug Resistance Transporters and their Role in Survival of Bacteria

5

Patrick J. Bakkes, H. Bart van den Berg van Saparoea and Arnold J.M. Driessen

Abstract

Multidrug transporters are membrane proteins that actively catalyse the extrusion of structurally and functionally unrelated drugs from the cell. They are considered the major contributors to multidrug resistance (MDR) of bacterial cells. The majority of the identified MDR transporters were shown to rely on ion motive forces for their extrusion activity, while only few systems characterized thus far use ATP hydrolysis to drive efflux. There is accumulating evidence that MDR transporters also have natural physiological roles in addition to the extrusion of man-made synthetic drugs. MDR efflux pumps have been implicated in the protection against toxic compounds present in the natural environment such as bile acids, host-defence molecules and hormones. Here we summarize recent advances in the characterization of ATP-binding cassette (ABC) type transporters involved in bacterial multidrug resistance with implications for their physiological functions. In addition, we will discuss possible pitfalls in the functional analysis of ABC-type MDR-like transporters in heterologous systems, and the need for functional characterization in the authentic host.

Introduction

Antibiotics were first discovered in the 1940s. This was an important step towards the treatment and control of bacterial infections. Antibiotics drastically reduced the lethal effects of infectious diseases thereby increasing the average life expectation of man (Cohen, 2000). However, bacteria have the astonishing ability to adapt to a rapidly changing environment. When exposed to toxic compounds, bacteria readily develop resistance against the detrimental effects of these compounds. Therefore, soon after the introduction of antibiotics it was observed that bacteria developed various ways to resist the bactericidal effects of these compounds (Neu, 1992; Levy, 1998; Wood and Moellering, Jr, 2003; Levy and Marshall, 2004). Especially in the past decade, the number of antibiotics to which bacteria have developed resistance has increased profoundly. Consequently, some antibiotics are no longer useful for the treatment of infections and diseases, which were thought to be under control, and these are re-emerging. Moreover, an increasing number of bacteria have become resistant to multiple antibiotics (Neu, 1992; Travis, 1994; Levy, 1998; Cohen, 2000; Levy and Marshall, 2004).

Bacteria employ a wide variety of mechanisms to resist the toxic effects of antimicrobial agents. In many cases, the resistance mechanisms are specific for a single drug or for a group of closely related drugs. A typical example of such a mechanism is the enzymatic inactivation of the toxic compound. For instance, β-lactamase hydrolyses the β-lactam ring of antibiotics thereby inactivating these compounds (Hayes and Wolf, 1990; Jenkinson, 1996; Walsh, 2000). Other mechanisms of resistance are the mutational modification of the drug target site such that the affinity of the target for the drug is reduced (Spratt, 1994), and cellular adaptations that reduce the permeability of the cell envelope for drugs.

The cytoplasmic membrane of bacteria consists of a lipid bilayer that acts as a general barrier for molecules to enter the cell. In Gram-positive bacteria, this lipid bilayer is surrounded by a thick extracellular wall made of peptidoglycan that offers only minor resistance to the diffusion of small molecules. Gram-negative bacteria on the other hand contain a second membrane in addition to the cytoplasmic membrane. This outer membrane constitutes a physical barrier that is relatively impermeable to lipophilic and amphiphilic compounds, which has been attributed to its phospholipid composition and the presence of lipopolysaccharides (Denyer and Maillard, 2002). In *Mycobacteria* a thick extracellular layer of mycolic acid functions as an effective diffusion barrier (Nikaido, 1994). The cell envelope of Gram-negative bacteria and *Mycobacteria* contains so-called porins, which are membrane proteins that serve as entry gates for small molecules and even facilitate the influx of certain nutrients. Since porins may also facilitate the influx of toxic compounds cells can increase their resistance to these compounds by decreasing the expression of porins (Nikaido, 1994).

A highly effective strategy to combat the cellular influx of drugs is the active extrusion of these compounds from the cytoplasmic membrane and the cytosol (Putman *et al.*, 2000b). Integral membrane proteins that use metabolic energy to transport drugs across the cytoplasmic membrane, against their concentration gradients, mediate drug extrusion. Based on bioenergetics criteria, drug efflux proteins can be classified into two main groups: transporters that use either ion motive forces or ATP-hydrolysis to drive efflux. In Gram-negative bacteria, drug transporters are located within the inner membrane and are usually connected via adaptor proteins to an outer membrane pore, thereby facilitating drug extrusion across the outer membrane. Transporter-based active extrusion can be specific for a given compound, i.e. mediated by so-called single drug or group-specific transporters, or can be multidrug-specific covering a wide range of toxic compounds that are structurally and functionally unrelated. These bacterial drug efflux systems have been implicated as the main contributors to clinical antibiotic resistance.

Multidrug resistance

The phenomenon of multidrug resistance (MDR) was first observed in mammalian cancer cells (Dano, 1972; Dano, 1973; Juliano and Ling, 1976; Paulsen and Lewis, 2001). Upon exposure to cytotoxic drugs used in cancer therapy, mammalian cancer cells developed a multidrug resistance phenotype; i.e. treated cancer cells became resistant not only to the drug of choice but also to other drugs (Tiirikainen and Krusius, 1991; Ling, 1997). The acquired multidrug resistance compromises successful cancer treatment and therefore poses a serious threat to public health. In mammalian cells, major contributors to multidrug resistance are ABC-type transporters like P-glycoprotein (Chen *et al.*, 1986; Ambudkar *et al.*, 2003), the multiple drug-resistant associated protein, MRP1 (Cole *et al.*, 1992; Kruh and Belinsky, 2003) and the breast cancer related protein, BCRP (Doyle *et al.*, 1998; Allikmets *et al.*, 1998; Miyake *et al.*, 1999; Doyle and Ross, 2003).

Also in bacteria, MDR is a widespread phenomenon. With the discovery and first clinical use of antibiotics to treat bacterial infections, antibiotic resistance emerged rapidly. Treatment of these resistant bacteria with antibiotics of different structural classes was often found to be ineffective, too. Apparently, these bacteria had acquired resistance to other drugs as well, even though they were never exposed to these compounds before. Thus, bacteria may rapidly develop resistance to multiple drugs.

Bacterial multidrug resistance is most frequently due to active transporters that extrude a broad spectrum of functionally and structurally different bactericidal agents from the cell. In 1980, Levy and co-workers described the active extrusion of tetracycline as the underlying mechanism of tetracycline specific resistance in *Escherichia coli* (McMurry *et al.*, 1980). Approximately 10 years later, a large number of MDR transporters were discovered in prokaryotes, as for instance EmrE (Lomovskaya and Lewis, 1992) and AcrAB (Ma *et al.*, 1993; Ma *et al.*, 1995) from *E. coli* and QacA from *Staphylococcus aureus* (Tennent *et al.*, 1989). At present, MDR transporters are found in virtually every organism and the MDR phenomenon caused by active transporters is recognized as an important mechanism contributing to clinical resistance against antibiotics in important pathogenic bacteria such as *Campylobacter jejuni, E. coli, Neisseria gonorrhoeae, Pseudomonas aeruginosa, Salmonella enterica* serovar Typhimurium, *S. aureus* and *Streptococcus pneumoniae* (Piddock, 2006a, b).

Classification, organization and distribution of MDR transporters

Active transport of solutes across the membrane requires metabolic energy. Based on how MDR transporters couple metabolic energy to the extrusion of dedicated substrates, MDR transporters can be classified into two main groups, i.e. the primary and secondary active transporters. Primary active transporters belong to the ATP binding cassette (ABC) transporter superfamily, which use the energy released upon the hydrolysis of ATP to extrude the drug against its concentration gradient from the cell (Fig. 5.1). Secondary transporters, on the other hand, use the proton or sodium motive force for drug expulsion. As such, these transporters function as drug/H^+ or drug/Na^+ antiporters (Fig. 5.1).

On the basis of their amino acid sequence homology, predicted secondary structure and experimental data, secondary MDR transporters are generally classified into four major families: (i) the major facilitator superfamily (MFS), which constitutes the largest family (Marger and Saier, Jr., 1993), (ii) the small multidrug resistance superfamily (SMR) (Paulsen *et al.*, 1996), (iii) the resistance-nodulation-cell division superfamily (RND) (Saier, Jr. *et al.*, 1994), and (iv) the multidrug and toxic compound extrusion superfamily (MATE) (Brown *et al.*, 1999) (Fig. 5.1).

In this chapter, we will focus on the ABC-type transporters that play a major role in the survival and multidrug resistance of bacteria and other prokaryotes. The first cloned and characterized MDR transporters were the eukaryotic

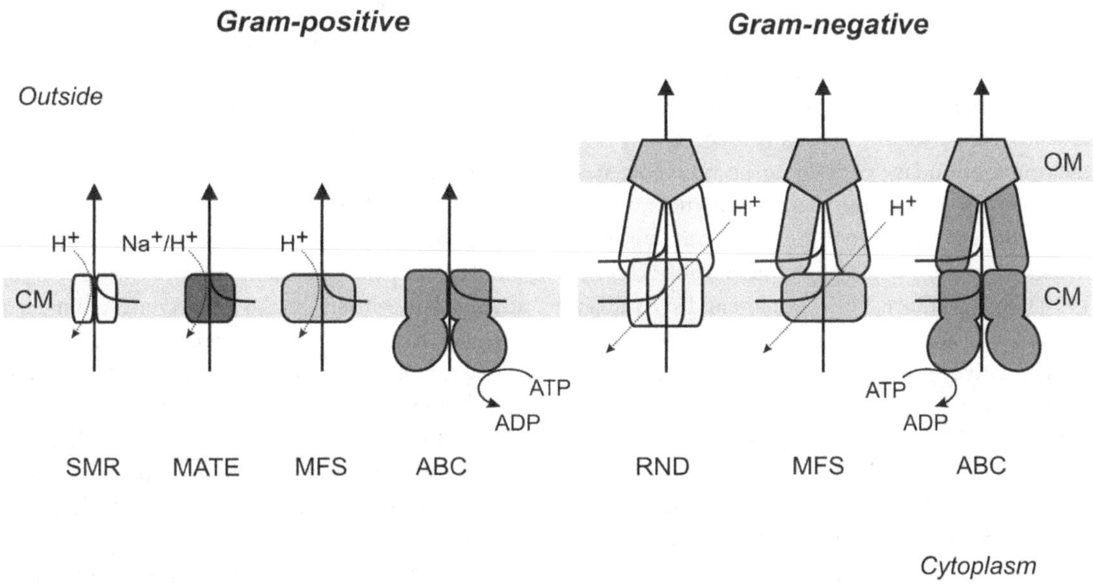

Figure 5.1 Schematic representation of the major families of MDR transporters in bacteria. Primary transporters that utilize the energy of ATP hydrolysis belong to the superfamily of ATP-Binding Cassette (ABC) transporters. The functional unit of these transporters consists of a dimer composed of two identical (homodimer) or two different (heterodimer) half transporters. Ion motive forces energize secondary transporters. Four major MDR transporter families rely on ion motive forces: the Major Facilitator Superfamily (MFS), which are monomeric; the Small Multidrug Resistance family (SMR), which form homo- or hetero-oligomers; the Multidrug and Toxic-compound Extrusion family (MATE); and the Resistance-Nodulation-cell Division family (RND), which are tripartite systems that consist of an integral membrane permease domain, a membrane fusion protein (MFP) and an outer membrane protein (OMP). The MDR efflux pumps in Gram-negative bacteria are generally tripartite systems that contain the outer membrane protein TolC. MDR efflux pumps of the ABC, MFS, MATE and SMR superfamily have been identified in Gram-positive organisms, while in Gram-negative organisms multidrug resistance has been attributed to members of the RND, ABC and MFS superfamily. OM, outer membrane; CM, cytoplasmic membrane.

ABC-type transporters P-glycoprotein (Chen *et al.*, 1986) and MRP1 (Cole *et al.*, 1992). In the early nineties, the first bacterial MDR transporters were described, which all belonged to the group of secondary active transporters (Paulsen and Lewis, 2001). The first bacterial ABC-type MDR transporter, called LmrA, was found in the non-pathogenic lactic acid bacterium *Lactococcus lactis* (van Veen *et al.*, 1996). However, the vast majority of bacterial MDR efflux pumps identified so far are secondary transporters suggesting that ABC-type MDR transporters are rather uncommon in bacteria (Li and Nikaido, 2004). Recent and on-going completion of genome sequences of many different bacteria may now provide a more complete view on the distribution of MDR transporters.

The wealth of sequence information available inspired M. Saier to classify membrane transport proteins in terms of transport mode, energy coupling mechanism, molecular phylogeny and substrate specificity (Saier, Jr, 2000).* Comparison of entire sequences of various bacterial genomes revealed that, in general, with increasing genome size also the number of transport proteins increases proportionally. In this respect, *E. coli* is at the upper boundary containing ~6 transport systems per 100 kb and the archaeon *Methanococcus jannaschii* is at the lower boundary containing ~2 transporters per 100 kb of genome (Paulsen *et al.*, 1998). Furthermore, both primary and secondary transporters are ubiquitous in prokaryotes, while their relative presence may vary considerably among species. The relative distribution of ABC-type and MFS transporters often correlates with the mode of metabolic energy generation by the bacterial host. Obligate aerobic bacteria rely on the proton motive force (pmf) as their primary source of energy, while strict fermentative bacteria employ substrate level phosphorylation to generate ATP. As a consequence, fermentative bacteria tend to rely more on primary transporters, whereas strict aerobic bacteria rely on secondary transporters (Paulsen *et al.*, 2000). The genome of the obligate aerobe *Mycobacterium tuberculosis*, for instance, encodes a total of 113 transporters of which 28 are ABC-type transporters and 24 are MFS transporters. The strict fermentative bacterium *Mycoplasma pneumoniae*, on the other hand, harbors a total of 22 transporters of which 11 are ABC-type transporters and just one MFS transporter (Paulsen *et al.*, 1998; Paulsen *et al.*, 2000).

Comparison of the genome sequences of more than 100 organisms, including a wide variety of prokaryotes, revealed that multidrug efflux pumps are found throughout nature. For prokaryotes, it was found that 3–24% of the total number of transporters constitute MDR transporters or proteins homologous to them (Paulsen *et al.*, 2000; Paulsen, 2003). The number and type of MDR transporters may differ considerably among these microorganisms. Nevertheless, ABC-type and MFS-type MDR transporters seem to be the most abundant drug efflux systems in prokaryotes (Paulsen *et al.*, 1998; Paulsen *et al.*, 2000).

Structure and function of prokaryotic ABC-type MDR transporters

The ATP-binding cassette (ABC) protein family constitutes one of the largest superfamilies of paralogous proteins and comprises transporters involved in either uptake or secretion but also includes ABC-ATPases that are involved in processes unrelated to transport, like DNA repair, translation, regulation of gene expression and also antibiotic resistance. The structural, functional and evolutionary aspects of bacterial ABC systems were excellently reviewed recently (Davidson *et al.*, 2008). In addition, comprehensive reviews on the functional classification of bacterial ABC systems are available (Dassa and Bouige, 2001; Bouige *et al.*, 2002)† (See also Chapter 1).

The first described ABC-type transporter, the histidine permease, HisP of *Salmonella enterica*

* The Transport Classification (TC) tool is available at http://www-biology.ucsd.edu/~msaier/transport/. The Transport classification (TC) system has been approved and recommended by the Nomenclature Committee of the International Union of Biochemistry and Molecular Biology). A relational database (TransportDB) is available at http://membranetransport.org/ that classifies and predicts the membrane transporters from organisms of which the genome has been sequenced (Ren *et al.*, 2004).

† Database is available at: http://www.pasteur.fr/recherche/unites/pmtg/abc/index.html.

serovar Typhimurium, was cloned and sequenced in 1982 (Higgins *et al.*, 1982). Around the same time the complete sequences of the maltose transporter MalK of *E. coli* and the oligopeptide transporter OppD of *S.* Typhimurium became available (Gilson *et al.*, 1982; Higgins *et al.*, 1985). It was observed that these three transporter proteins shared a consensus nucleotide binding motif, similar to that identified in ATP synthase, myosin and adenylate kinase (Walker *et al.*, 1982). Functional analysis of transporter proteins soon followed and revealed that these sequence motifs are responsible for ATP binding and that ATP hydrolysis is coupled to the transport event (Higgins *et al.*, 1985; Bishop *et al.*, 1989; Mimmack *et al.*, 1989).

In 1986, it became generally accepted that transporters containing these ATP binding motifs constitute a large family of proteins, which are now collectively called ABC-type (ATP binding cassette) transporters (Higgins *et al.*, 1986; Hyde *et al.*, 1990). The ATP binding cassette domain contains a number of short conserved sequence motifs as well as conserved structural elements, among which the Walker A and B motifs that form the nucleotide binding site, and the signature motif that is unique to ABC proteins (see also below). More than two decades of research has now resulted in the identification of thousands of ABC-type ATPases (Bouige *et al.*, 2002). It has become evident that the ABC protein superfamily is ubiquitously distributed in all three kingdoms of life (Higgins, 1992; Jones and George, 1999; Higgins, 2001; Kerr, 2002; Schmitt and Tampe, 2002; Lage, 2003; Holland *et al.*, 2003; Davidson and Chen, 2004; Higgins and Linton, 2004; van der Does C. and Tampe, 2004; Locher, 2004; Jones and George, 2004).

ABC-type transporters are membrane proteins that are responsible for the uptake and secretion of a wide range of substrates, from inorganic ions to small molecules such as amino acids, sugars, xenobiotics and vitamins up to polymers such as peptides, proteins and polysaccharides. Although the ABC-type transporters first described were nutrient uptake systems it is now evident that they play a role in a range of cellular processes, such as extrusion of cellular waste products, protection against xenobiotics, bacterial immunity and virulence, osmotic stress, lipid transport, and export of macromolecules during biogenesis, differentiation and pathogenesis (Higgins, 1992; Higgins, 2001; Jones and George, 2004).

The ABC-type transporter typically consists of four core domains, i.e. two nucleotide binding domains (NBD) and two transmembrane domains (TMD) (Higgins *et al.*, 1986), which, depending on the type of transporter, can be organized in a variety of ways. The 4 modules (2 NBDs and 2 TMDs) may be encoded by four different genes or by fewer genes that either code for polypeptides consisting of a fusion of (i) two NBDs, (ii) two TMDs, (iii) one NBD and one TMD, or (iii) all four modules comprising the entire ABC transporter. This remarkable modular organization further discriminates uptake from extrusion systems.

Uptake systems usually consist of four separate polypeptides that together constitute an active transporter, as is the case for the oligopeptide transporter OppBCDF of *S.* Typhimurium (Hiles *et al.*, 1987). Most uptake systems in addition require an extracytoplasmic substrate binding protein (BP) that delivers the substrate to the permease domain. In Gram-negative organisms, BPs are soluble proteins, which are located in the periplasm (Higgins, 2001). Gram-positive bacteria, on the other hand, lack an outer membrane and BPs are often lipoproteins that are anchored to the outer leaflet of the cytoplasmic membrane either through an N-terminal lipid anchor, a N-terminal transmembrane segment or even stably anchored to the transporter through fusion with the integral membrane domains (Biemans-Oldehinkel *et al.*, 2006).

Extrusion systems generally have a homo- or heterodimeric organization in which one protomer consists of one NBD fused to one TMD (constituting a half transporter). In eukaryotes, ABC-type transporters are often found as a single protein in which all four domains are present (full transporters). The overall organization of these four domains is both sufficient and necessary to form a functional ABC-type transporter (Higgins, 2001; Davidson *et al.*, 2008).

Proteins belonging to the ABC superfamily are specified by the presence of the nucleotide-binding domain (NBD), i.e. the ATP-binding cassette (See Chapters 2 and 3). In ABC-type transporters this domain acts as the motor domain that binds and hydrolyses ATP in order to elicit conformational changes in the TMDs, which facilitates transport of substrate across the cellular membrane. The NBD is evolutionarily the most conserved domain and contains highly conserved amino acid sequence motifs that are involved in nucleotide binding and co-ordination of the Mg^{2+} ions, i.e., the Walker A (GxxGxGKS/T, where x is any amino acid) and Walker B (ΦΦΦΦD, where Φ is a hydrophobic amino acid) motifs (Walker *et al.*, 1982; Davidson *et al.*, 2008).

The Walker A and B motifs determine the 3-D structural fold of the nucleotide-binding site and are characteristic for P-loop ATPases, which besides membrane transport are involved in diverse cellular processes such as receptor signaling, ATP synthesis, phosphoryl transfer reactions, motility, DNA translation as well as DNA maintenance and repair (Jones and George, 2004). The ABC-type transporter superfamily is further specified by the presence of the unique ABC signature motif, LSGGQ also known as the C-loop that also interacts with ATP. Other conserved structural elements are the Q-loop, which mediates interactions between the NBDs and TMDs (Diederichs *et al.*, 2000; Dawson and Locher, 2006; Hollenstein *et al.*, 2007) and the H-loop (or switch region) (Linton and Higgins, 1998).

At present, a large number of crystal structures of isolated NBDs, as well as of seven full-length ABC-type transporters, is available. The available full-length structures are of the ABC exporters Sav1866 (Dawson and Locher, 2006; Dawson and Locher, 2007) and MsbA (Ward *et al.*, 2007) and of the importers for vitamin B12, BtuCD (Locher *et al.*, 2002; Hvorup *et al.*, 2007) and the homologous HI1470/1471 (Pinkett *et al.*, 2007); for molybdate, ModBC (Hollenstein *et al.*, 2007; Gerber *et al.*, 2008); for maltose, MalFGK (Oldham *et al.*, 2007), and for methionine, MetNI (Kadaba *et al.*, 2008). A general feature of the isolated NBDs is that they have been crystallized in various dimeric forms (Kerr, 2002; Jones and George, 2004), some of which appear non-physiological. Interestingly, in the physiologically active dimeric state, the ABC signature motif of an NBD subunit complements the ATP binding site formed by the Walker A and B motifs of the neighbouring subunit. Co-crystallization studies with isolated NBDs defective in ATPase activity revealed that two molecules of ATP are bound at the interface of the dimer (Smith *et al.*, 2002; Zaitseva *et al.*, 2005). The head-to-tail arrangement of the NBD subunits was initially described for the Rad50 protein (Hopfner *et al.*, 2000), and more recently, also in the structures of complete ABC-type transporters (Locher *et al.*, 2002; Dawson and Locher, 2006; Hvorup *et al.*, 2007; Hollenstein *et al.*, 2007; Dawson and Locher, 2007; Pinkett *et al.*, 2007; Ward *et al.*, 2007; Oldham *et al.*, 2007; Gerber *et al.*, 2008; Kadaba *et al.*, 2008). This indicates that the overall fold and organization of the ABC-type NBDs is conserved. Biochemical evidence corroborates the typical organization of the NBD dimer and therefore it is considered to represent the physiological arrangement.

Other distinctive structural features of membrane transport proteins are the transmembrane domains (TMDs), which form a conduit that facilitates transport of dedicated substrate(s) across the cellular membrane. The TMDs are believed to contain binding sites for the solutes destined for transport. The dimeric transporter contains two closely associated TMDs each of which usually consist of six transmembrane α-helices. This general topology has been demonstrated biochemically and is found in the recently described crystal structures of the putative MDR transporter, Sav1866 from *Staphylococcus aureus* (Fig. 5.2B) (Dawson and Locher, 2006; Dawson and Locher, 2007), the lipid flippase, MsbA from *E. coli*, *Vibrio cholerae* and *Salmonella enterica* serovar Typhimurium (Fig. 5.2B) (Ward *et al.*, 2007), and a putative molybdate importer in complex with its binding protein, $ModB_2C_2A$ from *Archeoglobus fulgidus* (Hollenstein *et al.*, 2007). Some ABC-type transporters contain additional domains or auxiliary proteins such as the extracytoplasmic substrate binding proteins in bacterial uptake systems (Young and Holland, 1999; Locher *et al.*, 2002; Holland *et al.*, 2003; Hvorup *et al.*, 2007; Hollenstein *et al.*, 2007; Oldham *et al.*, 2007; Kadaba *et al.*, 2008), and/or contain additional transmembrane α-helices in the TMDs (Hvorup *et al.*, 2007; Pinkett *et al.*, 2007; Oldham *et al.*, 2007). Interestingly, there is little amino acid se-

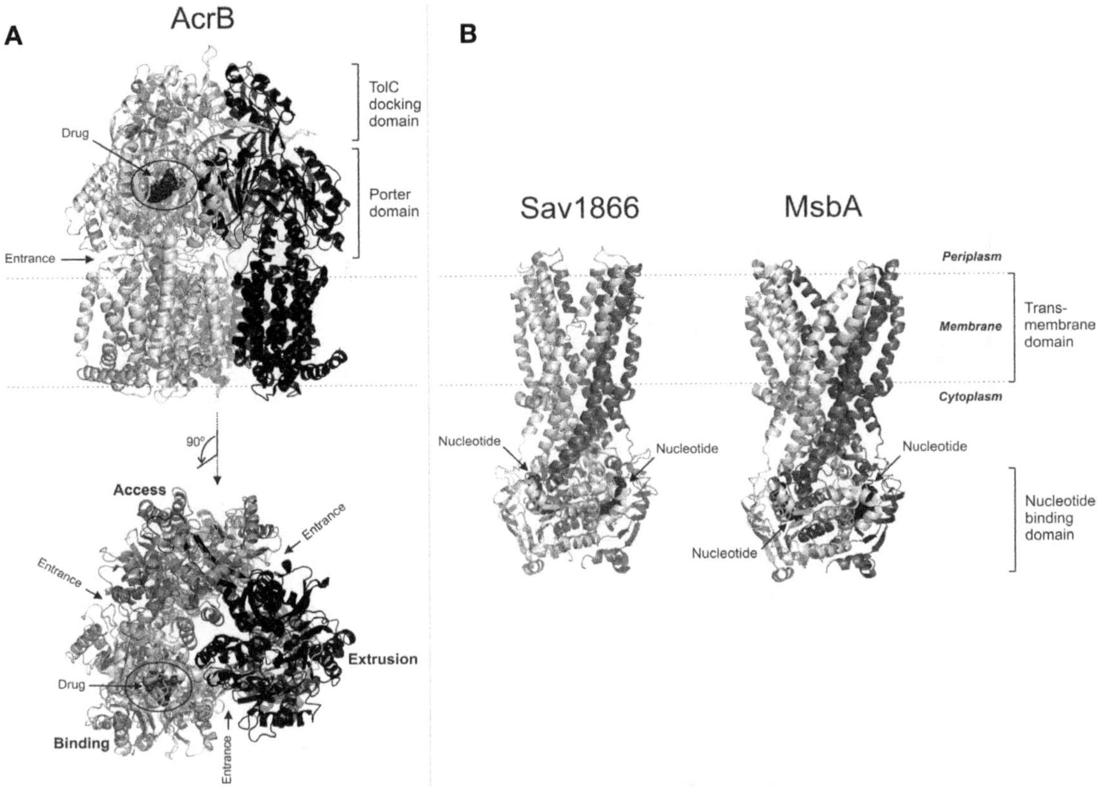

Figure 5.2 Structures of bacterial (putative) multidrug transporters. **A** Ribbons representation of the MDR protein AcrB of the RND efflux pump from *Escherichia coli* with the bound drug doxorubicin (2DR6); upper panel, side view; lower panel, top view. The doxorubicin molecule is depicted as black spheres. Side view shows the porter domain that carries the substrate molecule and the funnel-like domain that interacts with TolC. The top view shows the different structural conformations of the trimer subunits, each reflecting different stages of the transport cycle. AcrB-mediated drug efflux is thought to occur via a three-step transport cycle. Each subunit of the AcrB trimer undergoes sequential conformational changes in an alternating fashion, which facilitate access, binding and extrusion of the drug, respectively. In the access conformation, drug entrance is allowed from the periplasmic side, while the drug binding site is inaccessible. In the binding state, the entrance remains open and the binding pocket is expanded to allow substrate binding. At the final stage, the entrance is closed while the exit is opened. Drug extrusion into the top funnel part of AcrB is facilitated by shrinking of the drug binding pocket. It is anticipated that the conformational changes are coupled to proton translocation across the membrane. **B** Ribbons representation of the ABC-type transporters Sav1866 (left) from *Staphylococcus aureus* (2ONJ) and MsbA (right) from *Salmonella* Typhimurium (3B60). Sav1866 and MsbA form homodimers and share a very similar architecture. The Sav1866 and MsbA subunits are depicted in light and dark grey. The bound AMP-PNP nucleotides are shown as black spheres. This figure has been prepared using PyMol program from the respective PDB files.

quence identity between TMDs of different types of ABC transporters. This variability in amino acid sequence likely reflects the large diversity of the dedicated substrates. While most of the ABC-type transporters exhibit a high specificity for one dedicated substrate or for a group of closely related compounds, the ABC-type MDR transporters can accommodate a variety of unrelated substrates.

To date, ABC-type transporters involved in bacterial MDR are poorly characterized. Recent reviews on efflux based resistance mechanisms in bacteria mainly focus on the importance of MDR transporters driven by the proton/sodium motive force (Li and Nikaido, 2004; Poole, 2005). Here the focus will be on chromosomally encoded ABC-type transporters that are involved in bacterial multidrug resistance (Table 5.1). In addition, ABC-type transporters that are specific for a given drug or a closely related group of substrates and ABC transporters that are involved in self-immunity are discussed.

Table 5.1. Characterized bacterial MDR transporters belonging to the ABC-type superfamily.

Organism	Transporter	Origin[a]	Natural host[b] phenotype	Substrates[c]	References
Gram-positive					
Lactococcus lactis	LmrA	c	–	DA, EB, H, R6G, TPP^+, various antibiotics, Zn, $CdCl_2$	van Veen *et al.*, 1996; Putman *et al.*, 2000a
	LmrCD[d]	c	MDR	DA, EB, H, R6G, CA, GDCA, BCECF-AM	Lubelski *et al.*, 2004; Zaidi *et al.*, 2008
Lactobacillus brevis	HorA	p	Hop sensitive	EB, H, Hop-compounds	Sakamoto *et al.*, 2001
Bacillus subtilis	BmrA (YvcC)	c	EB sensitive	H, DO, AD	Steinfels *et al.*, 2002
Enterococcus faecalis	EfrAB[d]	c	–	AC, EB, SO, DAPI, DA, DO, NB, AK, DC, NF, CF, TPP^+	Davis *et al.*, 2001; Lee *et al.*, 2003b
	Abc7	c	MDR	DA, DO, EB, OF, CM	
	Abc11	c	MDR	PT, CH	
	Abc16	c	MDR	AZ, EM, CR	
	Abc23	c	MDR	CL, LM, VM, SC	
Oenococcus oeni	OmrA	c	ND	Ethanol, SL, $CdCl_2$, Zn, $HgCl_2$	Bourdineaud *et al.*, 2004
Streptococcus pneumoniae	PatA/PatB[d]	c	MDR	AC, BB, EB, CF, NF, NB, SM, CL, Reserpine	Robertson *et al.*, 2005; Marrer *et al.*, 2006a; Marrer *et al.*, 2006b
Bifidobacterium breve	BbmAB[d] (AbcAB)	c	ND	Nisin, PMB, H	Margolles *et al.*, 2006
Serratia marcescens	SmdAB[d]	c	ND	NF, TC, DAPI, H	Matsuo *et al.*, 2008
Staphylococcus aureus	Sav1866	c	ND	EB	Velamakanni *et al.*, 2008

Gram-negative					
Escherichia coli	MacAB	c	Macrolide sensitive	EM, CR, O, AZ	Sulavik *et al.*, 2001; Nishino and Yamaguchi, 2001
	MsbA	c	EB resistant	EB, H	Reuter *et al.*, 2003
Vibrio cholerae	VcaM	c	ND	H, TC, CF, NF, OF, DA, DO, DAPI, TPP^+, R6G, EB, AO	Huda *et al.*, 2003
Corynebacterium striatum	TetAB	p	ND	OX, TC, OTC	Tauch *et al.*, 1999
Mycoplasmas					
Mycoplasma hominis	MD1/MD2	c	ND	EB, CF	Raherison *et al.*, 2005

AC, Acriflavine; AD, 7-Aminoactinomysin D; AK, Arbekacin; AO, Acridine orange; AZ, Azithromycin; BB, Berberine; BCECF-AM, Acetoxymethyl ester 2′;7′-Bis-(2-carboxyethyl)-5(and 6)-carboxyfluorescein; CA, Sodium cholate; CF, Ciprofloxacin; CH, Chlorhexidine; CL, Clindamycin; CM, Chloramphenicol; CR, Clarithromycin; DA, Daunomycin (daunorubicin); DAPI, 4′,6-diamidino-2-phenylindole; DC, Doxycycline; DO, Doxorubicin (adriamycin); EB, Ethidium bromide; EM, Erythromycin; GDCA, Glycodexycholate; H, Hoechst 33342; LM, Lincomycin; NB, Novobiocin; NF, Norfloxacin; O, Oleandomycin; OF, Ofloxacin; OTC; Oxytetracycline; OX, Oxacillin; PMB, Polymyxin B; PT, Pentamidine; R6G, Rhodamine 6G; SC, Synercid; SL, Sodium laureate; SGB, Streptogramin B; SM, Streptomycin; SO, Safranin O; TC, Tetracycline; TPP^+= Tetraphenylposphonium; VM, Virginiamycin M.

a Transporter origin; c, chromosome, p, plasmid.

b Function of transporter assessed in the natural host either by overexpression or disruption of the transporter with the corresponding phenotype; –, no phenotype; ND, not determined.

c Substrates identified following expression in natural or heterologous systems.

d Two half-transporters that assemble in a heterodimeric functional unit.

ABC-type MDR transporters in Gram-negative bacteria

The key structural feature that distinguishes Gram-negative bacteria from Gram-positive bacteria is the presence of a second cellular membrane. The auxiliary outer membrane of Gram-negative organisms constitutes a physical barrier that is relatively impermeable to lipophilic and amphiphilic compounds, which has been attributed to its phospholipid composition and the presence of lipopolysaccharides (Denyer and Maillard, 2002). The presence of such an impermeable outer membrane does, however, not sufficiently explain the high intrinsic level of drug resistance observed in many Gram-negative bacteria (Nikaido, 1996). One also has to take into account the substantial contribution of efflux pumps to drug resistance in these bacteria.

It has been well established that Gram-negative bacteria contain multiple MDR transporters that catalyse the efflux of drugs across the cytoplasmic membrane. In most cases these transporters interact with a periplasmic partner protein, called the membrane fusion protein (MFP), which in turn interacts with an outer membrane channel protein (OMP) (Nikaido, 1996; Li and Nikaido, 2004). Of these tripartite transporter systems, the RND-type multidrug transporter AcrAB-TolC from *E. coli* is characterized in most detail (Ma *et al.*, 1993; Ma *et al.*, 1995). AcrB is the integral membrane part of the transporter, and, together with the MFP AcrA and the OMP TolC constitute a pmf-dependent MDR pump that plays a major role in both the intrinsic and acquired drug resistance of Gram-negative bacteria (Li and Nikaido, 2004; Piddock, 2006a; Piddock, 2006b). The phenotype caused by the deletion of AcrAB was already described in 1965 as a locus disruption which renders *E. coli* sensitive to a wide variety of unrelated compounds including basic dyes, detergents and hydrophobic antibiotics (George, 1996).

Recently, the 3D-structure of the membrane component AcrB with (Fig. 5.2A) or without drugs bound to it has been elucidated (Murakami *et al.*, 2006; Seeger *et al.*, 2006). These structures for the first time provided direct information on the actual mechanism of drug extrusion by an RND-type MDR pump. AcrB forms trimers in the cytoplasmic membrane with each subunit exhibiting a conformation different from the other (Fig. 5.2A). These three conformations are thought to represent consecutive states in a transport cycle, i.e. initial entry, tight binding and final release of the substrate (into the funnel TolC). From these structures it is also suggested that the RND-type transporters are likely to pick up the substrate from the periplasmic space and/or from the cytoplasmic membrane (Murakami *et al.*, 2006; Seeger *et al.*, 2006). In Gram-negative bacteria these RND-type efflux systems are considered the main contributors to clinically significant multidrug resistance (Li and Nikaido, 2004; Poole, 2004; Piddock, 2006a; Piddock, 2006b). In addition, various members of the p(s)mf-dependent superfamilies MFS, MATE and SMR have been implicated in drug resistance in Gram-negative bacteria (Poole, 2004; Piddock, 2006a; Piddock, 2006b). In contrast, only very few members of the ABC family have been reported to be involved in resistance to drugs in Gram-negative bacteria (Li and Nikaido, 2004; Poole, 2004). In this respect, a systematic gene disruption study (Sulavik *et al.*, 2001) and an expression study of all putative drug transporters (Nishino and Yamaguchi, 2001) did not identify a single ABC-type transporter involved in multidrug resistance in *E. coli*. However, one ABC-type transporter was implicated in single-drug resistance (see also further down in the chapter) (Nishino and Yamaguchi, 2001). Nevertheless, a small number of ABC-type transporters involved in single- or multidrug resistance has been characterized in Gram-negative bacteria and these will be discussed below.

Escherichia coli

MacAB

The *E. coli* genome encodes 37 putative drug transporters (19 MFS, 7 RND, 7 ABC, 3 SMR and 1 MATE) (Nishino and Yamaguchi, 2001). These predicted efflux pumps were functionally expressed in a drug sensitive *E. coli* strain that lacks the RND pump AcrAB. Of the pumps tested, twenty conferred considerable resistance to selected drugs. However, of the seven potential ATP-type drug transporters only YbjYZ conferred appreciable drug resistance to *E. coli*. YbjYZ specifically conferred resistance to macrolides composed of 14- and 15-membered lactones, such as erythromycin and clarithromycin (Nishino and Yamaguchi, 2001). The macrolide-specific YbjYZ pump was subsequently renamed MacAB (Kobayashi *et al.*, 2001). In addition, it was demonstrated that the periplasmic MFP MacA stimulates the ATPase activity of the integral membrane protein MacB by 45-fold (Tikhonova *et al.*, 2007). *In vivo*, MacAB requires the interaction with the outer membrane channel protein TolC for its full function (Kobayashi *et al.*, 2001; Tikhonova *et al.*, 2007). However, deletion of the *macAB* (*ybjYZ*) genes in *E. coli* failed to show a significant contribution of this transporter to the intrinsic drug resistance (Sulavik *et al.*, 2001). Since *E. coli* possesses multiple MDR pumps with overlapping substrate specificities it may very well be that these transporters may compensate for the lack of MacAB.

Recently, a role of MacAB in the secretion of heat-stable enterotoxin II (STII) by enterotoxigenic *E. coli* was demonstrated (Yamanaka *et al.*, 2008). Secretion of STII was largely impaired in a *macAB* knockout strain, while the toxin molecules were found to accumulate in the periplasm. Plasmid-based expression of *macAB* subsequently restored the secretion of STII, indicating that MacAB mediates the translocation of STII from the periplasm into the extracellular space. This process involves the OMP TolC. Homologues of MacAB have been identified and/or characterized in *Neisseria gonorrhoeae* (Rouquette-Loughlin *et al.*, 2005), *S.* Typhimurium (Nishino *et al.*, 2006) and *Bacillus subtilis* (Butcher and Helmann, 2006). Loss of *macAB* in a clinical isolate of *N. gonorrhoeae* exhibited decreased resistance to azithromycin and erythromycin. Interestingly, deletion of the *macAB* genes of *Salmonella* attenuated the virulence of this organism. The *macAB* homologue of *B. subtilis* is encoded by the operon *yknXYZ* and is involved in self-immunity to the endogenously produced antimicrobial peptide SdpC. These findings indicate that the ABC-type transporter MacAB might have physiological functions unrelated to the efflux of macrolides.

MsbA

The ABC-type transporter MsbA is found in Gram-negative bacteria and has an essential function in biogenesis of the outer membrane, i.e. it transports lipid A and lipopolysaccharide (LPS) from the inner leaflet to the outer leaflet of the inner (cytoplasmic) membrane (Karow and Georgopoulos, 1993; Zhou *et al.*, 1998; Doerrler *et al.*, 2004). MsbA shares significant sequence identity with the human multidrug transporter p-glycoprotein and LmrA from *L. lactis* (discussed below). The drug transport function of MsbA was studied by means of *E. coli* WD2, which harbours a temperature sensitive mutant of MsbA, resulting in ethidium sensitivity at elevated temperatures. Interestingly, at the non-permissive temperature of 42°C, the ethidium sensitivity could be reversed by the expression of wild-type MsbA, which was shown to support the efflux of ethidium from cells. In addition, MsbA was shown to interact with the DNA-intercalating dyes daunomycin and Hoechst 33342, indicating a possible role in drug resistance (Reuter *et al.*, 2003). The crystal structures of MsbA have been revisited (Ward *et al.*, 2007). Structures of MsbA were obtained from the closely related bacteria, *E. coli*, *V. cholerae* and *S.* Typhimurium and exhibited three different conformations; (i) a nucleotide-free open state, (ii) a nucleotide-free, closed state and (iii) a nucleotide-bound, closed state (Fig. 5.2B) (Ward *et al.*, 2007). The large structural differences observed in these crystal structures are thought to reflect various intermediate states of the transporter during a transport cycle *in vivo*. The nucleotide-bound state of MsbA proved to be very similar to that of Sav1866 (Fig. 5.2B). In the nucleotide-free open conformation of MsbA the NBDs are separated by approximately 50 Å, while the ATP-binding half sites still face each other. This structure exhibits a large wedge-shaped chamber that faces the cytoplasm, which may accommodate the substrate. It is of interest to determine whether this conformation of MsbA indeed represents the substrate acceptor state or that it is a physiologically irrelevant state of the transporter. Site-directed spin-labelling and electron paramagnetic spin resonance (EPR) studies to probe the conformation of MsbA are in line with the open, nucleotide-free structure (Dong *et al.*, 2005; Borbat *et al.*, 2007). These studies indicate a large change in the accessibility of the internal chamber to water molecules upon nucleotide binding, i.e. ATP-binding induces the chamber to close. In addition, ATP hydrolysis can drive a relative motion of the NBDs of about 30 Å in both detergent-solubilized and reconstituted MsbA, suggestive of a catalytic intermediate in which the NBD dimeric interface is disrupted and the two NBDs are well separated. The lipid flippase MsbA is required for the viability of *E. coli*, indicating that its primary function, i.e. lipid transport, is essential to the cell. The contribution of MsbA to drug resistance in the natural host *E. coli*, however, remains to be determined. The drug efflux activity of MsbA is likely a secondary and possibly opportunistic activity of this transporter that is likely to be more pronounced when the transporter is expressed at high level.

Vibrio cholerae

VcaM

Cholera is a severe diarrhoeal disease caused by toxigenic and drug-resistant strains of the enteric bacterium *Vibrio cholerae* (Huda *et al.*, 2003). The *V. cholera* genome contains an ORF, *vcaM*, which encodes a putative ABC-type transporter that has considerable amino acid sequence homology to human P-glycoprotein and lactococcal LmrA. Expression of *vcaM* in a drug-hypersensitive *E. coli* strain that lacks the MDR pumps AcrAB and YdhE conferred protection against a wide range of unrelated drugs including tetracycline, norfloxacin, ciprofloxacin, doxorubicin, daunomycin, DAPI, Hoechst 33342 and rhodamine 6G (Table 5.1). VcaM was shown to mediate the efflux of Hoechst 33342 and doxorubicin from *E. coli* cells, and this activity is inhibited by reserpine and vanadate (Huda *et al.*, 2003). To our knowledge, VcaM is the first and thus far only reported example of an ABC-type transporter from a Gram-negative bacterium that is truly multidrug specific. However, the role of VcaM in multidrug resistance of the natural host *V. cholerae* has thus far not been addressed. Since tetracycline and fluoroquinolones are commonly used for the treatment of patients infected with *V. cholerae*, VcaM mediated efflux of these antibiotics would be a serious threat to patients with cholera.

ABC-type MDR transporters in Gram-positive bacteria

Recent reports and functional genomics predictions indicate that the role of ABC-type transporters in bacterial drug resistance might be underestimated (Lubelski *et al.*, 2007; Davidson *et al.*, 2008). Thusfar, the majority of the MDR transporters from Gram-positive bacteria that have been characterized are secondary transporters belonging to the MFS superfamily such as QacA (Tennent *et al.*, 1989) and NorA (Yoshida *et al.*, 1990) from *Staphylococcus aureus*, the NorA homologues Bmr and Blt from *Bacillus subtilis* (Neyfakh *et al.*, 1991; Ahmed *et al.*, 1995), PmrA from *Streptococcus pneumoniae* (Gill *et al.*, 1999) and LmrP from *Lactococcus lactis* (Bolhuis *et al.*, 1995). However, ABC-type MDR pumps have also been identified in Gram-positive bacteria and several of them have been described in detail. One of the best-studied primary active MDR transporters is LmrA of *L. lactis*, which was the first prokaryotic ABC-type MDR transporter discovered (van Veen *et al.*, 1996).

Lactococcus lactis

Lactococcus lactis is a non-pathogenic lactic acid bacterium that has served as a convenient model organism to study bacterial multidrug drug resistance (MDR). The genome of *L. lactis* contains 40 putative drug transporter genes of which only a few have been characterized in detail. In 1994, Bolhuis *et al.* presented first evidence of an efflux based mechanism of drug resistance in *L. lactis* (Bolhuis *et al.*, 1994). When *L. lactis* MG1363 cells were grown in the presence of increasing sub-lethal concentrations of toxic compounds like ethidium bromide, rhodamine 6G or daunomycin they not only readily developed resistance to the drug used for selection, but they also exhibited cross-resistance to structurally unrelated drugs to which they were previously not exposed (Bolhuis *et al.*, 1994). The acquired MDR phenotype of these *L. lactis* mutants was found to be a stable trait and, at least in part, due to active efflux of the drugs. Moreover, bioenergetics studies indicate that the efflux activity involved both ATP- and pmf-dependent extrusion system(s) (Bolhuis *et al.*, 1994).

Subsequent studies were aimed at identifying the lactococcal transporters that are involved in (multi)drug resistance. At present, three transporters have been implicated in MDR in *L. lactis*. These are the ABC-type transporter LmrA (van Veen *et al.*, 1996), the MFS-type transporter, LmrP (Bolhuis *et al.*, 1995) and the recently described heterodimeric ABC-type MDR pump,

LmrCD (Lubelski *et al.*, 2004). In addition, sequence analysis of the genome of *L. lactis* IL1403 indicates that it encodes an additional 36 putative drug transporters (Lubelski *et al.*, 2006b), of which 16 are ABC-type proteins. A similar large number of putative MDR transporters are found in the recently completed genome of the closely related *L. lactis* MG1363 strain (Wegmann *et al.*, 2007).

LmrA

The first ABC-type MDR transporter discovered in bacteria is LmrA (Lactococcus multidrug resistance protein ATP) from *L. lactis*. LmrA is a structural and functional homologue of the human MDR transporter P-glycoprotein. The LmrA polypeptide constitutes a half-transporter that consists of 590 amino acids exhibiting a topology of six α-helical transmembrane segments in the N-terminal hydrophobic domain, followed by a hydrophilic domain containing a single ATP-binding site. While P-glycoprotein is encoded by a single polypeptide comprising the full transporter, LmrA constitutes a half transporter that functions as a homodimer. Fusion of two LmrA monomers by means of a 'linker peptide' derived from P-glycoprotein yields a functional protein (van Veen *et al.*, 2000). Substitution of a single amino acid residue important for ATPase activity in either one of the two half-transporter domains of the covalently linked LmrA dimer resulted in complete loss of ATPase activity. In addition, a lipid flippase assay demonstrated a dominant negative effect of an NEM-inactivated single cysteine mutant of LmrA when co-reconstituted with wild-type LmrA in proteolipsomes (van Veen *et al.*, 2000). These observations are consistent with the notion that LmrA functions as a homodimer.

Several lines of research indicate that LmrA is a transporter with a function and substrate spectrum similar to that of the human MDR transporter P-glycoprotein. Expression of LmrA in the drug hypersensitive strain of *E. coli* CS1562, which lacks the OMP TolC, confers protection against a remarkable wide range of toxic agents. These include drugs like ethidium bromide, daunomycin, rhodamine 6G, and TPP^+ and antibiotics like aminoglycosides, lincosamides, macrolides, quinolones, streptogramins, tetracyclines and chloramphenicol (Putman *et al.*, 2000a). Recently, expression of LmrA in *E. coli* strain CS1562 (Δ*tolC*) was shown to protect these cells from toxic effects of high concentrations of salt, sodium laurate, ethanol (Bourdineaud *et al.*, 2004), and cadmium (Achard-Joris *et al.*, 2005). Transient transfection of human lung fibroblast cells with LmrA increased the resistance of these cells to a variety of compounds which are typical substrates of P-glycoprotein, including anthracyclines such as daunomycin and doxorubicin, vinca-alkaloids such as vinblastine and vincristine, and cytotoxic agents such as ethidium bromide, rhodamine 123 (van Veen *et al.*, 1998). Complementary transport assays indicate that *E. coli* cells expressing LmrA exhibit a reduced rate of ethidium accumulation (van Veen *et al.*, 1996). In addition, LmrA-mediated transport of daunomycin and Hoechst 33342 could be demonstrated in inside-out membrane vesicles prepared from *E. coli* or *L. lactis* cells, respectively (van Veen *et al.*, 1996; Margolles *et al.*, 1999). This transport activity could be blocked by the established MDR-pump inhibitors verapamil and reserpine. Similarly, reconstituted purified LmrA in proteoliposomes showed transport activity of Hoechst 33342 and the fluorescently labelled lipid phosphatidylethanolamine (Margolles *et al.*, 1999). In addition, LmrA when expressed at 42°C has been shown to partially restore the growth of *E. coli* WD2, which harbours a temperature sensitive mutant of the lipid A transporter MsbA (Reuter *et al.*, 2003), suggesting that LmrA can also transport lipids. Conversely, MsbA expressed in *L. lactis* improved the resistance to erythromycin and inside-out vesicles containing MsbA showed a reduced rate of accumulation of Hoechst 33342 and ethidium (Woebking *et al.*, 2005).

It is important to note that the proposed role of LmrA in drug transport mostly originates from studies in heterologous systems when the system is overexpressed at a high level. The exact role of LmrA in multidrug resistance in *L. lactis* has however, remained unclear, as the protein expressed at wild-type levels does not appear to confer any significant level of resistance. Plasmid-based overexpression of LmrA in *L. lactis* yields levels up to 30% of total membrane protein and such cells have been reported to show an enhanced activity of ethidium efflux (Margolles *et al.*, 1999). In contrast, other studies failed to detect any significant activity above the already high endogenous ethidium extrusion levels (Bolhuis *et al.*, 1994; Lubelski *et al.*, 2004). The fluorescent compound Hoechst 33342 is a substrate for many MDR transporters and is widely used to demonstrate transport activity. Inside-out membrane vesicles containing high levels of LmrA show a low level of ATP-dependent Hoechst 33342 transport activity. However, *L. lactis* LmrA knockouts did not show an appreciable difference in drug susceptibility profiles (Davis *et al.*, 2001) and plasmid-based overexpression of LmrA in *L. lactis* cells lacking genomic *lmrA* did not confer protection against the toxic effect of Hoechst 33342 (van den Berg van Saparoea *et al.*, 2005).

Van den Berg van Saparoea *et al.* (2005) demonstrated that the LmrA dependent transport activity of Hoechst 33342 in membrane vesicles was not coupled directly to the hydrolysis of ATP by LmrA, but rather relates to the generation of a pH gradient driven by the hydrolysis of ATP by the F_1F_0-ATPase (van den Berg van Saparoea *et al.*, 2005). Because of the change in pH, Hoechst 33342 repartitions across the membrane in a LmrA-dependent manner but by a mechanism that does not require the ATPase activity of LmrA itself. Indeed, liposomes containing purified reconstituted LmrA exhibited a very low and barely detectable level of ATP-dependent Hoechst 33342 transport (Margolles *et al.*, 1999).

It is clear that the numerous studies on LmrA still leave us with many unanswered questions about its function. However, there is compelling evidence that LmrA does not function as a major MDR transporter in *L. lactis*. This is best illustrated by the fact that genomic deletion of *lmrA* does not render cells sensitive to drugs and that overexpression of LmrA in the *L. lactis* Δ*lmrA*Δ*lmrCD* strain, which is sensitive to drugs (see next section on LmrCD) does not confer protection against drugs. In contrast, high-level expression of LmrA in a drug sensitive strain of *E. coli* confers protection against a remarkable wide range of toxic agents (Putman *et al.*, 2000a). Several aspects concerning this discrepancy need to be considered. First, the apparent lack of a phenotype of the *L. lactis lmrA* deletion strain may be attributed to compensatory mechanisms of resistance as for instance the increase in expression levels of other lactococcal drug transporters, although this possibility has not been demonstrated to occur. Second, the observed drug extrusion activity of LmrA in heterologous hosts is observed only upon high-level overexpression and may therefore reflect only a relatively minor extrusion activity, or, may even relate to a secondary cellular responses to the high expression. The latter may for instance involve an increased expression of indigenous drug transporters. In this respect, the *E. coli* genome encodes 37 putative drug transporters. Summarizing, at the normal, physiological expression levels, LmrA does not seem to provide resistance to drugs in *L. lactis*. The natural substrate(s) of LmrA and its physiological role, therefore, remain to be determined.

Despite its poorly defined physiological role, the molecular mechanism by which LmrA mediates drug transport has been studied in detail. As such, LmrA has become a model protein and its structural and mechanistic properties are considered exemplary of ABC-type MDR transporters. Based on photoaffinity labelling and drug equilibrium binding experiments, LmrA was shown to contain two cooperative drug binding sites, i.e. a high affinity drug-binding site facing the cytoplasm and a low affinity drug-release site at the outer leaflet of the membrane (van Veen *et al.*, 2000). LmrA functions as a homodimer and the two binding sites of each subunit were suggested to be allosterically linked in an alternating fashion. The alternating two-site transport cycle, or so-called two-cylinder engine mechanism proposes a displacement of the high affinity binding site from the inside to the outside of the cell accompanied by a change from a high into a low affinity binding site, thereby forcing substrate release. This conformational transition of the LmrA half-transporter is driven by the hydrolysis of ATP by its NBD, which is accompanied by the converse transition in the adjacent LmrA subunit, i.e. the low affinity outward facing binding site reverts to a high affinity drug-binding site facing the cytoplasm, with concomitant release of ADP and P_i. The interconversion of the drug-binding and drug-release sites occurs via a transition state intermediate which is mimicked by the vanadate-trapped ADP-bound state of LmrA (van Veen *et al.*, 2000; Ecker *et al.*, 2004). Interestingly, only the low affinity site of this transition state intermediate was found to be accessible to drugs whereas the high affinity drug-binding site appeared to be occluded. Although binding sites are presented in separate transmembrane domains, it is important to note that the drug binding sites may very well be present at the interface between the transmembrane domains of the LmrA subunits. Thus, according to this model, the LmrA dimer would contain four drug-binding sites that function as pairs in parallel in an alternating fashion. This model implies a stoichiometry of one substrate molecule transported per one molecule of ATP hydrolysed and resembles the alternating catalytic sites mechanism of ATP hydrolysis suggested for ABC-type transporters (Senior and Gadsby, 1997).

In addition to the two-cylinder model, different transporter-based drug efflux mechanisms have been proposed. In 2006, Dawson and Locher solved the structure of the ABC-type transporter Sav1866 from the human pathogen *S. aureus* (Dawson and Locher, 2006). The amino acid sequence of Sav1866 shows a high sequence identity to the human multidrug resistance protein MDR1 and LmrA from *L. lactis*. Sav1866 may specify an MDR transporter as reconstitution studies demonstrated that its ATPase activity is stimulated by various agents including the anti-cancer drugs doxorubicin and vinblastin and the fluorescent dye Hoechst 33342. A recent study furthermore shows that overexpression of Sav1866 mediates ethidium efflux in *L. lactis* (Velamakanni *et al.*, 2008). It is suggested that drug transport by Sav1866 occurs by an 'alternating access and release' mechanism. Sav1866 was crystallized both in the presence of ADP and AMP-PNP (Dawson and Locher, 2006; Dawson and Locher, 2007). In both structures nucleotide is stacked between the Walker A motif (P-loop) of one subunit and the ABC signature motif of the opposite subunit, while the transmembrane domains of interacting Sav1866 subunits form a central chamber that is inaccessible from the inner leaflet of the lipid bilayer. The AMP-PNP structure (Fig. 5.2B) is in line with the notion that the ATP-bound state of the transporter is coupled to the outward-facing conformation of the transporter required for drug release. For drug binding, an inward-facing conformation of the transporter is proposed. ATP binding and hydrolysis would supply the energy to drive the conformational transition from the inward-facing drug binding state to the outward-facing drug release state, while the central chamber may function as a conduit for drug-translocation. This model differs from the two-cylinder engine model that supposes two re-orienting binding sites, whereas with Sav1866 there is only a single re-orienting binding site. An interesting structural feature of Sav1866 is the tight and intertwined association of the transmembrane domains of the two half-transporters (Fig. 5.2B). The structure of the lipid flippase MsbA when bound to nucleotide shows similar intimate interactions (Fig. 5.2B) (Ward *et al.*, 2007). Such architecture is thought to facilitate close association of the two NBDs during the catalytic cycle and may prevent complete dissociation of the NBDs. This view differs from the ATP-switch model that has been postulated for ABC-type transporters (Higgins and Linton, 2004). In the ATP-switch model, ATP binding induces NBD dimer formation, while ATP hydrolysis drives dissociation. Hydrolysis of the two ATP molecules is thought to occur sequentially. The structural rearrangements within the NBDs trigger further conformational changes in the TMDs that mediate transport.

LmrCD

L. lactis readily acquires an MDR phenotype when challenged with increasing concentrations of ethidium bromide, rhodamine 6G, daunomycin or cholate (Bolhuis *et al.*, 1994; Yokota *et al.*, 2000). An ATP-dependent transporter has been implicated with the acquired resistance but its identity initially remained obscured. However, transcriptome analysis revealed that these four MDR strains exhibited an increased expression of *lmrCD* as compared to the wild-type strain (Lubelski *et al.*, 2006b). LmrC and LmrD are ABC-type half transporters proteins that share considerable amino acid sequence identity to other members of the ABC-type MDR transporter family, including LmrA, human P-glycoprotein and BmrA from *Bacillus subtilis*. The deletion of the *lmrCD* genes renders *L. lactis* more susceptible to a wide range of unrelated toxic compounds, including ethidium bromide, rhodamine 6G, daunomycin, cholate and Hoechst 33342 than the wild-type cells (Lubelski *et al.*, 2006b; Zaidi *et al.*, 2008). Plasmid-based expression of *lmrCD* subsequently restored the drug resistant phenotype (Lubelski *et al.*, 2006b; Zaidi *et al.*, 2008). In contrast, mutants of LmrCD that lacked ATPase activity were unable to rescue the cells (Lubelski *et al.*, 2006a). These data demonstrate that LmrC and LmrD are involved in the intrinsic drug resistance of *L. lactis*.

LmrC and LmrD each are composed of a TMD consisting of 6 TMS, which is connected to the NBD located at the carboxyl-terminal. LmrC and LmrD can be co-purified in a 1:1 ratio as a complex that exhibits a high endogenous ATPase activity, while the separately purified subunits are essentially devoid of ATPase activity (Lubelski *et al.*, 2004). These data indicate that LmrC and LmrD are functionally interacting proteins. This is further corroborated by the fact *L. lactis* cells show only high level of drug extrusion when LmrC and LmrD are expressed together, whereas no increased activity is observed when these subunits are expressed separately. Transport assays using intact cells indicated that expression of LmrCD reduces the cellular accumulation of the fluorescent dyes BCECF, daunomycin, ethidium and Hoechst 33342 (Lubelski *et al.*, 2004). In addition, it was shown that LmrCD mediates Hoechst 33342 transport in membrane vesicles (van den Berg van Saparoea *et al.*, 2005). This process occurs independently of the pmf and is abrogated in ATPase mutants of the LmrCD complex.

Amino acid sequence analysis of LmrC and LmrD reveals that their nucleotide binding sites are distinct, i.e. the heterodimeric LmrCD contains partially degenerated NBD sequence motifs. Important amino acids that commonly interact – either directly or via a water molecule – with the γ-phosphate of ATP bound in the nucleotide active site, such as the glutamate following the Walker B motif, a histidine in the H-loop and the first glycine in the ABC signature motif, are conserved in the NBD of LmrD. However, in the NBD of LmrC, the aforementioned catalytic residues are substituted by aspartate, glutamine and valine, respectively. Thus, the LmrCD heterodimer contains structurally asymmetric ATP binding/hydrolysis sites (Lubelski *et al.*, 2006a).

Together with the histidine in the H-loop, the conserved glutamate directly following the Walker B motif is suggested to play an essential role in ATP hydrolysis as a catalytic base positioning the water molecule that attacks the γ-phosphate of bound ATP (Lubelski *et al.*, 2006a). In LmrD, this residue corresponds to Glu587, while the corresponding residue in LmrC is an aspartate at position 495. Both residues were substituted for their respective carboxamides, and the effect thereof on ATP binding and hydrolysis as well as drug-transport activity was determined. LmrCD complexes containing LmrC (D495N) exhibited significant ATPase activity, but were found to be instable, whereas LmrCD complexes containing LmrD(E587Q) completely lacked ATPase activity. The asymmetric nature of the NBDs in the LmrCD heterodimer was further investigated by azido-ATP photoaffinity studies. Interestingly, under conditions non-permissive for hydrolysis, this nucleotide analogue labelled predominantly LmrC, while in the presence of AlF_x to trap the hydrolysed azido-ATP at the nucleotide-binding site, LmrD was mostly labelled (Lubelski *et al.*, 2006a). Together these data indicate that the heterodimeric transporter LmrCD contains a canonical ATP active site in the NBD of LmrD, presumably responsible for a high ATPase activity, and a slow hydrolysing non-canonical ATP active site in LmrC. Transport studies subsequently demonstrated that both the single mutant LmrC/LmrD(E587Q) and the double mutant LmrC(D495N)/LmrD(E587Q) were defective in ethidium, BCECF and Hoechst 33342 transport. On the other hand, *L. lactis* cells expressing LmrC(D495N)/LmrD showed a transport activity similar to that of cells overexpressing the wild-type LmrCD proteins (Lubelski *et al.*, 2006a).

The heterodimeric organization of an active ABC-type transporter is a more common feature in eukaryotes. The *Drosophila* white, brown and scarlet genes (Ewart and Howells, 1998) and the fatty acid peroxisomal transporter genes (Liu *et al.*, 1999) all specify ABC-type half-transporters that typically associate to form various heterodimers with different properties. It can be envisioned that such hybrid ABC-type transporters may exist also in bacteria. Regarding MDR efflux pumps, recruitment of different half-transporters would represent an elegant means to allow the formation of efflux pumps with different substrate specificity.

Most transport studies on bacterial MDR efflux pumps are carried out with compounds that are usually not encountered by the bacteria in their natural environment, such as fluorescent dyes or detergents. Since drug transporters are ubiquitous in nature it is thought that many of them play an active role in the extrusion of host-derived compounds and/or offer protection against noxious agents present within their ecological niche. At present, our knowledge of natural substrates of MDR transporters is rather limited. Pioneering studies to identify natural substrates focused on enteric bacteria that inhabit the gastrointestinal (GI) tract of humans and animals. Bile acids are important toxic agents that are frequently encountered in the GI tract. In Gram-negative bacteria, MDR efflux pumps that belong especially to the RND family, such as AcrAB, have been implicated in bile acid resistance (Thanassi *et al.*, 1997; Piddock, 2006b). In Gram-positive bacteria the exact mechanisms of bile acid resistance are less clear, but the involvement of efflux pumps based on secondary transport have also been proposed (Gunn, 2000). However, recently it was established that in *L. lactis* the ABC-type MDR transporter LmrCD plays a major role in the resistance to cholate (Lubelski *et al.*, 2006b; Zaidi *et al.*, 2008), which is an important constituent of human bile. DNA microarray analysis indicated that a cholate resistant MDR strain of *L. lactis* (Yokota *et al.*, 2000) showed elevated expression of *lmrCD* (Lubelski *et al.*, 2006b), suggesting a role in cholate resistance. In line with this, cholate challenged wild-type *L. lactis* cells show a transient induction of *lmrCD* (Zaidi *et al.*, 2008). Moreover, deletion of *lmrCD* renders *L. lactis* sensitive to the bile acids cholate and glycodeoxycholate (Lubelski *et al.*, 2006b; Zaidi *et al.*, 2008), while plasmid-based expression of LmrCD in sensitive cells provides high-level resistance to these bile acids (Zaidi *et al.*, 2008). Transport assays using radiolabelled cholate demonstrated that only those cells containing chromosomally or plasmid-encoded *lmrCD* genes were able to actively extrude the cholate from the cell (Zaidi *et al.*, 2008). These data indicate that LmrCD fulfils a central role in bile acid resistance. *Lactococci* are not considered part of the human gut, however, homologues of LmrCD are widely distributed among gut bacteria such as *lactobacilli* and *bifidobacteria*. As these organisms are frequently exposed to bile acids, it is conceivable that the LmrCD-like transporters have evolved into dedicated bile acid transporters that may contribute to both colonization and survival in the intestine.

In addition to the cholate resistance, LmrCD plays a major role in drug resistance in *L. lactis*. However, it has remained unclear whether other putative MDR transporters contribute to resistance. For instance, DNA micro-array analysis demonstrated that the MDR strains selected by rhodamine 6G (Bolhuis *et al.*, 1994) or cholate (Yokota *et al.*, 2000) show besides increased levels of *lmrCD*, also increased expression of *llmg2513* (formerly named *yxbD*) (Lubelski *et al.*, 2006b), which encodes a putative MDR transporter of the MFS family. Attempts to re-isolate ethidium and daunomycin resistant strains from the drug sensitive *L. lactis* Δ*lmrCD* strain have been unsuccessful, suggesting that at least for these drugs there is no functional alternative extrusion route (Lubelski *et al.*, 2006b). In contrast, when the drug-sensitive *L. lactis* Δ*lmrCD* cells were challenged with cholate, a resistant strain (Δ*lmrCD*^{R}) was obtained that compared to the parental strain showed: (i) a significantly improved resistance to cholate, (ii) morphological changes, and (iii) an altered susceptibility to antimicrobial peptides (Zaidi *et al.*, 2008). Remarkably, transcriptome analysis of the Δ*lmrCD*^{R} strain indicated that none of the indigenous putative MDR transporters exhibited an altered gene expression, including *llmg2513* and the well-characterized drug transporters LmrA and LmrP. In contrast, a multitude of genes involved in stress responses, cell envelope biogenesis and cell metabolism were differentially expressed (Zaidi *et al.*, 2008). Moreover, the Δ*lmrCD*^{R} strain was shown to be unable to actively extrude cholate from the cell (Zaidi *et al.*, 2008). The data indicate that in the absence of LmrCD, cholate resistance relies on a multitude of transport unrelated cellular responses and that LmrCD is the main system responsible for cholate extrusion in *L. lactis*.

Taken together, it seems that LmrCD is a major determinant of both intrinsic and acquired drug resistance in *L. lactis*. While many bacteria depend on secondary active transporters for drug resistance, *L. lactis* seems to rely on ABC-type MDR transporters.

Lactobacillus brevis

Lactobacillus brevis is a spoilage bacterium that is able to resist the bactericidal hop compounds present in beer. Hop compounds consist predominantly of iso-α-acids that act as ionophores, which dissipate the pH gradient across the cytoplasmic membrane and reduce the proton motive force of gram-positive bacteria (Sakamoto and Konings, 2003). A major determinant in hop-resistance is the *horA* gene, which is localized on a plasmid in hop-tolerant strains (Sakamoto and Konings, 2003). *HorA* encodes an ABC-type half-transporter that shares 53% amino acid sequence identity with LmrA from *L. lactis* and may specify an MDR transporter. Expression of HorA in *L. lactis* conferred resistance to hop compounds and reduced the cellular accumulation of the fluorescent dyes ethidium and Hoechst 33342. In addition, HorA-mediated transport of Hoechst 33342 transport in *L. lactis* inside-out membrane vesicles was inhibited by hop-compounds (Sakamoto *et al.*, 2001). It appears that HorA is an MDR transporter that confers protection to antimicrobial compounds used as food-preservatives. Interestingly, when the HorA pump is not available hop-resistance of *L. brevis* relies on a variety of transport unrelated cellular responses like changes in cell metabolism and cell wall composition (Behr *et al.*, 2006). These data implicate HorA as the main efflux pump for hop-compounds.

Bacillus subtilis

Bacillus subtilis is a bacterium commonly found in soil and has the ability to form a protective endospore, allowing the organism to tolerate extreme environmental conditions. *B. subtilis* exhibits an exceptionally versatile metabolism and its genome encodes many putative MDR transporters of which only few have been characterized. *BmrA* (formerly known as *yvcC*) encodes an ABC-type half-transporter that shares about 42% and 40% amino acid sequence identity with LmrA and HorA, respectively. Initially BmrA was characterized in *E. coli* (Steinfels *et al.*, 2002). Inside-out membrane vesicles containing BmrA exhibit a high ATPase activity inhibited by vanadate and show efflux of Hoechst 33342, doxorubicin and 7-aminoactinomycin D (Steinfels *et al.*, 2002; Steinfels *et al.*, 2004). Although expression of BmrA in a heterologous host is associated with MDR, this is less clear in *B. subtilis*. In *B. subtilis*, the *bmrA* gene is constitutively expressed, while no significant increase in expression is observed when cells are challenged with typical MDR substrates. However, *B. subtilis* cells lacking the *bmrA* gene showed reduced ethidium efflux, suggesting a role of BmrA in ethidium resistance (Steinfels *et al.*, 2004).

Cryo-electron microscopy of purified BmrA reconstituted in lipid bilayer revealed that the protein forms large 40 nm wide ring-shaped structures consisting of 12 identical units identified as dimers of BmrA. The BmrA homodimer contains a central open chamber located between the two BmrA subunits while the two NBDs exhibit asymmetry (Chami *et al.*, 2002). Time-resolved FRET experiments are in line with the notion that BmrA forms stable homodimeric complexes (Dalmas *et al.*, 2005). The mechanistic aspects of the ATPase cycle of BmrA have been studied in some detail. The conserved glutamate residue next to the Walker B site, which is assumed to play a critical role as the catalytic base required for ATP hydrolysis, was subjected to site-directed mutagenesis. All mutants lacked ATPase activity and showed a trapping of azido-ATP at the nucleotide-binding site independent of vanadate. In the wild-type BmrA vanadate traps the nucleotide in its hydrolysed state. These observations strongly suggest that the conserved glutamate residue is involved in catalysis (Orelle *et al.*, 2003).

Recently it was found that the addition of Mg-ATP prevented the formation of the typical ring-shaped BmrA structures or disrupted preformed ring-structures (Orelle *et al.*, 2008). The molecular basis responsible for such an effect was studied by means of different hydrolysis-deficient mutants of BmrA. While wild-type (and glutamate mutants defective in ATP hydrolysis) showed ring disruption upon addition of Mg-ATP, lysine Walker A mutants remained unaffected by the addition of Mg-ATP (Orelle *et al.*, 2008). Although the latter mutants still bound Mg-ATP, they do not exhibit or exhibit a very slow ATPase activity. It was suggested that ATP-driven dimerization of the NBDs of these mutants does not occur properly; while with wild-type BmrA ATP binding drives dimerization of the NBDs, which in turn triggers conformational changes that destabilize the ring-structures. These large conformational changes might reflect the 'power stroke' that likely is coupled to substrate transport across the membrane.

Enterococcus faecalis

Enterococci are major constituents of the gastrointestinal tract and are increasingly recognized as important nosocomial pathogens because they cause serious diseases like bacteraemia, endocarditis, urinary tract infections and wound infections and are often found to be resistant to antibiotic treatment (Lee *et al.*, 2003b). *E. faecalis* is resistant to a variety of toxic compounds, which has been attributed to efflux pumps. Analysis of the genome of *E. faecalis* revealed the presence of 34 putative MDR transporters, which include 23 ABC-, 9 MFS-, 1 MATE- and 1 SMR-type transporters (Davis *et al.*, 2001). A systematic single gene disruption study revealed that only four of the ABC-type MDR transporters contributed significantly to the intrinsic drug resistance of *E. faecalis* (Table 5.1). These genes specify 4 separate transporters that show little overlap in their substrate specificities (Table 5.1) (Davis *et al.*, 2001). In addition, several MDR transporters have been characterized in more detail. These include amongst others the MFS-type transporter EmeA, which is a homologue of NorA from *Staphylococcus aureus* (Jonas *et al.*, 2001; Lee *et al.*, 2003a) and the heterodimeric ABC-type transporter EfrAB (Lee *et al.*, 2003b). Co-expression of *efrA* and *efrB* renders *E. coli* cells resistant to a variety of toxic agents including acriflavine, norfloxacin, ciprofloxacin, doxycycline, 4′,6-diamidino-2-phenylindole and tetraphenylphosphonium chloride. In addition it was shown that cells overexpressing EfrAB exhibited reduced cellular accumulation of acriflavine (Lee *et al.*, 2003b). EfrAB most likely functions as a heterodimeric complex since the increased drug resistance was observed only when *efrA* and *efrB* were jointly expressed. In this respect, EfrAB may be structurally and functionally related to LmrCD of *L. lactis*. Although LmrCD has been shown to play a major role in both intrinsic and acquired resistance, this seems however not to be the case for EfrAB, since single gene disruption of all ABC-type transporters in *E. faecalis* did not reveal involvement of heterodimeric ABC-type transporters in the intrinsic resistance of this organism (Davis *et al.*, 2001). EfrAB might therefore, contribute only to the acquired drug resistance.

Oenococcus oeni

The *omrA* gene of the wine lactic acid bacterium *Oenococcus oeni* encodes a protein that shares 54% amino acid sequence identity with LmrA. OmrA expression is increased in response to stress conditions such as high temperature and osmotic shock. Heterologous expression of OmrA in a drug sensitive *E. coli* strain provided resistance to wine, ethanol and sodium laureate, while in the presence of verapamil, a known inhibitor of ABC-type MDR pumps, this protection was lost (Bourdineaud *et al.*, 2004). Interestingly, expression of OmrA in *E. coli* rendered cells more resistant to the metal cadmium (Achard-Joris *et al.*, 2005). A similar phenomenon was observed when LmrA and human P-glycoprotein were expressed. In all cases resistance to cadmium could be blocked by the addition of verapamil (Achard-Joris *et al.*, 2005). It is suggested that cadmium is transported as a glutathione–derivative in a manner similar to that reported for human MRP1. Recently, it was shown that heterologous expression of OmrA, LmrA or human MDR1 in drug sensitive *E. coli* cells also conferred resistance against zinc and mercury (Achard-Joris and Bourdineaud, 2006).

Streptococcus pneumoniae

Streptococcus pneumoniae is an important human pathogen that was recognized as a major cause of pneumonia in the late 19th century (Butler *et al.*, 1999). Despite the name, this bacterium causes many types of infection other than pneumonia, including meningitis, bacteraemia, endocarditis, peritonitis, and pericarditis. In addition, mortality of persons infected with human immunodeficiency virus (HIV) is often associated with pneumococcal infections (Schneider and Rosen, 1999). Fluoroquinilones are often used to treat such pneumococcal infections.

The genome of *S. pneumoniae* encodes 14 putative drug transporters. Single deletion of the predicted MDR genes resulted in 13 knockout strains that were tested for drug susceptibility (Robertson *et al.*, 2005). Remarkably only the deletion of two closely spaced genes SP2073 and SP2075, both encoding ABC proteins, resulted in a significant decrease in resistance to compounds like acriflavine, ethidium bromide, ciprofloxacin, norfloxacin and the plant alkaloid berberine. The individual knockout strains of SP2073 or SP2075, as well as the double knockout strain, exhibited identical drug resistance profiles, indicating that these gene products do not function in an additive or synergistic manner. From this it was suggested that SP2073 and SP2075 may specify a heterodimeric ABC-type MDR transporter.

In parallel, Marrer *et al.* described two putative ABC-type transporters, PatA (SP2075) and PatB (SP2073) that confer multidrug resistance to a laboratory strain of *S. pneumoniae* (M22) that was selected for ciprofloxacin resistance (Marrer *et al.*, 2006a; Marrer *et al.*, 2006b). Transcriptome profiling revealed that this MDR strain constitutively overexpresses amongst others the *patA* and *patB* genes (Marrer *et al.*, 2006a; Marrer *et al.*, 2006b). Moreover, these genes were strongly induced by quinolone antibiotics and distamycin. Interestingly, the *patA* gene could not be disrupted in the MDR strain M22 suggesting that it is essential for growth in this strain. In contrast, disruption of *patB* did not affect growth and rendered cells more susceptible to norfloxacin and ciprofloxacin. Compared to the parental strain M22, the *patB* disruption strain showed also altered expression of genes involved in amino acid biosynthesis suggesting a possible role of PatB in metabolism.

Multidrug resistant *S. pneumoniae* was also obtained when specifically selected for resistance against the efflux pump inhibitor reserpine. The MDR phenotype was attributed to the increased expression of the *patA* and *patB* genes (Garvey and Piddock, 2008). Disruption of either gene or addition of the ABC transporter inhibitor sodium orthovanadate rendered these cells sensitive again to antibiotics. Interestingly, only in the case of *patA* disruption both MDR and reserpine resistance was lost, which suggests that PatB is not involved in reserpine resistance (Garvey and Piddock, 2008). The differences in phenotype observed when either *patA* or *patB* is disrupted and the lack of coordinate translation of the *patA* and *patB* genes indicates that PatA and PatB may function independently. It is hypothesized that PatA forms a homodimer and is responsible for reserpine resistance and that the heterodimer PatAB is responsible for multidrug resistance (Garvey and Piddock, 2008).

PatA and PatB are close homologues of the well-characterized multidrug efflux proteins LmrC and LmrD from *L. lactis* (Lubelski *et al.*, 2006b; Lubelski *et al.*, 2007). Analogous to drug sensitive *S. pneumonia* lacking *patA* and *patB*, the deletion of the *lmrC* and *lmrD* genes renders *L. lactis* cells sensitive to a wide range of drugs. Close homologues of these paired efflux proteins are found in a wide variety of gram-positive organisms, including *Bacillus, Clostridium, Lactobacillus, Listeria, Corynebacterium, Mycobacterium, Streptomyces, Treponema and Mycoplasma* (Marrer *et al.*, 2006b; Lubelski *et al.*, 2007). Homologues of these efflux proteins may therefore play an important role in clinical multidrug resistance of human pathogens. It would be interesting to determine the contributions of these efflux pumps to multidrug resistance and whether they function as heterodimers in a manner similar to LmrCD.

ABC-type transporters involved in self-immunity

Bacteria that produce and secrete bactericidal agents such as antibiotics, lantibiotics and bacteriocins require to resist the detrimental effects of these compounds. Often ABC-type transporters are involved in the self-protection of these organisms. A common feature of such transporters is that they are relatively specific for a subset of substrates. Many of the antibiotics and drugs used today are produced by a group of bacteria collectively called *Streptomycetes* (Chater, 2006). For instance, *Streptomyces peucetius* produces daunorubicin (daunomycin) and doxorubicin, which are important chemotherapeutic agents used in cancer treatment (Hutchinson and Colombo, 1999). The *drrAB* genes were shown to be involved in the intrinsic resistance of *S. peucetius* to these compounds (Guilfoile and Hutchinson, 1991). These genes are members of the large DRA family of ABC-type transporters, which are involved in drug and antibiotic resistance (Davidson *et al.*, 2008).

The *drrA* gene encodes an ABC-type NBD, while *drrB* encodes a transmembrane protein.

Together these proteins are thought to constitute an efflux pump involved in self-protection, with a likely stoichiometry of $DrrA_2B_2$ (Kaur and Russell, 1998). Homologues of the DrrAB pump have been identified in different species of drug-producing *Streptomyces*. Although these efflux pumps are commonly specific for substrates produced in the associated biosynthetic gene clusters, several of them have been associated with an MDR-like phenotype. For instance, expression of MsrAB in drug-sensitive *Streptomyces* strains conferred resistance to daunorubicin and doxorubicin and to unrelated compounds like erythromycin, spiramycin, and tetracycline (Fernandez-Moreno *et al.*, 1998).

The *drr* operon has also been identified in the genomes of several mycobacteria (Choudhuri *et al.*, 2002). The *drr* genes have been implicated in the virulence of the human pathogen *Mycobacterium tuberculosis*. The efflux pump DrrAB from *M. tuberculosis* was functionally expressed in *E. coli* and *Mycobacterium smegmatis* (Choudhuri *et al.*, 2002). In *E. coli*, DrrAB conferred resistance to ethidium bromide, daunorubicin, doxorubicin, puromycin and chloramphenicol. A similar MDR-phenotype was observed when DrrAB was expressed in *M. smegmatis*; DrrAB provided resistance to ethidium bromide, daunorubicin, ethambutol, doxorubicin, puromycin, chloramphenicol, erythromycin, norflaxin, streptomycin and tetracycline. The MDR phenotype could be reversed by reserpine and verapamil (Choudhuri *et al.*, 2002), two known inhibitors of ABC-type transporters. These data suggest a possible role of DrrAB in antibiotic resistance of *M. tuberculosis*.

A different class of ABC-type transporters involved in bacterial self-immunity is constituted by the drug resistance and immunity (DRI) protein family (Davidson *et al.*, 2008), which are involved in drug resistance and immunity to bacteriocins. Bacteriocins are protein-derived antimicrobial compounds secreted by bacteria to kill related bacteria that compete for the same nutrients. Bacteriocins are classified into three distinct groups (Chatterjee *et al.*, 2005). Class I bacteriocins or lantibiotics are small ribosomally synthesized peptides that are post-translationally modified, whilst class II bacteriocins are heat stable peptides that are not modified. Bacteriocins belonging to class III are generally large heat labile proteins. Many gram-positive organisms produce lantibiotics and the best studied lantibiotic is nisin, which is produced by *Lactococcus lactis*. Nisin is one of the oldest known antimicrobial compounds and is widely used as a food preservative with antimicrobial action directed predominantly against Gram-positive food borne pathogens like *Clostridium* and *Listeria* (Ross *et al.*, 2002; Chatterjee *et al.*, 2005). The cytotoxicity of nisin is a consequence of peptide-induced inhibition of cell wall synthesis and pore-formation in the cytoplasmic membrane (Hsu *et al.*, 2004). Interestingly even though nisin has been used extensively as a food preservative for several decades no substantial development of bacterial resistance against this peptide has emerged. The genes responsible for lantibiotic biosynthesis and immunity are organized in a large gene cluster. These gene clusters usually contain two ABC-type transporters; one that secretes an inactive precursor form of the lantibiotic peptide (*lanT*) and one that provides resistance to the mature active form thereof (*lanFEG*). The genes and gene products involved in nisin biosynthesis and self-immunity are schematically depicted in Fig. 5.3.

The nisin immunity genes *nisFEG* were first described by Siegers and Entian (Siegers and Entian, 1995). NisF and NisE are homologues of the ABC-type HisP transporter family. NisF is a cytoplasmic ATP binding protein, while NisE and NisG constitute transmembrane proteins that together likely form the permease domain of the transporter. Considering the structural similarity to the $HisMQP_2$ transporter and that many ABC-type transporters consist of two NBDs and two TMDs, it is suggested that the complete nisin efflux pump is comprised of $NisF_2EG$ (Peschel and Gotz, 1996; Chatterjee *et al.*, 2005; Lubelski *et al.*, 2008). Single disruption of the *nisF*, *nisE*, and *nisG* genes did not affect the biosynthesis of nisin, indicating that NisFEG is not involved in secretion of the inactive nisin precursor (Siegers and Entian, 1995). In contrast, these disruption mutants were found to be considerably more sensitive to mature nisin, suggesting an active role of NisFEG in the extrusion of nisin (Siegers and Entian, 1995). The extracellular lipoprotein NisI,

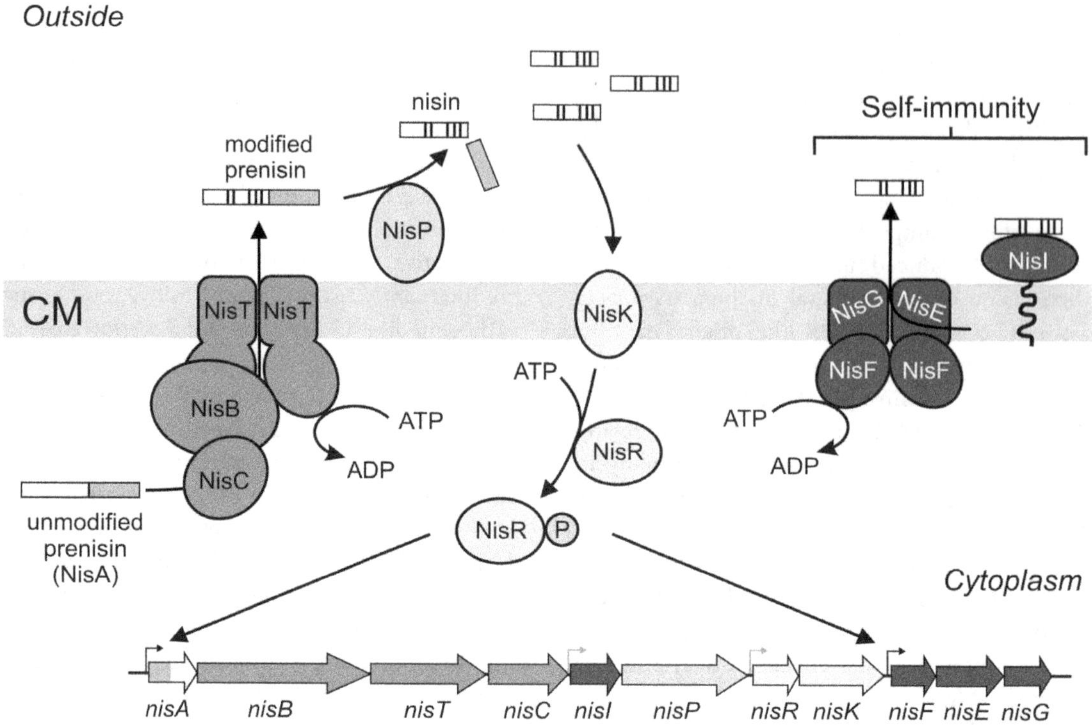

Figure 5.3 Schematic representation of nisin biosynthesis and self-immunity. Extracellular mature nisin binds to the receptor histidine kinase NisK, which induces autophosphorylation of NisK. The phosphate group is subsequently transferred to the response regulator NisR. The phosphorylated transcription factor NisR regulates the expression of two transcriptional units, *nisABTC* and *nisFEG* involved in biosynthesis and self-immunity, respectively The *nisI* gene is preceded by an internal promoter, which allows expression of the immunity protein NisI before full nisin production is established. A weak constitutive promoter precedes the two-component regulatory genes nisRK. A putative membrane-associated complex consisting of NisBTC acts upon the unmodified prenisin, NisA. The modification enzymes NisB and NisC post-translationally introduce dehydro amino acids and (Me) lanthionine residues, respectively. The modified prenisin is secreted by the dedicated ABC-type transporter NisT and subsequently processed outside the cell by the leader peptidase NisP, thereby releasing active nisin. NisFEG constitute an ABC-type transporter that is believed to confer self-protection against the produced nisin and acts in a co-operative manner with the nisin immunity protein NisI.

which is thought to function as a nisin scavenger, also contributes to immunity (Siegers and Entian, 1995; Ra *et al.*, 1996; Ra *et al.*, 1999; Stein *et al.*, 2003). Cells lacking either one of the two systems showed a 80 – 95% reduction in resistance. Maximum immunity is only observed when both NisFEG and NisI are present, indicating synergistic action.

Functional expression of NisFEG in the heterologous host *Bacillus subtilis,* which is sensitive to nisin, rendered these cells resistant to nisin. The highest level of nisin resistance was obtained when NisFEG and NisI were co-expressed, confirming their co-operative action (Stein *et al.*, 2003). *B. subtilis* expressing NisFEG showed greatly reduced levels of cell-associated nisin when compared to control cells, in support of the notion that NisFEG transports nisin from the membrane into the extracellular space. Remarkably, the acquired immunity was specific to nisin and offered no protection against the closely related lantibiotic subtilin. This suggests that this ABC-type lantibiotic efflux pump involved in self-immunity has a more narrow substrate specificity compared to the ABC transporter NisT which is involved in production (Kuipers *et al.*, 2004; van den Berg van Saparoea HB *et al.*, 2008).

Transcriptional regulation of ABC-type MDR transporters

Bacteria are frequently exposed to varying conditions in their natural environment, including changes in temperature, water and nutrient availability, and exposure to toxic compounds. Therefore bacteria rely on regulatory mechanisms that allow the cells to rapidly adapt to changes in their natural environment. These adaptive responses are commonly triggered by regulatory proteins that modulate the expression of other proteins required to alleviate the cellular stress (Ramos *et al.*, 2005).

In this respect, bacterial MDR transporters are commonly subject to regulation at the level of transcription, which generally involves both local and global transcriptional regulators (Ramos *et al.*, 2005). In many cases, genes encoding local regulators, either transcriptional activators or repressors, are present adjacent to the genes encoding MDR transporters. Local regulation of the expression of the drug efflux pump often depends on the presence of the drug in the extracellular milieu. The drug is thought to interact directly with the regulatory protein and trigger conformational changes. Due to these conformational changes the transcriptional activator or repressor respectively binds to, or is released from, the DNA, which in turn leads to enhanced expression of the drug transporter. Additionally, the expression of MDR transporters is controlled by global modulators in response to a global stress response (Grkovic *et al.*, 2001; Grkovic *et al.*, 2002). The complex transcriptional regulation of MDR transporters may reflect that these pumps potentially can extrude also important molecules required by the cell. As is the case for many other membrane proteins, the excessive expression of MDR transporters can be detrimental to the cell but it remains to be determined whether this relates to specific transporter-induced effects such as a compromised integrity of the cell membrane or the undesired secretion of important metabolites (Grkovic *et al.*, 2001; Grkovic *et al.*, 2002).

The transcriptional regulators of MDR proteins in bacteria identified thus far belong to one of four regulatory protein families; the AraC, MarR, MerR, or TetR family. Their respective members are characterized by similarities within their DNA-binding domains. Typically, the MDR regulators contain the α-helix-turn-α-helix DNA-binding motif, a conserved structural motif that is present in the majority of bacterial activators and repressors. To date, nearly all of the described MDR transporters that are subject to transcriptional regulation are pmf driven pumps belonging to the MFS or RND family, whereas the SMR-type transporters are usually constitutively expressed and not subject to transcriptional regulation (Grkovic *et al.*, 2002).

Several local transcriptional regulators have been characterized in detail (Grkovic *et al.*, 2001). Among these, BmrR and BltR from *B. subtilis*, which are transcriptional activators belonging to the MerR family that regulate the expression of the MFS-type efflux pumps Bmr and Blt, respectively (Ahmed *et al.*, 1994; Ahmed *et al.*, 1995; Zheleznova *et al.*, 1999). The DNA-binding domains of BmrR and BltR are similar, but their drug-binding domains are distinct, suggesting that Bmr and Blt expression is induced by different compounds. When drugs like tetraphenylphosphonium or rhodamine 6G bind to BmrR, this results in an increased affinity of the regulator for the *bmr* promoter, which, in turn, enhances the transcription of the *bmr* structural gene (Markham *et al.*, 1996). In contrast, rhodamine 6G did not induce *blt* expression even though this compound is a substrate for the transporter Blt, which supports the notion that spermidine efflux rather than multidrug efflux is the primary function of Blt (Woolridge *et al.*, 1997). Structural and biochemical analysis have given more insight into the molecular mechanism of drug binding and transcriptional activation by BmrR (Heldwein and Brennan, 2001).

A well-studied transcriptional repressor is QacR, which regulates the expression of the MFS-type transporter QacA from *Staphylococcus aureus*. QacR belongs to the TetR family and interacts with a wide variety of toxic compounds including dyes, quaternary ammonium compounds and plant alkaloids (Grkovic *et al.*, 2003). Under un-induced conditions, QacR is bound to the target promoter thereby blocking transcription of *qacA*. Under inducing conditions, drugs bind to QacR and promote its release from the target promoter, which, in turn, relieves the repression of *qacA* expression (Grkovic *et al.*, 1998). The structure of QacR has been solved and provided further insight into the mechanism of drug recognition (Schumacher *et al.*, 2001). Moreover, it was shown that QacR can bind simultaneously ethidium and proflavin demonstrating plasticity of the drug-binding pocket (Schumacher *et al.*, 2004). Recently, the TetR family of transcriptional repressors has been excellently reviewed (Ramos *et al.*, 2005).

The MarR family is another large family of transcriptional repressors and its members have been implicated in controlling the expression of either MFS- or RND-type efflux pumps. For instance, EmrR has been reported to regulate the expression of the MFS transporter EmrAB form *E. coli* (Lomovskaya *et al.*, 1995), while MexR has been shown to function as a repressor of the RND-type efflux pump MexAB-OprM from *Pseudomonas aeruginosa* (Evans et al., 2001).

The expression of several MDR transporters has been shown to be controlled by global regulators. The expression of the main MDR pump AcrAB-TolC from *E. coli* is regulated by the AraC-like global activators MarA, Rob and SoxS (Alekshun and Levy, 1997). In addition, expression of the *acrAB* genes is controlled by the local transcriptional regulator AcrR, which functions as a non-inducible repressor (Grkovic *et al.*, 2001). In *B. subtilis* expression of the transporters Bmr and Blt is regulated by the MerR-type global activator Mta in addition to the aforementioned local activators BmrR and BltR (Baranova *et al.*, 1999). Additionally, transcriptional regulation of drug transporters may involve two-component signal transduction systems, which consist of a sensory kinase and a response regulator. The kinase senses particular environmental conditions, such as exposure to toxic compounds, and accordingly phosphorylates the response regulator that controls the expression of the drug transporter. Typical examples of such regulatory systems are: ArlSR (Fournier *et al.*, 2000) which modulates the expression of NorA from *S. aureus*; BaeSR (Nishino *et al.*, 2007) from *Salmonella enterica* which induces the AcrD and MdtABC drug efflux systems. In *E. coli*, the two-component system EvgAS (Nishino and Yamaguchi, 2002) was shown to regulate the expression of the RND-type efflux pump YhiUV. Moreover, in the presence of a constitutively active mutant of EvgS, the level of transcription for various genes related to drug efflux increased, including *emrKY*, *yhiUV*, *acrAB*, *mdfA* and *tolC* (Eguchi *et al.*, 2003). These data suggest a central role of EvgAS in transcriptional regulation of drug transporters that contribute to MDR.

In clinical isolates, antibiotic resistance is often associated with elevated expression of efflux pumps. In many cases this increase in expression could be attributed to mutations in genes encoding local repressors or global regulators, while in some cases the promoter region of the transporter was mutated (Piddock, 2006a). For instance, various mutations in the local repressor AcrR have been described that cause a relief of repression of transcription of its cognate transporter AcrAB. Such mutations have been found in clinical isolates of *E. coli*, *E. aerogenes*, *H. influenza* and *S. enterica* (Piddock, 2006a).

Interestingly, a similar phenomenon is observed in the non-pathogenic bacterium *L. lactis*. DNA microarray analysis of MDR strains of *L. lactis* selected for daunomycin, ethidium or rhodamine 6G resistance (Bolhuis *et al.*, 1994) revealed that in these strains the expression level of the MDR transporter LmrCD is strongly increased (Lubelski *et al.*, 2006b). The elevated expression of LmrCD was attributed to a defect in the gene coding for the transcriptional regulator LmrR (formerly named YdaF), which is located immediately upstream of the *lmrCD* genes (Lubelski *et al.*, 2006b; Agustiandari *et al.*, 2008). It was found that the *lmrR* gene of the different *L. lactis* MDR strains contained either a frameshift mutation or a point mutation (Lubelski *et al.*, 2006b; Agustiandari *et al.*, 2008).

Based on amino acid sequence homology, LmrR belongs to the PadR family of transcriptional regulators, which in turn is related to the bacterial MarR (multiple antibiotic resistance) family of transcriptional regulators. PadR-like regulators have been shown to modulate diverse cellular processes such as virulence, detoxification of phenolic compounds and MDR (Agustiandari *et al.*, 2008). The characteristic PadR structural fold (De Silva *et al.*, 2005; Arita *et al.*, 2007) consists of a conserved N-terminal winged α-helix-turn-α-helix DNA binding domain (Aravind *et al.*, 2005) that is similar to that of MarR-type repressors (Alekshun *et al.*, 2001), and a highly diverse C-terminal helical domain which is important for dimerization. Recently the structure of LmrR was determined both in the absence and presence of drugs (Madoori *et al.*, 2008). Interestingly, the general architecture of the LmrR dimer is very similar to that of other PadR/MarR family members, with one major distinction; the LmrR dimer contains a large central pore at the dimer interface, which is absent in all structures of PadR/MarR-type regulators reported thus far. Co-crystallization of LmrR with Hoechst 33342 or daunomycin shows that the central pore acts as the drug-binding pocket, which permits binding of a single drug molecule (Madoori *et al.*, 2008). Hoechst 33342 and daunomycin exhibited a common mode of binding: the planar ring moieties of these drugs are wedged in between two tryptophan residues – one tryptophan located on each LmrR side chain – resulting in aromatic stacking interactions with each of the two indole groups. Hydrophobic amino acids lining the central pore further contribute to the stabilization of the bound drug. Substitution of the catalytic tryptophan residues by tyrosine or alanine abolished the binding of Hoechst 33342, indicating that the tryptophans in the central pore are essential for drug binding (Madoori *et al.*, 2008).

The structure of LmrR represents the first structure of a transcriptional regulator that modulates the expression of an ABC-type multidrug efflux pump. At present it has proven very difficult to obtain structures of MDR transporters. The mechanism by which these transporters can bind and extrude structurally and functionally unrelated compounds has remained largely illusive. The multidrug binding abilities of cytosolic regulators like LmrR may therefore also shed some light on how the cognate transporter (LmrCD) may recognize and bind different drugs.

The above biochemical and structural studies have provided insights in the molecular mechanism by which the MDR phenotype in *L. lactis* is regulated. When *L. lactis* cells are exposed to toxic compounds, these molecules may cross the cell membrane and bind to the drug binding pocket of the transcriptional repressor LmrR. Drug-binding likely induces a conformational change in LmrR such that its affinity for the *lmrCD* promoter is reduced, thereby allowing the RNA polymerase to initiate transcription. This relieve in repression is accompanied by the expression of the MDR transporter LmrCD that extrudes the toxic molecules from the cell.

Conclusions and future trends

Bacterial ABC-type membrane transport proteins fulfil very diverse cellular functions, including transport of ions, small molecules such as metabolites, nutrients and drugs, or macromolecules such as polysaccharides and polypeptides across the cellular membrane. It is estimated that less than 30% of the bacterial ABC systems have been identified so far and only few of these have been functionally characterized (Davidson *et al.*, 2008). A subset of these ABC-type transporters has been implicated in multidrug resistance (MDR), an important function in the survival of bacteria in their often hostile environment.

Until recently, research on bacterial MDR focused predominantly on RND- and MFS-type transporters, which are major determinants of clinical multidrug resistance (Piddock, 2006a). However, at present also the ABC-type transporters are under more scrutiny since several ABC-type MDR transporters have now been identified that are involved in intrinsic and acquired drug resistance in bacteria. In pathogenic Gram-positive bacteria such as *E. faecalis* and *S. pneumoniae*, ABC-type transporters are important contributors to multidrug resistance (Robertson *et al.*, 2005; Lubelski *et al.*, 2007). These organisms contain paired ABC-type MDR efflux genes. Homologues of these transporters are found in many Gram-positive bacteria, including clinically important pathogens. In this respect, the non-pathogenic Gram-positive bacterium *L. lactis*, LmrC and LmrD constitute a heterodimeric transporter that is the major determinant of lactococcal MDR. Thus, functional characterization of these transporters may reveal a wider involvement of ABC-type transporters in bacterial MDR (Table 5.1) (Lubelski *et al.*, 2007).

MDR transporters encoded by bacteria can provide protection against a wide range of structurally diverse toxic compounds, including clinically relevant antibiotics. However, there is accumulating evidence that these efflux pumps may have physiological functions other than the extrusion of drugs. MDR transporters have been shown to export natural compounds produced by the host organism, including bile acids, defence molecules and hormones. For instance, bacteria that are exposed to bile in their natural habitat employ predominantly RND-type pumps to expel the bile acids from the cell. Typical examples are the AcrAB-TolC pump of *E. coli* and *S.* Typhimurium and the CmeABC pump of *Campylobacter jejuni* (Piddock, 2006b). However, recently the ABC-type MDR transporter LmrCD was demonstrated to be the primary efflux system in *L. lactis* for the bile acids cholate and glycodeoxycholate (Zaidi *et al.*, 2008). Notably, ABC-type transporters are commonly responsible for bile secretion and transport within the liver. Homologues of LmrCD are widely distributed among human gut bacteria such as *lactobacilli* and *bifidobacteria.* Since these organisms are frequently exposed to high concentrations of bile acids, it is conceivable that in these bacteria the LmrCD-like ABC-type transporters are involved in bile acid resistance.

Besides the export of antimicrobial compounds and host-derived substrates, MDR transporters have also been implicated in the secretion of virulence determinants. Bacteria may secrete compounds such as adhesion molecules, toxins and proteins that are required for the colonization and infection of mammalian cells, or compounds for killing other bacteria competing for the same ecological niche. The AcrAB-TolC system of *S.* Typhimurium has been shown to be involved in pathogenesis as deletion of this efflux pump resulted in impaired host cell adherence and invasion. A role for ABC-type MDR transporters in bacterial virulence has yet to be established. However, the deletion of the ABC-type macrolide-specific efflux pump MacAB attenuated the virulence of *S.* Typhimurium (Nishino *et al.*, 2006). Deletion of the *macAB* genes of this organism did however not alter its sensitivity to erythromycin, while overexpression of these genes provided only a minor increase in resistance (Nishino *et al.*, 2006). In contrast, overexpression of *macAB* in *E. coli* lacking *acrAB* conferred significant resistance to erythromycin (Kobayashi *et al.*, 2001). It remains therefore to be determined whether the MacAB efflux pump has a significant contribution to clinical resistance to macrolide antibiotics.

Many Gram-positive bacteria secrete lantibiotics, which are antimicrobial peptides directed against closely related bacteria. Probably the best studied lantibiotic is nisin, which was discovered in 1928, one year prior to penicillin, and is secreted by several *L. lactis* strains. For several decades now, nisin has been used in the food industry as a preservative and remarkably without significant development of bacterial resistance. Nisin exerts its antimicrobial action by binding to the cell wall precursor molecule lipid II, which increases the efficiency of nisin pore formation and in addition interferes with cell wall synthesis. Since lipid II is also an important constituent of the *lactococcal* cell envelope *L. lactis* requires a defence system against the endogenously produced nisin. The nisin immunity genes *nisFEG* encode an ABC-type transporter that is a homologue of the HisP transporter family, which expels nisin from the cell membrane. Similar lantibiotic extrusion systems are found in *B. subtilis* and *Staphylococcus epidermidis* (Chatterjee *et al.*, 2005). These organisms produce and secrete the lantibiotics subtilin and epidermin, respectively, which are bacteriocins similar to nisin. Heterologous expression of *nisFEG* in *B. subtilis* imparted resistance to nisin while no increased cross-resistance to the closely related subtilin was observed (Stein *et al.*, 2003). This suggests that these ABC-type lantibiotic immunity efflux pumps have very narrow substrate specificity, which may be restricted to the dedicated lantibiotic.

B. subtilis produces many antimicrobial peptides in addition to subtilin. For instance, sporulating *B. subtilis* cells secrete the toxic peptide SdpC. This antimicrobial peptide induces lysis of sister cells thereby providing the sporulating cells with nutrients, which in turn allows them to keep growing rather than to complete spore formation (Gonzalez-Pastor *et al.*, 2003; Ellermeier *et al.*, 2006). To protect itself against this harmful peptide, *B. subtilis* expresses an ABC-type efflux pump encoded by the operon *yknXYZ* (Butcher and Helmann, 2006). This efflux system is homologous to the macrolide-specific pump MacAB and is responsible for the extrusion of the toxic peptide from the cell (Butcher and Helmann, 2006). These observations further illustrate that ABC-type drug transporters may have physiological functions unrelated to the extrusion of drugs.

It is now understood that ABC-type transporters play important roles in bacterial survival, which include functions in (multi)drug resistance and self-immunity. However, many important questions remain: Why do bacterial genomes specify so many MDR pumps? Do these pumps have redundant or overlapping functions? Which MDR transporters contribute to intrinsic and/or acquired resistance and do they contribute to clinical resistance? Do the MDR pumps fulfill functions other than conferring protection against xenobiotics? If so, what is their primary function and what is their natural substrate? What is the molecular mechanism by which the substrates are transported across the cell membrane?

To address (some of) these questions it is desirable to use a combination of genetic and biochemical approaches. Systematic gene deletion of putative MDR transporters in the natural host is a useful technique to identify the major determinants of the resistance phenotype (Hsieh *et al.*, 1998; Sulavik *et al.*, 2001). In addition, putative MDR transporters can be expressed in drug sensitive strains of the natural or a heterologous host to establish whether they are able to confer resistance to drugs. Besides the widely used drug-sensitive mutants of the Gram-negative bacterium *E. coli* (lacking e.g. AcrAB-TolC), drug-sensitive strains of Gram-positive bacteria would be similarly helpful. In this respect, the *L. lactis* Δ*lmrA* Δ*lmrCD* strain is very sensitive to a range of unrelated drugs and may therefore be a suitable host for characterizing transporters from Gram-positive organisms (Lubelski *et al.*, 2006b; Lubelski *et al.*, 2007).

Several aspects regarding the identification of an MDR transporter need to be further considered. Deletion of a putative drug efflux gene does not necessarily result in a drug-sensitive phenotype; the loss of drug export activity may be compensated by the increased expression of other indigenous transporters. Therefore, expression of the putative MDR transporter is often attempted in a heterologous host, for which in many cases *E. coli* is the preferred organism. When heterologous expression of the transporter confers protection against a range of drugs, a role in MDR seems apparent, but there still is a need to demonstrate this activity in the natural host. For instance, the drug-sensitive *E. coli* strains, which lack the major efflux pumps, still contain many uncharacterized putative MDR transporters. It is possible that the heterologous expression of a putative MDR transporter results in the expression of these cryptic MDR-like transporters, thus indirectly resulting in a MDR phenotype. Therefore, it is important to demonstrate a direct involvement of an expressed MDR transporter in drug resistance, for instance by expressing inactive variants. In addition, it has to be taken into account that transporters with a specific or restricted substrate range may exhibit an opportunistic transport of drugs when expressed at very high level. From an evolutionary point of view, such transporters may serve as a reservoir of potential resistance proteins and may therefore eventually contribute to the acquired drug resistance of bacteria. Furthermore, analysis of the expression profile of putative MDR transporter genes can substantiate the involvement of a transporter in drug resistance. DNA micro-array analysis may reveal specific cellular adaptations like increased expression of efflux pumps in response to a challenge with drug(s). In addition, the genomic context of the efflux gene may also provide clues its physiological function. Adjacent genes may encode regulatory proteins and/or proteins, which are involved in the same metabolic pathway. The various genetic and biochemical approaches discussed here are powerful tools, especially when they are combined, to correctly determine the function of the many uncharacterized bacterial ABC-type transporters, whether they are involved in general survival of the host, multidrug resistance or the extrusion of a single dedicated substrate.

Acknowledgements

This work was supported by the Dutch Technology Foundation STW (project number 06927).

References

Achard-Joris, M. and Bourdineaud, J.P. (2006). Heterologous expression of bacterial and human multidrug resistance proteins protect *Escherichia coli* against mercury and zinc contamination. Biometals *19*, 695–704.

Achard-Joris, M., van den Berg van Saparoea, H.B., Driessen, A.J.M., and Bourdineaud, J.P. (2005). Heterologously Expressed Bacterial and Human Multidrug Resistance Proteins Confer Cadmium Resistance to *Escherichia coli*. Biochemistry *44*, 5916–5922.

Agustiandari, H., Lubelski, J., van den Berg van Saparoea, H.B, Kuipers, O.P., and Driessen, A.J. (2008). LmrR is a transcriptional repressor of expression of the multidrug ABC transporter LmrCD in *Lactococcus lactis*. J. Bacteriol. *190*, 759–763.

Ahmed, M., Borsch, C.M., Taylor, S.S., Vazquez-Laslop, N., and Neyfakh, A.A. (1994). A protein that activates expression of a multidrug efflux transporter upon binding the transporter substrates. J. Biol. Chem. *269*, 28506–28513.

Ahmed, M., Lyass, L., Markham, P.N., Taylor, S.S., Vazquez-Laslop, N., and Neyfakh, A.A. (1995). Two highly similar multidrug transporters of Bacillus subtilis whose expression is differentially regulated. J. Bacteriol. *177*, 3904–3910.

Alekshun, M.N. and Levy, S.B. (1997). Regulation of chromosomally mediated multiple antibiotic resistance: the mar regulon. Antimicrob. Agents Chemother. *41*, 2067–2075.

Alekshun, M.N., Levy, S.B., Mealy, T.R., Seaton, B.A., and Head, J.F. (2001). The crystal structure of MarR, a regulator of multiple antibiotic resistance, at 2.3 A resolution. Nat. Struct. Biol. *8*, 710–714.

Allikmets, R., Schriml, L.M., Hutchinson, A., Romano-Spica, V., and Dean, M. (1998). A human placenta-specific ATP-binding cassette gene (ABCP) on chromosome 4q22 that is involved in multidrug resistance. Cancer Res. *58*, 5337–5339.

Ambudkar, S.V., Kimchi-Sarfaty, C., Sauna, Z.E., and Gottesman, M.M. (2003). P-glycoprotein: from genomics to mechanism. Oncogene 22, 7468–7485.

Aravind, L., Anantharaman, V., Balaji, S., Babu, M.M., and Iyer, L.M. (2005). The many faces of the helix–turn–helix domain: transcription regulation and beyond. FEMS Microbiol. Rev. *29*, 231–262.

Arita, K., Hashimoto, H., Igari, K., Akaboshi, M., Kutsuna, S., Sato, M., and Shimizu, T. (2007). Structural and biochemical characterization of a cyanobacterium circadian clock-modifier protein. J. Biol. Chem. *282*, 1128–1135.

Baranova, N.N., Danchin, A., and Neyfakh, A.A. (1999). Mta, a global MerR-type regulator of the Bacillus subtilis multidrug-efflux transporters. Mol. Microbiol. *31*, 1549–1559.

Behr, J., Ganzle, M.G., and Vogel, R.F. (2006). Characterization of a highly hop-resistant Lactobacillus brevis strain lacking hop transport. Appl. Environ. Microbiol. 72, 6483–6492.

Biemans-Oldehinkel, E., Doeven, M.K., and Poolman, B. (2006). ABC transporter architecture and regulatory roles of accessory domains. FEBS Lett. *580*, 1023–1035.

Bishop, L., Agbayani, R., Jr., Ambudkar, S.V., Maloney, P.C., and Ames, G.F. (1989). Reconstitution of a bacterial periplasmic permease in proteoliposomes and demonstration of ATP hydrolysis concomitant with transport. Proc. Natl. Acad. Sci. U. S. A.*86*, 6953–6957.

Bolhuis, H., Molenaar, D., Poelarends, G., van Veen, H.W., Poolman, B., Driessen, A.J., and Konings, W.N. (1994). Proton motive force-driven and ATP-dependent drug extrusion systems in multidrug-resistant *Lactococcus lactis*. J. Bacteriol. *176*, 6957–6964.

Bolhuis, H., Poelarends, G., van Veen, H.W., Poolman, B., Driessen, A.J., and Konings, W.N. (1995). The Lactococcal *lmrP* gene encodes a proton motive force-dependent drug transporter. J. Biol. Chem. *270*, 26092–26098.

Borbat, P.P., Surendhran, K., Bortolus, M., Zou, P., Freed, J.H., and Mchaourab, H.S. (2007). Conformational motion of the ABC transporter MsbA induced by ATP hydrolysis. PLoS. Biol. *5*, e271.

Bouige, P., Laurent, D., Piloyan, L., and Dassa, E. (2002). Phylogenetic and functional classification of ATP-binding cassette (ABC) systems. Curr. Protein Pept. Sci. *3*, 541–559.

Bourdineaud, J .P., Nehme, B., Tesse, S., and Lonvaud-Funel, A. (2004). A bacterial gene homologous to ABC transporters protect Oenococcus oeni from ethanol and other stress factors in wine. Int. J. Food Microbiol. *92*, 1–14.

Brown, M.H., Paulsen, I.T., and Skurray, R.A. (1999). The multidrug efflux protein NorM is a prototype of a new family of transporters. Mol. Microbiol. *31*, 394–395.

Butcher, B.G. and Helmann, J.D. (2006). Identification of Bacillus subtilis sigma-dependent genes that provide intrinsic resistance to antimicrobial compounds produced by Bacilli. Mol. Microbiol. *60*, 765–782.

Butler, J.C., Shapiro, E.D., and Carlone, G.M. (1999). Pneumococcal vaccines: history, current status, and future directions. Am. J. Med. *107*, 69S-76S.

Chami, M., Steinfels, E., Orelle, C., Jault, J.M., Di Pietro, A., Rigaud, J.L., and Marco, S. (2002). Three-dimensional structure by cryo-electron microscopy of YvcC, an homodimeric ATP-binding cassette transporter from Bacillus subtilis. J. Mol. Biol. *315*, 1075–1085.

Chater, K.F. (2006). Streptomyces inside-out: a new perspective on the bacteria that provide us with antibiotics. Philos. Trans. R. Soc. Lond B Biol. Sci. *361*, 761–768.

Chatterjee, C., Paul, M., Xie, L., and van der Donk, W.A. (2005). Biosynthesis and mode of action of lantibiotics. Chem. Rev. *105*, 633–684.

Chen, C.J., Chin, J.E., Ueda, K., Clark, D.P., Pastan, I., Gottesman, M.M., and Roninson, I.B. (1986). Internal duplication and homology with bacterial transport proteins in the mdrl (P-glycoprotein) gene from multidrug-resistant human cells. Cell 47, 381–389.

Choudhuri, B.S., Bhakta, S., Barik, R., Basu, J., Kundu, M., and Chakrabarti, P. (2002). Overexpression and functional characterization of an ABC (ATP-binding cassette) transporter encoded by the genes drrA and drrB of Mycobacterium tuberculosis. Biochem. J. *367*, 279–285.

Cohen, M.L. (2000). Changing patterns of infectious disease. Nature *406*, 762–767.

Cole, S.P., Bhardwaj, G., Gerlach, J.H., Mackie, J.E., Grant, C.E., Almquist, K.C., Stewart, A.J., Kurz, E.U., Duncan, A.M., and Deeley, R.G. (1992). Overexpression of a transporter gene in a multidrug-resistant human lung cancer cell line. Science *258*, 1650–1654.

Dalmas, O., Do Cao, M.A., Lugo, M.R., Sharom, F.J., Di, P.A., and Jault, J.M. (2005). Time-resolved fluorescence resonance energy transfer shows that the bacterial multidrug ABC half-transporter BmrA functions as a homodimer. Biochemistry *44*, 4312–4321.

Dano, K. (1972). Cross resistance between vinca alkaloids and anthracyclines in Ehrlich ascites tumor *in vivo*. Cancer Chemother. Rep. *56*, 701–708.

Dano, K. (1973). Active outward transport of daunomycin in resistant Ehrlich ascites tumor cells. Biochim. Biophys. Acta *323*, 466–483.

Dassa, E. and Bouige, P. (2001). The ABC of ABCS: a phylogenetic and functional classification of ABC systems in living organisms. Res. Microbiol. *152*, 211–229.

Davidson, A.L. and Chen, J. (2004). ATP-binding cassette transporters in bacteria. Annu. Rev. Biochem. *73*, 241–268.

Davidson, A.L., Dassa, E., Orelle, C., and Chen, J. (2008). Structure, function, and evolution of bacterial ATP-binding cassette systems. Microbiol. Mol. Biol. Rev. *72*, 317–64, table.

Davis, D.R., McAlpine, J.B., Pazoles, C.J., Talbot, M.K., Alder, E.A., White, C., Jonas, B.M., Murray, B.E., Weinstock, G.M., and Rogers, B.L. (2001). *Enterococcus faecalis* multi-drug resistance transporters: application for antibiotic discovery. J. Mol. Microbiol. Biotechnol. *3*, 179–184.

Dawson, R.J. and Locher, K.P. (2006). Structure of a bacterial multidrug ABC transporter. Nature *443*, 180–185.

Dawson, R.J. and Locher, K.P. (2007). Structure of the multidrug ABC transporter Sav1866 from *Staphylococcus aureus* in complex with AMP-PNP. FEBS Lett. *581*, 935–938.

De Silva, R.S., Kovacikova, G., Lin, W., Taylor, R.K., Skorupski, K., and Kull, F.J. (2005). Crystal structure of the virulence gene activator AphA from Vibrio cholerae reveals it is a novel member of the winged helix transcription factor superfamily. J. Biol. Chem. *280*, 13779–13783.

Denyer, S.P. and Maillard, J.Y. (2002). Cellular impermeability and uptake of biocides and antibiotics in Gram-negative bacteria. J. Appl. Microbiol. *92 Suppl*, 35S-45S.

Diederichs, K., Diez, J., Greller, G., Muller, C., Breed, J., Schnell, C., Vonrhein, C., Boos, W., and Welte, W. (2000). Crystal structure of MalK, the ATPase subunit of the trehalose/maltose ABC transporter of the archaeon Thermococcus litoralis. EMBO J. *19*, 5951–5961.

Doerrler, W.T., Gibbons, H.S., and Raetz, C.R. (2004). MsbA-dependent translocation of lipids across the inner membrane of *Escherichia coli*. J. Biol. Chem. *279*, 45102–45109.

Dong, J., Yang, G., and Mchaourab, H.S. (2005). Structural basis of energy transduction in the transport cycle of MsbA. Science *308*, 1023–1028.

Doyle, L.A. and Ross, D.D. (2003). Multidrug resistance mediated by the breast cancer resistance protein BCRP (ABCG2). Oncogene *22*, 7340–7358.

Doyle, L.A., Yang, W., Abruzzo, L.V., Krogmann, T., Gao, Y., Rishi, A.K., and Ross, D.D. (1998). A multidrug resistance transporter from human MCF-7 breast cancer cells. Proc. Natl. Acad. Sci. U. S. A.*95*, 15665–15670.

Ecker, G.F., Pleban, K., Kopp, S., Csaszar, E., Poelarends, G.J., Putman, M., Kaiser, D., Konings, W.N., and Chiba, P. (2004). A three-dimensional model for the substrate binding domain of the multidrug ATP binding cassette transporter LmrA. Mol. Pharmacol. *66*, 1169–1179.

Eguchi, Y., Oshima, T., Mori, H., Aono, R., Yamamoto, K., Ishihama, A., and Utsumi, R. (2003). Transcriptional regulation of drug efflux genes by EvgAS, a two-component system in *Escherichia coli*. Microbiology *149*, 2819–2828.

Ellermeier, C.D., Hobbs, E.C., Gonzalez-Pastor, J.E., and Losick, R. (2006). A three-protein signaling pathway governing immunity to a bacterial cannibalism toxin. Cell *124*, 549–559.

Evans, K., Adewoye, L., and Poole, K. (2001). MexR repressor of the mexAB-oprM multidrug efflux operon of Pseudomonas aeruginosa: identification of MexR binding sites in the mexA-mexR intergenic region. J. Bacteriol. *183*, 807–812.

Ewart, G.D. and Howells, A.J. (1998). ABC transporters involved in transport of eye pigment precursors in Drosophila melanogaster. Methods Enzymol. *292*, 213–224.

Fernandez-Moreno, M.A., Carbo, L., Cuesta, T., Vallin, C., and Malpartida, F. (1998). A silent ABC transporter isolated from Streptomyces rochei F20 induces multidrug resistance. J. Bacteriol. *180*, 4017–4023.

Fournier, B., Aras, R., and Hooper, D.C. (2000). Expression of the multidrug resistance transporter NorA from *Staphylococcus aureus* is modified by a two-component regulatory system. J. Bacteriol. *182*, 664–671.

Garvey, M.I. and Piddock, L.J. (2008). The efflux pump inhibitor reserpine selects multidrug-resistant *Streptococcus pneumoniae* strains that overexpress the ABC transporters PatA and PatB. Antimicrob. Agents Chemother. *52*, 1677–1685.

George, A.M. (1996). Multidrug resistance in enteric and other gram-negative bacteria. FEMS Microbiol. Lett. *139*, 1–10.

Gerber, S., Comellas-Bigler, M., Goetz, B.A., and Locher, K.P. (2008). Structural basis of trans-inhibition in a molybdate/tungstate ABC transporter. Science *321*, 246–250.

Gill, M.J., Brenwald, N.P., and Wise, R. (1999). Identification of an efflux pump gene, pmrA, associated with fluoroquinolone resistance in *Streptococcus pneumoniae*. Antimicrob. Agents Chemother. *43*, 187–189.

Gilson, E., Higgins, C.F., Hofnung, M., Ferro-Luzzi, A.G., and Nikaido, H. (1982). Extensive homology between membrane-associated components of histidine and maltose transport systems of *Salmonella typhimurium* and *Escherichia coli*. J. Biol. Chem. *257*, 9915–9918.

Gonzalez-Pastor, J.E., Hobbs, E.C., and Losick, R. (2003). Cannibalism by sporulating bacteria. Science *301*, 510–513.

Grkovic, S., Brown, M.H., Roberts, N.J., Paulsen, I.T., and Skurray, R.A. (1998). QacR is a repressor protein that regulates expression of the *Staphylococcus aureus* multidrug efflux pump QacA. J. Biol. Chem. *273*, 18665–18673.

Grkovic, S., Brown, M.H., and Skurray, R.A. (2001). Transcriptional regulation of multidrug efflux pumps in bacteria. Semin. Cell Dev. Biol. *12*, 225–237.

Grkovic, S., Brown, M.H., and Skurray, R.A. (2002). Regulation of bacterial drug export systems. Microbiol. Mol. Biol. Rev. *66*, 671–701.

Grkovic, S., Hardie, K.M., Brown, M.H., and Skurray, R.A. (2003). Interactions of the QacR multidrug-binding protein with structurally diverse ligands: implications for the evolution of the binding pocket. Biochemistry *42*, 15226–15236.

Guilfoile, P.G. and Hutchinson, C.R. (1991). A bacterial analog of the mdr gene of mammalian tumor cells is present in *Streptomyces peucetius*, the producer of daunorubicin and doxorubicin. Proc. Natl. Acad. Sci. U. S. A.*88*, 8553–8557.

Gunn, J.S. (2000). Mechanisms of bacterial resistance and response to bile. Microbes. Infect. *2*, 907–913.

Hayes, J.D. and Wolf, C.R. (1990). Molecular mechanisms of drug resistance. Biochem. J. *272*, 281–295.

Heldwein, E.E. and Brennan, R.G. (2001). Crystal structure of the transcription activator BmrR bound to DNA and a drug. Nature *409*, 378–382.

Higgins, C.F. (1992). ABC transporters: from microorganisms to man. Annu. Rev. Cell Biol. *8*, 67–113.

Higgins, C.F. (2001). ABC transporters: physiology, structure and mechanism--an overview. Res. Microbiol. *152*, 205–210.

Higgins, C.F., Haag, P.D., Nikaido, K., Ardeshir, F., Garcia, G., and Ames, G.F. (1982). Complete nucleotide sequence and identification of membrane components of the histidine transport operon of *S. typhimurium*. Nature *298*, 723–727.

Higgins, C.F., Hiles, I.D., Salmond, G.P., Gill, D.R., Downie, J.A., Evans, I.J., Holland, I.B., Gray, L., Buckel, S.D., Bell, A.W., and . (1986). A family of related ATP-binding subunits coupled to many distinct biological processes in bacteria. Nature 323, 448–450.

Higgins, C.F., Hiles, I.D., Whalley, K., and Jamieson, D.J. (1985). Nucleotide binding by membrane components of bacterial periplasmic binding protein-dependent transport systems. EMBO J. *4*, 1033–1039.

Higgins, C.F. and Linton, K.J. (2004). The ATP switch model for ABC transporters. Nat. Struct. Mol. Biol. *11*, 918–926.

Hiles, I.D., Gallagher, M.P., Jamieson, D.J., and Higgins, C.F. (1987). Molecular characterization of the oligopeptide permease of *Salmonella typhimurium*. J. Mol. Biol. *195*, 125–142.

Holland, I.B., Cole, S.P., Kuchler, K., and Higgins, C.F. (2003). ABC proteins, from Bacteria to Man. Academic Press.

Hollenstein, K., Frei, D.C., and Locher, K.P. (2007). Structure of an ABC transporter in complex with its binding protein. Nature *446*, 213–216.

Hopfner, K.P., Karcher, A., Shin, D.S., Craig, L., Arthur, L.M., Carney, J.P., and Tainer, J.A. (2000). Structural biology of Rad50 ATPase: ATP-driven conformational control in DNA double-strand break repair and the ABC-ATPase superfamily. Cell *101*, 789–800.

Hsieh, P.C., Siegel, S.A., Rogers, B., Davis, D., and Lewis, K. (1998). Bacteria lacking a multidrug pump: a sensitive tool for drug discovery. Proc. Natl. Acad. Sci. U. S. A.*95*, 6602–6606.

Hsu, S.T., Breukink, E., Tischenko, E., Lutters, M.A., de, K.B., Kaptein, R., Bonvin , A.M., and van Nuland, N.A. (2004). The nisin-lipid II complex reveals a pyrophosphate cage that provides a blueprint for novel antibiotics. Nat. Struct. Mol. Biol. *11*, 963–967.

Huda, N., Lee, E.W., Chen, J., Morita, Y., Kuroda, T., Mizushima, T., and Tsuchiya, T. (2003). Molecular cloning and characterization of an ABC multidrug efflux pump, VcaM, in Non-O1 *Vibrio cholerae*. Antimicrob. Agents Chemother. *47*, 2413–2417.

Hutchinson, C.R. and Colombo, A.L. (1999). Genetic engineering of doxorubicin production in
: a review. J. Ind. Microbiol. Biotechnol. *23*, 647–652.

Hvorup, R.N., Goetz, B.A., Niederer, M., Hollenstein, K., Perozo, E., and Locher, K.P. (2007). Asymmetry in the structure of the ABC transporter-binding protein complex BtuCD-BtuF. Science *317*, 1387–1390.

Hyde, S.C., Emsley, P., Hartshorn, M.J., Mimmack, M.M., Gileadi, U., Pearce, S.R., Gallagher, M.P., Gill, D.R., Hubbard, R.E., and Higgins, C.F. (1990). Structural model of ATP-binding proteins associated with cystic fibrosis, multidrug resistance and bacterial transport. Nature *346*, 362–365.

Jenkinson, H.F. (1996). Ins and outs of antimicrobial resistance: era of the drug pumps. J. Dent. Res. *75*, 736–742.

Jonas, B.M., Murray, B.E., and Weinstock, G.M. (2001). Characterization of emeA, a NorA homolog and multidrug resistance efflux pump, in *Enterococcus faecalis*. Antimicrob. Agents Chemother. *45*, 3574–3579.

Jones, P.M. and George, A.M. (1999). Subunit interactions in ABC transporters: towards a functional architecture. FEMS Microbiol. Lett. *179*, 187–202.

Jones, P.M. and George, A.M. (2004). The ABC transporter structure and mechanism: perspectives on recent research. Cell Mol. Life Sci. *61*, 682–699.

Juliano, R.L. and Ling, V. (1976). A surface glycoprotein modulating drug permeability in Chinese hamster ovary cell mutants. Biochim. Biophys. Acta *455*, 152–162.

Kadaba, N.S., Kaiser, J.T., Johnson, E., Lee, A., and Rees, D.C. (2008). The high-affinity *E. coli* methionine ABC transporter: structure and allosteric regulation. Science *321*, 250–253.

Karow, M. and Georgopoulos, C. (1993). The essential *Escherichia coli* msbA gene, a multicopy suppressor of null mutations in the htrB gene, is related to the universally conserved family of ATP-dependent translocators. Mol. Microbiol. *7*, 69–79.

Kaur, P. and Russell, J. (1998). Biochemical coupling between the DrrA and DrrB proteins of the doxorubicin efflux pump of *Streptomyces peucetius*. J. Biol. Chem. *273*, 17933–17939.

Kerr, I.D. (2002). Structure and association of ATP-binding cassette transporter nucleotide-binding domains. Biochim. Biophys. Acta *1561*, 47–64.

Kobayashi, N., Nishino, K., and Yamaguchi, A. (2001). Novel macrolide-specific ABC-type efflux transporter in *Escherichia coli*. J. Bacteriol. *183*, 5639–5644.

Kruh, G.D. and Belinsky, M.G. (2003). The MRP family of drug efflux pumps. Oncogene *22*, 7537–7552.

Kuipers, A., de, B.E., Rink, R., Fekken, S., Kluskens, L.D., Driessen, A.J., Leenhouts, K., Kuipers, O.P., and Moll, G.N. (2004). NisT, the transporter of the lantibiotic nisin, can transport fully modified, dehydrated, and unmodified prenisin and fusions of the leader peptide with non-lantibiotic peptides. J. Biol. Chem. *279*, 22176–22182.

Lage, H. (2003). ABC transporters: implications on drug resistance from microorganisms to human cancers. Int. J. Antimicrob. Agents 22, 188–199.

Lee, E.W., Chen, J., Huda, M.N., Kuroda, T., Mizushima, T., and Tsuchiya, T. (2003a). Functional cloning and expression of emeA, and characterization of EmeA, a multidrug efflux pump from *Enterococcus faecalis*. Biol. Pharm. Bull. *26*, 266–270.

Lee, E.W., Huda, M.N., Kuroda, T., Mizushima, T., and Tsuchiya, T. (2003b). EfrAB, an ABC multidrug efflux pump in *Enterococcus faecalis*. Antimicrob. Agents Chemother. *47*, 3733–3738.

Levy, S.B. (1998). The challenge of antibiotic resistance. Sci. Am. *278*, 46–53.

Levy, S.B. and Marshall, B. (2004). Antibacterial resistance worldwide: causes, challenges and responses. Nat. Med. *10*, S122-S129.

Li, X.Z. and Nikaido, H. (2004). Efflux-mediated drug resistance in bacteria. Drugs *64*, 159–204.

Ling, V. (1997). Multidrug resistance: molecular mechanisms and clinical relevance. Cancer Chemother. Pharmacol. *40 Suppl*, S3-S8.

Linton, K.J. and Higgins, C.F. (1998). The *Escherichia coli* ATP-binding cassette (ABC) proteins. Mol. Microbiol. *28*, 5–13.

Liu, L.X., Janvier, K., Berteaux-Lecellier, V., Cartier, N., Benarous, R., and Aubourg, P. (1999). Homo- and heterodimerization of peroxisomal ATP-binding cassette half-transporters. J. Biol. Chem. *274*, 32738–32743.

Locher, K.P. (2004). Structure and mechanism of ABC transporters. Curr. Opin. Struct. Biol. *14*, 426–431.

Locher, K.P., Lee, A.T., and Rees, D.C. (2002). The *E. coli* BtuCD structure: a framework for ABC transporter architecture and mechanism. Science *296*, 1091–1098.

Lomovskaya, O. and Lewis, K. (1992). Emr, an *Escherichia coli* locus for multidrug resistance. Proc. Natl. Acad. Sci. U. S. A.*89*, 8938–8942.

Lomovskaya, O., Lewis, K., and Matin, A. (1995). EmrR is a negative regulator of the *Escherichia coli* multidrug resistance pump EmrAB. J. Bacteriol. *177*, 2328–2334.

Lubelski, J., Konings, W.N., and Driessen, A.J. (2007). Distribution and physiology of ABC-type transporters contributing to multidrug resistance in bacteria. Microbiol. Mol. Biol. Rev. *71*, 463–476.

Lubelski, J., Mazurkiewicz, P., van Merkerk, R., Konings, W.N., and Driessen, A.J.M. (2004). *ydaG* and *ydbA* of *Lactococcus lactis* encode a heterodimeric ATP-binding cassette-type multidrug transporter. J. Biol. Chem. *279*, 34449–34455.

Lubelski, J., Rink, R., Khusainov, R., Moll, G.N., and Kuipers, O.P. (2008). Biosynthesis, immunity, regulation, mode of action and engineering of the model lantibiotic nisin. Cell Mol. Life Sci. *65*, 455–476.

Lubelski, J., van Merkerk, R., Konings, W.N., and Driessen , A.J.M. (2006a). Nucleotide-binding sites of the heterodimeric LmrCD ABC-Multidrug transporter of *Lactococcus lactis* are asymmetric. Biochemistry *45*, 648–656.

Lubelski, J., de Jong, A., van Merkerk, R., Agustiandari, H., Kuipers, O.P., Kok, J., and Driessen, A.J.M. (2006b). LmrCD is a major multidrug resistance transporter in *Lactococcus lactis*. Molecular Microbiology *61*, 771–781.

Ma, D., Cook, D.N., Alberti, M., Pon, N.G., Nikaido, H., and Hearst, J.E. (1993). Molecular cloning and characterization of acrA and acrE genes of *Escherichia coli*. J. Bacteriol. *175*, 6299–6313.

Ma, D., Cook, D.N., Alberti, M., Pon, N.G., Nikaido, H., and Hearst, J.E. (1995). Genes acrA and acrB encode a stress-induced efflux system of *Escherichia coli*. Mol. Microbiol. *16*, 45–55.

Madoori, P.K., Agustiandari, H., Driessen, A.J.M., and Thunnissen, A.M. (2008). Structure of the transcriptional regulator LmrR and its mechanism of multidrug recognition. MBO J. *28*, 156–166.

Marger, M.D. and Saier, M.H., Jr. (1993). A major superfamily of transmembrane facilitators that catalyse uniport, symport and antiport. Trends Biochem. Sci. *18*, 13–20.

Margolles, A., Florez, A.B., Moreno, J.A., van, S.D., and de Los Reyes-Gavilan, C.G. (2006). Two membrane proteins from *Bifidobacterium breve* UCC2003 constitute an ABC-type multidrug transporter. Microbiology *152*, 3497–3505.

Margolles, A., Putman, M., van Veen, H.W., and Konings, W.N. (1999). The purified and functionally reconstituted multidrug transporter LmrA of *Lactococcus lactis* mediates the transbilayer movement of specific fluorescent phospholipids. Biochemistry *38*, 16298–16306.

Markham, P.N., Ahmed, M., and Neyfakh, A.A. (1996). The drug-binding activity of the multidrug-responding transcriptional regulator BmrR resides in its C-terminal domain. J. Bacteriol. *178*, 1473–1475.

Marrer, E., Satoh, A.T., Johnson, M.M., Piddock, L.J., and Page, M.G. (2006a). Global transcriptome analysis of the responses of a fluoroquinolone-resistant *Streptococcus pneumoniae* mutant and its parent to ciprofloxacin. Antimicrob. Agents Chemother. *50*, 269–278.

Marrer, E., Schad, K., Satoh, A.T., Page, M.G., Johnson, M.M., and Piddock, L.J. (2006b). Involvement of the putative ATP-dependent efflux proteins PatA and PatB in fluoroquinolone resistance of a multidrug-resistant mutant of *Streptococcus pneumoniae*. Antimicrob. Agents Chemother. *50*, 685–693.

Matsuo, T., Chen, J., Minato, Y., Ogawa, W., Mizushima, T., Kuroda, T., and Tsuchiya, T. (2008). SmdAB, a heterodimeric ABC-Type multidrug efflux pump, in Serratia marcescens. J. Bacteriol. *190*, 648–654.

McMurry, L., Petrucci, R.E., Jr., and Levy, S.B. (1980). Active efflux of tetracycline encoded by four genetically different tetracycline resistance determinants in *Escherichia coli*. Proc. Natl. Acad. Sci. U. S. A.*77*, 3974–3977.

Mimmack, M.L., Gallagher, M.P., Pearce, S.R., Hyde, S.C., Booth, I.R., and Higgins, C.F. (1989). Energy coupling to periplasmic binding protein-dependent transport systems: stoichiometry of ATP hydrolysis during transport *in vivo*. Proc. Natl. Acad. Sci. U. S. A.*86*, 8257–8261.

Miyake, K., Mickley, L., Litman, T., Zhan, Z., Robey, R., Cristensen, B., Brangi, M., Greenberger, L., Dean, M., Fojo, T., and Bates, S.E. (1999). Molecular cloning of cDNAs which are highly overexpressed in mitoxantrone-resistant cells: demonstration of homology to ABC transport genes. Cancer Res. *59*, 8–13.

Murakami, S., Nakashima, R., Yamashita, E., Matsumoto, T., and Yamaguchi, A. (2006). Crystal structuRes. of a multidrug transporter reveal a functionally rotating mechanism. Nature *443*, 173–179.

Neu, H.C. (1992). The crisis in antibiotic resistance. Science *257*, 1064–1073.

Neyfakh, A.A., Bidnenko, V.E., and Chen, L.B. (1991). Efflux-mediated multidrug resistance in Bacillus subtilis: similarities and dissimilarities with the mammalian system. Proc. Natl. Acad. Sci. U. S. A.*88*, 4781–4785.

Nikaido, H. (1994). Prevention of drug access to bacterial targets: permeability barriers and active efflux. Science *264*, 382–388.

Nikaido, H. (1996). Multidrug efflux pumps of gram-negative bacteria. J. Bacteriol. *178*, 5853–5859.

Nishino, K., Latifi, T., and Groisman, E.A. (2006). Virulence and drug resistance roles of multidrug efflux systems of Salmonella enterica serovar Typhimurium. Mol. Microbiol. *59*, 126–141.

Nishino, K., Nikaido, E., and Yamaguchi, A. (2007). Regulation of multidrug efflux systems involved in multidrug and metal resistance of Salmonella enterica serovar Typhimurium. J. Bacteriol. *189*, 9066–9075.

Nishino, K. and Yamaguchi, A. (2001). Analysis of a complete library of putative drug transporter genes in *Escherichia coli*. J. Bacteriol. *183*, 5803–5812.

Nishino, K. and Yamaguchi, A. (2002). EvgA of the two-component signal transduction system modulates production of the yhiUV multidrug transporter in *Escherichia coli*. J. Bacteriol. *184*, 2319–2323.

Oldham, M.L., Khare, D., Quiocho, F.A., Davidson, A.L., and Chen, J. (2007). Crystal structure of a catalytic intermediate of the maltose transporter. Nature *450*, 515–521.

Orelle, C., Dalmas, O., Gros, P., Di Pietro, A., and Jault, J.M. (2003). The conserved glutamate residue adjacent to the Walker-B motif is the catalytic base for ATP hydrolysis in the ATP-binding cassette transporter BmrA. J. Biol. Chem. *278*, 47002–47008.

Orelle, C., Gubellini, F., Durand, A., Marco, S., Levy, D., Gros, P., Di, P.A., and Jault, J.M. (2008). Conformational change induced by ATP binding in the multidrug ATP-binding cassette transporter BmrA. Biochemistry *47*, 2404–2412.

Paulsen, I.T. (2003). Multidrug efflux pumps and resistance: regulation and evolution. Curr. Opin. Microbiol. *6*, 446–451.

Paulsen, I.T. and Lewis, K. (2001). Microbial multidrug efflux: introduction. J. Mol. Microbiol. Biotechnol. 3, 143–144.

Paulsen, I.T., Nguyen, L., Sliwinski, M.K., Rabus, R., and Saier, M.H., Jr. (2000). Microbial genome analyses: comparative transport capabilities in eighteen prokaryotes. J. Mol. Biol. *301*, 75–100.

Paulsen, I.T., Skurray, R.A., Tam, R., Saier, M.H., Jr., Turner, R.J., Weiner, J.H., Goldberg, E.B., and Grinius, L.L. (1996). The SMR family: a novel family of multidrug efflux proteins involved with the efflux of lipophilic drugs. Mol. Microbiol. *19*, 1167–1175.

Paulsen, I.T., Sliwinski, M.K., and Saier, M.H., Jr. (1998). Microbial genome analyses: global comparisons of transport capabilities based on phylogenies, bioenergetics and substrate specificities. J. Mol. Biol. *277*, 573–592.

Peschel, A. and Gotz, F. (1996). Analysis of the Staphylococcus epidermidis genes epiF, -E, and -G involved in epidermin immunity. J. Bacteriol. *178*, 531–536.

Piddock, L.J. (2006a). Clinically relevant chromosomally encoded multidrug resistance efflux pumps in bacteria. Clin. Microbiol. Rev. *19*, 382–402.

Piddock, L.J. (2006b). Multidrug-resistance efflux pumps – not just for resistance. Nat. Rev. Microbiol. *4*, 629–636.

Pinkett, H.W., Lee, A.T., Lum, P., Locher, K.P., and Rees, D.C. (2007). An inward-facing conformation of a putative metal-chelate-type ABC transporter. Science *315*, 373–377.

Poole, K. (2004). Efflux-mediated multiresistance in Gram-negative bacteria. Clin. Microbiol. Infect. *10*, 12–26.

Poole, K. (2005). Efflux-mediated antimicrobial resistance. J. Antimicrob. Chemother. *56*, 20–51.

Putman, M., van Veen, H.W., Degener, J.E., and Konings, W.N. (2000a). Antibiotic resistance: era of the multidrug pump. Mol. Microbiol. *36*, 772–773.

Putman, M., van Veen, H.W., and Konings, W.N. (2000b). Molecular properties of bacterial multidrug transporters. Microbiol. Mol. Biol. Rev. *64*, 672–693.

Ra, R., Beerthuyzen, M.M., de Vos, W.M., Saris, P.E., and Kuipers, O.P. (1999). Effects of gene disruptions in the nisin gene cluster of *Lactococcus lactis* on nisin production and producer immunity. Microbiology *145 (Pt 5)*, 1227–1233.

Ra, S.R., Qiao, M., Immonen, T., Pujana, I., and Saris, E.J. (1996). Genes responsible for nisin synthesis, regulation and immunity form a regulon of two operons and are induced by nisin in Lactoccocus lactis N8. Microbiology *142 (Pt 5)*, 1281–1288.

Raherison, S., Gonzalez, P., Renaudin, H., Charron, A., Bebear, C., and Bebear, C.M. (2005). Increased expression of two multidrug transporter-like genes is associated with ethidium bromide and ciprofloxacin resistance in Mycoplasma hominis. Antimicrob. Agents Chemother. *49*, 421–424.

Ramos, J.L., Martinez-Bueno, M., Molina-Henares, A.J., Teran, W., Watanabe, K., Zhang, X., Gallegos, M.T., Brennan, R., and Tobes, R. (2005). The TetR family of transcriptional repressors. Microbiol. Mol. Biol. Rev. *69*, 326–356.

Ren, Q., Kang, K.H., and Paulsen, I.T. (2004). TransportDB: a relational database of cellular membrane transport systems. Nucleic Acids Res. 32, D284-D288.

Reuter, G., Janvilisri, T., Venter, H., Shahi, S., Balakrishnan, L., and van Veen, H.W. (2003). The ATP binding cassette multidrug transporter LmrA and lipid transporter MsbA have overlapping substrate specificities. J. Biol. Chem. *278*, 35193–35198.

Robertson, G.T., Doyle, T.B., and Lynch, A.S. (2005). Use of an efflux-deficient *Streptococcus pneumoniae* strain panel to identify ABC-class multidrug transporters involved in intrinsic resistance to antimicrobial agents. Antimicrob. Agents Chemother. *49*, 4781–4783.

Ross, R.P., Morgan, S., and Hill, C. (2002). Preservation and fermentation: past, present and future. Int. J. Food Microbiol. *79*, 3–16.

Rouquette-Loughlin, C.E., Balthazar, J.T., and Shafer, W.M. (2005). Characterization of the MacA-MacB efflux system in *Neisseria gonorrhoeae*. J. Antimicrob. Chemother. *56*, 856–860.

Saier, M.H., Jr. (2000). A functional-phylogenetic classification system for transmembrane solute transporters. Microbiol. Mol. Biol. Rev. *64*, 354–411.

Saier, M.H., Jr., Tam, R., Reizer, A., and Reizer, J. (1994). Two novel families of bacterial membrane proteins concerned with nodulation, cell division and transport. Mol. Microbiol. *11*, 841–847.

Sakamoto, K. and Konings, W.N. (2003). Beer spoilage bacteria and hop resistance. Int. J. Food Microbiol. *89*, 105–124.

Sakamoto, K., Margolles, A., van Veen, H.W., and Konings, W.N. (2001). Hop resistance in the beer spoilage bacterium *Lactobacillus brevis* is mediated by the ATP-binding cassette multidrug transporter HorA. J. Bacteriol. *183*, 5371–5375.

Schmitt, L. and Tampe, R. (2002). Structure and mechanism of ABC transporters. Curr. Opin. Struct. Biol. *12*, 754–760.

Schneider, R.F. and Rosen, M.J. (1999). Pneumococcal infections in HIV-infected adults. Semin. Respir. Infect. *14*, 237–242.

Schumacher, M.A., Miller, M.C., and Brennan, R.G. (2004). Structural mechanism of the simultaneous binding of two drugs to a multidrug-binding protein. EMBO J. *23*, 2923–2930.

Schumacher, M.A., Miller, M.C., Grkovic, S., Brown, M.H., Skurray, R.A., and Brennan, R.G. (2001). Structural mechanisms of QacR induction and multidrug recognition. Science *294*, 2158–2163.

Seeger, M.A., Schiefner, A., Eicher, T., Verrey, F., Diederichs, K., and Pos, K.M. (2006). Structural asymmetry of AcrB trimer suggests a peristaltic pump mechanism. Science *313*, 1295–1298.

Senior, A.E. and Gadsby, D.C. (1997). ATP hydrolysis cycles and mechanism in P-glycoprotein and CFTR. Semin. Cancer Biol. *8*, 143–150.

Siegers, K. and Entian, K.D. (1995). Genes involved in immunity to the lantibiotic nisin produced by *Lactococcus lactis* 6F3. Appl. Environ. Microbiol. *61*, 1082–1089.

Smith, P.C., Karpowich, N., Millen, L., Moody, J.E., Rosen, J., Thomas, P.J., and Hunt, J.F. (2002). ATP binding to the motor domain from an ABC transporter drives formation of a nucleotide sandwich dimer. Mol. Cell *10*, 139–149.

Spratt, B.G. (1994). Resistance to antibiotics mediated by target alterations. Science *264*, 388–393.

Stein, T., Heinzmann, S., Solovieva, I., and Entian, K.D. (2003). Function of *Lactococcus lactis* nisin immunity genes nisI and nisFEG after coordinated expression in the surrogate host Bacillus subtilis. J. Biol. Chem. *278*, 89–94.

Steinfels, E., Orelle, C., Dalmas, O., Penin, F., Miroux, B., Di Pietro, A., and Jault, J.M. (2002). Highly efficient over-production in *E. coli* of YvcC, a multidrug-like ATP-binding cassette transporter from *Bacillus subtilis*. Biochim. Biophys. Acta *1565*, 1–5.

Steinfels, E., Orelle, C., Fantino, J.R., Dalmas, O., Rigaud, J.L., Denizot, F., Di Pietro, A., and Jault, J.M. (2004). Characterization of YvcC (BmrA), a multidrug ABC transporter constitutively expressed in *Bacillus subtilis*. Biochemistry *43*, 7491–7502.

Sulavik, M.C., Houseweart, C., Cramer, C., Jiwani, N., Murgolo, N., Greene, J., DiDomenico, B., Shaw, K.J., Miller, G.H., Hare, R., and Shimer, G. (2001). Antibiotic susceptibility profiles of *Escherichia coli* strains lacking multidrug efflux pump genes. Antimicrob. Agents Chemother. *45*, 1126–1136.

Tauch, A., Krieft, S., Puhler, A., and Kalinowski, J. (1999). The tetAB genes of the Corynebacterium striatum R-plasmid pTP10 encode an ABC transporter and confer tetracycline, oxytetracycline and oxacillin resistance in Corynebacterium glutamicum. FEMS Microbiol. Lett. *173*, 203–209.

Tennent, J.M., Lyon, B.R., Midgley, M., Jones, I.G., Purewal, A.S., and Skurray, R.A. (1989). Physical and biochemical characterization of the qacA gene encoding antiseptic and disinfectant resistance in *Staphylococcus aureus*. J. Gen. Microbiol. *135*, 1–10.

Thanassi, D.G., Cheng, L.W., and Nikaido, H. (1997). Active efflux of bile salts by *Escherichia coli*. J. Bacteriol. *179*, 2512–2518.

Tiirikainen, M.I. and Krusius, T. (1991). Multidrug resistance. Ann. Med. 23, 509–520.

Tikhonova, E.B., Devroy, V.K., Lau, S.Y., and Zgurskaya, H.I. (2007). Reconstitution of the *Escherichia coli* macrolide transporter: the periplasmic membrane fusion protein MacA stimulates the ATPase activity of MacB. Mol. Microbiol. *63*, 895–910.

Travis, J. (1994). Reviving the antibiotic miracle? Science *264*, 360–362.

van den Berg van Saparoea, H.B., Bakkes, P.J., Moll, G.N., and Driessen, A.J. (2008). Distinct contributions of the nisin biosynthesis enzymes NisB and NisC and transporter NisT to prenisin production by *Lactococcus lactis*. Appl. Environ. Microbiol. *74*, 5541–5548.

van den Berg van Saparoea, H., Lubelski, J., van Merkerk, R., Mazurkiewicz, P.S., and Driessen, A.J.M. (2005). Proton motive force-dependent Hoechst 33342 transport by the ABC transporter LmrA of *Lactococcus lactis*. Biochemistry *44*, 16931–16938.

van der Does, C. and Tampe, R. (2004). How do ABC transporters drive transport? Biol. Chem. *385*, 927–933.

van Veen, H.W., Callaghan, R., Soceneantu, L., Sardini, A., Konings, W.N., and Higgins, C.F. (1998). A bacterial antibiotic-resistance gene that complements the human multidrug-resistance P-glycoprotein gene. Nature *391*, 291–295.

van Veen, H.W., Margolles, A., Muller, M., Higgins, C.F., and Konings, W.N. (2000). The homodimeric ATP-binding cassette transporter LmrA mediates multidrug transport by an alternating two-site (two-cylinder engine) mechanism. EMBO J. *19*, 2503–2514.

van Veen, H.W., Venema, K., Bolhuis, H., Oussenko, I., Kok, J., Poolman, B., Driessen, A.J., and Konings, W.N. (1996). Multidrug resistance mediated by a bacterial homolog of the human multidrug transporter MDR1. Proc. Natl. Acad. Sci. U. S. A. *93*, 10668–10672.

Velamakanni, S., Yao, Y., Gutmann, D.A., and van Veen, H.W. (2008). Multidrug transport by the ABC transporter Sav1866 from *Staphylococcus aureus*. Biochemistry *47*, 9300–9308.

Walker, J.E., Saraste, M., Runswick, M.J., and Gay, N.J. (1982). Distantly related sequences in the alpha- and beta-subunits of ATP synthase, myosin, kinases and other ATP-requiring enzymes and a common nucleotide binding fold. EMBO J. *1*, 945–951.

Walsh, C. (2000). Molecular mechanisms that confer antibacterial drug resistance. Nature *406*, 775–781.

Ward, A., Reyes, C.L., Yu, J., Roth, C.B., and Chang, G. (2007). Flexibility in the ABC transporter MsbA: Alternating access with a twist. Proc. Natl. Acad. Sci. U. S. A.*104*, 19005–19010.

Wegmann, U., O'Connell-Motherway, M., Zomer, A., Buist, G., Shearman, C., Canchaya, C., Ventura, M., Gotesmann, A., Gasson, M.J., Kuipers, O.P., van, S.D., and Kok, J. (2007). Complete genome sequence of the prototype lactic acid bacterium *Lactococcus lactis* subsp. cremoris MG1363. J. Bacteriol. *189*, 3256–3270.

Woebking, B., Reuter, G., Shilling, R.A., Velamakanni, S., Shahi, S., Venter, H., Balakrishnan, L., and van Veen, H.W. (2005). Drug-lipid A interactions on the *Escherichia coli* ABC transporter MsbA. J. Bacteriol. *187*, 6363–6369.

Wood, M.J. and Moellering, R.C., Jr. (2003). Microbial resistance: bacteria and more. Clin. Infect. Dis. *36*, S2–S3.

Woolridge, D.P., Vazquez-Laslop, N., Markham, P.N., Chevalier, M.S., Gerner, E.W., and Neyfakh, A.A. (1997). Efflux of the natural polyamine spermidine facilitated by the Bacillus subtilis multidrug transporter Blt. J. Biol. Chem. *272*, 8864–8866.

Yamanaka, H., Kobayashi, H., Takahashi, E., and Okamoto, K. (2008). MacAB Is Involved in the Secretion of *Escherichia coli* Heat-Stable Enterotoxin II. J. Bacteriol. *190*:7693–7698..

Yokota, A., Veenstra, M., Kurdi, P., van Veen, H.W., and Konings, W.N. (2000). Cholate resistance in *Lactococcus lactis* is mediated by an ATP-dependent multispecific organic anion transporter. J. Bacteriol. *182*, 5196–5201.

Yoshida, H., Bogaki, M., Nakamura, S., Ubukata, K., and Konno, M. (1990). Nucleotide sequence and characterization of the *Staphylococcus aureus* norA gene, which confers resistance to quinolones. J. Bacteriol. *172*, 6942–6949.

Young, J. and Holland, I.B. (1999). ABC transporters: bacterial exporters-revisited five years on. Biochim. Biophys. Acta *1461*, 177–200.

Zaidi, A.H., Bakkes, P.J., Lubelski, J., Agustiandari, H., Kuipers, O.P., and Driessen, A.J. (2008). The ABC-type MDR transporter LmrCD is responsible for an extrusion based mechanism of bile acid resistance in *Lactococcus lactis*. J. Bacteriol. *190*: 7357–7366.

Zaitseva, J., Jenewein, S., Wiedenmann, A., Benabdelhak, H., Holland, I.B., and Schmitt, L. (2005). Functional characterization and ATP-induced dimerization of the isolated ABC-domain of the haemolysin B transporter. Biochemistry *44*, 9680–9690.

Zheleznova, E.E., Markham, P.N., Neyfakh, A.A., and Brennan, R.G. (1999). Structural basis of multidrug recognition by BmrR, a transcription activator of a multidrug transporter. Cell *96*, 353–362.

Zhou, Z., White, K.A., Polissi, A., Georgopoulos, C., and Raetz, C.R. (1998). Function of *Escherichia coli* MsbA, an essential ABC family transporter, in lipid A and phospholipid biosynthesis. J. Biol. Chem. *273*, 12466–12475.

ABC Transporters in *Plasmodium falciparum* and their Involvement in Resistance to Antimalarial Drugs

6

Bruno Pradines, Véronique Parquet and Eve Orlandi-Pradines

Abstract

Malaria is by far the world's most important parasitic disease. Unfortunately, the extensive use of drugs led to the emergence of drug resistance. Quinoline resistance is associated with decreased accumulation of drugs in intracellular vacuoles, due to reduced drug uptake into the cell, increased efflux from the cell or a combination of both. A number of candidate transporters in *P. falciparum*, including ABC transporters, such as Pgh1 (PfMDR1) and PfMRP, have been proposed to be involved in antimalarial resistance.

Targeting these transport proteins with specific drugs may permit to reverse quinoline resistance. Compounds, belonging to different pharmacological classes have demonstrated in the past decade a promising capability to reverse the antimalarial resistance *in vitro* of parasite isolates, as well as in animal models and human malaria. Herein we will present the progress made in biochemical and genetic basis of antimalarial resistance, emphasizing the recent developments on drugs, which interfere with transmembrane proteins involved in drug efflux, or uptake.

Introduction

Malaria is by far the most prominent infectious disease, affecting 300–500 million people and killing three million people every year, including 200–300 children per hour. It is endemic in most tropical regions, placing 2 billion people at risk. From its virtual eradication in the 1960s, malaria has re-emerged to such an extent that more than 41% of the world's population is now at risk. Of the four *Plasmodium* species that infect humans, *P. falciparum* causes the most serious pathology, and the overwhelming parasitaemia results in high mortality rates if left untreated.

Until now, chemotherapy has been the main line of defence against malaria. The current expansion of the disease is due to the emergence and rapid spread of multidrug resistant forms of the parasites. In fact, *P. falciparum* resistance to chloroquine (CQ) creates a severe and increasing public health threat. This inexpensive and widely used drug has been the main defence tool used against malaria, and its ineffectiveness go with re-emergence of malaria-related morbidity and mortality. Initially reported from independent regions in South America in 1961 (Young *et al.*, 1961) and Southeast Asia in 1962 (Harinasuta *et al.*, 1965), CQ resistance is now found in all areas where malaria is endemic (Cowman *et al.*, 1990). Furthermore, failure of antimalarial prophylaxis based on the combination of CQ and proguanil (Barnes *et al.*, 1991) or mefloquine (Gari-Toussaint *et al.*, 2002) in addition to clinical failures with halofantrine (Basco *et al.*, 1991) and quinine (Molinier *et al.*, 1994) are now common in Africa.

One strategy to reduce the prevalence of malaria is the use of drugs in combination. In fact, since 2001, more than 60 countries have officially adopted artemisinin-based combination therapies (ACTs) for the treatment of falciparum malaria. Still, individual *P. falciparum* isolates resistant to artemisinin *in vitro* – whether associated with clinical failures or not – have been described in Cambodia (Jambou *et al.*, 2005).

An enormous scientific effort has been initiated to elucidate the mechanisms underlying resistance to antimalarial drugs with the hope of restoring and improving the efficacy of existing drugs and developing new drugs that can bypass resistance mechanisms. Among the various mechanisms identified until now that are related to drug resistance, those based on drug transport appear to play an important role by pumping drugs out of target sites and therefore decreasing their therapeutic concentrations.

Proteins that belong to the ATP-binding cassette (ABC) family provide the basis for multidrug resistance in mammalian cancer cells and in pathogenic yeasts, fungi, bacteria and parasites.

The ABC transporters are generally composed of four domains: two membrane domains (MDs), each usually with six putative α-helical transmembrane segments (TMS), and two nucleotide-binding domains (NBDs). The NBDs contain the 'Walker A' and 'Walker B' motifs and a short but highly conserved sequence, which are typical for members of the ABC family (see Chapter 1).

This chapter focuses in the involvement of ATP-binding cassette transporters in *Plasmodium* in drug resistance, emphasizing recent insights into the molecular interactions between drug compounds and transmembrane proteins involved in drug efflux or uptake.

ABC transporters and multidrug resistance in *Plasmodium* spp.

Resistance is associated with CQ concentration in vacuoles, which results from reduced drug uptake, increased drug efflux or a combination of both processes (Wellems, 2001). A number of *P. falciparum* genes that are involved in membrane transport have been proposed to play a role in CQ resistance. Restoration *in vitro* of the drug's efficacy by agents like verapamil is a characteristic feature of CQ-resistant strains that share characteristics with those present in cancer cells. Indeed, decreased CQ accumulation in resistant parasites (Krogstad *et al.*, 1987) and *in vitro* modulation of resistance by verapamil suggest that *P. falciparum* multidrug resistance could be similar to the mechanisms described for cancer cells. Thus, homologous proteins might be involved in both cases.

Four ABC transporter genes have been identified in *P. falciparum.*: one is a member of the P-gp cluster (P-glycoprotein), *pfmdr1* (Cowman *et al.*, 1991; Foote *et al.*, 1989; Wilson *et al.*, 1989) and a second one a member of the TAP cluster (Transporter associated with Antigen Processing), *Pfmdr2* (Zalis *et al.*, 1993; Rubio *et al.*, 1994). The last two are members of the MRP cluster (Multidrug Resistance associated Protein), *Pfgcn20* (Bozdech *et al.*, 1996; Bozdech *et al.*, 1998), which is a homologue of the yeast *Gcn20* gene, and *Pfa0590w* (Klokousas *et al.*, 2004) (Fig. 6.1).

Plasmodial ABC transporters included in P-gp cluster: PfMDR1 or Pgh1

The gene *pfmdr1* is overexpressed in chloroquine-resistant strains of *P. falciparum* (Wilson *et al.*, 1989). This gene encodes a 162 kDa protein named *P. falciparum* homologue of the P-glycoprotein (Pgh1) and is located on chromosome 5 (http://plasmodb.org, systematic name: PFE1150w, previous Id: MAL5P1.230). Pgh1 contains the four usual domains, two MDs and two NBDs (Figs 6.1 and 6.2), and is located in the membrane of the parasite digestive vacuole (Wilson *et al.*, 1989). Competition experiments carried out with photoaffinity-labelling of purified parasite digestive vacuole preparations associated with ATP analogues, indicate that ATP binds to a nucleotide site located in the NH_2 terminal moiety of the molecule, which is exposed on the cytoplasmic side of the vacuolar membrane (Karcz *et al.*, 1993).

Involvement of Pgh1 in CQ resistance

Three possible functions have been proposed for Pgh1. It was originally reported that Pgh1 pumped CQ out of the food vacuole and was expressed in CQ-resistant parasites (Krogstad *et al.*, 1987); however, the role of Pgh1 in CQ resistance is still debated. Initial studies demonstrated the relationship between *pfmdr1* amplification and overexpression in resistant parasites (Foote *et al.*, 1989). Nevertheless, subsequent studies failed to correlate gene overexpression and CQ resistance (Cowman *et al.*, 1991) and demonstrated that selection of CQ resistance was associated with a decrease in copy number of *pfmdr1* (Barnes *et al.*, 1992). Pretreatment with the anticonvulsant phe-

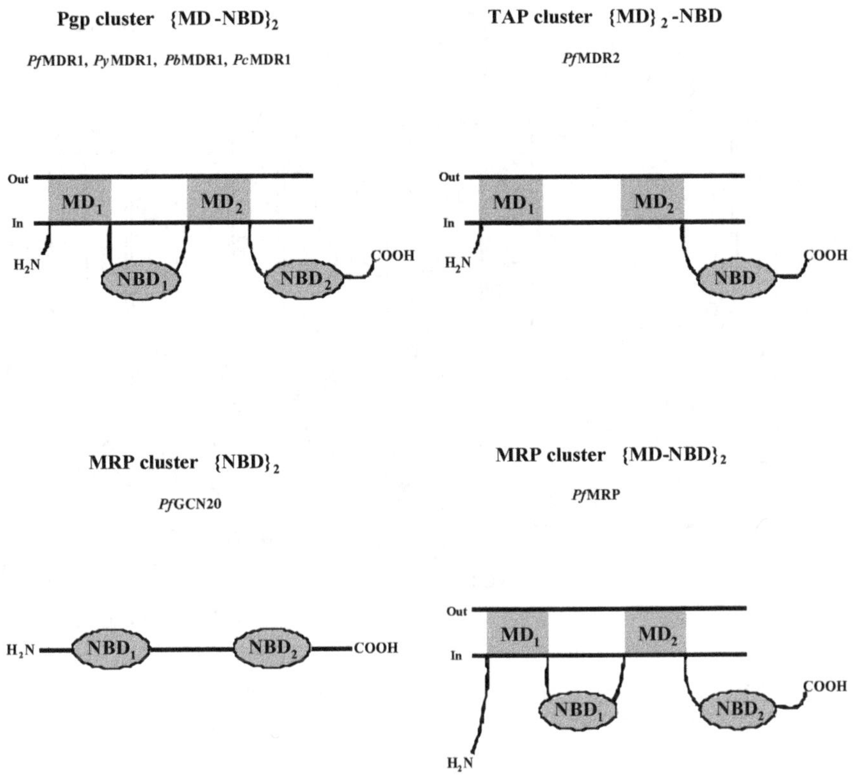

Figure 6.1 Predicted topological configuration of ATP-binding cassette (ABC) transporters in *Plasmodium* spp. ABC proteins in *Plasmodium spp*. belong to P-gp, MRP and TAP clusters. Members of the TAP cluster contain two membrane domains (MD_1 and MD_2) and only one nucleotide-binding domain (NBD). Typical members of the MRP cluster and the atypical ones, which contain two NBDs, but no MDs, are found in *Plasmodium spp*.

nobarbitone elicited a decreased susceptibility to CQ in both CQ-resistant and CQ-susceptible parasites, and this was associated with increased expression of Pgh1 (Johnson *et al.*, 2008).

Based on the previously discussed information it was then speculated that Pgh1 played an important role in CQ import into the parasites food vacuole (Van Es *et al.*, 1994b). In particular, there was a considerable controversy as to whether specific mutations in *pfmdr 1* actually conferred clinical drug resistance. However, work made in the field demonstrated the predictive value for CQ resistance of point mutations found in the *pfmdr1* sequence, resulting in amino acid changes that varied depending on the geographic area (Duraisingh *et al.*, 1997; Basco *et al.*, 1998). For example, point mutations, especially at codon 86, have been associated with a decrease in the CQ susceptibility (Foote *et al.*, 1990; Basco *et al.*, 1995; Adagu *et al.*, 1996). In some of these epidemiological studies, however, the number of CQ-susceptible samples was too limited to provide statistically a meaningful analysis (Duraisingh *et al.*, 1997; Grobusch *et al.*, 1998). In fact, no relationship, or only a weak relationship, was found in *P. falciparum* between CQ resistance and mutations in *pfmdr1* (Basco *et al.*, 1996; Basco *et al.*, 1998) in large epidemiological studies. Additionally, polymorphisms in *pfmdr1* such as point mutations at codons 184, 1034, 1042 and 1246 have been found and identified in both CQ-susceptible and -resistant parasites (Povoa *et al.*, 1998; Pickard *et al.*, 2003). Altogether, from these results it is tempting to conclude that Pgh1 is not involved in CQ import.

In fact, Pgh1 could act as a chloride channel or as a modulator of such a channel (van Es *et al.*, 1994a). As a chloride channel, Pgh1 may be constitutively expressed to maintain a vacuolar acidic pH *via* the proton motive force of the vacuolar

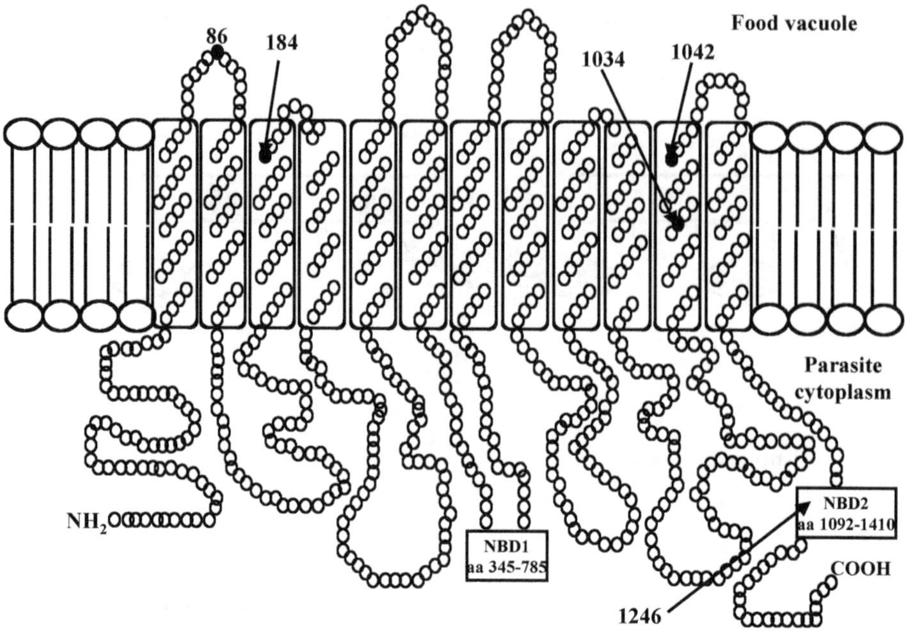

Figure 6.2 Predicted topology of the membrane ABC transporter Pgh1. Numbers indicate the mutated amino acids related to antimalarial drug resistance described in the literature. NBD, nucleotide-binding domain.

H^+-pump. Since the pH of the vacuole plays an important role in CQ action and resistance, it has been analysed whether CQ reaches the parasite food vacuole. The results document and confirm that CQ reaches and accumulates in the parasite food vacuole due to its weak base properties; this is even stronger in resistant parasites due to the locally acidic pH, which is higher than in susceptible parasites (Bray *et al.*, 1992b). CQ resistance is associated with increased vacuolar pH. Recently it was however revealed that the pH of the CQ-resistant parasites is significantly more acidic than in CQ-susceptible parasites (Saliba *et al.*, 1998; Dzekunov *et al.*, 2000). These findings diverge from earlier ideas that suggested that the reduced vacuolar pH leads to an increase in CQ accumulation (Bray *et al.*, 1992b). What is clear is that vacuoles isolated from CQ-susceptible strains actively accumulate significantly more CQ (3-fold increase) than those isolated from the CQ-resistant parasites (Yayon *et al.*, 1985). CQ reaches the parasite digestive vacuole, where it accumulates due to the local pH, binds to freely soluble ferriprotoporphyrin IX (FPIX), and prevents its detoxification process to insoluble dimeric haemozoin. The solubility of FPIX in the vacuole and the binding to CQ depend on vacuolar pH. Changes in pH are said to modify the conversion of soluble haem to insoluble aggregates (Dzekunov *et al.*, 2000). The aggregate form of FPIX dimers, which are promoted by the lower pH, is still capable of crystallization as haemozoin (Dorn *et al.*, 1998; Pagola 2000), but does not bind CQ as avidly as the freely soluble FPIX does (Dzekunov 2000). The lower vacuolar pH present in CQ-resistant parasites, which is close to the critical pH where alteration occurs, creates better conditions for the formation of FPIX (Ursos *et al.*, 2001). Indeed, acidification of the digestive vacuole would significantly decrease the free haem concentration available for the formation of toxic complexes with CQ.

Involvement of Pgh1 in resistance to other quinolines

Expression of *pfmdr1* gene by yeast cells causes resistance to quinine, mefloquine and halofantrine in these cells (Ruetz *et al.*, 1996). Amplification and overexpression of *pfmdr1* has been associated with mefloquine resistance and decreased halofantrine susceptibility in *P. falciparum* (Volkman *et al.*, 1993; Wilson *et al.*, 1993; Cowman *et al.*, 1994; Peel *et al.* 1994; Zalis *et al.*1998). In contrast, two additional studies describe that both the copy number and level of expression of *pfmdr1* are not altered in mefloquine or halofantrine resistant parasites (Lim *et al.*, 1996; Ritchie *et al.*, 1996). However, recently, the group of Price demonstrated that amplification of *pfmdr1* was the main cause of resistance to mefloquine in *P. falciparum* (Price *et al.*, 2004). Additionally they also argue that the *pfmdr1* copy number could also be used as a molecular marker to monitor mefloquine resistance in areas of emerging resistance (Price *et al.*, 1999; Pickard *et al.*, 2003).

P. falciparum strains with an Asn-to-Tyr mutation at codon 86 need significantly lower mefloquine concentrations to decrease their growth by 50% (IC50) than wild-type cells (Price *et al.*, 1999). This information correlates with analyses of *P. falciparum* isolates that have demonstrated an association between mutations at codon 86 and an increase in susceptibility to mefloquine, halofantrine or artemisinin derivatives (Price *et al.*, 1999; Duraisingh *et al.*, 2000a; Duraisingh *et al.*, 2000b; Ngo *et al.*, 2003). Substitution of wild-type *pfmdr1* gene by *pfmdr1* gene encoding the mutations at codons 1034 and 1042 abolish or reduce the level of resistance to mefloquine. Moreover, transfection with a wild-type *pfmdr1* allele at codons 1034, 1042 and 1246 confers mefloquine resistance to susceptible parasites (Reed *et al.*, 2000). Mutations at codons 1034, 1042 and 1246 in *P. falciparum pfmdr1* isolates, however, are not sufficient to explain the variation in mefloquine susceptibility because associations between polymorphism in *pfmdr1* and mefloquine resistance are not systematically observed (Basco *et al.*, 2002; Mawili *et al.*, 2002; Pillai *et al.*, 2003).

Pfmdr1 homologues have been identified in *P. chabaudi*, *pcmdr1* (Carlton *et al.*, 1998; Cravo *et al.*, 2003; Hunt *et al.*, 2004); in *P. yoelii*, *pymdr1* (Ferrer-Rodriguez *et al.*, 2004); and in *P. berghei*, *pbmdr1* (Gervais *et al.*, 1999). As has so far been described for *pfmdr1*, no association has been found between CQ susceptibility and *pcmdr1* (Carlton *et al.*, 1998; Cravo *et al.*, 2003; Hunt *et al.*, 2004), while overexpression of *pcmdr1*, *pbmdr1* or *pymdr1* gene products associates with mefloquine and artemisinin resistance (Carlton *et al.*, 1998; Gervais *et al.*, 1999; Ferrer-Rodriguez *et al.*, 2004).

The role of Pgh1 in quinine resistance is still unclear. Amplification of the *pfmdr1* gene has already been associated with quinine resistance (Sidhu *et al.*, 2006), but polymorphisms within *pfmdr1* had no effect on quinine accumulation in the food vacuole (Sanchez *et al.*, 2008b). Nevertheless, the same authors showed that wild type Pgh1 transported quinine, whereas polymorphic Pgh1 variants did not transport quinine (Sanchez *et al.*, 2008a). However, a recent study demonstrated that Pgh1 transport dye into the parasite food vacuole and that this transport activity was inhibited by quinine (Rohrbach *et al.*, 2006). An analogous finding was reported for human MDR in which QN also inhibited MDR transport function (Weaver *et al.*, 1993). Quinine can be both a substrate and an inhibitor of Pgh1.

Plasmodial ABC transporters included in TAP cluster: PfMDR2

With regard to other ABC transporters present in Plasmodia, *pfmdr2* is located in chromosome 14 and encodes a protein of 110 kDa termed *Pf*MDR2 (Rubio *et al.*, 1994; Bozdech *et al.*, 1996) (http://plasmodb.org, systematic name: PF14_0455). *Pf*MDR2 contains two membrane domains (MDs), each with five transmembrane domains and a single NBD, signature of the TAP cluster (Fig. 6.1). Amplification and overexpression of *pfmdr2* are still under debate: some studies have reported increased transcription of *pfmdr2* in CQ-resistant parasites (Ekong *et al.*, 1993), while other studies have demonstrated neither amplification nor overexpression of the gene occur in CQ-resistant parasites (Rubio *et al.*, 1994; Bozdech *et al.*, 1996). In conclusion, to date, and with the accumulated information, there is not a clear evidence to support that *Pf*MDR2 is involved in drug resistance.

Plasmodial ABC transporters included in MRP cluster: PfMRP or CAB63558 and PfGCN20

Pfa0590w encodes a protein of 210–215 kDa termed *Pf*MRP or CAB63558 (Klokousas *et al.*, 2004) (http://plasmodb.org, systematic name: PF0590w, previous Id: MAL1P3.03). This protein contains two MDs and two NBDs and is a member of the MRP cluster (Fig. 6.1). It shows structural similarity with the human MRP2 and contains the same structural domains described for Pgh1: TMD1, NBD1, TMD2 and NBD2. The NBDs are made up of the conserved motifs Walker A, Walker B and the ABC signature. The predicted protein lacks the additional NH2-terminal TMD (TMD0) that exists in human MRP2. Eleven probable transmembrane helices are predicted by the amino acids 121–142, 175–197, 318–340, 350–372, 426–448, 1158–1180, 1218–1240, 1250–1272, 1293–1310, 1314–1333 and 1392–1414. This transporter is localized in the digestive vacuole membrane of *Plasmodium*, similar to the transporters involved in drug resistance described above, and is expressed in both CQ-susceptible and CQ-resistant strains. Its structural similarities with human MRP2 suggest that PfMRP functions as an organic anion transporter in *P. falciparum* with substrates that include oxidized glutathione (GSSG) and glutathione conjugates. In fact, glutathione, which is known to guard *P. falciparum* from oxidative damage, may have an additional protective role by promoting haem catabolism. Ferriprotoporphyrin IX is demonstrably degraded by reduced glutathione (GSH) in a radical-mediated mechanism. Moreover, this degradation is inhibited by CQ. Increased GSH levels in *P. falciparum* infected cells increase resistance to CQ (Ginsburg *et al.*, 1998). Previous studies described two mutations in *pfmrp* gene that correlated with CQ resistance (Mu *et al.*, 2003; Ursing *et al.*, 2006; Henry *et al.*, 2008b). In fact, the *pfmrp* mutated haplotype 191Y and 437A was associated with a decrease in the CQ response.

Pfgcn20 is located in chromosome 11 and encodes a protein of 95.5 kDa termed *Pf*GCN20 (Bozdech *et al.*, 1996; Bozdech *et al.*, 1998) (http://plasmodb.org, systematic name: PF11_0255). This protein contains two NBDs but no MD and is a member of the MRP cluster (Fig. 6.1). PfGCN20 is localized to multiple regions of the infected erythrocyte, including the membrane and cytosolic compartments of the parasite (Bozdech *et al.*, 1998; Bozdech *et al.*, 1999). *Pf*GCN20 could be an ATP-binding subunit of a multimeric ABC transporter (Bozdech *et al.*, 1998; Bozdech *et al.*, 1999).

ABC transporter inhibitors and modulators of CQ resistance in *Plasmodium* spp.

Many molecules – related and unrelated to multidrug resistance mechanisms – have shown a capacity to modulate the level of resistance (see Chapter 9). This is especially true for the CQ response of CQ resistant plasmodial strains *in vitro*, without any effect on CQ susceptible strains. These molecules include calcium channel blockers, calmodulin inhibitors, antidepressant drugs, histamine H_1-receptor antagonists and synthetic surfactants.

Calcium channel blockers

Classical chemosensitizers such as calcium channel blockers (e.g., verapamil, diltiazem, tiapamil, nifedipine and nicardipine) and β-blockers (e.g., propranolol) are known to effectively overcome CQ resistance *in vitro* (Kyle *et al.*, 1990; Kyle *et al.*, 1993; Bray *et al.*, 1996, Menezes *et al.*, 2003; Henry *et al.* 2008a). Verapamil (VP) was the first active compound in reversing CQ resistance (Krogstad *et al.*, 1987; Martin *et al.*, 1987). Verapamil reversed the CQ resistance of laboratory strains but more interestingly of clinical isolates (Bayoumi *et al.*, 1993; Millet *et al.*, 2004b). Calcium channels are stereospecific for the S-(-)-isomer of the drug, but the reversal of resistance could be achieved by both the (R)-(+) and (S)-(−) isomers (Ye and Van Dyke, 1988).

Differences in CQ uptake versus efflux in CQ-susceptible and CQ-resistant parasites have been proposed to explain the lower CQ accumulation that occurs in resistant parasites, but since the available information is not yet clear, this debate continues (Bray *et al.*, 1992a; Bayoumi *et al.*, 1994, Pradines *et al.*, 2002b). Some authors have concluded, after evaluating CQ influx into infected erythrocytes that a decrease in CQ uptake could account for drug resistance (Verdier *et al.*, 1985; Ferrari *et al.*, 1991; Bray *et al.*, 1992a), while others propose that CQ-resistant parasites rapidly expel the drug via a pump-mediated mechanism (Krogstad *et al.*, 1987). When resistant parasites are treated with VP, CQ accumulation levels increase in parasites to reach CQ concentrations obtained in CQ susceptible parasites (Martiney *et al.*, 1995; Taylor *et al.*, 2000). When used at concentrations between 0.5 µM to 1 µM, VP was found to reverse CQ resistance from 40% to 85% (Martin *et al.*, 1987; Gerena *et al.*, 1992; Oduola *et al.*, 1998; Pradines *et al.*, 2002a). Moreover, VP at concentrations ranging from 2 to 10 µM was able to increase the level of CQ accumulation in CQ-resistant parasites by two- to six-fold (Martiney *et al.*, 1995; Bray *et al.*, 1999; Pradines *et al.*, 2002b).

CQ has little morphological effects on resistant parasites, while the combination of CQ and VP results in typical CQ-related food vacuolar swelling with increased amounts of granular matrix (Jacobs *et al.*, 1988). These effects are similar to those observed in drug-susceptible parasites treated with CQ alone or with a CQ-VP combination. Additionally, it has been found that VP normalizes vacuolar pH in CQ-resistant parasites to a value close to that measured in susceptible parasites (Ursos *et al.*, 2000). However, as pointed out before, the capacity to reverse resistance of VP cannot be explained by its effects on global drug accumulation alone.

With regard to malaria models in mice, daily injections of VP, nicardipine or diltiazem in combination with CQ reverse resistance in CQ-resistant *P. chabaudi* parasites (Tanabe *et al.*, 1990); this reversal is dose-dependent between 5 and 50 mg/kg of VP. Amlodipine increases CQ accumulation in infected mouse erythrocytes and potentiates CQ action against CQ-resistant *P. yoelii* strains (Deloron *et al.*, 1991). VP and nicardipine display dose-independent reversal activity against CQ-resistant *P. berghei* parasites *in vivo* in opposition to the other *Plasmodium*, and unlike *P. falciparum*, the CQ-susceptible strains are the most susceptible to these modulators (Platel *et al.*, 1998). This result supports the hypothesis that *P. berghei* has an innate low-level efflux mechanism that is susceptible to VP. The loss of the efficacy of VP during the increase of the level of CQ resistance could indicate that this mechanism of resistance is rapidly swamped, and that one (or more) mode(s) of resistance take over. Resistance in *P. berghei* seems to be incomparable with *P. falciparum*. In addition, the efficacy of VP and nicardipine decreases as CQ resistance increases. VP also sensitizes resistant parasites to quinine and amodiaquine *in vitro* (Bray *et al.*, 1996; Sidhu *et al.*, 2002; Johnson *et al.*, 2004).

Calmodulin inhibitors (phenothiazines)

Antipsychotic agents such as phenothiazine drugs (e.g., chlorpromazine, trifluoperazine, prochlorperazine) enhance the potency of chloroquine against *P. falciparum* CQ-resistant strains *in vitro* (Guan *et al.*, 2002; Kalkanidis *et al.*, 2002; Menezes *et al.* 2002). It now seems increasingly likely that phenothiazines exert their CQ resistance reversal activity by interacting with Pgh1 in *P. falciparum* (Reed *et al.*, 2000). Like verapamil, phenothiazine drugs are less effective to reverse CQ-resistance in South American parasites than in African and Southeast Asian isolates (Mehlotra *et al.*, 2001; Menezes *et al.*, 2002; Sidhu *et al.*, 2002). Chlorpromazine has been shown to bind to human P-glycoprotein and to modulate multidrug resistance in cancer cells (Zamora *et al.*, 1988). For this reason, it was undeniable to test chlorpromazine in *P. falciparum* resistant parasites. Indeed, combinations of chlorpromazine or prochlorperazine with CQ confirm the reversal effect of these drugs on CQ resistance as shown by results in *Aotus* monkeys infected with CQ-resistant *P. falciparum* strains (Kyle *et al.*, 1993). Moreover, daily injection of chlorpromazine or trifluoperazine reverses resistance to chloroquine in *P. chabaudi* resistant parasites (Miki *et al.*, 1992).

Tricyclic antidepressant drugs

Desipramine and other tricyclic antidepressant drugs (e.g., imipramine, fluoxetine, oxaprotiline, amitriptyline or citalopram) reverse CQ resistance in *P. falciparum in vitro* at concentrations observed in the plasma of human patients treated for depression (Bitonti *et al.*, 1988; Basco *et al.*, 1990b; Boulter *et al.* 1993). Although desipramine reverses CQ resistance *in vitro* (Menezes *et al.*, 1997), it does not enhance the efficacy of CQ in human clinical trials (Warsame *et al.*, 1992). Daily injection of desipramine, imipramine or citalopram in combination with CQ inhibited the growth of CQ-resistant *P. chabaudi* parasites in a dose-dependent manner (Miki *et al.*, 1992; Evans *et al.*, 1998). Combinations of desipramine with CQ reversed resistance in *Aotus* monkeys infected with CQ-resistant *P. falciparum* strains (Bitonti *et al.*, 1988). It has been shown that an aminoalkyl substitution at the N5-position of the heterocycle of the tricyclic antidepressants and a secondary or tertiary aliphatic aminoalkyl nitrogen with a two or three carbon long side-chain branched to the heteroatomic nitrogen (N5) were both required for reversal resistance (Bhattacharjee *et al.*, 2001; Bhattacharjee *et al.*, 2002). Tricyclic antidepressant drugs with different reuptake inhibition properties on various neurotransmitter systems were examined as potential reversal agents for CQ resistance. Unfortunately, CQ resistance reversal properties in *P. falciparum* with these compounds were not associated with their previous activities as transporter blockers (Taylor *et al.*, 2000).

Tricyclic histamine (H-1) receptor antagonists

Tricyclic histamine (H-1) receptor antagonists (e.g., cyproheptadine, promethazine, ketotifen and azatadine) also reverse CQ resistance of *P. falciparum in vitro* (Basco *et al.*, 1990a; Basco *et al.*, 1991). These compounds produce a marked reversal of the CQ resistance in *P. berghei* and *P. yoelii* strains *in vivo* (Peters *et al.*, 1990; Singh *et al.*, 2000a; Oduola *et al.*, 2004). *Aotus* monkeys infected with CQ-resistant *P. falciparum* parasites cured following treatment with CQ and promethazine (Oduola *et al.*, 1998). The combination of CQ with chlorpheniramine enhanced the efficacy of the former in *P. falciparum* malaria and was effective in CQ- and pyrimethamine/sulfadoxine-resistant malaria in Nigerian children (Sowunmi *et al.*, 1997; Sowunmi *et al.*, 1998, Okonkwo *et al.*, 1999). In addition, co-administration of cyproheptadine or ketotifen with halofantrine significantly improved the efficacy of the combination over treatment with halofantrine alone in halofantrine-resistant *P. yoelii* strains (Singh *et al.*, 2000b).

Synthetic surfactants

Nonylphenolethoxylates (NPEs) are synthetic surfactants used as intestinal permeability enhancers to improve oral drug delivery and have synergistic effects in combination with CQ (Crandall *et al.*, 2000). However, there is a subset of *P. falciparum* isolates that display polyethoxylated nonylphenol (NP30)-insensitive CQ resistance (Ciach *et al.*, 2003). The difference between NP30 activities against these isolates could not be explained by point mutations in *pfmdr1* since these mutations were present in all of the parasites used in the study. In addition, only poor interactions have been identified between mammalian P-glycoproteins and long-chain NPEs (Loe *et al.* 1993).

Synthetic 9,10-dihydroethanoanthracene derivatives

Synthetic 9,10-dihydroethanoanthracene derivatives have been shown to reverse the resistance to CQ of several strains and isolates (Alibert *et al.*, 1999; Alibert *et al.*, 2002; Pradines *et al.*, 2002c; Millet *et al.*, 2004a; Henry *et al.* 2008). Some of these compounds increase CQ accumulation in CQR *P. falciparum* parasites by five to eight times more than promethazine, while other dihydroethanoanthracenic reversal agents have no effect (Pradines *et al.*, 2002b).

In conclusion, all these functionally and structurally different compounds previously described share structural characteristics that may be important in chemosensitization. Most have a secondary or tertiary nitrogen atom, two planar rings and significant lipophilicity. Indeed, the presence of the amino group suggests that their reversal activity depends on the ability to have a protonatable nitrogen at physiological pH (Alibert *et al.*, 2002), i.e., it depends on their pKa value, which is weaker for inactive molecules. Substitution of the amine group is also a determining factor. In fact, secondary and tertiary amines are more active than primary and quaternary amines (Bhattacharjee *et al.*, 2002; Alibert *et al.*, 2002). Moreover, the reversal activity depends on the distance of the amino group from the two planar rings. A lack of aromatic pharmacophores also decreases activity drastically (Alibert *et al.*, 2002; Guan *et al.*, 2002).

ABC transporter inhibitors and modulators of quinine resistance in *Plasmodium* spp.

The mode of action of quinine is less documented than that of CQ. Like CQ, quinine is a quinoline drug, and in some cases, shows cross-resistance to CQ. Quinine shares some common characteristics with CQ. A parasite's resistance to CQ and quinine is directly related to its relative content of acid phospholipids and inversely related to the level of cholesterol in red blood cell membranes (Shalmiev *et al.*, 1993).

The role of mutations at residues 1034, 1042 and 1246 of the *pfmdr1* gene has been investigated using a transfection-based approach (Reed *et al.*, 2000). The results showed that these mutations affected quinine susceptibility, suggesting that *pfmdr1* mutations contribute to quinine resistance (Sidhu *et al.*, 2005). Recent findings have suggested that other loci such as the *pfmrp* gene may be involved in modulating quinine responses (Mu *et al.*, 2003; Klokousas *et al.*, 2004).

Although selection and transfection experiments have implicated these genes in quinine response, their relative contributions to drug responses and the way in which they interact with each other and additional genes to generate a specific drug response have not been elucidated.

The quinine response in *P. falciparum* can be modulated by molecules often found to modulate the CQ response in *P. falciparum*: namely, calcium channel blockers (e.g., verapamil, diltiazem and tiapamil) (Kyle *et al.*, 1990; Sidhu *et al.*, 2002; Menezes *et al.*, 2003), calmodulin inhibitors (e.g., chlorpromazine) (Kyle *et al.*, 1990) and synthetic surfactants (e.g., NP30) (Ciach *et al.*, 2003).

ABC transporter inhibitors and modulators of mefloquine resistance in *Plasmodium* spp.

The emergence of mefloquine-resistant *P. falciparum* has been documented over two decades. In preliminary studies, resistance to mefloquine and halofantrine was linked to mutations in *pfmdr1* (Peel *et al.*, 1994; Price *et al.*, 1999) and to increases in the *pfmdr1* gene copy number (Price *et al.*, 2004; Nelson *et al.*, 2005). The tyrosine-86 allele of *pfmdr1* has been associated with increased susceptibility to mefloquine and increased resistance to chloroquine (Duraisingh *et al.*, 2000a; Duraisingh *et al.*, 2000b). Mefloquine interacts with the ABC proteins MRP1 (ABCC1) and MRP4 (ABCC4) present in human erythrocyte membranes (Wu *et al.*, 2005). Mefloquine stimulates ATPase activity although may itself be a substrate for transport of the ABC proteins.
Interestingly, mefloquine efflux from resistant parasites is blocked by penfluridol, a calcium channel blocker (Peters *et al.*, 1991; Oduola *et al.*, 1993; Mungthin *et al.*, 1999) but not by VP. Few molecules are able to modulate the mefloquine response in *P. falciparum*. They are generally different from those known to modulate CQ responses and include penfluridol and the synthetic surfactant NP30 (Ciach *et al.*, 2003).

Conclusions and future trends

Efflux transporters play a key role in the emergence and dissemination of resistant parasites and in the acquisition of additional mechanisms of drug resistance caused by a decrease in intracellular drug concentration. Despite their noticeable divergence in structure and membrane topology, the major drug efflux systems depend on specific key parameters including (i) the functional assembly of a membrane transporter, (ii) the energy required (e.g., ATP, ion antiport or membrane potential) for active transport, and (iii) the presence of affinity sites inside the transporter that are involved in substrate recognition and transport.
The identification of functional domains and the characterization of various interactions with the transported drug may elucidate key parameters that govern efflux activity. At present, some 3D structures have been solved for bacterial drug transporters, and these have allowed the proposal of dynamic and mechanical models for drug transport (Seeger et al, 2006). The same approach must be used for malaria. Drug–transporter interactions have recently been shown to be an important part of multidrug resistance. Modelling efforts are currently being undertaken using both ligand- and transporter-based methods such as structure–activity relationship studies (SAR), quantitative-SAR (QSAR) studies, hologram QSAR (HQSAR), comparative molecular field analysis (CoMFA) and comparative molecular similarity index analysis (CoMSIA) studies, pharmacophore modelling, homology modelling and molecular dynamics studies. This approach has been carried out in the case of human ATP-transporter Multidrug-Resistance Associated Protein 2 (MRP2; ABCC2) (Pedersen et al, 2008).

The aim of these efforts is to decipher the molecular basis of drug transport, to explain how differences in chemical structures modify interactions with the transporter or to elucidate how the transporter functions in general. In addition, original molecules have been demonstrated to restore the antimalarial activity of drugs that are pump substrates and these studies make it possible to identify pharmacophoric groups that are involved in efflux inhibition.

These data are crucial for the design of (i) new anti-malarial molecules that are devoid of efflux-substrate characteristics and can reach a normal intracellular accumulation level, and (ii) new compounds that have strong efflux pump affinity associated with a high inhibitor capability and block the pump, restor the intracellular concentration of antimalarial drugs.

The question about the effective use of chemosensitizers as antiplasmodial therapeutic is yet to be answered. A CQ and VP analogue combination has been administered to *Aotus* monkeys but failed to safely clear falciparum parasitaemias. This could be due to unfavourable pharmacokinetics or bioavaibility (Williams et al, 1992). Combinations of chlorpromazine, prochlorperazine, desipramine or propethazine with CQ confirm the reversal effect of these drugs on the CQ resistant parasites as showed by cures obtain in *Aotus* monkeys (Kyle et al, 1993; Oduola et al, 1998). An important factor to the lack of activity *in vivo* is the binding of the modulators to plasma proteins (see Chapter 9) (Boulter et al, 1993). In this regard, it has been suggested that the reason for the more successful trial of CQ combined with chlorpheniramine in humans was the reduced capacity of this drug to bind plasma proteins.

In vivo assays in animal models are in progress to evaluate the efficacy of different modulators, such as synthetic 9,10-dihydroethanoanthracene derivatives, on CQ, amodiaquine, quinine and mefloquine resistance. New human clinical studies in Nigerian children are in progress to evaluate the efficacy of promethazine or chlorpheniramine in association with several antimalarial drugs.

References

Adagu, I.S., Dias, F., Pinheiro, L., Rombo, L., do Rosario, V., and Warhurst, D.C. (1996). Guinea Bissau: association of chloroquine resistance of *Plasmodium falciparum* with the Tyr86 allele of the multiple drug-resistance gene *pfmdr1*. Trans. R. Soc. Trop. Med. Hyg. *90*, 90–91.

Alibert, S., Santelli-Rouvier, C, Pradines, B., Houdoin, C., Parzy, D., Karolak-Wojciechowska, J., and Barbe, J. (2002). Synthesis and effects on chloroquine susceptibility in *Plasmodium falciparum* of a series of new dihydroanthracene derivatives. J. Med. Chem. *45*, 3195–3209.

Alibert-Franco, S., Santelli-Rouvier, C., Barbe, J., Pradines, B., Houdoin, C., and Parzy, D. (1999). 9, 10-(3′, 4′-pyrrolino)-9, 10-dihydroanthracene and structurally related compounds as synergistic antimalarial drugs. Heterocycl. Commun. *5*, 235–240.

Barnes, D.A., Foote, S.J., Galatis, D., Kemp, D.J., and Cowman, A.F. (1992). Selection for high-level chloroquine resistance results in deamplification of the *pfmdr1* gene and increased sensitivity to mefloquine in *Plasmodium falciparum*. E.M.B.O J. *11*, 3067–3075.

Barnes, A.J., Ong, E.L., Dunbar, E.M., Mandal, B.K., and Wilkins, E.G.L. (1991). Failure of chloroquine and proguanil prophylaxis in travellers to Kenya. Lancet *338*, 1338–1339.

Basco, L.K., and Le Bras, J. (1990a). Desipramine or cyproheptadine for reversing chloroquine resistance? Lancet *335*, 422.

Basco, L.K., and Le Bras, J. (1990b). Reversal of chloroquine resistance with desipramine in isolates of *Plasmodium falciparum* from Central and West Africa. Trans. R. Soc. Trop. Med. Hyg. *84*, 479–481.

Basco, L.K., Le Bras, J., Gillotin, C., Ringwald, P., Rabenjarson, E., Gimenez, F., Bouchaud, O., Farinotti, R., and Coulaud, J.P. (1991). Type RI resistance to halofantrine in West Africa. Trop. Med. Parasitol. *42*, 413–414.

Basco, L.K., Le Bras, J., Rhoades, Z., and Wilson, C.M. (1995). Analysis of *pfmdr1* and drug susceptibility in fresh isolates of *Plasmodium falciparum* from subsaharan Africa. Mol. Biochem. Parasitol. *74*, 157–1966.

Basco, L.K., de Pecoulas, P.E., Le Bras, J., and Wilson, C.M. (1996). *Plasmodium falciparum*: molecular characterization of multidrug-resistant Cambodian isolates. Exp. Parasitol. *82*, 97–103.

Basco, L.K., and Ringwald, P. (1998). Molecular epidemiology of malaria in Yaoundé, Cameroon. III. Analysis of chloroquine resistance and point mutations in the multidrug resistance 1 (pfmdr1) gene of *Plasmodium falciparum*. Am. J. Trop. Med. Hyg. *59*, 577–581.

Basco, L.K., and Ringwald, P. (2002). Molecular epidemiology of malaria in Cameroon. X. Evaluation of *pfmdr1* mutations as genetic markers for resistance to amino alcohols and artemisinin derivatives. Am. J. Trop. Med. Hyg. *66*, 667–671.

Basco, L.K., Ringwald, P., and Le Bras, J. (1991). Chloroquine-potentiating action of antihistaminics in *Plasmodium falciparum in vitro*. Ann. Trop. Med. Parasitol. *85*, 223–228.

Bayoumi, R.A., Creasey, A.M., Babiker, H.A., Carlton, J.M., Sultan, A.A., Satti, G., Sohal, A.K., Walliker, D., Jensen, J.B., and Arnot, D.E. (1993). Drug response and genetic characterization of *Plasmodium falciparum* clones recently isolated from a Sudanese village. Trans. R. Soc. Trop. Med. Hyg. *87*, 454–458.

Bayoumi, R.A.L., Babiker, H.A., and Arnot, D.E. (1994). Uptake and efflux of chloroquine by chloroquine-resistant *Plasmodium falciparum* clones recently isolated in Africa. Acta Trop. *58*, 141–149.

Bhattacharjee, A.K., Kyle, D.E., and Vennerstrom, J.L. (2001). Structural analysis of chloroquine resistance reversal by imipramine analogs. Antimicrob. Agents Chemother. 45, 2655–2657.

Bhattacharjee, A.K., Kyle, D.E., Vennerstrom, J.L., and Milhous, W.K. (2002). A 3D QSAR pharmacophore model and quantum chemical structure--activity analysis of chloroquine(CQ)-resistance reversal. J. Chem. Inf. Comput. Sci. 42, 1212–1220.

Bitonti, A.J., Sjoerdsma, A., McCann, P.P., Kyle, D.E., Oduola, A.M.J., Rossan, R.N., Milhous, W.K., and Davidson, D.E. (1988). Reversal of chloroquine resistance in malaria parasite *Plasmodium falciparum* by desipramine. Science 242, 1301–1303.

Boulter, M.K., Bray, P.G., Howells, R.E., and Ward, S.A. (1993). The potential of desipramine to reverse chloroquine resistance of *Plasmodium falciparum* is reduced by its binding to plasma protein. Trans. R. Soc. Trop. Med. Hyg. 87, 303.

Bozdech, Z., Delling, U., Volkman, S.K., Cowman, A.F., and Schurr, E. (1996). Cloning and sequence analysis of a novel member of the ATP-binding cassette (ABC) protein gene family from *Plasmodium falciparum*. Mol. Biochem. Parasitol. *81*, 41–51.

Bozdech, Z., and Schurr, E. (1999). Protein transport in the host cell cytoplasm and ATP-binding cassette proteins in *Plasmodium falciparum*-infected erythrocytes. Novartis Found. Symp. 226, 231–241.

Bozdech, Z., Van Wye, J., Haldar, K., and Schurr, E. (1998). The human malaria parasite *Plasmodium falciparum* exports the ATP-binding cassette protein PFGCN20 to membrane structuRes. in the host red blood cell. Mol. Biochem. Parasitol. 97, 81–95.

Bray, P.G., Hawley, S.R., Mungthin, M., and Ward, S.A. (1996). Physicochemical properties correlated with drug resistance and the reversal of drug resistance in *Plasmodium falciparum*. Mol. Pharmacol. 50, 1559–1566.

Bray, P.G., Howells, R.E., Pitchie, G.Y., and Ward, S.A. (1992a). Rapid chloroquine efflux phenotype in both chloroquine-sensitive and chloroquine-resistant *Plasmodium falciparum*. A correlation of chloroquine sensitivity with energy-dependent drug accumulation. Biochem. Pharmacol. 44, 1317–1324.

Bray, P.G., Howells, R.E., and Ward, S.A. (1992b). Vacuolar acidification and chloroquine sensitivity in *Plasmodium falciparum*. Biochem. Pharmacol. 43, 1219–1227.

Bray, P.G., Jenneh, O., Raynes, K.J., Mungthin, M., Ginsburg, H., and Ward, S.A. (1999). Cellular uptake of chloroquine is dependent on binding to ferriprotoporphyrin IX and is independent of NHE activity in *Plasmodium falciparum*. J. Cell Biol. 145, 363–376.

Carlton, J., Mackinnon, M., and Walliker, D. (1998). A chloroquine resistance locus in the rodent malaria parasite *Plasmodium chabaudi*. Mol. Biochem. Parasitol. 93, 57–72.

Ciach, M., Zong, K., Kain, K.C., and Crandall, I. (2003). Reversal of mefloquine and quinine resistance in *Plasmodium falciparum* with NP30. Antimicrob. Agents Chemother. 47, 2393–2396.

Cowman, A.F., and Foote, S. (1990). Chemotherapy and drug resistance in malaria. Int. J. Parasitol. 20, 503–513.

Cowman, A.F., Galatis, D., and Thompson, J.K. (1994). Selection for mefloquine resistance in *Plasmodium falciparum* is linked to amplification of the *pfmdr1* gene and cross-resistance to halofantrine and quinine. Proc. Natl. Acad. Sci. U.S.A. *91*, 1143–1147.

Cowman, A.F., Karze, S., Galatis, D., and Culvenor, J.G. (1991). A P-glycoprotein homologue of *Plasmodium falciparum* is localized on the digestive vacuole. J. Cell Biol. *113*, 1033–1042.

Crandall, I., Charuk, J., and Kain, K.C. (2000). Nonylphenolethoxylates as malarial chloroquine resistance reversal agents. Antimicrob. Agents Chemother. 44, 2431–2434.

Cravo, P.V.L., Carlton, J.M.R., Hunt, P., Bisoni, L., Padua, R.A., and Walliker, D. (2003). genetics of mefloquine resistance in the rodent malaria parasite *Plasmodium chabaudi*. Antimicrob. Agents Chemother. 47, 709–718.

Deloron, P., Basco, L.K., Dubois, B., Gaudin, C., Clavier, F., Le Bras, J., and Verdier, F. (1991). *In vitro* and *in vivo* potentiation of chloroquine against malaria parasites by an enantiomer of amlodipine. Antimicrob. Agents Chemother. 35, 1338–1342.

Dorn, A., Vippagunta, S.R., Matile, H., Jaquet, C., Vennerstrom, J.L., and Ridley, R.G. (1998). An assessment of drug-haematin binding as a mechanism for inhibition of haematin polymerisation by quinoline antimalarials. Biochem. Pharmacol. 55, 727–736.

Duraisingh, M.T., Drakeley, C.J., Muller, O., Bailey, R., Snounou, G., Targett, G.A.T., Greenwood, B.M., and Warhurst, D.C. (1997). Evidence for selection for the tyrosine-86 allele of the pfmdr1 gene of *Plasmodium falciparum* by chloroquine and amodiaquine. Parasitology *114*, 205–211.

Duraisingh, M.T., Jones, P., Sambou, I., von Seidlein, L., Pinder, M., and Warhurst, D.C. (2000a). The tyrosine-86 allele of the *pfmdr1* gene of *Plasmodium falciparum* is associated with increased sensitivity to the anti-malarials mefloquine and artemisinin. Mol. Biochem. Parasitol. *108*, 13–23.

Duraisingh, M.T., Roper, C., Walliker, D., and Warhurst, D.C. (2000b). Increased sensitivity to the antimalarials mefloquine and artemisinin is conferred by mutations in the *pfmdr1* gene of *Plasmodium falciparum*. Mol. Microbiol. 36, 955–961.

Dzekunov, S.M., Ursos, L.M.B., and Roepe, P.D. (2000). Digestive vacuolar pH of intact intraerythrocytic *P. falciparum* either sensitive or resistant to chloroquine. Mol. Biochem. Parasitol. *110*, 107–124.

Ekong, R.M., Robson, K.J., Baker, D.A., and Warhurst, D.C. (1993). Transcripts of the multidrug resistance genes in chloroquine-sensitive and chloroquine-resistant *Plasmodium falciparum*. Parasitology *106*, 107–115.

Evans, S.G., Butkow, N., Stilwell, C., Berk, M., Kirchmann, N., and Havlik, I. (1998). Citalopram enhances the activity of chloroquine in resistant *Plasmodium in vitro* and *in vivo*. J. Pharmacol. Exp. Ther. *286*, 172–174.

Ferrari, V., and Cutler, D.J. (1991). Simulation of kinetic data on the influx and efflux of chloroquine by erythrocytes infected with *Plasmodium falciparum*. Evidence for a drug-importer in chloroquine-sensitive strains. Biochem. Pharmacol. 42, S167–179.

Ferrer-Rodriguez, I., Perez-Rosado, J., Gervais, G.W., Peters, W., Robinson, B.L., and Serrano, A.E. (2004). *Plasmodium yoelii*: identification and partial characterization of an MDR1 gene in an artemisinin-resistant line. J. Parasitol. *90*, 152–160.

Foote, S.J., Kyle, D.E., Martin, R.K., Oduola, A.M., Forsyth, K., Kemp, D.J., and Cowman, A.F. (1990). Several alleles of the multidrug-resistance gene are closely linked to chloroquine resistance in *Plasmodium falciparum*. Nature *345*, 255–258.

Foote, S.J., Thompson, J.K., Cowman, A.F., and Kemp, D.J. (1989). Amplification of the multidrug resistance gene in some chloroquine-resistant isolates of *P. falciparum*. Cell *57*, 921–930.

Gari-Toussaint, M., Pradines, B., Mondain, V., Keundjian, A., Dellamonica, P., and Le Fichoux, Y. (2002). Sénégal et paludisme. Echec prophylactique vrai de la méfloquine. Presse Med. *31*, 1136.

Gerena, L., Bass, G.T., Kyle, D.E., Oduola, A.M., Milhous, W.K., and Martin, R.K. (1992). Fluoxetine hydrochloride enhances *in vitro* susceptibility to chloroquine in resistant *Plasmodium falciparum*. Antimicrob. Agents Chemother. *36*, 2761–2765.

Gervais, G.W., Trujillo, K., Robinson, B.L., Peters, W., and Serrano, A.E. (1999). *Plasmodium berghei*: identification of an mdr-like gene associated with drug resistance. Exp. Parasitol. *91*, 86–92.

Ginsburg, H., Famin, O., Zhang, J., and Krugliak, M. (1998). Inhibition of glutathione-dependent degradation of heme by chloroquine and amodiaquine as a possible basis for their antimalarial mode of action. Biochem. Pharmacol. *56*, 1305–1313.

Grobusch, M.P., Adagu, I.S., Kremsner, P.G., and Warhurst, D.C. (1998). *Plasmodium falciparum*: *in vitro* chloroquine susceptibility and allele-specific PCR detection of *pfmdr1* Asn86Tyr polymorphism in Lambarene, Gabon. Parasitology *116*, 211–217.

Guan, J., Kyle, D.E., Gerena, L., Zhang, Q., Milhous, W.K., and Lin, A.J. (2002). Design, synthesis, and evaluation of new chemosensitizers in multi-drug-resistant *Plasmodium falciparum*. J. Med. Chem. *45*, 2741–2748.

Harinasuta, T., Suntharasamai, P., and Vivaran, C. (1965) Chloroquine-resistant falciparum malaria in Thailand. Lancet *2*, 657–660.

Henry, M., Alibert, S., Baragatti, M., Mosnier, J., Baret, E., Amalvict, R., Legrand, E., Fusai, T., Barbe, J., Rogier, C., Pages, J.M., and Pradines, B. (2008a). Dihydroethano-anthracene derivatives reverse *in vitro* quinoline resistance in *Plasmodium falciparum* malaria. Med. Chem. *4*, 426–437.

Henry, M., Briolant, S., Fontaine, A., Mosnier, J., Baret, E., Amalvicy, R., Fusai, T., Fraisse, L., Rogier, C., and Pradines, B. (2008b). In vitro activity of ferroquine is independent of polymorphisms in transport protein genes implicated in quinoline resistance in *Plasmodium falciparum*. Antimicrob. Agents Chemother. *52*, 2755–2759.

Hunt, P., Cravo, P.V.L., Donleavy, P., Carlton, J.M.R., and Walliker, D. (2004). Chloroquine resistance in *Plasmodium chabaudi*: are chloroquine-resistance transporter (*crt*) and multidrug resistance (*mdr1*) orthologues involved ? Mol. Biochem. Parasitol. *133*, 27–35.

Jacobs, G.H., Oduola, A.M.J., Kyle, D.E., Milhous, W.K., Martin, S.K., and Aikawa M. (1988). Ultrastructural study of the effects of chloroquine and verapamil on *Plasmodium falciparum*. Am. J. Trop. Med. Hyg. *39*, 15–20.

Jambou, R., Legrand, E., Niang, M., Khim, L., Lim, P., Volney, B., Rkala, M.T., Vouchier, C., Esterre, P., Fandeur, T., and Mercereau-Puijalon, O. (2005). Resistance of *Plasmodium falciparum* field isolates to in-vitro artemether and point mutations of the SERCA-type PfATPase6. Lancet *366*, 1960–1963.

Johnson, D.J., Fidock, D.A., Mungthin, M., Lakshmanan, V., bir Singh Sidhu, A., Bray, P.G., and Ward, S.A. (2004). Evidence for a central role for PfCRT in conferring *Plasmodium falciparum* resistance to diverse antimalarial agents. Mol. Cell *15*, 867–877.

Johnson, D.J., Owen, A., Plant, N., Bray, P.G., and Ward, S.A. (2008). Drug-regulated expression of *Plasmodium falciparum* P-Glycoprotein homologue 1: a putative role for nuclear receptors. Antimicrob. Agents Chemother. *52*, 1438–1445.

Kalkanidis, M., Klonis, N., Tilley, L., and Deady, L.W. (2002). Novel phenothiazine antimalarials: synthesis, antimalarial activity, and inhibition of the formation of beta-haematin. Biochem. Pharmacol. *63*, 833–842.

Karcz S.R., Galatis, D., and Cowman, A.F. (1993). Nucleotide binding properties of a P-glycoprotein homologue from *Plasmodium falciparum*. Mol. Biochem. Parasitol. *58*, 269–276.

Klokousas, A., Tiffert, T., van Schalkwyk, D., Wu, C.P., van Veen, H.W., Barrand, M.A., and Hladky, S.B. (2004). *Plasmodium falciparum* expresses a multidrug resistance-associated protein. Biochem. Biophys. Res. Commun. *321*, 197–201.

Krogstad, D.J., Gluzman, I.Y., Kyle, D.E., Oduola, A.M.J., Martin, S.K., Milhous, W.K., and Schlesinger, P.H. (1987). Efflux of chloroquine from *Plasmodium falciparum*: mechanism of chloroquine resistance. Science *238*, 1283–1285.

Kyle, D.E., Milhous, W.K., and Rossan, R.N. (1993). Reversal of *Plasmodium falciparum* resistance to chloroquine in Panamanian *Aotus* monkeys. Am. J. Trop. Med. Hyg. *48*, 126–133.

Kyle, D.E., Oduola, A.M., Martin, S.K., and Milhous, W.K. (1990). *Plasmodium falciparum*: modulation by calcium antagonists of resistance to chloroquine, desethylchloroquine, quinine, and quinidine *in vitro*. Trans. R. Soc. Trop. Med. Hyg. *84*, 474–478.

Lim, A.S., Galatis, D., and Cowman, A.F. (1996). *Plasmodium falciparum*: amplification and overexpression of *pfmdr1* is not necessary for inreased mefloquine resistance. Exp. Parasitol. *86*, 295–303.

Loe, W., and Sharom, F.J. (1993). Interaction of multidrug-resistant Chinese hamster ovary cells with amphiphiles. Br. J. Cancer *68*, 342–351.

Martin, S.K., Oduola, A.M., and Milhous, W.K. (1987). Reversal of chloroquine resistance in *Plasmodium falciparum* by verapamil. Science *235*, 899–901.

Martiney, J.A., Cerami, A., and Slater, A.F.G. (1995). Verapamil reversal of chloroquine resistance in the malaria parasite *Plasmodium falciparum* is specific for resistant parasites and independent of the weak base effect. J. Biol. Chem. *270*, 22393–22398.

Mawili-Mboumba, D.P., Kun, J.F.J., Lell, B., Kremsner, P.G., and Ntoumi, F. (2002). *Pfmdr1* alleles and response to ultralow-dose mefloquine treatment in Gabonese patients. Antimicrob. Agents Chemother. *46*, 166–170.

Mehlotra, R.K., Fusioka, H., Roepe, P.D., Janneh, O., Ursos, L.M.B., Jacobs-Lorena, V., Mcnamara, D.T., Bockarie, M.J., Kasura, J.W., Kyle, D.E., and Fidock, D.A. (2001). Evolution of a unique *Plasmodium falciparum* chloroquine-resistance phenotype in association with *pfcrt* polymorphism in Papua New Guinea and South America. Proc. Natl. Acad. Sci. U.S.A. *98*, 12689–12694.

Menezes, C.M., Kirchgatter, K., Di Santi, M., Savalli, C., Monteiro, F.G., Paula, G.A., and Ferreira, E.I. (2003). *In vitro* evaluation of verapamil and other modulating agents in Brazilian chloroquine-resistant *Plasmodium falciparum* isolates. Rev. Soc. Bras. Med. Trop. *36*, 5–9.

Menezes, C.M.S., Kirchgatter, K, di Santi, S.M., Savalli, C., Monteiro, F.G., Paula, G.A., and Ferreira, E.I. (1997). Antimalarial effect *in vitro* and lack of modulating effect of desipramine and imipramine. Trans. R. Soc. Trop. Med. Hyg. *91*, 697–700.

Menezes, C.M.S., Kirchgatter, K., di Santi, S.M., Savall, C., Monteiro, F.G., Paula, G.A., and Ferreira, E.I. (2002). *In vitro* chloroquine resistance modulation study on fresh isolates of Brazilian *Plasmodium falciparum*: intrinsic antimalarial activity of phenothiazine drugs. Mem. Inst. Oswaldo Cruz Rio Janeiro *97*, 1033–1039.

Miki, A., Tanabe, K., Nakayama, T., Kiryon, C., and Ohsawa, K. (1992). *Plasmodium chabaudi*: association of reversal of chloroquine resistance with increased accumulation of chloroquine in resistant parasites. Exp. Parasitol. *74*, 134–142.

Millet, J., Alibert, S., Torrentino-Madamet, M., Rogier, C., Santelli-Rouvier, C., Bigot, P., Mosnier, J., Baret, E., Barbe, J., Parzy, D., and Pradines, B. (2004a). Polymorphisms in *Plasmodium falciparum* drug transporter proteins and reversal of *in vitro* chloroquine resistance by a 9, 10-dihydroethanoanthracene derivative. Antimicrob. Agents Chemother. *48*, 4869–4872.

Millet, J., Torrentino-Madamet, M., Alibert, S., Rogier, C., Santelli-Rouvier, C., Mosnier, J., Baret, E., Barbe, J., Parzy, D., and Pradines, B. (2004b). Dihydroethanoanthracene derivatives as *in vitro* malarial chloroquine resistance reversal agents. Antimicrob. Agents Chemother. *48*, 2753–2756.

Molinier, S., Imbert, P., Verrot, D., Morillon, M., Parzy, D., and Touze, J.E. (1994). Paludisme à *Plasmodium falciparum* : résistance de type RI à la quinine en Afrique de l'Est. Presse Med. *23*, 1484.

Mu, J., Ferdig, M.T., Feng, X., Joy, D.A., Duan, J., Furuya, T., Subramanian, G., Aravind, L., Cooper, R.A., Wootton, J.C., Xiong, M., and Su, X.Z. (2003). Multiple transporters associated with malaria parasite responses to chloroquine and quinine. Mol. Microbiol. *49*, 977–989.

Mungthin, M., Bray, P.G., and Ward, S.A. (1999). Phenotypic and genotypic characteristics of recently adapted isolates of *Plasmodium falciparum* from Thailand. Am. J. Trop. Med. Hyg. *60*, 469–474.

Nelson, A.L., Purfield, A., McDaniel, P., Uthaimongkol, N., Buathong, N., Sriwichai, S., Miller, R.S., Wongsrichanalai, C., and Meshnick, S.R. (2005). *Pfmdr1* genotyping and *in vivo* mefloquine resistance on the Thai-Myanmar border. Am. J. Trop. Med. Hyg. *72*, 586–592.

Ngo, T., Duraisingh, M., Reed, M., Hipgrave, D., Biggs, B., and Cowman, A.F. (2003). Analysis of *pfcrt*, *pfmdr1*, *dhfr*, and *dhps* mutations and drug sensitivities in *Plasmodium falciparum* isolates from patients in Vietnam before and after treatment with artemisinin. Am. J. Trop. Med. Hyg. *68*, 350–356.

Oduola, A.M., Omitowoju, G.O., Gerena, L., Kyle, D.E., Milhous, W.K., Sowunmi, A., and Salako, L.A. (1993). Reversal of mefloquine resistance with penfluridol in isolates of *Plasmodium falciparum* from south-west Nigeria. Trans. R. Soc. Trop. Med. Hyg. *87*, 81–83.

Oduola, A.M., Sowunmi, A., Milhous, W.K., Brewer, T.G., Kyle, D.E., Gerena, L., Rossan, R.N., Salako, A., and Schuster, B.G. (1998). *In vitro* and *in vivo* reversal of chloroquine resistance in *Plasmodium falciparum* with promethazine. Am. J. Trop. Med. Hyg. *58*, 625–629.

Oduola, O.O., Happi, T.C., Gbotosho, G.O., Ogundahunsi, O.A., Falade, C.O., Akinboye, D.O., Sowunmi, A., and Oduola, A.M. (2004). *Plasmodium berghei*: efficacy and safety of combinations of chloroquine and promethazine in chloroquine resistant infections in gravid mice. Afr. J. Med. Sci. *33*, 77–81.

Okonkwo, C.A., Coker, H.A.B., Agomo, P.U., Ogunbanwo, J.A., Mafe, A.G., Agomo, C.O., and Afolabi, B.M. (1999). Effect of chlorpheniramine on the pharmacokinetics of and response to chloroquine of Nigerian children with *falciparum* malaria. Trans. R. Soc. Trop. Med. Hyg. *93*, 306–311.

Pagola, S., Stephens, P.W., Bohle, D.S., Kosar, A.D., and Madsen, S.K. (2000). The structure of malaria pigment beta-haematin. Nature *404*, 307–310.

Pedersen, J.M., Matsson, P., Bergstrom, C.A., Norinder, U., Hoogstraate, J., and Artursson, P. (2008). Prediction and identification of drug interactions with the human ATP-binding cassette transporter multidrug-resistance associated protein 2 (MRP2; ABCC2). J. Med. Chem. *51*, 3275–3287.

Peel, S.A., Bright, P., Yount, B., Handy, J., and Baric, R.S. (1994). A strong association between mefloquine and halofantrine resistance and amplification, overexpression, and mutation in the P-glycoprotein gene homolog (*pfmdr*) of *Plasmodium falciparum in vitro*. Am. J. Trop. Med. Hyg. *51*, 648–658.

Peters, W., Ekong, R., Robinson, B.L., and Warhurst, D.C. (1990). The chemotherapy of rodent malaria. XLV. Reversal of chloroquine resistance in rodent and human *Plasmodium* by antihistaminic agents. Ann. Trop. Med. Parasitol. *84*, 541–551.

Peters, W., and Robinson, B.L. (1991). The chemotherapy of rodent malaria. XLVI. Reversal of mefloquine resistance in rodent *Plasmodium*. Ann. Trop. Med. Parasitol. *85*, 5–10.

Pickard, A.L., Wongsrichanalai, C., Purfield, A., Kamwendo, D., Emery, K., Zalewski, C., Kawamoto, F., Miller, R.S., and Meshnick, S.R. (2003). Resistance to antimalarials in Southeast Asia and genetic polymorphisms in *pfmdr1*. Antimicrob. Agents Chemother. *47*, 2418–2423.

Pillai, D.R., Hijar, G., Montoya, I., Marquino, W., Ruebush, T.K., Wrongsrichanalai, C., and Kain, K.C. (2003). Lack of prediction of mefloquine and mefloquine-artesunate treatment outcome by mutations in the *Plasmodium falciparum* multidrug resistance 1 (*pfmdr1*) gene for *P. falciparum* malaria in Peru. Am. J. Trop. Med. Hyg. *68*, 107–110.

Platel, D.F., Mangou, F., and Tribouley-Duret, J. (1998). High-level chloroquine resistance of *Plasmodium berghei* is associated with multiple drug resistance and loss of reversal by calcium antagonists. Int. J. Parasitol. *28*, 641–651.

Povoa, M.M., Adagu, I.S., Oliveira, S.G., Machado, R.L.D., Miles, M.A., and Warhurst, D.C. (1998). *Pfmdr1* Asn1042Asp and Asp1246Tyr polymorphisms, thought to be associated with chloroquine resistance, are present in chloroquine-resistant and -sensitive Brazilian field isolates of *Plasmodium falciparum*. Exp. Parasitol. *88*, 64–68.

Pradines, B., Alibert, S., Houdoin, C., Mosnier, J., Santelli-Rouvier, C., Papa, V., Rogier, C., Fusai, T., Barbe, J., and Parzy, D. (2002a). *In vitro* reversal of chloroquine resistance in *Plasmodium falciparum* with dihydroethanoanthracene derivatives. Am. J. Trop. Med. Hyg. *66*, 661–666.

Pradines, B., Alibert, S., Houdoin, C., Santelli-Rouvier, C., Mosnier, J., Fusai, T., Rogier, C., Barbe, J., and Parzy, D. (2002b). *In vitro* increase in chloroquine accumulation induced by dihydroethano- and ethenoanthracene derivatives in *Plasmodium falciparum*-parasitized erythrocytes. Antimicrob. Agents Chemother. *46*, 2061–2068.

Pradines, B., Fusai, T., Rogier, C., Sinou, V., Millet, J., Henry, M., and Parzy, D. (2002c). Progress in the development of antimalarial drugs: iron chelation and reversal chloroquine resistance as therapy for malaria. Recent Res. Devel. Antimicrob. Agents Chemother. *5*, 24–43.

Price, R.C., Uhlemann, A.C., Brockman, A., McReady, R., Asley, E., Phaipun, L., Patel, R., Laing, K., Looareesuwan, S., White, N.J., Nosten, F., and Krishna, S. (2004). Mefloquine resistance in *Plasmodium falciparum* and increased *pfmdr1* gene copy number. Lancet *364*, 438–447.

Price, R.N., Cassar, C., Brockman, A., Duraisingh, M., van Vugt, M., White, N.J., Nosten, F., and Krishna, S. (1999). The *pfmdr1* gene is associated with a multi-drug-resistant phenotype in *Plasmodium falciparum* from the western border of Thailand. Antimicrob. Agents Chemother. *43*, 2943–2949.

Reed, M.B., Saliba, K.J., Caruana, S.R., Kirk, K., and Cowman, A.F. (2000). Pgh1 modulates sensitivity and resistance to multiple antimalarials in *Plasmodium falciparum*. Nature *403*, 906–909.

Ritchie, G.Y., Mungthin, M., Green, J.E., Bray, P.G., Hawley, S.R., and Ward, S.A. (1996). In vitro selection of halofantrine resistance in *Plasmodium falciparum* is not associated with increased expression of Pgh1. Mol. Biochem. Parasitol. *83*, 35–46.

Rohrbach, P., Sanchez, C.P., Hayton, K., Friedrich, O., Patel, J., Sidhu, A.B., Ferdig, M.T., Fidock, D.A., and Lanzer, M. (2006). Genetic linkage of *pfmdr1* with food vacuolar solute import in *Plasmodium falciparum*. EMBO J. *25*, 3000–3011.

Rubio, J.P., and Cowman, A.F. (1994). *Plasmodium falciparum*: the pfmdr2 protein is not overexpressed in chloroquine-resistant isolates of the malaria parasite. Exp. Parasitol. *79*, 137–147.

Ruetz, S., Delling, U., Brault, M., Schurr, E., and Gros, P. (1996). The pfmdr1 gene of *Plasmodium falciparum* confers cellular resistance to antimalarial drugs in yeast cells. Proc. Natl. Acad. Sci. U.S.A. *93*, 9942–9947.

Saliba, K.J., Folb, P.I., and Smith, P.J. (1998). Role for the *Plasmodium falciparum* digestive vacuole in chloroquine resistance. Biochem. Parasitol. *56*, 313–320.

Sanchez, P.C., Rotmann, A., Stein, W.D., and Lanzer, M. (2008a). Polymorphisms within PfMDR1 alter the substrate specificity for antimalarial drugs in *Plasmodium falciparum*. Mol. Microbiol. (postprint doi: 10.1111/j.1365-2958.2008.06413.x).

Sanchez, P.C., Stein, W.D., and Lanzer, M. (2008b). Dissecting the components of quinine accumulation in *Plasmodium falciparum*. Mol. Microbiol. *67*, 1081–1093.

Seeger, M.A., Schiefner, A., Eicher, T., Verrey, F., Diederichs, K., and Pos, K.M. (2006). Structural asymmetry of AcrB trimer suggests a peristaltic pump mechanism. Science *313*, 1295–1298.

Shalmiev, G., and Ginsburg, H. (1993). The susceptibility of the malarial parasite *Plasmodium falciparum* to quinoline-containing drugs is correlated to the lipid composition of the infected erythrocyte membranes. Biochem. Pharmacol. *46*, 365–374.

Sidhu, A.B., Uhlemann, A.C., Valderramos, S.G., Valderramos, J.C., Krishna, S., and Fidock, D.A. (2006). Decreasing *pfmdr1* copy number in *Plasmodium falciparum* malaria heightens susceptibility to mefloquine, lumefantrine, halofantrine, quinine, and artemisinin. J. Infect. Dis. *194*, 528–535.

Sidhu, A.B., Valderramos, S.G., and Fidock, D.A. (2005). *Pfmdr1* mutations contribute to quinine resistance and enhance mefloquine and artemisinin sensitivity in *Plasmodium falciparum*. Mol. Microbiol. *57*, 913–926.

Sidhu, A.B.S., Verdier-Pinard, D., and Fidock, D.A. (2002). Chloroquine resistance in *Plasmodium falciparum* malaria parasites conferred by *pfcrt* mutations. Science *298*, 210–213.

Singh, N., and Puri, S.K. (2000a). Interaction between chloroquine and diverse pharmacological agents in chloroquine resistant *Plasmodium yoelii nigeriensis*. Acta Trop. *77*, 185–193.

Singh, N, and Puri, SK. (2000b). Modulation of halofantrine resistance after coadministration of halofantrine with diverse pharmacological agents in a rodent malaria model. Life Sci. *67*, 1345–1354.

Sowunmi, A., Oduola, A.M.J., Ogundahunsi, O.A.T., and Salako, L.A. (1998). Comparative efficacy of chloroquine plus chlorpheniramine and pyrimethamine/sulfadoxine in acute uncomplicated *falciparum* malaria in Nigerian children. Trans. R. Soc. Trop. Med. Hyg. *92*, 77–81.

Sowunmi, A., Oduola, A.M.J., Ogundahunsi, O.A.T., Falade, C.O., Gbotosho, G.O., and Salako, L.A. (1997). Enhanced efficacy of chloroquine-chlorpheniramine combination in acute uncomplicated *falciparum* malaria in children. Trans. R. Soc. Trop. Med. Hyg. *91*, 63–67.

Tanabe, K., Kato, M., Izumo, A., Hagiwara, A., and Doi, S. (1990). *Plasmodium chabaudi*: *in vivo* effects of Ca2+ antagonists on chloroquine-resistant and chloroquine-sensitive parasites. Exp. Parasitol. *70*, 419–426.

Taylor, D., Walden, J.C., Robins, A.H., and Smith P.J. (2000). Role of the neurotransmitter reuptake-blocking activity of antidepressants in reversing chloroquine resistance *in vitro* in *Plasmodium falciparum*. Antimicrob. Agents Chemother. *44*, 2689–2692.

Ursing, J., Zakeri, S., Gil, J.P., and Bjorkman, A. (2006). Quinoline resistance associated polymorphisms in the *pfcrt, pfmdr1* and *pfmrp* genes of *Plasmodium falciparum* in Iran. Acta Trop. 97, 352–356.

Ursos, L.M.B., DuBay, K F., and Roepe, P.D. (2001). Antimalarial drugs influence the pH dependent solubility of heme via apparent nucleation phenomena. Mol. Biochem. Parasitol. *112*, 11–17.

Ursos, L.M.B.; Dzekunov, S M., and Roepe, P.D. (2000). The effects of chloroquine and verapamil on digestive vacuolar pH of *P. falciparum* either sensitive or resistant to chloroquine. Mol. Biochem. Parasitol. *110*, 125–134.

van Es, H.H.G., Karcz, S., Chu, F., Cowman, A.F., Vidal, S., Gros, P., and Schurr, E. (1994b). Expression of the plasmodial pfmdr1 gene in mammalian cells is associated with increased susceptibility to chloroquine. Mol. Cell Biol. *14*, 2419–2428.

van Es, H.H., Renkema, H., Aerts, H., and Schurr, E. (1994a). Enhanced lysosomal acidification leads to increased chloroquine accumulation in CHO cells expressing the *pfmdr1* gene. Mol. Biochem. Parasitol. *6*, 209–219.

Verdier, F., Le Bras, J., Clavier, F., Hatin, I., and Blayo, M. (1985). Chloroquine uptake by *Plasmodium falciparum*-infected human erythrocytes during *in vitro* culture and its relationship to chloroquine resistance. Antimicrob. Agents Chemother. *27*, 561–564.

Volkman, S.K., Wilson, C.M., and Wirth, D.F. (1993). Stage-specific transcripts of the *Plasmodium falciparum pfmdr1* gene. Mol. Biochem. Parasitol. *57*, 203–211.

Warsame, M., Wernsdorfer, W.H., and Björkman, A. (1992). Lack of effect of desipramine on the response to chloroquine of patients with chloroquine-resistant falciparum malaria. Trans. R. Soc. Trop. Med. Hyg. *86*, 235–236.

Weaver, J.L., Szabo, G., Pine, P. S., Gottesman, M.M., Goldenberg, S., and Aszalos, A. (1993). The effect of ion channel blockers, immunosuppressive agents, and other drugs on the activity of the multi-drug transporter. Int. J. Cancer *54*, 456–461.

Wellems, T.E., and Plowe, C.V. (2001). Chloroquine-resistant malaria. J. Infect. Dis. *18*, 770–776.

Williams, H.L., Johnson, D.J., Okoye, V.C., and Martin, S.K. (1992). Treatment of chloroquine-resistant malaria in monkeys with a drug combination that reverses resistance *in vitro*. Ann. Trop. Med. Parasitol. *86*, 467–473.

Wilson, C.M., Serrano, A.E., Wasley, A., Bogenschutz, M.P., Shankar, A.H. and Wirth, D.F. (1989). Amplification of a gene related to mammalian mdr genes in drug-resistant *Plasmodium falciparum*. Science *244*, 1184–1186.

Wu, C.P., Klokouzas, A., Hladky, S.B., Ambudkar, S.V., and Barrand, M.A. (2005). Interactions of mefloquine with ABC proteins, MRP1 (ABCC1) and MRP4 (ABCC4) that are present in human red cell membranes. Biochem. Pharmacol. *70*, 500–510.

Yayon, A., Cabantchik, Z.I., and Ginsburg, H. (1985). Susceptibility of human malaria parasites to chloroquine is pH dependent. Proc. Natl. Acad. Sci. U.S.A. *82*, 2784–2788.

Ye, Z.G., and Van Dyke, K. (1988). Reversal of chloroquine in falciparum independent of calcium channels. Biochem. Biophys. Res. Commun. *155*, 476–481.

Young, M.D. and Moore, D.V. (1961). Chloroquine resistance in *Plasmodium falciparum*. Am. J. Trop. Med. Hyg. *10*, 317–320.

Zalis, M.G., Pang, L., Silveira, M.S., Milhous, W.K., and Wirth, D.F. (1998). Characterization of *Plasmodium falciparum* isolated from the Amazon region of Brazil: evidence for quinine resistance. Am. J. Trop. Med. Hyg. *58*, 630–637.

Zamora, J.M., Pearce, H.L., and Beck, W.T. (1988). Physical–chemical properties shared by compounds that modulate multidrug resistance in human leukemic cells. Mol. Pharmacol. 33, 454–462.

Zalis, M.G., Wilson, C.M., Zhang, Y., and Wirth, D.F. (1993). Characterization of the *pfmdr2* gene for *Plasmodium falciparum*. Mol. Biochem. Parasitol. *62*, 83–92.

Cellular Functions of ABC Proteins in *Trypanosomatidae*

7

Philippe Leprohon, Danielle Légaré and Marc Ouellette

Abstract

The ATP-binding cassette (ABC) protein superfamily is a ubiquitous and functionally versatile family of proteins that is conserved from archaea to humans. In eukaryotes, most of these proteins are implicated in the transport of a variety of molecules across cellular membranes, whereas others are involved in biological processes unrelated to transport. ABC proteins have been reported in several parasitic protozoa, including parasites of the *Trypanosomatidae* family, which gathers evolutionary related protozoa of medical relevance like *Leishmania* spp., *Trypanosoma cruzi* and *Trypanosoma brucei*. A recent survey of the genome sequences of three trypanosomatid parasites indicated the presence of a complete set of ABC genes, with representative members of every subfamily described in eukaryotes (ABCA-ABCH). The biological functions of several members of the ABCA, ABCB, ABCC, ABCE and ABCG subfamilies have been described in protozoan parasites and include vesicular trafficking, phospholipids movement, translation and drug resistance. This chapter will review the ABC proteins that have been described in *Trypanosomatidae* and will present our current understanding on their functions in the biology of these medically important protozoan parasites.

Introduction

The <u>A</u>TP-<u>b</u>inding <u>c</u>assette (ABC) protein superfamily is a family of proteins that combine the hydrolysis of ATP with several biological functions. Whereas most of these proteins are involved in the ATP-dependent transport of a variety of molecules across biological membranes, including amino acids, sugars, peptides, lipids, ions and chemotherapeutic drugs (Higgins, 1992; Gottesman *et al.*, 2002), some ABC proteins are devoid of transmembrane domains and are associated with functions unrelated to transport. In prokaryotes, ABC transporters are primarily involved in the import of nutrients from the extracellular milieu (Davidson *et al.*, 2008), whereas in eukaryotes they are mainly associated with the export of molecules from the cytoplasm to the extracellular space or to intracellular compartments (Borst *et al.*, 2002). The functional significance of this group of proteins is reflected by the diversity of Mendelian and complex disorders associated with mutations and/or overexpression of many of the 48 human ABC genes (Dean, 2005).

Proteins can be classified as part of the ABC superfamily based on the presence in their sequence of a strongly conserved <u>n</u>ucleotide-<u>b</u>inding <u>d</u>omain (NBD) with three major motifs. Along with the Walker A and B motifs (Walker *et al.*, 1982) common to most families of ATPases, the NBD of ABC proteins is composed of a characteristic ABC signature, the 'C' motif (Hyde *et al.*, 1990), which is located between the two Walker boxes and which is a diagnostic feature of the ABC superfamily (see Chapter 1). In addition to the NBDs, ABC proteins involved in transport also contain <u>t</u>rans<u>m</u>embrane <u>d</u>omains (TMD) composed of multiple transmembrane α-helices. Eukaryotic ABC transporters are organized either as full transporters containing two TMDs and two NBDs, or as half transporters containing only one of each domain. Given that a functional transporter appears to require a minimum of at least two NBDs coupled with two TMDs, the half transporter molecules need to assemble as homodimers or heterodimers in the membrane to create a functional transporter. Soluble ABC proteins do not possess any transmembrane domains and are generally composed of two NDBs fused on the same molecule. On the basis of homology in the NBD sequence, eukaryotic ABC proteins can be divided into eight major subfamilies (ABCA to ABCH), seven of which (ABCA to ABCG) are found in the human genome (Allikmets *et al.*, 1996). For plants, a further subfamily (ABC1) has been proposed to incorporate the 'prokaryotic'-type ABCs that are absent from most animal genomes (Verrier *et al.*, 2008). Few proteins do not fit into these subfamilies and the subfamilies can be further divided into subgroups.

The ABC protein superfamily is one of the largest evolutionary conserved protein families and is found in all kingdom of life. Among others, ABC proteins have been reported in parasites of the *Trypanosomatidae* family, a group of protozoan flagellates morphologically characterized by a rounded or elongated form, a unique elongated mitochondrion and an anteriorly directed flagellum (in some genera, it borders an undulating membrane). The *Trypanosomatidae* family gathers together asexual blood or tissue parasites including the genera *Crithidia, Herpetomonas, Leptomonas* and *Blastocrithidia,* all of which are monogenetic and are parasites of invertebrates, and the genera *Leishmania* and *Trypanosoma,* which are digenetic and include important pathogens of man and animals (Simpson *et al.*, 2006). *Leishmania* spp. are protozoan pathogens transmitted by *Phlebotomus* and *Lutzomyia* spp. sand flies. They infect and persist in phagolysosomes of human macrophages and of other mammalian and reptile hosts and are responsible for a variety of clinical manifestations commonly referred as leishmaniasis. *Trypanosoma cruzi* is an intracellular parasite transmitted by the feces of reduvidae bugs and is responsible for Chagas disease in South America. The haemoflagellate *Trypanosoma brucei* is an extracellular parasite transmitted by the bite of the tsetse fly and is the aetiological agent of sleeping sickness in Africa.

Despite intensive attempts, there are no effective vaccines for the prevention of parasitic diseases of human. Accordingly, drugs (often toxic) are required for the control and treatment of these infections when simple prevention measures (such as pesticide-impregnated bed-nets) fail or prove impractical. The available arsenal of antiprotozoal drugs is limited and often relies on antiquated drugs with several drawbacks. The treatment of parasitic diseases is further complicated by drug resistance now described on a frequent basis in several endemic regions (Lira *et al.*, 1999; Hadighi *et al.*, 2006; Rojas *et al.*, 2006). Several mechanisms of drug resistance have been described in protozoan parasites, some of which imply transport-related mechanisms mediated by ABC proteins.

The genome sequences of a number of trypanosomatid parasites are either completed or under way (www.genedb.org) and numerous ABC proteins are being revealed. Extensive work is being carried out to characterize the biological function of many of these ABC proteins. Since most data related to ABC proteins were reported for parasites belonging to the digenetic branch, we will present and summarize our current understanding of the function and the phylogenetic classification of this class of ubiquitous proteins in *Trypanosoma cruzi, Leishmania* spp. and the African trypanosome.

A complete set of ABC proteins in trypanosomatids

The availability of the genome sequences of *T. brucei* (Berriman *et al.*, 2005), *T. cruzi* (El-Sayed *et al.*, 2005) and three *Leishmania* species (*L. major, L. infantum* and *L. braziliensis*) (Ivens *et al.*, 2005; Peacock *et al.*, 2007) enables database mining and the identification of entire protein families. Systematic BLAST analyses using the NBD consensus sequence motif (Pfam00005) as a query, followed by phylogenetic analyses of the retrieved NBDs, allowed the recovery and classification of every ABC protein encoded in the genome of these clinically important parasites (Leprohon *et al.*, 2006). The analysis revealed a complete ABC protein dataset in trypanosomatid parasites, with representative members of all subfamilies described in eukaryotes (ABCA-ABCH). The analysis also revealed a relatively stable ABC protein dataset among the *Leishmania* species, with *L. major* (GeneDB V5), *L. infantum*

(GeneDB V3) and *L. braziliensis* (GeneDB V2) having 42 (Leprohon *et al.*, 2006), 43 (Leprohon *et al.*, submitted) and 39 (Table 7.1) ABC proteins, respectively. Conversely, a slightly more pronounced difference was observed between the two *Trypanosoma* species, with *T. cruzi* (GeneDB V4) and *T. brucei* (GeneDB V4) having 28 and 22 ABC proteins, respectively (Leprohon *et al.*, 2006).

Comparative phylogenetic studies between *Leishmania* and *Trypanosoma* indicated that an expansion of the ABC gene family occured in *Leishmania* after its split from the *Trypanosoma* lineage and probably explain the difference in the ABC gene content between the two genera. Interestingly, the expansion was only seen for subfamilies frequently involved with redundant functions among the subfamily (i.e. ABCA, ABCC and ABCG). Furthermore, whereas most ABC genes are found alone on several chromosomes in *Leishmania,* some are grouped in a head-to-tail fashion on clusters of two or three repeated ABC genes. It is noteworthy that these clusters contain members of the expanded subfamilies in *Leishmania* and are consistent with gene duplication events (i.e. ABCA genes on chromosomes 11 and 27, ABCC genes on chromosomes 23 and 31 or ABCG genes on chromosome 6) (Leprohon *et al.*, 2006). The duplication events seem to have occured before speciation of the *Leishmania* genera since the synteny of ABC genes (including those clustered in tandem) is very well conserved between the three *Leishmania* species analysed despite over fifteen million years of proposed divergence (Momen *et al.*, 2000; Kerr, 2006; Lukes *et al.*, 2007).

The functions of several ABC proteins in the biology of trypanosomatid parasites have been assessed, mainly by gene overexpression and gene deletion experiments, and include drug resistance, transport of phospholipids and translation.

ABC proteins and drug resistance in trypanosomatids

In order to better understand the role of ABC proteins in drug resistance, we shall first describe the drugs currently used in therapy against parasites of the *Trypanosomatidae* family. Pentavalent antimonial (SbV)-containing drugs have been considered to be the first-line treatment against all clinical forms of leishmaniasis (i.e. visceral, cutaneous, and mucosal) for more than a century (Herwaldt, 1999; Berman, 2003). Although used in medicine for several decades as anti-infective in the treatment of tropical diseases, especially those of protozoan origin like leishmaniasis, the mechanism of action of SbV-containing drugs remains poorly understood. It is clear that the SbV anti-leishmanial activity is dependent on its reduction to its trivalent state (SbIII), but the mechanism of conversion and where it occurs, i.e. whether it takes place in the macrophage or in the parasite or in both, is still an opened question (Sereno *et al.*, 2001; Shaked-Mishan *et al.*, 2001). Second-line drugs against *Leishmania* include the diamidine pentamidine (Soto *et al.*, 1994) and the polyene antibiotic amphotericin B (Sundar *et al.*, 2004), but severe side effects and high cost (when part of liposomes) have limited their widespread use. The aminoglycoside paromomycin is also more frequently used (Jha *et al.*, 1998; Sundar *et al.*, 2007) and the first oral drug miltefosine shows promising efficacy (Croft *et al.*, 2006) despite evidence suggesting that resistance could develop rapidly (Perez-Victoria *et al.*, 2003). As for *Leishmania,* a number of old drugs are also available against *T. brucei* infections. However, in the late-stage of the disease, when the parasites have crossed the blood–brain barrier, the trivalent arsenical melarsoprol is the only drug with therapeutic potential (Pepin *et al.*, 1994). Despite more than 50 years of use, the mode of action of arsenicals is uncertain and several primary targets have been considered. Finally, drugs active against the acute phase of *T. cruzi* infections include the nitrofuran nifurtimox, which is involved in the generation of hydroxyl radicals, and the nitroimidazole benznidazole, which mode of action remains unclear (de Castro, 1993).

Members of the ABCB, ABCC and ABCG subfamilies of the ABC proteins superfamily have been implicated in drug resistance in the *Trypanosomatidae* and will be discussed in the following sections.

The ABCB subfamily

The ABCB subfamily is unique in eukaryotes in that it contains both full transporters and half transporters. In most cell types analysed, full transporters are usually found at the plasma

Table 7.1 ABC proteins in *Leishmania*

Gene nomenclature	*Leishmania major* (GeneDB V5)	*Leishmania infantum* (GeneDB V3)	*Leishmania braziliensis* (GeneDB V2)	Gene topology
ABCA1	LmjF2.0300	LinJ02_V3.0270	LbrM02_V2.0350	[TM-NBD]2
ABCA2	LmjF11.1220	LinJ11_V3.1210	LbrM11_V2.0960	[TM-NBD]2
ABCA3	LmjF11.1240	LinJ11_V3.1230	LbrM11_V2.1000	[TM-NBD]2
ABCA4	LmjF11.1250	LinJ11_V3.1240	LbrM11_V2.1010	[TM-NBD]2
ABCA5	LmjF11.1270	LinJ11_V3.1260	LbrM11_V2.1020	[TM-NBD]2
ABCA6	LmjF11.1290	LinJ11_V3.1280	LbrM11_V2.1040	[TM-NBD]2
ABCA7	LmjF15.0760	LinJ15_V3.0800	LbrM15_V2.0820	[TM-NBD]2
ABCA8	LmjF27.0970	LinJ27_V3.0830	LbrM27_V2.1050	[TM-NBD]2
ABCA9	LmjF27.0980	LinJ27_V3.0840	LbrM27_V2.1060	[TM-NBD]2
ABCA10	LmjF29.0620	LinJ29_V3.0640	LbrM29_V2.0630	[TM-NBD]2
ABCB1	LmjF25.0530	LinJ25_V3.0540	LbrM25_V2.0480	TM-NBD
ABCB2	LmjF26.2670	LinJ26_V3.2700	LbrM26_V2.2600	[TM-NBD]2
ABCB3	LmjF32.3080	LinJ32_V3.3280	LbrM32_V2.3380	TM-NBD
ABCB4	LmjF34.0990	LinJ34_V3.1060	*Absent?*	[TM-NBD]2
ABCC1	LmjF23.0210	LinJ23_V3.0230	*Absent?*	[TM-NBD]2
ABCC2	LmjF23.0220	LinJ23_V3.0240	LbrM23_V2.0230	[TM-NBD]2
ABCC3	LmjF23.0250	LinJ23_V3.0290	LbrM23_V2.0280	[TM-NBD]2
ABCC4	LmjF31.1270	LinJ31_V3.1290	LbrM31_V2.1500	[TM-NBD]2
ABCC5	LmjF31.1280	LinJ31_V3.1300	LbrM31_V2.1510	[TM-NBD]2
ABCC6	LmjF31.1290	LinJ31_V3.1310	LbrM31_V2.1520	[TM-NBD]2
ABCC7	LmjF31.1430	LinJ31_V3.1460	LbrM31_V2.1650	[TM-NBD]2
ABCC8	LmjF34.0670	LinJ34_V3.0690	LbrM20_V2.0600	[TM-NBD]2
ABCC9	*Absent*	LinJ24_V3.1510	*Pseudogene*	[TM-NBD]2
ABCD1	LmjF27.0470	LinJ27_V3.0480	LbrM27_V2.0580	TM-NBD
ABCD2	LmjF31.0540	LinJ31_V3.0560	LbrM31_V2.0700	TM-NBD
ABCD3	LmjF33.1860	LinJ33_V3.1960	LbrM33_V2.2130	TM-NBD
ABCE1	LmjF21.0710	LinJ21_V3.0770	LbrM21_V2.0770	[NBD]2
ABCF1	LmjF3.0160	LinJ03_V3.0150	LbrM03_V2.0170	[NBD]2
ABCF2	LmjF19.0800	LinJ19_V3.0800	LbrM19_V2.1110	[NBD]2
ABCF3	LmjF33.0310	LinJ33_V3.0340	LbrM33_V2.0320	[NBD]2
ABCG1	LmjF06.0080	LinJ06_V3.0080	LbrM06_V2.0010	NBD-TM
ABCG2	LmjF06.0090	LinJ06_V3.0090	LbrM06_V2.0080	NBD-TM
ABCG3	LmjF06.0100	LinJ06_V3.0100	*Absent?*	NBD-TM

Gene nomenclature	*Leishmania major* (GeneDB V5)	*Leishmania infantum* (GeneDB V3)	*Leishmania braziliensis* (GeneDB V2)	Gene topology
ABCG4	LmjF15.0890	LinJ15_V3.0950	LbrM15_V2.0930	NBD-TM
ABCG5	LmjF23.0380	LinJ23_V3.0430	LbrM23_V2.0420	NBD-TM
ABCG6	LmjF36.2890	LinJ36_V3.3040	LbrM35_V2.3110	NBD-TM
ABCH1	LmjF11.0040	LinJ11_V3.0040	LbrM19_V2.0280	NBD
ABCH2	LmjF29.1640	LinJ29_V3.1780	LbrM29_V2.1750	NBD
ABCH3	LmjF30.1330	LinJ30_V3.1390	LbrM30_V2.1450	NBD
Unclassified	LmjF12.1190	LinJ12_V3.0790	LbrM12_V2.0870	NBD
	LmjF32.2060	LinJ32_V3.2190	LbrM32_V2.2270	TM-NBD
	LmjF33.3040	LinJ33_V3.3200	LbrM33_V2.3320	NBD
	LmjF33.3260	LinJ33_V3.3410	LbrM33_V2.3540	NBD

TM, Transmembrane domain; NBD, Nucleotide-binding domain.

Gene nomenclature from www.genedb.org.

membrane whereas half transporters are localized to the membrane of intracellular organelles like the mitochondria, the endoplasmic reticulum and lysosomes. Accordingly, the ABCB subfamily is associated with heterogenous functions like phospholipids (van Helvoort *et al.*, 1996) and peptides transport (Herget *et al.*, 2007), Fe/S clusters biogenesis (Kispal *et al.*, 1999; Mitsuhashi *et al.*, 2000) and drug resistance. Interestingly, it has recently been reported that a half transporter of the ABCB subfamily is required for efficient RNA interference in *Caenorhabditis elegans* (Sundaram *et al.*, 2006).

In *Leishmania,* the ABCB subfamily is composed of two full transporters and two half transporters in *L. major* (GeneDB V5) and *L. infantum* (GeneDB V3) (Leprohon *et al.*, 2006), whereas only one full transporter and two half transporters have been found in the *L. braziliensis* genome (GeneDB V2) (Table 7.1). The gene coding for the multidrug resistance protein 1 (*MDR1/P-gp*) was the first ABC gene to be cloned in human (Gottesman *et al.*, 2006) and its *Leishmania* homologue was the first ABCB gene to be identified in the *Trypanosomatidae* family (Chow *et al.*, 1993; Henderson *et al.*, 1992). MDR1/P-gp is responsible for the multidrug resistance phenotype (MDR) and is characterized by its ability to confer resistance to a plethora of structurally unrelated hydrophobic compounds when overexpressed. The *Leishmania MDR1* gene product is homologous to the mammalian MDR1 protein and it has been characterized in several *Leishmania* species. Analysis of the genome sequences available allowed the recovery of *MDR1* homologues in *L. donovani* (Henderson *et al.*, 1992), *L. enrietti* (Chow *et al.*, 1993), *L. amazonensis* (Gueiros-Filho *et al.*, 1995; Katakura *et al.*, 1999), *L. tropica* (Chiquero et *al.*, 1998), *L. major* and *L. infantum* (GeneDB V3). Surprisingly, *MDR* homologues could not be found in a recent version of the *L. braziliensis* genome (GeneDB V2). It is also intriguing that MDR1/Pgp homologues are absent from the genome of trypanosomes (Leprohon *et al.*, 2006). Since MDR1/P-gp homologues are often implicated in drug resistance, it seems that these parasites will resort to other resistance mechanisms in order to survive to the drugs part of the MDR spectrum.

In *Leishmania,* the homologue of *MDR1* was found to be amplified in mutants selected for daunomycin or vinblastine resistance *in vitro* and was shown to confer the MDR phenotype by transfection experiments. Although vinblastine and daunomycin are not used in therapy against *Leishmania*, the overexpression of MDR1 was also associated with cross-resistance to miltefosine (Perez-Victoria *et al.*, 2001) and could putatively be involved in clinical treatment failure to this alkyl-lysophospholipid highly active against *Leishmania*. The degree of similarity between the *Leishmania* and human MDR1 homologues suggested that the former could confer resistance by an active extrusion of the drugs. Consistent with this proposal, initial drug accumulation experiments showed an absence of accumulation of puromycin (Henderson *et al.*, 1992), a reduced accumulation of daunomycin (Perez-Victoria *et al.*, 1999) and an increased efflux of rhodamine 123 (Gueiros-Filho *et al.*, 1995) in MDR1-overexpressing drug-resistant parasites. However, subsequent subcellular localization studies did not support the plasma membrane efflux pump hypothesis and revealed that the majority of MDR1 is located in a number of endocytic and secretory compartments including the Golgi apparatus, the endoplasmic reticulum (ER) and the multivesicular tubule (MVT)-lysosome in *L. enrietti* and *L. mexicana* (Dodge *et al.*, 2004). Pulse-chase labelling experiments later helped to understand the reduced drug accumulation observed in MDR1-overexpressing parasites as they suggested that the Golgi and ER pools were active in pumping fluorescent markers outside the cells and that MDR1 substrates are probably exported from the parasite by exocytosis (Dodge *et al.*, 2004). Finally, the function of MDR1 in mediating drug resistance in *Leishmania* appears to be iron-dependent (Wong *et al.*, 2006).

The other full lengh ABCB transporter that has been described in *Leishmania* has also been identified following drug resistance studies. Indeed, the MDR2 protein of *L. amazonensis* (LaMDR2) exhibit 47% amino acids identity to its most closely related protein, LaMDR1, and parasites overexpressing its gene showed a significant decrease in accumulation of radioactively-labelled 5-fluorouracil (5-FU) (Katakura *et al.*, 2004). Although the overexpression of LaMDR2 resulted in low-level resistance to 5-FU, it could not confer resistance to the MDR1 substrates vinblastine and puromycine, which suggests that LaMDR2 is clearly functionally different from LaMDR1.

Conversely to the full transporters of the ABCB subfamily, the genome of every trypanosomatid parasites analysed (*L. major*, *L. infantum*, *L. braziliensis*, *T. brucei* and *T. cruzi*) encodes two half ABCB transporters (Leprohon *et al.*, 2006) (Table 7.1). In eukaryotes, the half transporters of the ABCB subfamily are usually involved in important biological processes and are often well conserved among organisms (Burke *et al.*, 2007; Herget *et al.*, 2007). Although no experimental data have been reported regarding the functions of half ABCB proteins in the biology of trypanosomatid parasites, phylogenetic studies indicated that the half transporter ABCB3 in *Leishmania* and *Trypanosoma* is one of the few clear orthologues that have been identified while comparing the ABC dataset of trypanosomatids to those of other eukaryotes (Leprohon *et al.*, 2006). Indeed, comparative analyses indicated that ABCB3 is an orthologue of the yeast mitochondrial transporter ATM1, a half ABCB transporter located to the mitochondrial inner membrane (Leighton *et al.*, 1995) and involved in the transport of Fe/S cluster precursors to the cytosol (Kispal *et al.*, 1999). Given the orthologous relationship between the yeast ATM1 protein and ABCB3 in trypanosomatid parasites, it is likely that the trypanosomatid orthologue of ATM1 is involved with very similar processes. Interestingly, the function of ATM1 is essential to the biogenesis of cytosolic Fe/S cluster proteins and the knock down of its mRNA is associated with a severe growth defect in yeast (Leighton *et al.*, 1995). Since *ABCB3* is the only homologue of *ATM1* in trypanosomatids (Leprohon *et al.*, 2006), it might be of interest to assess the usefulness of this gene as a therapeutic target by gene deletion experiments.

The ABCC subfamily

The ABCC subfamily is defined as a group of proteins involved in the ATP-dependent transport of thiol conjugates. The physiological functions of these proteins vary from the transport of molecules implicated in the immune response (Leukotriene C4) to drug resistance (Deeley *et al.*, 2006). Several groups have reported drug resistance mechanisms incriminating ABCC proteins in parasites and the first ABCC protein was described following drug resistance studies in *Leishmania* (Ouellette *et al.*, 1990). Indeed, our work on metal-resistant mutants of *Leishmania* has led to a coherent SbV resistance model involving ABCC genes. Analysis of *Leishmania* antimony-resistant mutants indicated that the gene coding for the ABC protein MRPA/ABCC3 (formerly known as PGPA) is frequently amplified following selection for a SbIII resistance phenotype. The isolated amplicons are usually circular and are generated by recombination events between direct or inverted DNA repeats (Ouellette *et al.*, 1991; Grondin *et al.*, 1993), a common mechanism of gene amplification in *Leishmania* (Beverley, 1991). Transfection experiments clearly indicated that MRPA/ABCC3 overexpression was associated with arsenite (AsIII) and antimonite (SbIII) resistance (Callahan *et al.*, 1991; Papadopoulou *et al.*, 1994), although the level of resistance conferred by MRPA/ABCC3 varied according to the species into which the gene was transfected (Papadopoulou *et al.*, 1994; Legare *et al.*, 1997). This led to the suggestion that the ability of MRPA/ABCC3 to confer metal resistance requires other factors and that the availability of these factors differs in various *Leishmania* species. By analogy with the GS-X pump (Ishikawa, 1992), it was proposed that MRPA/ABCC3 might recognize metals conjugated to the bisglutathione–spermidine conjugate trypanothione (TSH), the major reduced thiol in *Leishmania* (Fairlamb *et al.*, 1992). Supporting this hypothesis, MRPA/ABCC3 has been located to an intracellular vesicule by fusion with the green fluorescent protein (GFP) and transport experiments have confirmed its role in the sequestration of antimony conjugated to thiols (Legare *et al.*, 2001).

The enzymatic activities of γ-glutamylcysteine synthetase (γ-GCS) (Grondin *et al.*, 1997; Haimeur *et al.*, 2000), the rate-limiting step in glutathione (GSH) synthesis, and of ornithine decarboxylase (ODC) (Haimeur *et al.*, 1999), the rate-limiting step in spermidine synthesis, are frequently increased in *Leishmania* metal-resistant mutants and a dual increase in GSH and spermidine levels, the two building blocks of TSH, ultimately leads to increased TSH levels in these mutants (Mukhopadhyay *et al.*, 1996; Legare *et al.*, 1997; Haimeur *et al.*, 2000). Experiments using γ-GCS and ODC specific inhibitors have confirmed that TSH is essential for metal resistance (Haimeur *et al.*, 1999; Haimeur *et al.*, 2000), although elevated levels of TSH alone are usually not sufficient to sustain parasite survival in the presence of AsIII or SbIII (Grondin *et al.*, 1997; Haimeur *et al.*, 1999). Transfection of *GSH1* alone can, however, lead to resistance to antimonials in some species (Mukherjee *et al.*, 2007) and can lead to resistance to arsenicals in the African trypanosome (Shahi *et al.*, 2002). Co-transfection experiments indicated that a strong synergy occurs between *MRPA/ABCC3* and the gene coding for γ-GCS (*GSH1*) when both genes are co-transfected in a revertant strain (a SbIII-resistant mutant grown in the absence of drug for prolonged periods) of *L. tarentolae* but not when they are overexpressed in a wildtype background (Grondin *et al.*, 1997; Haimeur *et al.*, 1999). Since the synergy only occurs in revertant cells, it is clear that at least one other stable mutation involved in resistance is present in the metal-resistant mutants. By analogy to the GS-X system (Ishikawa, 1992), it has been proposed that the missing mutation could correspond to an increased trypanothione-S-transferase (TST) activity, although this seems unlikely in light of recent results reporting no changes of TST activity in SbIII-resistant *L. tarentolae* mutants (Wyllie *et al.*, 2008).

As opposed to metal-resistant mutants selected *in vitro*, few studies have dealt with antimony resistance mechanisms in clinical isolates. Some reports have indicated the involvement of *MRPA*, *GSH1* and *ODC* (Mukherjee *et al.*, 2007), which falls in line with the model elaborated while studying strains selected for SbIII resistance *in vitro*, while others revealed new mechanisms like a decreased propensity to undergo apoptosis in drug-resistant cells (Vergnes *et al.*, 2007). In other field strains, however, resistance mechanisms seem to differ (Decuypere *et al.*, 2005).

Since the discovery of MRPA/ABCC3, several other ABCC proteins have been identified in *Leishmania* (Legare *et al.*, 1994). The analysis of the genome indicated that the ABCC subfamily is composed of eight members in *L. major* (Leprohon *et al.*, 2006) and of nine members in *L. infantum*. *ABCC1* and *ABCC2* are clustered in tandem on chromosome 23, just upstream of *MRPA/ABCC3* (Legare *et al.*, 1994). The genes

ABCC4 to ABCC7 have been mapped to chromosome 31, whereas *ABCC8* and *ABCC9* are found alone on chromosome 34 and 24, respectively (Leprohon *et al.*, 2006). Apart from MRPA/ABCC3, three other ABCC proteins have been implicated in drug resistance in *Leishmania* using two different molecular approaches. The first two proteins were recovered by overexpression experiments of the entire ABCC subfamily, followed by SbIII resistance assays. These experiments revealed that the tandemly duplicated genes *ABCC4* and *ABCC5* are able to confer low-level resistance to SbIII when overexpressed in a revertant background of *L. tarentolae* (Leprohon *et al.*, submitted). The overexpression of these proteins in wildtype strains of *Leishmania* could not confer an increased tolerance to antimony, however, and the role of these proteins in resistance might require a putative mutation selected in the resistant mutant as described above for *MRPA/ABCC3*. The two proteins have been localized to an intracellular compartment oriented along the longitudinal axis of the parasite and are probably involved in the sequestration of the drug.

The last ABCC protein involved in resistance in *Leishmania* was recovered by functional cloning experiments. Functional cloning is a whole genome approach ideally suited for the isolation of drug resistance genes. Indeed, a genomic bank is transfected in *Leishmania* drug-sensitive cells and the transfectants are then tested for a change in drug susceptibility. If a gain of function is suspected, it is appropriate to transfect the cosmid bank in a wild-type background, to select with the marker of the cosmid (e.g. hygromycin B) and then to plate cells with varying concentration of the selective drug for which resistance is being studied. Functional cloning allowed the recovery of the ABC transporter PRP1/ABCC7 while selecting for resistance against the second-line drug pentamidine (Coelho *et al.*, 2003). In addition to pentamidine resistance, it has been shown that overexpression of *PRP1/ABCC7* confers low-level cross-resistance to SbIII. PRP1/ABCC7 has also been localized to an intracellular compartment by GFP fusion (Coelho *et al.*, 2006; Coelho *et al.*, 2007) and could be involved in a drug sequestration phenomenon.

Drug resistance mediated by the sequestration of drugs appears to be a frequent mechanism involved in metal tolerance involving ABC proteins. Indeed, whereas the *S. pombe* ABC transporter HTM1p confers cadmium tolerance by mediating the sequestration of its phytochelatin (a GSH-like molecule) conjugates (Ortiz *et al.*, 1992; Ortiz *et al.*, 1995), the *S. cerevisiae* ABCC protein Ycf1p confers resistance to cadmium, arsenite and antimonite by mediating the vacuolar transport of their GSH conjugates (Ghosh *et al.*, 1999; Li *et al.*, 1997; Li *et al.*, 1996). However, even if several intracellular ABCC proteins have been associated with metal resistance in *Leishmania*, it would be surprising that this parasite has evolved or conserved genes specifically involved in AsIII or SbIII sequestration, given the unlikely possibility that this parasite encounters metalloids during its parasitic life cycle.

The localization of the ABCC proteins in *Leishmania* is of interest. Indeed, it has previously been suggested that an ATP-dependent antimony efflux system located at the plasma membrane of the parasite could be encoded by one of the ABCC homologues. However, this assumption could not be supported, since confocal microscopy analyses of GFP fusions revealed that the entire family of ABCC proteins is intracellularly localized in *Leishmania* (Leprohon, P., submitted). While ABCC1-GFP, ABCC2-GFP and ABCC6-GFP are located to a network of intracellular membranes, ABCC3-GFP is located in an intracellular vesicule near the flagellar pocket of *Leishmania*, ABCC4-GFP, ABCC5-GFP and ABCC7-GFP are located to an intracellular organelle oriented along the longitudinal axis of the parasite and ABCC8-GFP has been localized to a punctuated organelle near the posterior end of the parasite. Although the physiological functions of any of the *Leishmania* ABCC protein have been described, it is intriguing that every of these proteins are located in intracellular compartments while several ABCC proteins have been localized to the plasma membrane in many other cell types, including the related parasite *T. brucei* (Shahi *et al.*, 2002). In eukaryotes, the ABCC subfamily is often implicated in secretion of toxic metabolites or in cellular detoxification and for an intracellular parasite like *Leishmania*, it might be advantageous to transport toxic compounds and waste metabolites inside intracellular compartments. The elucidation of the subcellular localization of

ABCC proteins in other intracellular parasites, like *T. cruzi*, could help to resolve this question.

The implication of ABC transporters in antimony resistance in *Leishmania* subsequently led to the search for ABC transporters involved in resistance to the related drug melarsoprol in *T. brucei*. Several ABC transporters have been described in *T. brucei* and three were classified as part of the ABCC subfamily by phylogenetic studies (Leprohon *et al.*, 2006). Among the three ABCC proteins identified, two were correlated with drug resistance by transfection experiments. Overexpression of the *T. brucei MRPA/ABCC3* homologue *TbMRPA* was found to increase the IC_{50} of melarsoprol by 10-fold *in vitro* (Luscher *et al.*, 2006; Shahi *et al.*, 2002), although its role in clinical resistance remains to be clarified (Alibu *et al.*, 2006). Immunofluorescence studies localized a GFP-tagged version of TbMRPA to the plasma membrane and suggested that this transporter proceeds by extrusion of the drug outside the cell. Transfection experiments using a second ABCC gene, *TbMRPE*, had little effect on susceptibility to melarsoprol but gave a two- to three fold resistance to suramin, a dye-related drug used against the early stage of human african trypanosomiasis infections (Shahi *et al.*, 2002). Immunofluorescence studies localized a tagged version of TbMRPE to intracellular vesicules and to the flagellar pocket, a specialized region of the plasma membrane where endocytosis and exocytosis take place. The function of the third and most divergent ABCC protein in *T. brucei* remains unknown.

Finally, an ABC transporter of the ABCC subfamily, TcPGP2, has been characterized in *T. cruzi* and is most probably the *Leishmania* MRPA/ABCC3 homologue. However, a role for TcPGP2 in resistance could not be confirmed by gene transfection experiments and susceptibility assays against several drugs (Dallagiovanna *et al.*, 1996). Another gene of the ABCC subfamily has been described in *T. cruzi* (*TcPGP1*) but it is interrupted by the insertion of a retrotransposon and has been annotated as a pseudogene

The ABCG subfamily

The ABCG proteins are unique in their architecture and consist of half transporters with a single NDB located at the amino terminal end of the proteins followed by six putative transmembrane domains (reverse topology). Transporters of the ABCG subfamily are mainly associated with cellular lipids homeostasis and drug resistance. Indeed, while most functionally characterized ABCG proteins are involved in lipid and sterol transport (Velamakanni *et al.*, 2007), elevated levels of ABCG2 in mammals have been associated with physiologically and pharmacologically important roles in resistance against a variety of anti-cancer drugs (Doyle *et al.*, 1998).

The analysis of the genome sequences of *Leishmania* spp. and the trypanosomes revealed ABCG subfamilies composed of six and four proteins, respectively (Leprohon *et al.*, 2006), and gene transfection experiments indicated a role in phospholipid transport and in drug resistance for two of the *Leishmania* proteins. Indeed, overexpression of *ABCG4* in *L. infantum* caused a reduced accumulation of a fluorescent analogue of phosphatidylcholine (NBD-PC) and of [^{14}C]-miltefosine (Castanys-Munoz *et al.*, 2007). Furthermore, parasites overexpressing *ABCG4* showed a two-fold increase in resistance towards the 8-quinoline sitamaquine and the alkyl-phospholipid analogues miltefosine, edelfosine and perifosine. Since a functional GFP-tagged version of ABCG4 localized to the plasma membrane and the flagellar pocket of the parasite (Castanys-Munoz *et al.*, 2007), it appears that the *Leishmania* ABCG4 protein is most likely involved in the outward transport of phospholipid analogues across the plasma membrane.

The second *Leishmania* ABCG protein was first described while studying mechanisms of resistance to camptothecin (BoseDasgupta *et al.*, 2008), a topoisomerase I inhibitor active against the parasite *in vitro*. Indeed, *L. donovani* camptothecin-resistant mutants selected in a stepwise fashion showed a three-fold increase in expression of *ABCG6* at the RNA level and gene overexpression experiments confirmed the role of *ABCG6* in camptothecin resistance. Shortly after, it was reported that *ABCG6* overexpression also confers significant resistance to the leishmanicidal agents miltefosine and sitamaquine and revealed a link between *ABCG6* overexpression and a reduced accumulation of various fluorescent analogues of phosphatidylcholine (NBD-PC), phosphatidylethanolamine

(NBD-PE) and phosphatidylserine (NBD-PS) (Castanys-Munoz *et al.*, 2008). Moreover, drug accumulation experiments in intact cells revealed a reduced accumulation of camptothecin, miltefosine and sitamaquine in *ABCG6*-overexpressing strains that correlated with an increased efflux of these drugs. Finally, a GFP-tagged version of ABCG6 has been localized to the plasma membrane and flagellar pocket of the parasite (BoseDasgupta *et al.*, 2008; Castanys-Munoz *et al.*, 2008) and transport experiments using everted plasma membrane vesicules confirmed the role of ABCG6 in the ATP-dependent efflux of camptothecin (BoseDasgupta *et al.*, 2008).

Drug resistance reversers

The MDR phenotype, due to the overexpression of ABC transporters, has been documented as a very efficient mechanism to reduce intracellular drug accumulation in cancer cells (Szakacs *et al.*, 2006) and parasites (Chiquero *et al.*, 1998; Foote *et al.*, 1989; Henderson *et al.*, 1992; Hendrickson *et al.*, 1993; Legare *et al.*, 2001; Ouellette *et al.*, 1991; Wilson *et al.*, 1989). Numerous modulators or chemosensitizers are known to alter the ability of such transporters to maintain intracellular drug concentrations to subtoxic levels (Ponte-Sucre, 2007; see Chapter 9). Flavonoids constitute a well-known class of natural inhibitors of different ATP-binding proteins and interact with both the ATP-binding site and a vicinal hydrophobic region of the proteins (Conseil *et al.*, 1998). In *Leishmania*, it has been reported that the binding of flavonoids to the C-terminal NBD of the *L. tropica* MDR1 correlated with a decreased daunomycin resistance mediated by the inhibition of its efflux in a *L. tropica* daunomycin-resistant mutant overexpressing MDR1 (Perez-Victoria *et al.*, 1999). Furthermore, synthetic flavonoid dimers have been shown to induce a dose-dependent sensitization to pentamidine and sodium stibogluconate (SbV) in resistant strains of *L. enrietti* and *L. donovani*, respectively, although the sensitization seemed to be independent of the MDR1 protein (Wong *et al.*, 2007).

It has been reported that several compounds of the dihydro-β-agarofuran sesquiterpenes class were able to induce a reversion of daunomycin resistance in a *L. tropica* daunomycin-resistant mutant overexpressing MDR1 (Kennedy *et al.*, 2001; Perez-Victoria *et al.*, 2001; Perez-Victoria *et al.*, 1999). Interestingly, a combinatorial strategy using suboptimal doses of one of the most potent dihydro-β-agarofuran sesquiterpenes, targeting the transmembrane domains of MDR1, in combination with a potent flavonoid derivative targeting the cytosolic NBD with high affinity, caused an increase in miltefosine intracellular accumulation and allowed the sensitization to daunomycin of a miltefosine-resistant *L. tropica* mutant overexpressing MDR1, without any side effects on different mammalian cell lines (Perez-Victoria *et al.*, 2006).

Finally, it has been shown that glibenclamide, an anti-diabetic drug part of a class of medications known as sulfonylureas, enhances glucantime's efficacy in decreasing the infection rate of *Leishmania*-infected macrophages. Glibenclamide targets a compartment associated with the multivesicular system in *L. major* that may represent the organelle where antimonials are sequestered (Padron-Nieves *et al.*, 2009).

ABC proteins involved in the transport of phospholipids

The ABCA subfamily

The ABCA subfamily contains some of the largest ABC proteins and several are over 2000 amino acids long. This subfamily is mainly involved in the transport of lipids and sterols and is composed of full transporters that can be located to the plasma membrane or to intracellular compartments (Albrecht *et al.*, 2007). The ABCA subfamily is absent from the yeast genome (Decottignies *et al.*, 1997) and it was previously thought to be restricted to multicellular organisms. However, several ABCA proteins have now been reported in unicellular eukaryotes like *Leishmania* (Parodi-Talice *et al.*, 2003; Araujo-Santos *et al.*, 2005) and *Trypanosoma* (Torres *et al.*, 2004).

The first ABCA protein to be described in a unicellular eukaryote has been isolated in *Leishmania* by the screen of a genomic bank using an ABCA-like transporter homologue as a probe (Parodi-Talice *et al.*, 2003). Among the ten ABCA genes that were later recovered from the *Leishmania* genome sequence (Leprohon *et al.*, 2006), two have been functionally characterized and might have a role in phospholipids trafficking.

Indeed, while *LtrABC1.1/ABCA8* was mapped on a cluster of two tandemly duplicated genes flanked by inverted repeats on chromosome 27 (Parodi-Talice *et al.*, 2003), *LtrABCA2/ABCA4* was found on a locus of chromosome 11 along with four other ABCA genes (Araujo-Santos *et al.*, 2005). Both proteins are mainly located to the flagellar pocket and internal vesicules of the parasite, where they appear to be involved in lipid movement across the plasma membrane. Indeed, phospholipid accumulation experiments in *L. tropica* indicated that overexpression of *LtrABC1.1/ABCA8* and *LtrABCA2/ABCA4* is associated with a decreased retention of fluorescent glycerophospholipid analogues. Furthermore, the exocytosis pathway appeared significantly altered in *LtrABCA1.1/ABCA8*- and *LtrABCA2/ABCA4*-overexpressing parasites as measured by the secreted acid phosphatase, a marker of the secretory pathway in *Leishmania*, and suggested that vesicular trafficking might be altered in the transfected parasites (Parodi-Talice *et al.*, 2003; Araujo-Santos *et al.*, 2005).

In trypanosomes, the only ABCA protein with a characterized function has also been isolated by a screen of a genomic bank of *T. cruzi* using a partial cDNA sequence encoding a putative ABCA-like transporter as a probe. TcABC1 shares 41% amino acid identity with the *Leishmania* ABCA4 and ABCA8 proteins and has been located to the plasma membrane and flagellar pocket of the parasite by immunolocalization experiments (Torres *et al.*, 2004). Gene expression studies indicated that TcABC1 is differentially expressed through the life cycle of the parasite, being absent at the mRNA and protein levels in the extracellular trypomastigote stage. As ABCA proteins in *Leishmania* were involved in lipid movement, the role of TcABC1 in such process was assessed using fluorescent phospholipid analogues. In contrast to *Leishmania*, no statistically significant differences in glycerophospholipid analogues accumulation were observed in the TcABC1-overexpressing parasites. However, a small but significant increase in accumulation of a sphingomyelin analogue (NBD-SM), which uptake depends mainly on endocytosis, was noted in the TcABC1-overexpressing parasites and suggested a relationship between the ABCA proteins and endocytosis. This was supported by experiments showing an increased accumulation of the endocytic marker FM4–64 and an increased ectophosphatase activity in TcABC1-overexpressing parasites, suggesting a role for this protein in endocytosis and in vesicular trafficking in *T. cruzi* (Torres *et al.*, 2004).

Finally, the ABCA subfamily has undergone several gene duplication and gene deletion events during evolution (Annilo *et al.*, 2006) and the *Trypanosomatidae* family is no exception. Trypanosomatid parasites contain a large ABCA subfamily showing extensive gene duplication events that occurred in *Leishmania* after its split from the *Trypanosoma* lineage (Leprohon *et al.*, 2006). The presence of an expanded family of ABCA transporters in a unicellular parasite like *Leishmania* is striking. Indeed, ABCA1 in human is an ABCA protein involved in severe disorders of cholesterol transport (i.e. Tangier disease) and in high density lipoprotein biosynthesis (Zarubica *et al.*, 2007). Interestingly, this protein has also been implicated in the engulfment of apoptotic cells by macrophages (Luciani *et al.*, 1996). Consequently, the presence of an expanded ABCA subfamily in a parasite engulfed by and living within macrophages like *Leishmania* is noteworthy and may suggest a role for these proteins in host–pathogen interactions. The paralogous relationship of ABC transporters in *Leishmania* did not allowed the identification of a clear ABCA1 orthologue by phylogenetic analyses (Leprohon *et al.*, 2006). However, it is noteworthy that the overexpression of the two functionally characterized *Leishmania* ABCA proteins (LtrABCA1.1/ABCA8 and LtrABCA2/ABCA4) has been correlated with an altered infectivity of the parasites (Araujo-Santos *et al.*, 2005; Parodi-Talice *et al.*, 2003).

ABC proteins involved in translation in trypanosomatids

Whereas most ABC proteins are involved in transport-related processes, the ABCE and ABCF subfamilies are devoid of transmembrane domains and are made up of two fused NBDs. These soluble ABC proteins are very conserved among eukaryotes and are mainly implicated in the regulation of mRNA translation in several organisms, including trypanosomatid parasites.

The ABCE subfamily

The ABCE subfamily is represented by the ABCE1 protein, which is one of the most conserved proteins throughout evolution. It is present in single copy in every archea and eukaryote studied, apart from *Arabidopsis thaliana* which possesses two ABCE genes. ABCE1 is composed of two NBDs fused to two Fe/S cluster domains located at the amino teminal end of the protein. The ABCE1 protein was first characterized as the oligo-adenylate-binding protein which is produced in response to infection by certain viruses (Bisbal *et al.*, 1995). It was later shown that this essential protein is also associated with ribosomes and has a more conserved role in translation, being involved in the assembly of the pre-initiation complex in yeast (Dong *et al.*, 2004; Kispal *et al.*, 2005), *Caenorhabditis* (Zhao *et al.*, 2004) and mammals (Chen *et al.*, 2006). Similarly, it was shown that depletion of *ABCE1* mRNA in *T. brucei* resulted in a severe growth arrest that correlated with a decrease in translation initiation as measured by the incorporation of [^{35}S]methionine (Estevez *et al.*, 2004). These results suggest that ABCE1 is very likely to be essential in trypanosomes and preliminary experiments indicated that this might also be the case in *Leishmania* (Leprohon, P., unpublished observations). Being essential to translation initiation, ABCE1 might prove to be of interest as a therapeutic target against parasites, although the high degree of conservation of this protein in eukaryotes might be challenging while designing a specific inhibitor.

The ABCF subfamily

The ABCF subfamily is composed of soluble proteins involved in the control of mRNA translation. Three ABCF protein coding genes have been identified in the genomes of *Leishmania* spp., *T. brucei* and *T. cruzi* (Leprohon *et al.*, 2006). No functional studies have been reported for any of these proteins in Trypanosomatids but the orthologous relationship with the human homologues allows tentative speculation on their role in the biology of the parasites (Leprohon *et al.*, 2006). Indeed, phylogenetic studies indicated that ABCF1 in *Trypanosomatidae* is an orthologue of the human ABCF3 protein which is involved in the biogenesis of ribosomes (Dong *et al.*, 2005). Furthermore, the ABCF2 protein in *Trypanosomatidae* seems to be an orthologue of the human ABCF2 and the yeast GCN20 proteins which are involved in the stress response occurring in amino acid-starved cells (Vazquez de Aldana *et al.*, 1995; Marton *et al.*, 1997). Finally, ABCF3 in *Trypanosomatidae* is an orthologue of the human ABCF1 protein which is involved in translation initiation (Tyzack *et al.*, 2000). Interestingly, microarray expression profiling of the ABC gene family in *Leishmania* revealed that the expression of *ABCF3* is downregulated in the mammalian stage (amastigote), compared to the insect stage (promastigote) of the parasite (Leprohon *et al.*, 2006) and might explain the decrease in mRNA translation rates that is occurring in the amastigote stage as observed by the profiling of polyribosomes (Papadopoulou, B., personal communication).

Other ABC proteins

The ABCD subfamily

The ABCD subfamily is composed of half-transporters which probably assemble as homodimers or heterodimers to create full transporters in the membrane of peroxisomes (Theodoulou *et al.*, 2006). Although ABCD proteins are predicted to be present in a variety of eukaryotic organisms, they have only been characterized to a significant extent in the yeast *Saccharomyces cerevisiae* (Shani *et al.*, 1996; Shani *et al.*, 1995), the plant *Arabidopsis thaliana* (Footitt et al., 2002) and different mammalian species. The ABCD proteins appear to be implicated in the transport of very-long-chain fatty acids inside the peroxisomes for their subsequent processing by the β-oxydation pathway (Cartier *et al.*, 1995; Hettema *et al.*, 1996; Braiterman *et al.*, 1998; Braiterman *et al.*, 1999).

The ABCD subfamily is composed of three members in *Leishmania* and *Trypanosoma* (Leprohon *et al.*, 2006). Trypanosomatid parasites do not possess peroxisomes *per se* but contain unique peroxisome-like organelles designated as glycosomes which sequester enzymes involved in several metabolic pathways including glycolysis (Parsons, 2004). The function of any of the ABCD proteins in the biology of Trypanosomatids has been reported. However, the ABCD proteins of *T. brucei*, termed GAT1–3 for Glycosomal ABC Transporter 1–3, have been localized to the glycosomes by immunofluorescence microscopy using GFP fusion constructs (Yernaux *et al.*, 2006). Furthermore, the role of these proteins in glycosomal functions has been suggested by the identification of the glycosome-targeting determinant of GAT1 and GAT2 and by experiments showing its interaction with the *T. brucei* homologue of the peroxin PEX19, a chaperonin and/or receptor for proteins synthetized in the cytosol and designated for insertion in the peroxisomal membrane.

The ABCH subfamily and unclassified ABC proteins

The genome of the *Trypanosomatidae* parasites encode several ABC proteins not clustering with any of the above-mentioned subfamilies and were referred as members of the ABCH subfamily or as unclassified ABC proteins whether they clustered or not with representatives of the *Drosophila* ABCH proteins by phylogenetic studies. The ABCH subfamily has been described for the first time in insects (*Drosophila* and *Anopheles*) (Dean *et al.*, 2001) but has later been observed in several organisms like *Dictyostelium* (Anjard *et al.*, 2002), *Caenorhabditis* (Sheps *et al.*, 2004), *Toxoplasma* (Sauvage *et al.*, 2006), *Trypanosoma* and *Leishmania* (Leprohon *et al.*, 2006) but not in mammals. The ABCH proteins are often devoid of transmembrane domains and are mainly found as a unique soluble NBD. The function of members of the ABCH subfamily has yet to be described.

The unclassified ABC proteins are part of a heterologous group of proteins and are usually more closely related to prokaryotic ABC proteins. It is interesting to note the good conservation of the ABCH subfamily and of the unclassified ABC proteins among the *Trypanosomatidae* (Leprohon *et al.*, 2006), especially between *Leishmania* and *T. cruzi*, and studies regarding their function in the biology of parasites seem of interest given their apparent absence in humans.

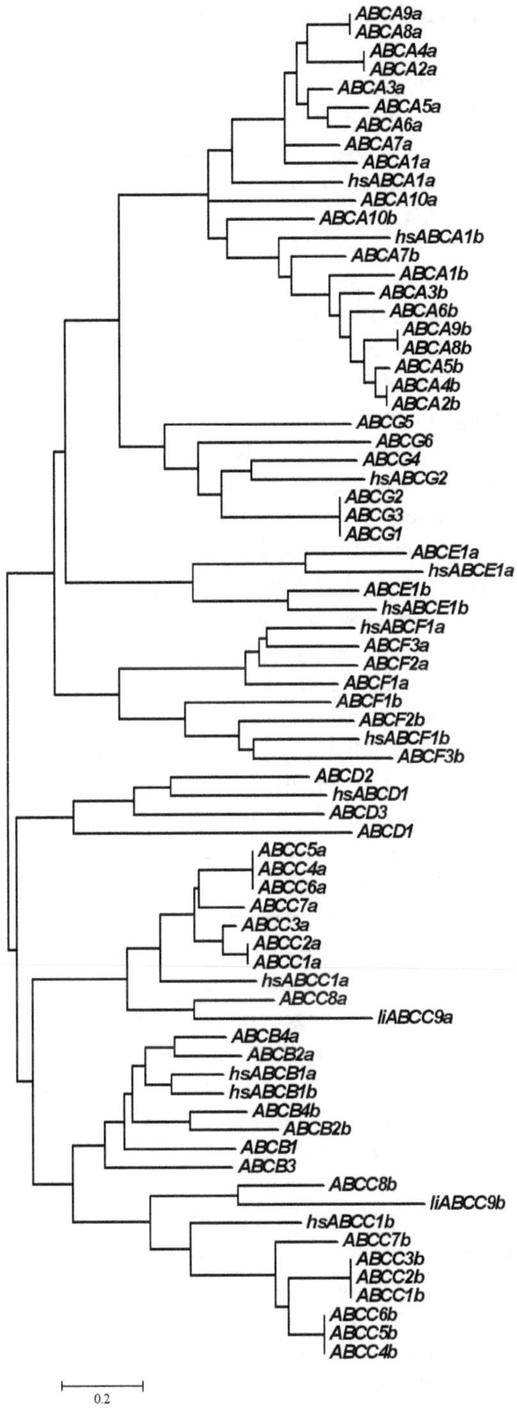

Figure 7.1 Phylogenetic analysis of the *Leishmania* ABC proteins. Nucleotide-binding domains were aligned using ClustalW. The resulting multiple alignment was subjected to phylogenetic analysis using the neighbour-joining algorithm of the MEGA software (Kumar *et al.*, 2004). The reliabilities of each branch point was assessed by the analysis of 1000 bootstrap replicates. A human representative of each mammalian subfamily was incorporated in the analysis to define each subfamily. The letters a and b represent the N- and C-terminal NBD of full length ABC proteins, respectively. li, *Leishmania infantum*; hs, *Homo sapiens*.

Concluding remarks and future trends

ABC transporters were discovered in the 1960s while studying bacterial nutrients uptake systems, revealing the histidine and maltose transporters in *Salmonella enterica* serovar Typhimurium and *Escherichia coli* (Gilson *et al.*, 1982; Higgins *et al.*, 1982), and were later discovered in every cell type studied. Since then, significant progresses have been made in understanding the structure, mechanism, biosynthesis and physiological roles of ABC proteins in several model organisms, although further studies are required to assess more fully the role of ABC proteins in trypanosomatid parasites. The high degree of sequence conservation between members of the same subfamily suggests however that the function of such proteins in parasites might be not so divergent from their orthologues described in other model organisms, i.e. to protect the parasites from cytotoxic compounds present in their environment, to help in detoxifying metabolites and to maintain cellular homeostasis. All this remains to be proven.

Inventories of ABC transporters are frequently being made in organisms for which the genome sequence has been completed and considerable experimental work is required in order to assess the biological roles of the recovered ABC proteins. One of the first steps that can be done in elucidating the functions of these transporters is to perform phylogenetic analyses with orthologues from different organisms. In parasites, episomal transfection, mutational approaches, gene disruption by homologous recombination, protein localization using GFP fusions and microarrays or proteomic analyses are available tools to provide insight into the function of ABC proteins and to study their putative involvement in host–pathogen interactions, homeostasis and drug resistance. Furthermore, post-translation modifications are well known to regulate the function of ABC proteins and might constitute an interesting line of research in trypanosomatids, given their lack of transcriptional control of gene expression. Reconstitution of purified ABC transporters in proteoliposomes is also useful to address their physiological substrates and to study their transport kinetics. Conversely, yeast two-hybrid assays and immunoprecipitation experiments may help to study the function of proteins not involved in transport activities and to reveal interaction partners of TMD-less ABC proteins. These studies would help to better understand the biology of protozoan parasites that are medically important and for which there are very few drugs available.

References

Albrecht, C., and Viturro, E. (2007). The ABCA subfamily--gene and protein structures, functions and associated hereditary diseases. Pflugers Arch. *453*, 581–589.

Alibu, V. P., Richter, C., Voncken, F., Marti, G., Shahi, S., Renggli, C. K., Seebeck, T., Brun, R., and Clayton, C. (2006). The role of *Trypanosoma brucei* MRPA in melarsoprol susceptibility. Mol. Biochem. Parasitol. *146*, 38–44.

Allikmets, R., Gerrard, B., Hutchinson, A., and Dean, M. (1996). Characterization of the human ABC superfamily: isolation and mapping of 21 new genes using the expressed sequence tags database. Hum. Mol. Genet. *5*, 1649–1655.

Anjard, C., and Loomis, W. F. (2002). Evolutionary analyses of ABC transporters of Dictyostelium discoideum. Eukaryot. Cell *1*, 643–652.

Annilo, T., Chen, Z.Q., Shulenin, S., Costantino, J., Thomas, L., Lou, H., Stefanov, S., and Dean, M. (2006). Evolution of the vertebrate ABC gene family: Analysis of gene birth and death. Genomics. *88*, 1–11.

Araujo-Santos, J.M., Parodi-Talice, A., Castanys, S., and Gamarro, F. (2005). The overexpression of an intracellular ABCA-like transporter alters phospholipid trafficking in *Leishmania*. Biochem. Biophys. Res. Commun. *330*, 349–355.

Berman, J. (2003). Current treatment approaches to leishmaniasis. Curr Opin Infect Dis. *16*, 397–401.

Berriman, M., Ghedin, E., Hertz-Fowler, C., Blandin, G., Renauld, H., Bartholomeu, D.C., Lennard, N.J., Caler, E., Hamlin, N.E., Haas, B., Bohme, U., Hannick, L., Aslett, M. A., Shallom, J., Marcello, L., Hou, L., Wickstead, B., Alsmark, U.C., Arrowsmith, C., Atkin, R. J., Barron, A.J., Bringaud, F., Brooks, K., Carrington, M., Cherevach, I., Chillingworth, T.J., Churcher, C., Clark, L.N., Corton, C.H., Cronin, A., Davies, R.M., Doggett, J., Djikeng, A., Feldblyum, T., Field, M.C., Fraser, A., Goodhead, I., Hance, Z., Harper, D., Harris, B.R., Hauser, H., Hostetler, J., Ivens, A., Jagels, K., Johnson, D., Johnson, J., Jones, K., Kerhornou, A. X., Koo, H., Larke, N., Landfear, S., Larkin, C., Leech, V., Line, A., Lord, A., Macleod, A., Mooney, P.J., Moule, S., Martin, D.M., Morgan, G.W., Mungall, K., Norbertczak, H., Ormond, D., Pai, G., Peacock, C. S., Peterson, J., Quail, M. A., Rabbinowitsch, E., Rajandream, M. A., Reitter, C., Salzberg, S. L., Sanders, M., Schobel, S., Sharp, S., Simmonds, M., Simpson, A.J., Tallon, L., Turner, C. M., Tait, A., Tivey, A. R., Van Aken, S., Walker, D., Wanless, D., Wang, S., White, B., White, O., Whitehead, S., Woodward, J., Wortman, J., Adams, M. D., Embley, T.M., Gull, K., Ullu, E., Barry, J. D., Fairlamb, A.H., Opperdoes, F., Barrell, B.G., Donelson, J.E., Hall, N., Fraser, C.M., Melville, S.E., and El-Sayed, N. M. (2005). The genome of the African trypanosome *Trypanosoma brucei*. Science. *309*, 416–422.

Beverley, S. M. (1991). Gene amplification in *Leishmania*. Annu Rev. Microbiol. *45*, 417–444.

Bisbal, C., Martinand, C., Silhol, M., Lebleu, B., and Salehzada, T. (1995). Cloning and characterization of a RNAse L inhibitor. A new component of the interferon-regulated 2–5A pathway. J. Biol. Chem. *270*, 13308–13317.

Borst, P., and Elferink, R.O. (2002). Mammalian ABC transporters in health and disease. Annu Rev. Biochem. *71*, 537–592.

Bose Dasgupta, S., Ganguly, A., Roy, A., Mukherjee, T., and Majumder, H.K. (2008). A novel ATP-binding cassette transporter, ABCG6 is involved in chemoresistance of *Leishmania*. Mol. Biochem. Parasitol. *158*, 176–188.

Braiterman, L.T., Watkins, P.A., Moser, A.B., and Smith, K.D. (1999). Peroxisomal very long chain fatty acid beta-oxidation activity is determined by the level of adrenodeukodystrophy protein (ALDP) expression. Mol. Genet. Metab. *66*, 91–99.

Braiterman, L.T., Zheng, S., Watkins, P.A., Geraghty, M.T., Johnson, G., McGuinness, M.C., Moser, A. B., and Smith, K. D. (1998). Suppression of peroxisomal membrane protein defects by peroxisomal ATP binding cassette (ABC) proteins. Hum. Mol. Genet. *7*, 239–247.

Burke, M.A., and Ardehali, H. (2007). Mitochondrial ATP-binding cassette proteins. Transl Res. *150*, 73–80.

Callahan, H.L., and Beverley, S.M. (1991). Heavy metal resistance: a new role for P-glycoproteins in *Leishmania*. J. Biol. Chem. *266*, 18427–18430.

Cartier, N., Lopez, J., Moullier, P., Rocchiccioli, F., Rolland, M.O., Jorge, P., Mosser, J., Mandel, J.L., Bougneres, P.F., Danos, O., and Aubourg, P. (1995). Retroviral-mediated gene transfer corrects very-long-chain fatty acid metabolism in adrenoleukodystrophy fibroblasts. Proc. Natl. Acad. Sci. U.S.A. *92*, 1674–1678.

Castanys-Munoz, E., Alder-Baerens, N., Pomorski, T., Gamarro, F., and Castanys, S. (2007). A novel ATP-binding cassette transporter from *Leishmania* is involved in transport of phosphatidylcholine analogues and resistance to alkyl-phospholipids. Mol. Microbiol. *64*, 1141–1153.

Castanys-Munoz, E., Perez-Victoria, J.M., Gamarro, F., and Castanys, S. (2008). Characterization of an ABCG-like transporter from the protozoan parasite *Leishmania* with a role in drug resistance and transbilayer lipid movement. Antimicrob. Agents Chemother.

de Castro, S.L. (1993). The challenge of Chagas' disease chemotherapy: an update of drugs assayed against Trypanosoma cruzi. Acta Trop. *53*, 83–98.

Chen, Z.Q., Dong, J., Ishimura, A., Daar, I., Hinnebusch, A.G., and Dean, M. (2006). The essential vertebrate ABCE1 protein interacts with eukaryotic initiation factors. J. Biol. Chem. *281*, 7452–7457.

Chiquero, M.J., Perez-Victoria, J.M., O'Valle, F., Gonzalez-Ros, J.M., del Moral, R.G., Ferragut, J.A., Castanys, S., and Gamarro, F. (1998). Altered drug membrane permeability in a multidrug-resistant *Leishmania tropica* line. Biochem. Pharmacol. *55*, 131–139.

Chow, L.M., Wong, A.K., Ullman, B., and Wirth, D.F. (1993). Cloning and functional analysis of an extrachromosomally amplified multidrug resistance-like gene in *Leishmania enriettii*. Mol. Biochem. Parasitol. *60*, 195–208.

Coelho, A.C., Beverley, S.M., and Cotrim, P.C. (2003). Functional genetic identification of PRP1, an ABC transporter superfamily member conferring pentamidine resistance in *Leishmania major*. Mol. Biochem. Parasitol. *130*, 83–90.

Coelho, A.C., Messier, N., Ouellette, M., and Cotrim, P.C. (2007). Role of the ABC transporter PRP1 (ABCC7) in pentamidine resistance in *Leishmania amastigotes*. Antimicrob. Agents Chemother. *51*, 3030–3032.

Coelho, A.C., Yamashiro-Kanashiro, E.H., Bastos, S.F., Mortara, R.A., and Cotrim, P.C. (2006). Intracellular location of the ABC transporter PRP1 related to pentamidine resistance in *Leishmania major*. Mol. Biochem. Parasitol. *150*, 378–383.

Conseil, G., Baubichon-Cortay, H., Dayan, G., Jault, J. M., Barron, D., and Di Pietro, A. (1998). Flavonoids: a class of modulators with bifunctional interactions at vicinal ATP- and steroid-binding sites on mouse P-glycoprotein. Proc. Natl. Acad. Sci. U.S.A. *95*, 9831–9836.

Croft, S.L., Seifert, K., and Yardley, V. (2006). Current scenario of drug development for leishmaniasis. Indian J. Med. Res. *123*, 399–410.

Dallagiovanna, B., Gamarro, F., and Castanys, S. (1996). Molecular characterization of a P-glycoprotein-related tcpgp2 gene in Trypanosoma cruzi. Mol. Biochem. Parasitol. 75, 145–157.

Davidson, A.L., Dassa, E., Orelle, C., and Chen, J. (2008). Structure, function, and evolution of bacterial ATP-binding cassette systems. Microbiol. Mol. Biol. Rev. 72, 317–364.

Dean, M. (2005). The genetics of ATP-binding cassette transporters. Methods Enzymol. *400*, 409–429.

Dean, M., Hamon, Y., and Chimini, G. (2001). The human ATP-binding cassette (ABC) transporter superfamily. J. Lipid Res. 42, 1007–1017.

Decottignies, A., and Goffeau, A. (1997). Complete inventory of the yeast ABC proteins. Nat. Genet. *15*, 137–145.

Decuypere, S., Rijal, S., Yardley, V., De Doncker, S., Laurent, T., Khanal, B., Chappuis, F., and Dujardin, J.C. (2005). Gene expression analysis of the mechanism of natural Sb(V) resistance in *Leishmania donovani* isolates from Nepal. Antimicrob. Agents Chemother. *49*, 4616–4621.

Deeley, R.G., Westlake, C., and Cole, S.P. (2006). Transmembrane Transport of Endo- and Xenobiotics by Mammalian ATP-Binding Cassette Multidrug Resistance Proteins. Physiol. Rev. *86*, 849–899.

Dodge, M.A., Waller, R.F., Chow, L.M., Zaman, M.M., Cotton, L.M., McConville, M.J., and Wirth, D.F. (2004). Localization and activity of multidrug resistance protein 1 in the secretory pathway of *Leishmania* parasites. Mol. Microbiol. *51*, 1563–1575.

Dong, J., Lai, R., Jennings, J.L., Link, A.J., and Hinnebusch, A.G. (2005). The novel ATP-binding cassette protein ARB1 is a shuttling factor that stimulates 40S and 60S ribosome biogenesis. Mol. Cell Biol. *25*, 9859–9873.

Dong, J., Lai, R., Nielsen, K., Fekete, C.A., Qiu, H., and Hinnebusch, A.G. (2004). The essential ATP-binding cassette protein RLI1 functions in translation by promoting preinitiation complex assembly. J. Biol. Chem. 279, 42157–42168.

Doyle, L.A., Yang, W., Abruzzo, L.V., Krogmann, T., Gao, Y., Rishi, A.K., and Ross, D.D. (1998). A multidrug resistance transporter from human MCF-7 breast cancer cells. Proc. Natl. Acad. Sci. U.S.A. 95, 15665–15670.

El-Sayed, N.M., Myler, P.J., Bartholomeu, D.C., Nilsson, D., Aggarwal, G., Tran, A.N., Ghedin, E., Worthey, E.A., Delcher, A.L., Blandin, G., Westenberger, S.J., Caler, E., Cerqueira, G. C., Branche, C., Haas, B., Anupama, A., Arner, E., Aslund, L., Attipoe, P., Bontempi, E., Bringaud, F., Burton, P., Cadag, E., Campbell, D. A., Carrington, M., Crabtree, J., Darban, H., da Silveira, J.F., de Jong, P., Edwards, K., Englund, P.T., Fazelina, G., Feldblyum, T., Ferella, M., Frasch, A.C., Gull, K., Horn, D., Hou, L., Huang, Y., Kindlund, E., Klingbeil, M., Kluge, S., Koo, H., Lacerda, D., Levin, M. J., Lorenzi, H., Louie, T., Machado, C. R., McCulloch, R., McKenna, A., Mizuno, Y., Mottram, J. C., Nelson, S., Ochaya, S., Osoegawa, K., Pai, G., Parsons, M., Pentony, M., Pettersson, U., Pop, M., Ramirez, J. L., Rinta, J., Robertson, L., Salzberg, S.L., Sanchez, D.O., Seyler, A., Sharma, R., Shetty, J., Simpson, A. J., Sisk, E., Tammi, M.T., Tarleton, R., Teixeira, S., Van Aken, S., Vogt, C., Ward, P.N., Wickstead, B., Wortman, J., White, O., Fraser, C.M., Stuart, K. D., and Andersson, B. (2005). The genome sequence of *Trypanosoma cruzi*, etiologic agent of Chagas disease. Science 309, 409–415.

Estevez, A.M., Haile, S., Steinbuchel, M., Quijada, L., and Clayton, C. (2004). Effects of depletion and overexpression of the *Trypanosoma brucei* ribonuclease L inhibitor homologue. Mol. Biochem. Parasitol. 133, 137–141.

Fairlamb, A.H., and Cerami, A. (1992). Metabolism and functions of trypanothione in the Kinetoplastida. Ann. Rev. Microbiol. 46, 695–729.

Foote, S.J., Thompson, J.K., Cowman, A.F., and Kemp, D.J. (1989). Amplification of the multidrug resistance gene in some chloroquine-resistant isolates of P. falciparum. Cell 57, 921–930.

Footitt, S., Slocombe, S.P., Larner, V., Kurup, S., Wu, Y., Larson, T., Graham, I., Baker, A., and Holdsworth, M. (2002). Control of germination and lipid mobilization by COMATOSE, the *Arabidopsis* homologue of human ALDP. EMBO J. 21, 2912–2922.

Ghosh, M., Shen, J., and Rosen, B. P. (1999). Pathways of As(III) detoxification in *Saccharomyces cerevisiae*. Proc. Natl. Acad. Sci. U.S.A. 96, 5001–5006.

Gilson, E., Higgins, C.F., Hofnung, M., Ferro-Luzzi Ames, G., and Nikaido, H. (1982). Extensive homology between membrane-associated components of histidine and maltose transport systems of *Salmonella typhimurium* and *Escherichia coli*. J. Biol. Chem. 257, 9915–9918.

Gottesman, M..M., Fojo, T., and Bates, S.E. (2002). Multidrug resistance in cancer: role of ATP-dependent transporters. Nat. Rev. Cancer. 2, 48–58.

Gottesman, M..M., and Ling, V. (2006). The molecular basis of multidrug resistance in cancer: the early years of P-glycoprotein research. FEBS Lett. 580, 998–1009.

Grondin, K., Haimeur, A., Mukhopadhyay, R., Rosen, B.P., and Ouellette, M. (1997). Co-amplification of the gamma-glutamylcysteine synthetase gene gsh1 and of the ABC transporter gene pgpA in arsenite-resistant *Leishmania* tarentolae. EMBO J. 16, 3057–3065.

Grondin, K., Papadopoulou, B., and Ouellette, M. (1993). Homologous recombination between direct repeat sequences yields P-glycoprotein containing amplicons in arsenite resistant *Leishmania*. Nucleic Acids Res. 21, 1895–1901.

Gueiros-Filho, F.J., Viola, J.P., Gomes, F.C., Farina, M., Lins, U., Bertho, A.L., Wirth, D.F., and Lopes, U.G. (1995). *Leishmania amazonensis*: multidrug resistance in vinblastine-resistant promastigotes is associated with rhodamine 123 efflux, DNA amplification, and RNA overexpression of a *Leishmania* mdr1 gene. Exp. Parasitol. 81, 480–490.

Hadighi, R., Mohebali, M., Boucher, P., Hajjaran, H., Khamesipour, A., and Ouellette, M. (2006). Unresponsiveness to Glucantime treatment in Iranian cutaneous leishmaniasis due to drug-resistant *Leishmania tropica* parasites. PLoS Med. 3, e162.

Haimeur, A., Brochu, C., Genest, P., Papadopoulou, B., and Ouellette, M. (2000). Amplification of the ABC transporter gene PGPA and increased trypanothione levels in potassium antimonyl tartrate (SbIII) resistant *Leishmania tarentolae*. Mol. Biochem. Parasitol. 108, 131–135.

Haimeur, A., Guimond, C., Pilote, S., Mukhopadhyay, R., Rosen, B. P., Poulin, R., and Ouellette, M. (1999). Elevated levels of polyamines and trypanothione resulting from overexpression of the ornithine decarboxylase gene in arsenite-resistant *Leishmania*. Mol. Microbiol. 34, 726–735.

Henderson, D.M., Sifri, C.D., Rodgers, M., Wirth, D.F., Hendrickson, N., and Ullman, B. (1992). Multidrug resistance in *Leishmania donovani* is conferred by amplification of a gene homologous to the mammalian mdr1 gene. Mol. Cell Biol. 12, 2855–2865.

Hendrickson, N., Sifri, C.D., Henderson, D.M., Allen, T., Wirth, D.F., and Ullman, B. (1993). Molecular characterization of the *ldmdr1* multidrug resistance gene from *Leishmania donovani*. Mol. Biochem. Parasitol. 60, 53–64.

Herget, M., and Tampe, R. (2007). Intracellular peptide transporters in human--compartmentalization of the 'peptidome'. Pflugers Arch. 453, 591–600.

Herwaldt, B.L. (1999). Leishmaniasis. Lancet 354, 1191–1199.

Hettema, E.H., van Roermund, C.W., Distel, B., van den Berg, M., Vilela, C., Rodrigues-Pousada, C., Wanders, R.J., and Tabak, H.F. (1996). The ABC transporter proteins Pat1 and Pat2 are required for import of long-chain fatty acids into peroxisomes of *Saccharomyces cerevisiae*. EMBO J. 15, 3813–3822.

Higgins, C.F. (1992). ABC transporters: from microorganisms to man. Annu Rev. Cell Biol. 8, 67–113.

Higgins, C.F., Haag, P.D., Nikaido, K., Ardeshir, F., Garcia, G., and Ames, G.F. (1982). Complete nucleotide sequence and identification of membrane components of the histidine transport operon of *S. typhimurium*. Nature. 298, 723–727.

Hyde, S.C., Emsley, P., Hartshorn, M.J., Mimmack, M.M., Gileadi, U., Pearce, S.R., Gallagher, M.P., Gill, D.R., Hubbard, R.E., and Higgins, C.F. (1990). Structural model of ATP-binding proteins associated with cystic fibrosis, multidrug resistance and bacterial transport. Nature. *346*, 362–365.

Ishikawa, T. (1992). The ATP-dependent glutathione S-conjugate export pump. Trends Biochem. Sci. *17*, 463–468.

Ivens, A.C., Peacock, C.S., Worthey, E.A., Murphy, L., Aggarwal, G., Berriman, M., Sisk, E., Rajandream, M.A., Adlem, E., Aert, R., Anupama, A., Apostolou, Z., Attipoe, P., Bason, N., Bauser, C., Beck, A., Beverley, S.M., Bianchettin, G., Borzym, K., Bothe, G., Bruschi, C.V., Collins, M., Cadag, E., Ciarloni, L., Clayton, C., Coulson, R.M., Cronin, A., Cruz, A.K., Davies, R.M., De Gaudenzi, J., Dobson, D.E., Duesterhoeft, A., Fazelina, G., Fosker, N., Frasch, A.C., Fraser, A., Fuchs, M., Gabel, C., Goble, A., Goffeau, A., Harris, D., Hertz-Fowler, C., Hilbert, H., Horn, D., Huang, Y., Klages, S., Knights, A., Kube, M., Larke, N., Litvin, L., Lord, A., Louie, T., Marra, M., Masuy, D., Matthews, K., Michaeli, S., Mottram, J. C., Muller-Auer, S., Munden, H., Nelson, S., Norbertczak, H., Oliver, K., O'Neil, S., Pentony, M., Pohl, T.M., Price, C., Purnelle, B., Quail, M.A., Rabbinowitsch, E., Reinhardt, R., Rieger, M., Rinta, J., Robben, J., Robertson, L., Ruiz, J.C., Rutter, S., Saunders, D., Schafer, M., Schein, J., Schwartz, D. C., Seeger, K., Seyler, A., Sharp, S., Shin, H., Sivam, D., Squares, R., Squares, S., Tosato, V., Vogt, C., Volckaert, G., Wambutt, R., Warren, T., Wedler, H., Woodward, J., Zhou, S., Zimmermann, W., Smith, D.F., Blackwell, J.M., Stuart, K.D., Barrell, B., and Myler, P. J. (2005). The genome of the kinetoplastid parasite, *Leishmania major*. Science *309*, 436–442.

Jha, T.K., Olliaro, P., Thakur, C.P., Kanyok, T.P., Singhania, B.L., Singh, I.J., Singh, N.K., Akhoury, S., and Jha, S. (1998). Randomised controlled trial of aminosidine (paromomycin) v sodium stibogluconate for treating visceral leishmaniasis in North Bihar, India. BMJ *316*, 1200–1205.

Katakura, K., Fujise, H., Takeda, K., Kaneko, O., Torii, M., Suzuki, M., Chang, K.P., and Hashiguchi, Y. (2004). Overexpression of LaMDR2, a novel multidrug resistance ATP-binding cassette transporter, causes 5-fluorouracil resistance in *Leishmania amazonensis*. FEBS Lett. *561*, 207–212.

Katakura, K., Iwanami, M., Ohtomo, H., Fujise, H., and Hashiguchi, Y. (1999). Structural and functional analysis of the LaMDR1 multidrug resistance gene in *Leishmania amazonensis*. Biochem. Biophys. Res. Commun. *255*, 289–294.

Kennedy, M.L., Cortes-Selva, F., Perez-Victoria, J.M., Jimenez, I.A., Gonzalez, A.G., Munoz, O. M., Gamarro, F., Castanys, S., and Ravelo, A.G. (2001). Chemosensitization of a multidrug-resistant *Leishmania tropica* line by new sesquiterpenes from *Maytenus magellanica* and *Maytenus chubutensis*. J. Med. Chem. *44*, 4668–4676.

Kerr, S.F. (2006). Molecular trees of trypanosomes incongruent with fossil records of hosts. Mem. Inst. Oswaldo Cruz. *101*, 25–30.

Kispal, G., Csere, P., Prohl, C., and Lill, R. (1999). The mitochondrial proteins Atm1p and Nfs1p are essential for biogenesis of cytosolic Fe/S proteins. EMBO J. *18*, 3981–3989.

Kispal, G., Sipos, K., Lange, H., Fekete, Z., Bedekovics, T., Janaky, T., Bassler, J., Aguilar Netz, D. J., Balk, J., Rotte, C., and Lill, R. (2005). Biogenesis of cytosolic ribosomes requires the essential iron-sulphur protein Rli1p and mitochondria. EMBO J. *24*, 589–598.

Kumar, S., Tamura, K., and Nei, M. (2004). MEGA3: Integrated software for Molecular Evolutionary Genetics Analysis and sequence alignment. Brief Bioinform. *5*, 150–163.

Legare, D., Hettema, E., and Ouellette, M. (1994). The P-glycoprotein-related gene family in *Leishmania*. Mol. Biochem. Parasitol. *68*, 81–91.

Legare, D., Papadopoulou, B., Roy, G., Mukhopadhyay, R., Haimeur, A., Dey, S., Grondin, K., Brochu, C., Rosen, B. P., and Ouellette, M. (1997). Efflux systems and increased trypanothione levels in arsenite-resistant *Leishmania*. Exp. Parasitol. *87*, 275–282.

Legare, D., Richard, D., Mukhopadhyay, R., Stierhof, Y. D., Rosen, B. P., Haimeur, A., Papadopoulou, B., and Ouellette, M. (2001). The *Leishmania* ATP-binding cassette protein PGPA is an intracellular metal-thiol transporter ATPase. J. Biol. Chem. *276*, 26301–26307.

Leighton, J., and Schatz, G. (1995). An ABC transporter in the mitochondrial inner membrane is required for normal growth of yeast. EMBO J. *14*, 188–195.

Leprohon, P., Legare, D., Girard, I., Papadopoulou, B., and Ouellette, M. (2006). Modulation of *Leishmania* ABC protein gene expression through life stages and among drug-resistant parasites. Eukaryot Cell *5*, 1713–1725.

Li, Z.S., Lu, Y.P., Zhen, R.G., Szczypka, M., Thiele, D.J., and Rea, P.A. (1997). A new pathway for vacuolar cadmium sequestration in *Saccharomyces cerevisiae*: YCF1-catalyzed transport of bis(glutathionato)cadmium. Proc. Natl. Acad. Sci. U.S.A. *94*, 42–47.

Li, Z.S., Szczypka, M., Lu, Y.P., Thiele, D.J., and Rea, P.A. (1996). The yeast cadmium factor protein (YCF1) is a vacuolar glutathione S-conjugate pump. J. Biol. Chem. *271*, 6509–6517.

Lira, R., Sundar, S., Makharia, A., Kenney, R., Gam, A., Saraiva, E., and Sacks, D. (1999). Evidence that the high incidence of treatment failuRes. in Indian kala-azar is due to the emergence of antimony-resistant strains of *Leishmania donovani*. J. Infect. Dis. *180*, 564–567.

Luciani, M.F., and Chimini, G. (1996). The ATP binding cassette transporter ABC1, is required for the engulfment of corpses generated by apoptotic cell death. EMBO J. *15*, 226–235.

Lukes, J., Mauricio, I.L., Schonian, G., Dujardin, J.C., Soteriadou, K., Dedet, J.P., Kuhls, K., Tintaya, K.W., Jirku, M., Chocholova, E., Haralambous, C., Pratlong, F., Obornik, M., Horak, A., Ayala, F.J., and Miles, M.A. (2007). Evolutionary and geographical history of the *Leishmania donovani* complex with a revision of current taxonomy. Proc. Natl. Acad. Sci. U.S.A. *104*, 9375–9380.

Luscher, A., Nerima, B., and Maser, P. (2006). Combined contribution of TbAT1 and TbMRPA to drug resistance in *Trypanosoma brucei*. Mol. Biochem. Parasitol. *150*, 364–366.

Marton, M.J., Vazquez de Aldana, C.R., Qiu, H., Chakraburtty, K., and Hinnebusch, A.G. (1997). Evidence that GCN1 and GCN20, translational regulators of GCN4, function on elongating ribosomes in activation of eIF2alpha kinase GCN2. Mol. Cell Biol. *17*, 4474–4489.

Mitsuhashi, N., Miki, T., Senbongi, H., Yokoi, N., Yano, H., Miyazaki, M., Nakajima, N., Iwanaga, T., Yokoyama, Y., Shibata, T., and Seino, S. (2000). MTABC3, a novel mitochondrial ATP-binding cassette protein involved in iron homeostasis. J. Biol. Chem. *275*, 17536–17540.

Momen, H., and Cupolillo, E. (2000). Speculations on the origin and evolution of the genus *Leishmania*. Mem Inst Oswaldo Cruz. *95*, 583–588.

Mukherjee, A., Padmanabhan, P.K., Singh, S., Roy, G., Girard, I., Chatterjee, M., Ouellette, M., and Madhubala, R. (2007). Role of ABC transporter MRPA, gamma-glutamylcysteine synthetase and ornithine decarboxylase in natural antimony-resistant isolates of *Leishmania donovani*. J. Antimicrob. Chemother. *59*, 204–211.

Mukhopadhyay, R., Dey, S., Xu, N., Gage, D., Lightbody, J., Ouellette, M., and Rosen, B. P. (1996). Trypanothione overproduction and resistance to antimonials and arsenicals in *Leishmania*. Proc. Natl. Acad. Sci. U.S.A. *93*, 10383–10387.

Ortiz, D.F., Kreppel, L., Speiser, D.M., Scheel, G., McDonald, G., and Ow, D.W. (1992). Heavy metal tolerance in the fission yeast requires an ATP-binding cassette-type vacuolar membrane transporter. EMBO J. *11*, 3491–3499.

Ortiz, D. F., Ruscitti, T., McCue, K. F., and Ow, D. W. (1995). Transport of metal-binding peptides by HMT1, a fission yeast ABC-type vacuolar membrane protein. J. Biol. Chem. *270*, 4721–4728.

Ouellette, M., and Borst, P. (1991). Drug resistance and P-glycoprotein gene amplification in the protozoan parasite *Leishmania*. Res. Microbiol. *142*, 737–746.

Ouellette, M., Fase-Fowler, F., and Borst, P. (1990). The amplified H circle of methotrexate-resistant leishmania tarentolae contains a novel P-glycoprotein gene. EMBO J. *9*, 1027–1033.

Ouellette, M., Hettema, E., Wust, D., Fase-Fowler, F., and Borst, P. (1991). Direct and inverted DNA repeats associated with P-glycoprotein gene amplification in drug resistant *Leishmania*. EMBO J. *10*, 1009–1016.

Padron-Nieves, M., Diaz, E., Machuca, C., Romero, A., and Ponte-Sucre, A. (2009). Glibenclamide modulates glucantime activity and disposition in *Leishmania*. Experimental parasitology. *in press*.

Papadopoulou, B., Roy, G., Dey, S., Rosen, B. P., and Ouellette, M. (1994). Contribution of the *Leishmania* P-glycoprotein-related gene ltpgpA to oxyanion resistance. J. Biol. Chem. *269*, 11980–11986.

Parodi-Talice, A., Araujo, J.M., Torres, C., Perez-Victoria, J.M., Gamarro, F., and Castanys, S. (2003). The overexpression of a new ABC transporter in *Leishmania* is related to phospholipid trafficking and reduced infectivity. Biochim Biophys. Acta. *1612*, 195–207.

Parsons, M. (2004). Glycosomes: parasites and the divergence of peroxisomal purpose. Mol. Microbiol. *53*, 717–724.

Peacock, C.S., Seeger, K., Harris, D., Murphy, L., Ruiz, J.C., Quail, M.A., Peters, N., Adlem, E., Tivey, A., Aslett, M., Kerhornou, A., Ivens, A., Fraser, A., Rajandream, M.A., Carver, T., Norbertczak, H., Chillingworth, T., Hance, Z., Jagels, K., Moule, S., Ormond, D., Rutter, S., Squares, R., Whitehead, S., Rabbinowitsch, E., Arrowsmith, C., White, B., Thurston, S., Bringaud, F., Baldauf, S.L., Faulconbridge, A., Jeffares, D., Depledge, D.P., Oyola, S. O., Hilley, J.D., Brito, L.O., Tosi, L.R., Barrell, B., Cruz, A.K., Mottram, J.C., Smith, D.F., and Berriman, M. (2007). Comparative genomic analysis of three *Leishmania* species that cause diverse human disease. Nat. Genet. *39*, 839–847.

Pepin, J., and Milord, F. (1994). The treatment of human African trypanosomiasis. Adv. Parasitol. *33*, 1–47.

Perez-Victoria, J.M., Chiquero, M.J., Conseil, G., Dayan, G., Di Pietro, A., Barron, D., Castanys, S., and Gamarro, F. (1999). Correlation between the affinity of flavonoids binding to the cytosolic site of *Leishmania tropica* multidrug transporter and their efficiency to revert parasite resistance to daunomycin. Biochemistry *38*, 1736–1743.

Perez-Victoria, J.M., Cortes-Selva, F., Parodi-Talice, A., Bavchvarov, B.I., Perez-Victoria, F.J., Munoz-Martinez, F., Maitrejean, M., Costi, M.P., Barron, D., Di Pietro, A., Castanys, S., and Gamarro, F. (2006). Combination of suboptimal doses of inhibitors targeting different domains of LtrMDR1 efficiently overcomes resistance of *Leishmania* spp. to Miltefosine by inhibiting drug efflux. Antimicrob. Agents Chemother. *50*, 3102–3110.

Perez-Victoria, F.J., Gamarro, F., Ouellette, M., and Castanys, S. (2003). Functional cloning of the miltefosine transporter. A novel P-type phospholipid translocase from *Leishmania* involved in drug resistance. J. Biol. Chem. *278*, 49965–49971.

Perez-Victoria, J.M., Perez-Victoria, F.J., Parodi-Talice, A., Jimenez, I.A., Ravelo, A.G., Castanys, S., and Gamarro, F. (2001). Alkyl-lysophospholipid resistance in multidrug-resistant *Leishmania tropica* and chemosensitization by a novel P-glycoprotein-like transporter modulator. Antimicrob. Agents Chemother. *45*, 2468–2474.

Perez-Victoria, J.M., Tincusi, B.M., Jimenez, I.A., Bazzocchi, I.L., Gupta, M.P., Castanys, S., Gamarro, F., and Ravelo, A. G. (1999). New natural sesquiterpenes as modulators of daunomycin resistance in a multidrug-resistant *Leishmania tropica* line. J. Med. Chem. *42*, 4388–4393.

Ponte-Sucre, A. (2007). Availability and applications of ATP-binding cassette (ABC) transporter blockers. Appl Microbiol. Biotechnol. *76*, 279–286.

Rojas, R., Valderrama, L., Valderrama, M., Varona, M.X., Ouellette, M., and Saravia, N.G. (2006). Resistance to antimony and treatment failure in human *Leishmania* (Viannia) infection. J. Infect. Dis. *193*, 1375–1383.

Sauvage, V., Millot, J.M., Aubert, D., Visneux, V., Marle-Plistat, M., Pinon, J. M., and Villena, I. (2006). Identification and expression analysis of ABC protein-encoding genes in *Toxoplasma gondii Toxoplasma gondii* ATP-binding cassette superfamily. Mol. Biochem. Parasitol. *147*, 177–192.

Sereno, D., Holzmuller, P., Mangot, I., Cuny, G., Ouaissi, A., and Lemesre, J.L. (2001). Antimonial-mediated DNA fragmentation in *Leishmania* infantum amastigotes. Antimicrob. Agents Chemother. *45*, 2064–2069.

Shahi, S.K., Krauth-Siegel, R.L., and Clayton, C.E. (2002). Overexpression of the putative thiol conjugate transporter TbMRPA causes melarsoprol resistance in *Trypanosoma brucei*. Mol. Microbiol. *43*, 1129–1138.

Shaked-Mishan, P., Ulrich, N., Ephros, M., and Zilberstein, D. (2001). Novel Intracellular SbV reducing activity correlates with antimony susceptibility in *Leishmania donovani*. J. Biol. Chem. *276*, 3971–3976.

Shani, N., and Valle, D. (1996). A *Saccharomyces cerevisiae* homolog of the human adrenoleukodystrophy transporter is a heterodimer of two half ATP-binding cassette transporters. Proc. Natl. Acad. Sci. U.S.A. *93*, 11901–11906.

Shani, N., Watkins, P.A., and Valle, D. (1995). PXA1, a possible *Saccharomyces cerevisiae* ortholog of the human adrenoleukodystrophy gene. Proc. Natl. Acad. Sci. U.S.A. *92*, 6012–6016.

Sheps, J.A., Ralph, S., Zhao, Z., Baillie, D.L., and Ling, V. (2004). The ABC transporter gene family of *Caenorhabditis elegans* has implications for the evolutionary dynamics of multidrug resistance in eukaryotes. Genome Biol. *5*, R15.

Simpson, A.G., Stevens, J.R., and Lukes, J. (2006). The evolution and diversity of kinetoplastid flagellates. Trends Parasitol. *22*, 168–174.

Soto, J., Buffet, P., Grogl, M., and Berman, J. (1994). Successful treatment of Colombian cutaneous leishmaniasis with four injections of pentamidine. Am J Trop Med Hyg. *50*, 107–111.

Sundar, S., Jha, T.K., Thakur, C.P., Sinha, P.K., and Bhattacharya, S.K. (2007). Injectable paromomycin for visceral leishmaniasis in India. N Engl J Med. *356*, 2571–2581.

Sundar, S., Mehta, H., Suresh, A.V., Singh, S.P., Rai, M., and Murray, H.W. (2004). Amphotericin B treatment for Indian visceral leishmaniasis: conventional versus lipid formulations. Clin Infect Dis. *38*, 377–383.

Sundaram, P., Echalier, B., Han, W., Hull, D., and Timmons, L. (2006). ATP-binding cassette transporters are required for efficient RNA interference in *Caenorhabditis elegans*. Mol. Biol. Cell *17*, 3678–3688.

Szakacs, G., Paterson, J.K., Ludwig, J.A., Booth-Genthe, C., and Gottesman, M.M. (2006). Targeting multidrug resistance in cancer. Nat. Rev. Drug Discov. *5*, 219–234.

Theodoulou, F.L., Holdsworth, M., and Baker, A. (2006). Peroxisomal ABC transporters. FEBS Lett. *580*, 1139–1155.

Torres, C., Barreiro, L., Dallagiovanna, B., Gamarro, F., and Castanys, S. (1999). Characterization of a new ATP-binding cassette transporter in *Trypanosoma cruzi* associated to a L1Tc retrotransposon. Biochim Biophys. Acta. *1489*, 428–432.

Torres, C., Perez-Victoria, F.J., Parodi-Talice, A., Castanys, S., and Gamarro, F. (2004). Characterization of an ABCA-like transporter involved in vesicular trafficking in the protozoan parasite Trypanosoma cruzi. Mol. Microbiol. *54*, 632–646.

Tyzack, J.K., Wang, X., Belsham, G.J., and Proud, C.G. (2000). ABC50 interacts with eukaryotic initiation factor 2 and associates with the ribosome in an ATP-dependent manner. J. Biol. Chem. *275*, 34131–34139.

van Helvoort, A., Smith, A.J., Sprong, H., Fritzsche, I., Schinkel, A.H., Borst, P., and van Meer, G. (1996). MDR1 P-glycoprotein is a lipid translocase of broad specificity, while MDR3 P-glycoprotein specifically translocates phosphatidylcholine. Cell *87*, 507–517.

Vazquez de Aldana, C.R., Marton, M.J., and Hinnebusch, A.G. (1995). GCN20, a novel ATP binding cassette protein, and GCN1 reside in a complex that mediates activation of the eIF-2 alpha kinase GCN2 in amino acid-starved cells. EMBO J. *14*, 3184–3199.

Velamakanni, S., Wei, S.L., Janvilisri, T., and van Veen, H.W. (2007). ABCG transporters: structure, substrate specificities and physiological roles: a brief overview. J. Bioenerg. Biomembr. *39*, 465–471.

Vergnes, B., Gourbal, B., Girard, I., Sundar, S., Drummelsmith, J., and Ouellette, M. (2007). A proteomics screen implicates HSP83 and a small kinetoplastid calpain-related protein in drug resistance in *Leishmania donovani* clinical field isolates by modulating drug-induced programmed cell death. Mol. Cell Proteomics. *6*, 88–101.

Verrier, P.J., Bird, D., Burla, B., Dassa, E., Forestier, C., Geisler, M., Klein, M., Kolukisaoglu, U., Lee, Y., Martinoia, E., Murphy, A., Rea, P.A., Samuels, L., Schulz, B., Spalding, E.J., Yazaki, K., and Theodoulou, F.L. (2008). Plant ABC proteins--a unified nomenclature and updated inventory. Trends Plant Sci. *13*, 151–159.

Walker, J.E., Saraste, M., Runswick, M.J., and Gay, N.J. (1982). Distantly related sequences in the alpha- and beta-subunits of ATP synthase, myosin, kinases and other ATP-requiring enzymes and a common nucleotide binding fold. EMBO J. *1*, 945–951.

Wilson, C.M., Serrano, A.E., Wasley, A., Bogenschutz, M.P., Shankar, A.H., and Wirth, D.F. (1989). Amplification of a gene related to mammalian mdr genes in drug-resistant *Plasmodium falciparum*. Science. *244*, 1184–1186.

Wong, I.L., Chan, K.F., Burkett, B.A., Zhao, Y., Chai, Y., Sun, H., Chan, T.H., and Chow, L.M. (2007). Flavonoid dimers as bivalent modulators for pentamidine and sodium stiboglucanate resistance in leishmania. Antimicrob. Agents Chemother. *51*, 930–940.

Wong, I.L., and Chow, L.M. (2006). The role of *Leishmania enriettii* multidrug resistance protein 1 (LeMDR1) in mediating drug resistance is iron-dependent. Mol. Biochem. Parasitol. *150*, 278–287.

Wyllie, S., Vickers, T.J., and Fairlamb, A.H. (2008). Roles of trypanothione S-transferase and tryparedoxin peroxidase in resistance to antimonials. Antimicrob. Agents Chemother.

Yernaux, C., Fransen, M., Brees, C., Lorenzen, S., and Michels, P.A. (2006). *Trypanosoma brucei* glycosomal ABC transporters: identification and membrane targeting. Mol. Membr Biol. 23, 157–172.

Zarubica, A., Trompier, D., and Chimini, G. (2007). ABCA1, from pathology to membrane function. Pflugers Arch. *453*, 569–579.

Zhao, Z., Fang, L.L., Johnsen, R., and Baillie, D.L. (2004). ATP-binding cassette protein E is involved in gene transcription and translation in *Caenorhabditis elegans*. Biochem. Biophys. Res. Commun. 323, 104–111.

ATP-Binding Cassette (ABC) Transporters in Yeasts: Their Role in Multidrug Resistance and Survival

8

Hina Sanwal, Sneh Lata Panwar and Rajendra Prasad

Abstract

Complete genome sequencing of the model yeast *Saccharomyces cerevisiae*, and many other pathogenic fungi such as *Candida albicans*, has led to the identification and classification of proteins belonging to the ATP-binding cassette (ABC) superfamily. Usually these ABC proteins encode transporters. These transporters provide an advantage to yeasts living in competitive environments by actively secreting toxic compounds, thereby serving as a first-line of defence. Amongst the different subfamilies of the ABC transporters, proteins of the multi drug resistance (MDR) and the pleiotropic drug resistance (PDR) subfamilies are well characterized. Most members of the PDR subfamily are drug transporters involved in the development of antifungal drug resistance caused by the over expression of genes encoding these drug efflux pump proteins. As compared to any other yeast, the role of efflux pumps in drug resistance has been extensively studied in *Saccharomyces cerevisiae.* However, several studies in the past decade have given momentum to the research pertaining to ABC drug transporters in pathogenic yeasts as well. The presence of large number of ABC members in yeast genomes suggests indispensability of these proteins and suggests that they are physiologically relevant.

Introduction

Members of the ABC superfamily constitute perhaps the most diverse and ubiquitous family of proteins spread throughout the evolutionary scale. Their representatives are localized in the membranes of all the intracellular organelles ranging from the plasma membrane to the membranes of lysosomes, peroxisomes, Golgi, mitochondria and endoplasmic reticulum. The ABC proteins, also known as 'traffic ATPases' are made up of an assorted group of transporters. The role of ABC proteins in Multidrug Resistance (MDR) has drawn considerable attention of researchers to this superfamily of transporters. However, the ABC transporters, besides being drug transporters, participate in a plethora of physiological roles like pheromone transport, ion homeostasis, weak organic acid tolerance, heavy metal detoxification, sterol homeostasis, etc. (Prasad *et al.*, 2002).

ABC proteins in yeasts

Identification of proteins from the ABC superfamily became easier with the complete sequencing of the yeast (*Saccharomyces cerevisiae*), genome (Decottignies and Goffeau, 1997). Gradually as the genome of other yeasts and fungi became available, it turned out that in each organism a family of ABC proteins performs a host of different functions. The ABC superfamily includes full transporters with two transmembrane domains (TMDs) and two nucleotide-binding domains (NBDs) as well as half transporters comprising one TMD and one NBD of varying topology. Those in which the TMD is predicted to be located N-terminal to the NBD are known as having forward orientation (TMD-NBD), while those in which the NBD precedes the TMD are defined as having reverse orientation (NBD-TMD) (discussed later). Irrespective of the orientation, each TMDs comprise of α-helices of six TMS while the NBDs have α-helices and β-sheets arranged to form a Rossman-fold architecture (Higgins *et al.*, 2001). Table 8.1 lists members of the ABC superfamily belonging to different subfamilies, which exist in pathogenic and non-pathogenic yeasts. This chapter will focus on the analysis of the subfamilies within ABC superfamily of the well-documented non-pathogenic bakers' yeast, *S. cerevisiae,* and of pathogenic yeast *Candida albicans.*

ABC proteins of *S. cerevisiae*

A complete inventory of the ABC proteins of *S. cerevisiae* revealed the presence of 31 putative ABC proteins in its genome (Decottignies and Goffeau, 1997). The family includes members with different topology and composition. These ABC proteins can be full transporters (NBD-TMD)$_2$, or half transporters (NBD-TMD); alternatively there are members which have no TMD and possess only a NBD. The number of amino acid residues in the ABC transporters of *S. cerevisiae* ranges between 289 and 1661 residues. On the basis of phylogenetic analysis, the yeast ABC proteins have been classified into distinct subfamilies (Table 8.1) (Decottignies and Goffeau, 1997). The MDR (Multi Drug Resistance), MRP (Multi Drug Resistance-associated Protein), and ALDp (Adrenoleukodystrophy protein) families have members with forward orientation (TMD-NBD) (Fig.1).

In the MDR family only ScSte6p (Sterile 6), which transports the **a** pheromone is a full transporter (TMD-NBD)$_2$ (McGrath and Vershavsky, 1989). The second member of the MDR family, ScAtm1p (ABC transporters of mitochondria 1), is a half-size transporter (TMD-NBD) localized in the inner mitochondrial membrane with its NBD facing the matrix (Leighton and Schatz, 1995). The last two half transporter members (TMD-NBD), ScMdl1p (Multidrug resistance like protein 1) and ScMdl2p (Multidrug resistance like protein 2) also localize in the mitochondrial inner membrane and share their topology with ScAtm1p (Dean *et al.*, 1994). The majority of the members of the MRP/CFTR (Cystic Fibrosis Transmembrane conductance Regulator) family are full transporters with (TMD-NBD)$_2$ topology. The exception is ScYcf1p (Yeast Cadmium factor1), that has a regulatory domain between the two halves (TMD-R-NBD)$_2$ (Szczypka *et al.*, 1994). The ALDp family comprises two half transporters ScPxa1p (Peroxisomal ATP-binding cassette transporter 1) and ScPxa2p (Peroxisomal ATP-binding cassette transporter 2) with the topology (TMD-NBD).

The PDR (Pleitropic Drug Resistance) is the largest ABC family in yeast with 10 members, all of which have reverse topological orientation and are predicted to be full transporters (NBD-TMD)$_2$ (Fig. 8.1). ScPdr5p and ScSnq2p (Sensitivity to 4-nitroquinoline-*N*-oxide 2), the most extensively studied proteins of the PDR subfamily are localized at the plasma membrane (Servos *et al.*, 1993; Balzi *et al.*, 1994; Bissinger and Kuchler, 1994). ScPdr10p and ScPdr15p share more than 65% overall sequence identity with ScPdr5p (Wolfger *et al.*, 1997). Another member ScPdr12p shares about 60% sequence identity with ScSnq2p (Piper *et al.*, 1998). Additional homologues of ScSnq2p included in the PDR family are ScPdr11p, ScAus1p (Aerobic up regulation of sterol 1) and YNR070w (Wilcox *et al.*, 2002c). YOL075c of the PDR subfamily encodes a putative full transporter. ScAdp1p (ATP-dependent permease 1), differs in its topology from other members of PDR subfamily. It has a NBD flanked by a TMD comprising of two TMSs on the N-terminal and another TMD with seven TMSs on the C-terminal (TMD (TMS$_2$)-NBD-TMD (TMS$_7$)) (Purnelle *et al.*, 1991). The YEF3 (Yeast Elongation Factor 3), RL1 family (RNase L inhibitor 1) members and 'Others' are soluble proteins as they do not have any TMDs and are made up of only one or two NBDs (Sandbaken *et al.*, 1990; Vazquez de Aldana *et al.*, 1995; Sarthy *et al.*, 1998; Maurice *et al.*, 1998; Liu *et al.*, 2001).

Study of yeast ABC transporters have also provided an insight into the functioning of their homologues in humans, that have been implicated in various diseases such as Zellweger syndrome, Dubin–Johnson syndrome, Tangier disease, cystic fibrosis, adrenoleukodystrophy, MDR in cancer therapy etc. (Decottignies and Goffeau, 1997; Taglicht and Michaelis, 1998; Klein *et al.*, 1999; Gottesman and Ambudkar, 2001; Suzuki *et al.*, 2001).

ABC proteins of *C. albicans*

The release of the genome sequence of the diploid human pathogenic fungus *C. albicans* made it pos-

Table 8.1 ABC transporters of fungi

Organism	Subfamily[a]	ORF names	Gene name	Chr.no.	Size[b]	Location	Topology[c]	Function	Reference
Saccharomyces cerevisiae	PDR		*ScPDR5*	XV	1511	PM	$(NBD\text{-}TMS_6)_2$	Drug efflux pump, phospholipid translocator	Balzi *et al.*, 1994; Bissinger and Kuchler, 1994;Decottignies and Goffeau, 1997
			ScPDR10	XV	1564	PM	$(NBD\text{-}TMS_6)_2$	?	Wolfger *et al.*, 1997
			ScPDR11	IX	1411	PM	$(NBD\text{-}TMS_6)_2$	Sterol uptake	Wilcox *et al.*, 2002
			ScPDR12	XVI	1511	PM	$(NBD\text{-}TMS_6)_2$	Resistance to water-soluble, monocarboxylic acids with chain lengths of from C-1 to C-7	Piper *et al.*, 1998 Kren *et al.*, 2003b
			ScPDR15	IV	1529	PM	$(NBD\text{-}TMS_6)_2$	Drug efflux pump, stress inducible	Wolfger *et al.*, 1997
			ScSNQ2	IV	1501	PM	$(NBD\text{-}TMS_6)_2$	Drug efflux pump	Servos *et al.*, 1993
			ScAUS1	XV	1394	PM	$(NBD\text{-}TMS_6)_2$	Sterol uptake	Wilcox *et al.*, 2002
			ScADP1	III	1049	?	$TMS_2\text{-}NBD\text{-}TMS_7$	?	Purnelle *et al.*, 1991
		YNR070w	***?***	XIV	1333	?	$(NBD\text{-}TMS_6)_2$	?	Decottignies and Goffeau, 1997
		YOL075c	***?***	XV	1294	?	$(NBD\text{-}TMS_6)_2$	?	Decottignies and Goffeau, 1997
	ALDp		***ScPXA1***	XVI	758	Peroxisomes	$TMS_6\text{-}NBD$	Required for β-oxidation of fatty acids	Shani *et al.*, 1995
			ScPXA2	XI	853	Peroxisomes	$TMS_6\text{-}NBD$	Required for β-oxidation of fatty acids	Shani *et al.*, 1995 Shani and Valle, 1996
	MRP/ CFTR		***ScYCF1***	IV	1515	Vacuoles	$(TMS_6\text{-}NBD)_2$	Required for glutathione conjugated cadmium resistance	Szczypka *et al.*, 1994

Table 8.1 Continued

Organism	Subfamily[a]	ORF names	Gene name	Chr.no.	Size[b]	Location	Topology[c]	Function	Reference
			ScYOR1/ ScYRS1	VII	1477	PM	$(TMS_6\text{-}NBD)_2$	Oligomycin resistance, phospholipid translocator	Katzmann *et al.*, 1995; Decottignies *et al.*, 1998
			ScBPT1	XII	1559	Vacuoles	$(TMS_6\text{-}NBD)_2$	Cellular detoxification	Petrovic *et al.*, 2000
			ScYBT1	XII	1661	Vacuoles	$(TMS_6\text{-}NBD)_2$	Bile acid transport	Ortiz *et al.*, 1997
		YHL035c	***?***	VIII	1218	?	$(TMS_6\text{-}NBD)_2$	?	Decottignies and Goffeau, 1997
		YKR103w & YKR104w	***?***	XI	1524	?	$(TMS_6\text{-}NBD)_2$	?	Decottignies and Goffeau, 1997
	MDR		***ScATM1***	XIII	690	Mitochondria	$TMS_6\text{-}NBD$	Essential mitochondrial protein involved in mitochondrial iron homeostatsis	Leighton and Schatz, 1995
			ScSTE6	XI	1290	PM	$(TMS_6\text{-}NBD)_2$	**a**-factor export	McGrath and Vershavsky, 1989
			ScMDL1	XII	695	Mitochondria	$TMS_6\text{-}NBD$	peptide transporter	Dean *et al.*, 1994
			ScMDL2	XVI	773	Mitochondria	$TMS_6\text{-}NBD$	?	Dean *et al.*, 1994
	YEF3		***ScYEF3***	XII	1044	?	$(NBD)_2$	Interaction with aminoacyl-tRNA	Sandbaken *et al.*, 1990
			ScGCN20	VI	752	Polysomes	$(NBD)_2$	Interaction with tRNA and Gcn2p	Vazquez de Aldana *et al.*, 1995
		YNL014w	***ScHef3p***	XIV	1044	?	$(NBD)_2$	stimulates the binding of aminoacyl tRNA (AA-tRNA) to ribosomes by releasing EF-1 alpha from the ribosomal complex	Maurice *et al.*, 1998 Sarthy *et al.*, 1998

		YPL226	***?***	XVI	1196	?	$(NBD)_2$	?	Decottignies and Goffeau, 1997
		YER036	***?***	V	610	?	$(NBD)_2$	?	Decottignies and Goffeau, 1997
	RL1	***YDR091c***	***?***	IV	608	?	$(NBD)_2$	?	Decottignies and Goffeau, 1997
	Others	***YDR061w***	***?***		539	?	NBD	?	Decottignies and Goffeau, 1997
			ScCAF16	VI	289	?	NBD	?	Liu *et al.*, 2001b
Schizosaccharomyces pombe	PDR		*Sppdr1*	I	1396	E.R	$(NBD\text{-}TMS_6)_2$	?	Bauer *et al.*, 1999
			Spbfr1	III	1530	P.M	$(NBD\text{-}TMS_6)_2$	Brefeldin A transport	Nagao *et al.*, 1995
	MDR		*Sppmdl1*	III	1362	P.M	$(TMS_6\text{-}NBD)_2$	Drug efflux pump	Nishi *et al.*, 1993
			Spmam1	II	1336	Endosome & Golgi	$(TMS_6\text{-}NBD)_2$	M-factor transport	Christensen *et al.*, 1997a
			Spabc1	I	1427	E.R	$(TMS_6\text{-}NBD)_2$	Unknown	Christensen *et al.*, 1997b
			Spabc2	I	1463	Vacuoles	$(TMS_6\text{-}NBD)_2$	Cellular detoxification	Iwaki *et al.*, 2006
			Spabc3	II	1465	Vacuoles	$(TMS_6\text{-}NBD)_2$	?	Iwaki *et al.*, 2006
			Spabc4	I	1469	Vacuoles	$(TMS_6\text{-}NBD)_2$	?	Iwaki *et al.*, 2006
			Spmdl1	II	726	Mitochondria	$TMS_6\text{-}NBD$	Mitochondrial –peptide transport	Iwaki *et al.*, 2006
			Spatm1	III	693	Mitochondria	$TMS_6\text{-}NBD$	Cellular iron homeostatsis	Iwaki *et al.*, 2006
	ALDp		*Sphmt1*	I	830	Vacuoles	$TMS_6\text{-}NBD$	Cellualar detoxification	Ortiz *et al.*, 1992
Candida albicans	**PDR**		***CaCDR1***	III	1502	PM	$(NBD\text{-}TMS_6)_2$	Drug efflux pump, phospholipid translocator	Prasad *et al.*, 1995 Sanglard *et al.*, 1995 Smriti *et al.*, 2002
			CaCDR2	III	1499	PM	$(NBD\text{-}TMS_6)_2$	Drug efflux pump, phospholipid translocator	Sanglard *et al.*, 1997 Smriti *et al.*, 2002

Table 8.1 Continued

Organism	Subfamily[a]	ORF names	Gene name	Chr.no.	Size[b]	Location	Topology[c]	Function	Reference
			CaCDR3	IV	1501	PM	$(NBD\text{-}TMS_6)_2$	Phospholipid translocator	Balan *et al.*, 1997 Smriti *et al.*, 2002
			CaCDR4	I	1490	?	$(NBD\text{-}TMS_6)_2$	Phospholipid translocator?	Franz *et al.*, 1998 Sanglard *et al.*, 1997
			CaCDR99	III	683	?	$NBD\text{-}TMS_6$	Drug efflux pump ?	Gaur *et al.*, 2005
		CaO.19.919	***?***	III	819	?	$NBD\text{-}TMS_6$	?	Gaur *et al.*, 2005
		CaO.19.5759	***?***	VI	1493	?	$(NBD\text{-}TMS_6)_2$	?	Gaur *et al.*, 2005
		CaO.19.4531	***?***	I	1245	?	$(NBD\text{-}TMS_6)_2$	?	Gaur *et al.*, 2005
		CaO.19.3120	***?***	IV	584	?	$NBD\text{-}TMS_6$	?	Gaur *et al.*, 2005
		CaO.19.459	***?***	R	719	?	$TMS_1\text{-}NBD\text{-}TMS_7$	?	Gaur *et al.*, 2005
	MDR		***CaHST6***	III	1323	?	$(TMS_6\text{-}NBD)_2$	Transport of **a**-factor, drugs?	Raymond *et al.*, 1998
			CaMDL1	R	659		$TMS_6\text{-}NBD$	?	Gaur *et al.*, 2005
		CaO19.13043	***?***	VI	726	?	$TMS_6\text{-}NBD$	?	Gaur *et al.*, 2005
		CaO19.1077	***?***	VI	730	?	$TMS_6\text{-}NBD$	?	Gaur *et al.*, 2005
	MRP/ CFTR		***CaYOR1***	II	711	PM	$TMS_6\text{-}NBD$	Drug efflux pump	Ogawa *et al.*, 1998
			CaYCF1	VII	1512	Vacuole?	$(TMS_6\text{-}NBD)_2$	Drug efflux pump	Gaur *et al.*, 2005
			CaMLT1	I	1544	Vacuole?	$(TMS_6\text{-}NBD)_2$	Transport/involved in virulence	Theiss *et al.*, 2002
		CaO19.6382	***?***	R	1254	?	$TMS_6\text{-}NBD$	?	Gaur *et al.*, 2005
		CaO19.1784	***?***	II	763	?	$TMS_6\text{-}NBD$	?	Gaur *et al.*, 2005
	ALDp	***CaO19.7500***	***?***	R	652	?	$TMS_6\text{-}NBD$	?	Gaur *et al.*, 2005
		CaO19.5255	***?***	I	615	?	$TMS_6\text{-}NBD$	?	Gaur *et al.*, 2005

	EF3	***CaO19.6060***	***?***	I	751	?	NBD-NBD	?	Gaur *et al.*, 2005
		CaO19.2183	***?***	II	585	?	NBD-NBD	?	Gaur *et al.*, 2005
			CaELF1	R	1097		NBD-NBD	m-RNA export factor	Sturtevant *et al.*, 1998
			CaCEF3	V	978		NBD-NBD	?	Myers *et al.*, 1992
	RL1	***CaO19.3034***	***?***	I	602	?	NBD-NBD	?	Gaur *et al.*, 2005
	Others	***CaO19.5029***	***?***	I	545	?	NBD-NBD	?	Gaur *et al.*, 2005
		CaO19.388	***?***	I	276	?	NBD	?	Gaur *et al.*, 2005
Candida glabrata	**PDR**		***CgCDR1***	?	1499	PM	$(NBD\text{-}TMS_6)_2$	Drug efflux pump	Sanglard *et al.*, 1999 Sanglard *et al.*, 2001
			CgCDR2	?	Pcds[d]	?	$(NBD\text{-}TMS_6)_2$	Drug efflux pump	Sanglard *et al.*, 2001
			CgPDH1	?	1542	?	$(NBD\text{-}TMS_6)_2$	Drug efflux pump	Miyazaki *et al.*, 1998
Candida dubliniensis	**PDR**		*CdCDR1*	III	1500	?	$(NBD\text{-}TMS_6)_2$	Drug efflux pump?	Moran *et al.*, 1998
			CdCDR2	III	1502	**?**	$(NBD\text{-}TMS_6)_2$	Drug efflux pump?	Moran *et al.*, 1998
Candida krusei	**PDR**		***CkABC1***	?	Pcds*	?	?	Drug efflux pump	Katiyar and Edlind, 2001
			CkABC2	?	Pcds*	?	?	Drug efflux pump	Katiyar and Edlind, 2001

a Names of subfamilies based on sequence similarity with human ALDP, MRP/CFTR, MDR and yeast (PDR, YEF3) ABC transporters.

b Number of amino acids residues.

c NBD- nucleotide binding domain, TMS- transmembrane segment.

* Pcds- partial coding sequence available in the data base.

The information in the above table has been obtained from the following sources:
Candida Genome Database: www.candidagenome.org
Saccharomyces Genome Database: www.yeastgenome.org
Comprehensive Yeast Genome Database: http://mips.gsf.de/genre/proj/yeast/

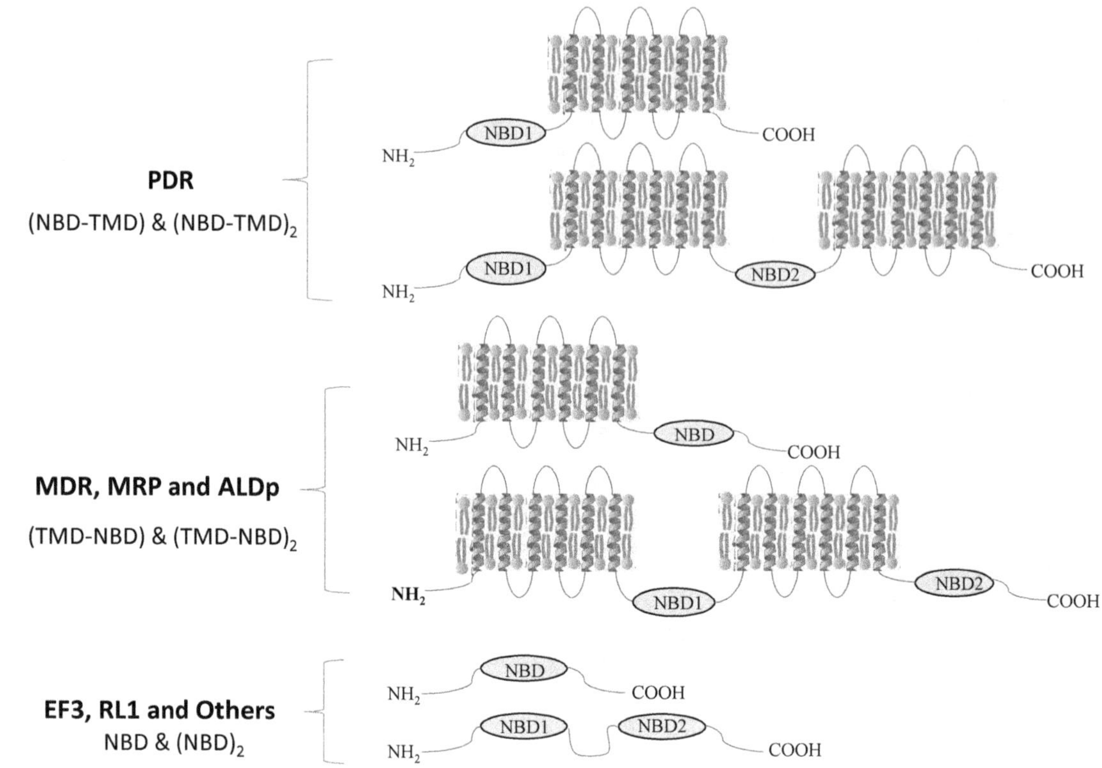

Figure 8.1 Predicted membrane topology of ABC proteins. In the PDR subfamily, the NBD (depicted by ovals) is followed by the TMD (usually comprising of 6 TMS depicted by helices). The PDR subfamily has transporters with forward orientation NBD-TMD. The subfamilies of MDR, MRP and ALDp have reverse topology (TMD-NBD). The EF3 and RL1 subfamilies are characterized by the absence of TMD and the presence of soluble cytoplasmic NBD/NBDs. NBD, nucleotide binding domain; TMD, transmembrane domain; TMS, transmembrane segment (Decottignies and Goffeau, 1997; (Gaur *et al.*, 2005).

sible to analyse the superfamily of ABC proteins (Jones *et al.*, 2004; Braun *et al.*, 2005). The inventory of *C. albicans* revealed that it has twenty-eight putative ABC proteins (Gaur *et al.*, 2005). All the putative ABC proteins ranging in sizes between 276 and 1544 amino acid residues segregated into six major clusters based on domain organization, resemblance to the ABC proteins from other organisms and the neighbour-joining tree.

Almost all of the *C. albicans* ABC proteins could be assigned to one of the subfamilies originally identified for *S. cerevisiae*, namely: PDR, MDR, MRP, ALDp, EF3 and RLI. However, the sixth cluster is represented by ABC proteins that do not belong to any known subfamily and were thus placed in the 'others' category. The putative ABC proteins identified in *C. albicans* follow the expected domain organization for each subfamily (Fig. 8.1). Those in which the TMD is predicted to be N-terminal to the NBD $(TMD\text{-}NBD)_2$ belong to MDR, MRP and ALDp subfamilies, while those in which the NBD precedes the TMD $(NBD\text{-}TMD)_2$ are members of the PDR subfamily. The putative ABC genes that lack TMD and possess only NBD belong to the EF3 and RLI subfamilies.

The MDR subfamily of *C. albicans* has four members which include both the full $(TMD\text{-}NBD)_2$ and half (TMD-NBD) transporters. These members bear a strong homology to their counterparts in mammalian cells where their over expression confers the multidrug resistance phenotype (Gottesman and Pastan, 1993). Only one ABC transporter, CaHst6p (Homologue of STE6), of the MDR sub-family is well characterized and is known to be a functional homologue of *S. cerevi-*

siae ScSte6p (Raymond *et al.*, 1998). The other members of this subfamily include homologues of the *S. cerevisiae* mitochondrial half transporters ScAtm1p (Leighton and Schatz, 1995), ScMdl1p (Dean *et al.*, 1994) and ScMdl2p (Dean *et al.*, 1994). In *C. albicans*, the ORF CaO19.1077 is the closest homologue of *S. cerevisiae* ScAtm1p which also has significant homology to the human mitochondrial half ABC transporter Abcb7 (ATP-binding cassette sub-family B).

Similar to *S. cerevisiae*, the PDR subfamily of *C. albicans* forms the largest subfamily. All the members of this subfamily display 'reverse' topology. The members include both the full $(NBD\text{-}TMD)_2$ and half (NBD-TMD) transporters. The well characterized members of this subfamily include the full $(NBD\text{-}TMD)_2$ transporters: CaCdr1p (*Candida* drug resistance protein 1) (Prasad *et al.*, 1995), CaCdr2p (Sanglard *et al.*, 1997), CaCdr3p (Balan *et al.*, 1997) and CaCdr4p (Franz *et al.*, 1998). They bear a close resemblance to *S. cerevisiae* pleiotropic drug resistance protein, ScPdr5p, which is the prototype of this subfamily (Balzi *et al.*, 1994). The half transporters (NBD-TMD) in the PDR subfamily of *C. albicans* comprise of CaCdr99p, CaO19.919, CaO19.459 and CaO19.3120. Interestingly, CaCdr99p and the ORFs CaO19.919 bear significant homology with both N- and C-terminal halves of ScPdr5p, respectively. The ORFs CaO19.459 which is a homologue of *S. cerevisiae* ScAdp1p, has the membrane topology $TMD_1\text{-}NBD_1\text{-}TMD_2$ with one TMS in TMD_1 and seven TMSs in TMD_2 (Gaur *et al.*, 2005).

Two important points emerge from the sequence analysis of different subfamily members. Firstly, the N- and C-terminal NBDs of *Candida* ABC proteins within each cluster segregate independently, implying that full transporters arose from the duplication and fusion of the half transporters (Saier, 1994; Saurin *et al.*, 1999). Secondly, N-terminal NBDs (N-NBDs) of different proteins within a subfamily are found to be more similar to one another as compared to their own corresponding C-terminal NBDs (C-NBDs). This information suggests that full proteins within the same subfamily share a common full progenitor and also points to the possibility that the two halves are functionally distinct (Daumke and Knittler, 2001).

Yeast ABC proteins in multidrug resistance

MDR is defined as resistance of an organism against a spectrum of drugs that share neither a common target nor a common structure (Skovsgaard *et al.*, 1994; Ruetz and Gros, 1994; Gottesman *et al.*, 1996, see also Chapter 9). The MDR phenotype is not only limited to humans but is seen in various other organisms (Higgins, 1992; Borst and Ouelette, 1995; Dupont *et al.*, 1996; Blanchard, 1996; Verrier *et al.*, 2008). A general theme that has emerged over the years suggests that the development of the drug-resistant phenotype is a step-wise phenomenon and mechanisms of MDR are the result of modification of normal pathways of cellular homeostasis (Gottesman *et al.*, 1995; Gottesman *et al.*, 1996).

Some of the major mechanisms that contribute to the development of MDR are: decreased accumulation of drugs, changes in expression of some cellular proteins, changes in cellular physiology affecting the structure of plasma membrane, cytosolic pH, lysosomal structure and function (Simon *et al.*, 1994; Prasad and Kapoor, 2005) and decreased permeability of the plasma membrane to the drugs. Various resistance mechanisms found in many microbes, including yeast, bacteria, parasitic pathogens and cancerous cells often arize out of different synergistic combinations of the above-mentioned mechanisms (Higgins, 1992).

In recent times reports of incidence of pathogenic *C. albicans* cells acquiring resistance to antifungals have been appearing (Odds *et al.*, 2003; Anderson, 2005). Candida infections are predominantly treated with azole antifungals, particularly with the triazole derivatives fluconazole (FLC), itraconazole (ITR) and ketoconazole (KET). Due to a prolonged exposure to these antifungals, *Candida* has evolved a variety of strategies to circumvent the lethal effects of these drugs. The resistance to azoles in *C. albicans* occurs primarily through an alteration or an over expression of the target enzyme 14α-lanosterol demethylase (P45014DM) involved in ergosterol biosynthesis (Marichal *et al.*, 1999). However, the observation that ATP-Binding Cassette (ABC) efflux proteins CaCdr1p and CaCdr2p (Prasad *et al.*, 1995; Sanglard *et al.*, 1997; Krishnamurthy *et al.*, 1998b) and CaMdr1p, a transporter of the major facilitator superfamily (MFS) (Fling *et al.*, 1991; Ben-Yaacov *et al.*, 1994; Morschhauser, 2002) are over expressed in azole resistant clinical isolates of *C. albicans* established drug extrusion mediated by efflux pump proteins as one of the prominent mechanisms of resistance.

Reduced intracellular accumulation of drugs mediated by efflux pump proteins in azole resistant clinical isolates has not only been observed in *C. albicans* but has also been reported for other pathogenic species such as *C. tropicalis, C. glabrata, C. krusei* and *C. dubliniensis* (Table 8.1). *C. glabrata* has emerged as an important nosocomial pathogen amongst the non-*albicans* species. Homologues of ABC transporter *CaCDR1* such as *CgCDR1* and *CgCDR2* have also been characterized from *C. glabrata* and implicated in azole resistance (Sanglard *et al.*, 1999). *CgPDH1* (Pleomorphic Drug resistance Homologue 1), another putative ABC transporter of *C. glabrata* has also been implicated in azole resistance (Miyazaki *et al.*, 1998). *C. dubliniensis* is phylogenetically closely related to *C. albicans* and is associated with oral candidasis. Over expression of two ABC transporters encoded by *CdCDR1* and *CdCDR2*, of *C. dubliniensis*, culminates in reduced accumulation of FLC in resistant isolates when compared to susceptible strains (Moran *et al.*, 1998). A potential role of two putative ABC transporters *CkABC1* and *CkABC2* in drug resistance has also been suggested for *C. krusei* (Katiyar and Edlind, 2001).

The major proteins that contribute to the drug-resistant phenotype in *S. cerevisiae* belong to the PDR subfamily. The PDR network comprises of the drug transporters of the PDR subfamily and the transcription factors that regulate the expression of these transporters. The expression of PDR genes encoding for ABC transporters, is controlled mainly by two transcription factors, ScPdr1p and ScPdr3p. While ScPdr5p and ScYor1p (Yeast oligomycin resistance 1) are strictly controlled by the transcription factors ScPdr1p/ScPdr3p, the expression of ScSnq2p is also regulated by another transcription factor, ScYrr1p (Yeast reveromycin resistance 1) (Cui *et al.*, 1998).

In the PDR subfamily, *ScPDR5* and *ScSNQ2* have been well documented for their roles in drug resistance. Amongst the two, ScPdr5p is the major drug transporter in *S. cerevisiae* (Balzi *et al.*, 1994; Bissinger and Kuchler, 1994; Decottignies and Goffeau, 1997). The Δ*Scpdr5* cells are sensitive to a range of compounds displaying remarkable broad substrate specificity. *ScPDR5, ScSNQ2* and *ScYOR1* have both specific and overlapping substrate profiles. The construction of mutants of ScPdr5p, ScSnq2p and ScYor1p in various combinations and screening them for growth inhibition by several toxic compounds has provided evidence for the existence of a largely overlapping substrate profile (Rogers *et al.*, 2001). Analysis of the null mutant cells namely Δ*Scpdr5* and Δ*Scsnq2* also revealed their susceptibility to cations like NaCl, $MnCl_2$ and LiCl. The expression of these genes is also induced by these above mentioned salts (Miyahara *et al.*, 1996). The variation in the phenotypes of these deletion mutants indicates that in addition to drug extrusion, they affect other cellular functions (discussed below). On the other hand, the ScYcf1p protein confers resistance to cadmium, hence the Δ*Scycf1* cells are hypersensitive to this cation (Szczypka *et al.*, 1994). The other members contributing to drug resistance in this yeast belong to the MRP/CFTR subfamily of transporters (Katzmann *et al.*, 1995; Petrovic *et al.*, 2000).

Structure–function analysis of yeast ABC drug transporters

ABC transporters are potential targets for the development of new drugs that aim at preventing the normal functioning of these pumps. In this context, considerable attention is being paid to the structural and functional aspects of these proteins, which in turn can lead to the development of better strategies for designing modulators/inhibitors of the efflux pumps. An understanding of the regulatory circuits controlling the expression of efflux pump encoding genes is also exploited as an alternate strategy, for the development of drugs that could block the expression of these genes. In this chapter, we restrict our discussion only to the structure and function of yeast ABC proteins followed by a section on their physiological roles. Additional information about structure and function of ABC transporters in general, as well as the characteristics of structure–function relationships in ABC multidrug transporters can be found in Chapters 1 and 3 within this volume.

It appears that in these transporters several TMSs associate together to form the substrate binding site(s) but this alone is probably not sufficient for substrate transport across the membrane bilayer. The extrusion of substrates requires energy that is derived from the hydrolysis of ATP carried out at the NBDs. Despite the overall conservation of the domain architecture of TMDs, their primary sequences are significantly different. On the other hand, NBDs of ABC transporters which power drug export are highly conserved both in terms of primary structure and architecture (Higgins and Linton, 2004).

CaCdr1p is the major drug transporter of *C. albicans*. Notably, out of 28 ABC members, only two, *CaCDR1* and *CaCDR2*, encode drug transporters. The other members of this family apparently do not contribute to MDR in this pathogenic yeast. The relevance of ABC members that do not contribute to efflux drugs remains an open question (discussed later). Both CaCdr1p and CaCdr2p comprise two homologous halves, each made up of a hydrophilic, cytoplasmic NBD followed by TMD represented by six TMSs. In comparison to their human counterparts, studies pertaining to the identification of drug and ATP binding sites, ATP hydrolysis and its coupling to drug extrusion of yeast ABC drug transporters have only recently been initiated (Egner *et al.*, 1998; Egner *et al.*, 2000; Jha *et al.*, 2003b; Shukla *et al.*, 2003; Shukla *et al.*, 2004; Jha *et al.*, 2004; Rai *et al.*, 2005; Saini *et al.*, 2005; Rai *et al.*, 2006; Rai *et al.*, 2008).

Drug binding

The promiscuity towards substrates is a characteristic feature of most ABC-type drug transporters and hence makes it difficult to study their functions (Rogers *et al.*, 2001; Prasad *et al.*, 2002). The range of substrates extruded by CaCdr1p varies enormously and includes structurally unrelated compounds such as azoles, lipids and steroids (Krishnamurthy *et al.*, 1998c; Dogra *et al.*, 1999; Smriti *et al.*, 2002). Expectedly, it is a challenge to predict the residues involved in substrate binding without high-resolution structural data to support these claims. Yet using deletion studies and site-directed mutagenesis followed by biochemical assays of the mutant variants thus generated, it has been possible to map the substrate binding pockets of these proteins with some success (Krishnamurthy *et al.*, 1998a; Shukla *et al.*, 2003; Shukla *et al.*, 2004; Pleban *et al.*, 2005; Saini *et al.*, 2005; N. Puri, unpublished).

In order to perform structural and functional studies, CaCdr1p was over expressed as a GFP-tagged fusion protein in *S. cerevisiae* wherein its expression was driven by the *ScPDR5* promoter (AD1–8u$^-$) (Shukla *et al.*, 2003). The expression system AD1–8u$^-$ which was developed by Goffeau's group (Decottignies *et al.*, 1998), lacks the seven major ABC transporters ScYor1p, ScSnq2p, ScPdr5p, ScPdr10p, ScPdr11p, ScYcf1p, ScPdr15p thereby rendering it hypersensitive to various drugs. The constitutive hyper-expression of heterologous proteins in this strain is facilitated by the presence of a hyperactive allele of Pdr1p (*pdr1-3*), one of the transcription factor that regulates expression of ScPdr5p (Nakamura *et al.*, 2002). In this system, the drug and nucleotide binding properties of CaCdr1p have been analysed (Shukla *et al.*, 2003; Shukla *et al.*, 2004; Jha *et al.*, 2004; Rai *et al.*, 2005; Saini *et al.*, 2005; Rai *et al.*, 2006; N. Puri, unpublished; A. Kumar, unpublished) (Fig. 8.2).

Iodoarylazido prazosin (IAAP, a photoaffinity analogue of human P-glycoprotein (P-gp) substrate, prazosine) and azidopine (a dihydropyridine photoaffinity analogue of its modulator, verapamil) were shown to specifically bind with CaCdr1p-GFP. Interestingly, IAAP binding with CaCdr1p-GFP was competed by nystatin (NYS) while azidopine binding could only be competed by miconazole (MIC), thus demonstrating the possibility of different drug binding sites for the two analogues (Shukla *et al.*, 2003).

The alanine scanning of TMS 5 and 11 of CaCdr1p identified interesting interactive residues. On one hand selected residues of TMS 11 when replaced with alanine decreased resistance to drugs and were found to bunch on one face of the helix in a pep wheel projection, on the other hand, substitution of all the residues of TMS 5 resulted in CaCdr1p variants with decreased resistance wherein these residues are distributed throughout the helix (Fig. 8.3). TMS 5 mutant variants also show uncoupling between ATP hydrolysis and drug extrusion. Studies suggest that TMS 5 could be a communication helix between NBDs and TMDs (N. Puri, unpublished).

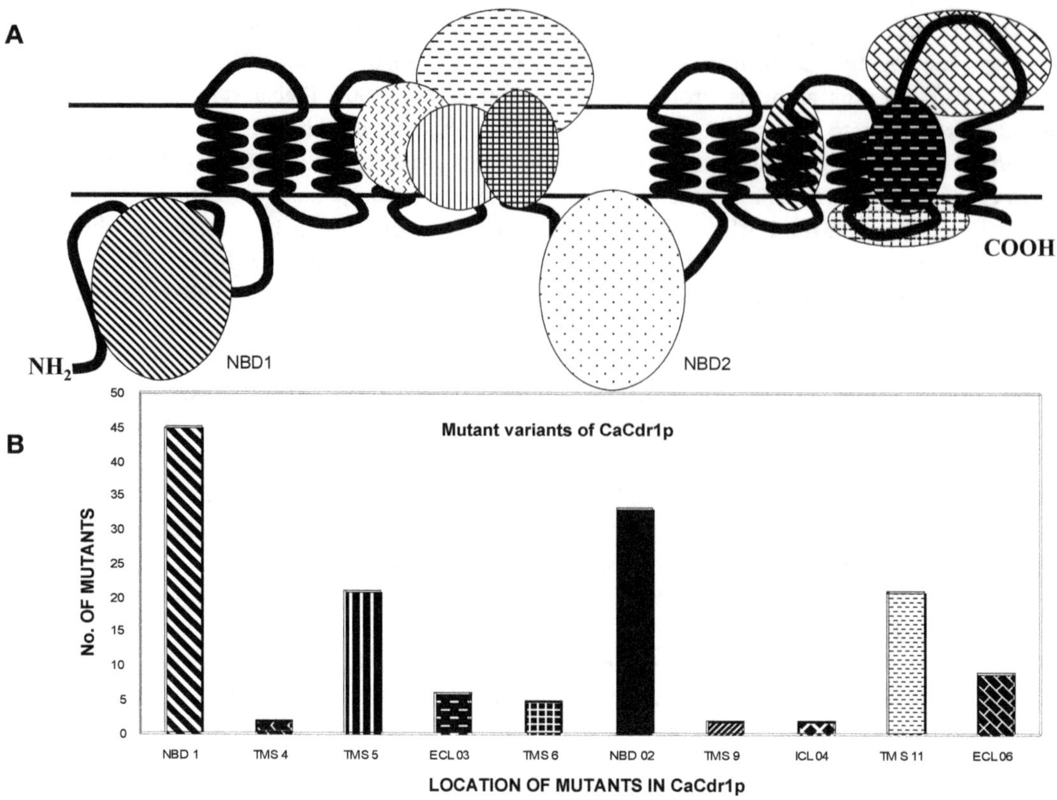

Figure 8.2 The location of mutated amino acid residues in CaCdr1p of *C. albicans*. **A** In the cartoon the position of various amino acid residues that have been studied in CaCdr1p by site directed mutagenesis are marked by ovals. **B** The bar diagram shows the number of mutants generated by mutating amino acid residues by site directed mutagenesis at different NBDs (nucleotide binding domain), TMS (transmembrane segment), ECL (extracellular loop) and ICL (intracellular loop). Every single amino acid residue in TM5 and TM11 has been subjected to alanine scanning mutagenesis, while only a select few have been mutated at the NBDs, ECLs, and ICLs (Shukla *et al.*, 2003; Jha *et al.*, 2003b; Shukla *et al.*, 2004; Jha *et al.*, 2004; Rai *et al.*, 2005; Saini *et al.*, 2005; Rai *et al.*, 2006; Rai *et al.*, 2008).

Deletion and site directed mutagenesis studies have implicated the contribution of TMS 5, 6, 11 and 12 in forming the substrate binding pocket(s) of the protein (Krishnamurthy *et al.*, 1998a; Shukla *et al.*, 2003; Shukla *et al.*, 2004; Saini *et al.*, 2005; N. Puri, unpublished).

Raymond's group has also shown that membranes prepared from CaCdr1p and CaCdr2p overexpressing cells are capable of binding the photoaffinity analogue of rhodamine 123 (^{125}I) iodoaryl azido-rhodamine 123 (IAARh123) and that both N-terminal and C-terminal halves of CaCdr2p contribute to rhodamine binding (Gauthier *et al.*, 2003). Coupled together, studies done so far suggest that the drug binding sites in CaCdr1p are scattered throughout the protein and probably more than one residue of different helices are involved in binding and extrusion of drugs. However, there is still insufficient information available to predict where exactly the most common antifungals, such as azoles bind prior to their extrusion.

Various groups are also gradually performing structure–function analysis of ABC drug transporters of other yeasts. In *S. cerevisiae*, ScPdr5p has been subjected to random mutational analysis wherein Egner and coworkers (Egner *et al.*, 1998; Egner *et al.*, 2000) identified several important amino acid residues critical for drug binding and transport. Additionally, imidazole and trialkyltin chloride derivatives have been used to determine the minimum prerequisite of ScPdr5p substrates interaction and transport (Hanson *et al.*, 2005). They observed that ScPdr5p has at least three

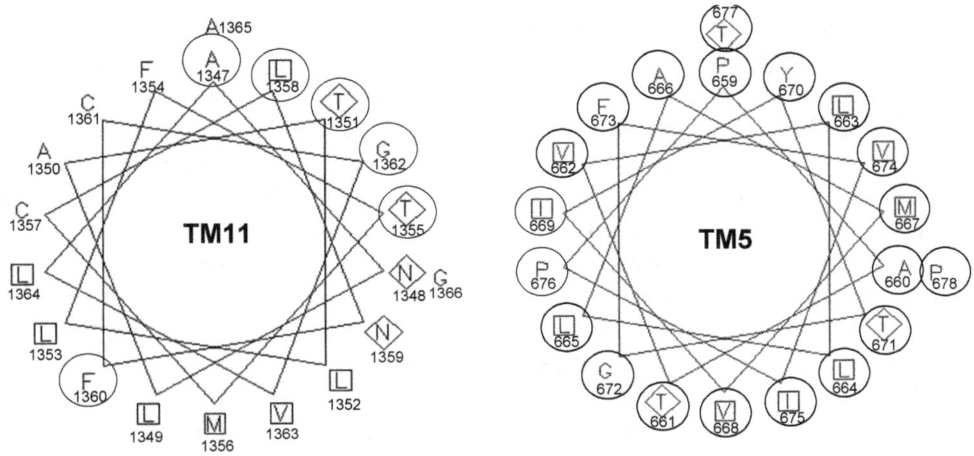

Figure 8.3 Helical wheel projection of TM5 and TM11 of CaCdr1p. Helical wheel projection of the primary amino acid sequence of CaCdr1p was constructed using 3.6 amino acids per turn of the helix by the EMBOSS PEPWHEEL program. The circled residues are the ones that play a role in drug binding and transport of CaCdr1p (Saini *et al.*, 2005; N. Puri, unpublished).

drug binding sites and some substrates might interact at more than one site. Thus, similar to CaCdr1p and its other mammalian homologues, ScPdr5p also seems to have multiple drug binding sites scattered throughout the protein.

ATP catalysis

ABC drug transporters utilize the energy derived from ATP hydrolysis to power drug extrusion across the plasma membrane. This is a key characteristic of their function. The conserved NBDs located at the cytoplasmic periphery are the hub of such an activity. The NBDs of all ABC transporters, irrespective of their origin and nature of the substrate that they transport, share extensive amino acid sequence identity and typical motifs (Higgins and Linton, 2004). For example, NBDs of the non-fungal ABC transporters have a β-sheet sub-domain containing the typical Walker A and Walker B motifs, as an essential feature of all ATP requiring enzymes (Walker *et al.*, 1982) along with an α-helical sub-domain that possesses the conserved ABC signature sequence. NBD protein sequences possess certain conserved amino acid stretches, which are considered to be critical for the domain's functionality (Schneider and Hunke, 1998). The Walker A motif has a consensus sequence GxxGxG<u>K</u>S/T, where 'x' represents any amino acid, the Walker B motif has the sequence ΦΦΦΦD, where 'Φ' is any aliphatic residue, and the ABC signature has the LSGGQQ/R/KQR sequence.

Structural and biochemical analyses of non-fungal NBDs show that the typical lysine residue of Walker A motif binds to the β- and γ-phosphates of ribonucleotides and plays a critical role in ATP hydrolysis (Schneider and Hunke, 1998). Mutation of this lysine residue in NBD reduces or abolishes the hydrolysis activity and in some other cases impairs nucleotide binding (Schneider and Hunke, 1998).

Interestingly, though N-NBD of CaCdr1p contains the conserved Walker A (GRPGAG<u>C</u>ST) and Walker B (IQCWD) motifs, and the ABC signature sequence (VSGGERKRVSIA) (Prasad *et al.*, 1995), it has got some unique and typical replacement of residues which exist in all fungal ABC transporters. Thus, both Walker A and Walker B of N-NBD have unique substitutions. While C-NBD has conserved Walker A and B unlike N-NBD it has a degenerate signature sequence. In the fungal Walker A motif, the commonly occurring lysine residue is replaced by a cysteine. This replacement appears to be a unique feature of N-NBDs of almost all the known fungal ABC transporters. The significance of this replacement from a charged residue to a sulfhydryl amino acid for the *in vivo* function of the protein is not apparent. Of note, Walker A in the C-NBD of most of yeast ABC transporters contains the evolutionary conserved lysine (Jha *et al.*, 2003a) (Fig. 8.4).

Recently, in an attempt to examine the role of the N-NBD of CaCdr1p independent of its C-terminal counterpart and to ascertain the role of the uncommon cysteine of Walker A, N-NBD of CaCdr1p was purified and characterized. It was demonstrated that this purified domain elicits a cation dependent general ribonucleotide triphosphatase activity (Jha *et al.*, 2003a). It was observed that an evolutionarily divergent Cys193 of Walker A of N-NBD was critical for ATP hydrolysis. In a recent *in vivo* study, the relative contribution of both the N- and C-terminal NBDs in ATP binding, hydrolysis and transporter activity of native CaCdr1p (full protein) was examined wherein the atypical Cys193 of Walker A of N-NBD (C193K) and conserved Lys901 (K901C) in the Walker A of C-NBD were replaced (Jha *et al.*, 2004). The drug resistance profile of CaCdr1p mutant variants viz C193K or K901C provided interesting insights into the functioning of the two NBDs. The cells expressing K901C showed hypersensitivity to drugs as compared to the C193K variant. These observations established that the two NBDs respond asymmetrically to the substitution of conserved residues of their respective Walker A motifs.

Two other residues of the N-NBD from CaCdr1p are also found to be important for domain functioning (Fig. 8.4). The unusual Trp326 in the Walker B motif of N-NBD, which is conserved in all fungal transporters, appears to be important for ATP binding and for the subsequent conformational change (Rai *et al.*, 2005). Thus although the mutant with W326A appears capable of ATP hydrolysis, it does so with a much higher K_M value, indicating that the

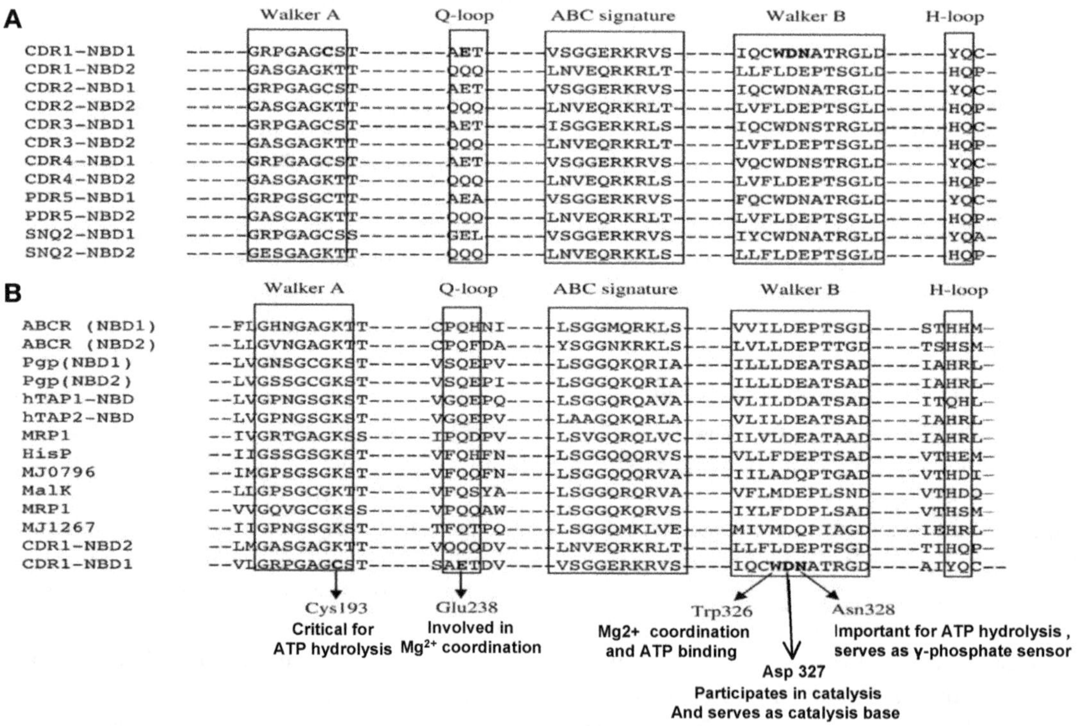

Figure 8.4 Sequence alignment of NBDs of fungal and non-fungal ABC transporters and the summarized role of uniquely placed, conserved but uncommon residues in N-NBD of CaCdr1p. The figure highlights the presence of conserved motifs like Walker A, Walker B, Q-loop, ABC signature, H-loop in **A** fungal and **B** non-fungal ABC transporters. The arrows mark the position of unique residues Cys193, Glu238, Trp326, Asp327 and Asn328 of N-NBD of CaCdrp1. These residues are also conserved in all fungal ABC transporters. C193 in Walker **A** is critical for ATP hydrolysis, E238 in the Q-Loop is involved in Mg^{2+} ion coordination. In the Walker **B** motif W326 is involved in Mg^{2+} coordination and ATP binding, D327 participates in catalysis and serves as catalysis base and N328 is important for ATP hydrolysis and serves as γ-phosphate sensor (Jha *et al.*, 2004; Rai *et al.*, 2005; Saini *et al.*, 2005; Rai *et al.*, 2006; Rai *et al.*, 2008).

docking of the substrate in the binding pocket has been altered by the mutation. However, the protein appears capable of near-normal function in cells expressing the full length protein carrying W326A mutation, implying that the conformational change that normally occurs upon ATP docking cannot by itself be responsible for the cross-talk by the domain with the TMDs. While the highly conserved Asp327 of N-NBD is shown to be the catalytic carboxylate in the context of other ABC transporters, in N-NBD of CaCdr1p, it does not appear to mediate catalysis via interaction with Mg^{2+} as is normally expected for similar transporters (Rai *et al.*, 2006). The conserved substitution of typical residues in fungal ABC transporters reveals their critical roles in ATP catalysis (Fig. 8.4). Since these substitutions are unique to fungal ABC transporters, they represent novel mechanism of ATP hydrolysis.

Another study performed to analyse the role of nucleotide binding domains (NBDs) of CaCdr1p, used the strategy of constructing chimeras comprising of either two N- or C-terminal NBDs (Saini *et al.*, 2006). It was observed that the variant protein comprising either two N-NBDs or two C- NBDs of CaCdr1p were non-functional. However, as compared to C-NBD, N-NBDs variant, which also showed poor cell surface localization, could be rescued to cell surface as primarily a non-functional protein, if cells were grown in presence of drug substrates. These results also demonstrated that the two NBDs of CaCdr1p are asymmetric and non-exchangeable. In-depth analysis of the catalytic cycle, namely nucleotide binding, hydrolysis and substrate binding/efflux is required to ascertain the reasons behind the dichotomy in functioning of the NBDs.

Recently, on addressing the question of functional asymmetry of NBDs in ScPdr5p it was found that catalysis at only one of the two NBD composite sites is crucial for protein function (Ernst *et al.*, 2008). Although a mutation of the glutamate residue (E1036) abolished ATPase activity and substrate transport, mutation of the catalytic dyad histidine (H1068) had no influence on ATP consumption. However, the H1068A mutation abolished rhodamine transport, while leaving the transport of other substrates unaffected. The study suggests that TMSs alone are not determinants for substrate selection. Notably, CaCdr1p and its homologue in Baker's yeast, ScPdr5p do not show substrate stimulated ATPase activity. It is suggested that yeast ABC drug transporters which show high specific activity of ATPase, may constantly hydrolyse ATP to ensure active substrate transport.

ScYor1p acts as a drug pump to clear toxic substances from the cytosol, and is required for cellular resistance to the mitochondrial poison, oligomycin. ScYor1p shares the same domain arrangement and topology as human CFTR and P-gp (Pagant *et al.*, 2007). Deletion of a phenylalanine residue within NBD1, equivalent to the CFTR-ΔF508 deletion, causes ScYor1p-ΔF670 to be retained in the ER. Notably, ΔF670 mutation of ScYor1p destabilized the interface between TMS6 and TMS12. In a recent study, the quaternary structure of Yor1p has been examined using biochemical and genetic approaches. The results demonstrate that ScYor1p presents a Sav1866-like architecture (Pagant *et al.*, 2008). By employing extensive cysteine cross linking, it was demonstrated that interactions between specific intracellular loops are influential in shaping the active quaternary structure of ABC transporters. (Pagant *et al.*, 2008).

Physiological roles of ABC transporters

Besides, playing an important role in drug transport, additional roles of ABC transporters in absorption, excretion, signal transduction and bacterial pathogenesis reflect their diverse functions (Chakraburtty, 2001; Detmers *et al.*, 2001; Mendez and Salas, 2001; Goldman and Kranz, 2001; Schneider, 2001;) (Fig. 8.5). Considering the existence of large numbers of ABC transporters in yeasts genome, there has been a focus on research related to understanding their physiological relevance. Notably, the abundance of the drug transporters and their wider specificity suggests that these ABC transporters may not exclusively export drugs in yeasts (Wolfger *et al.*, 2001; Jungwirth and Kuchler, 2006). Several studies confirm the multifunctional nature of the drug transporters belonging to the ABC superfamily of yeasts. Below we discuss some of the well known physiological roles of yeast ABC drug transporters and some of the other ABC proteins.

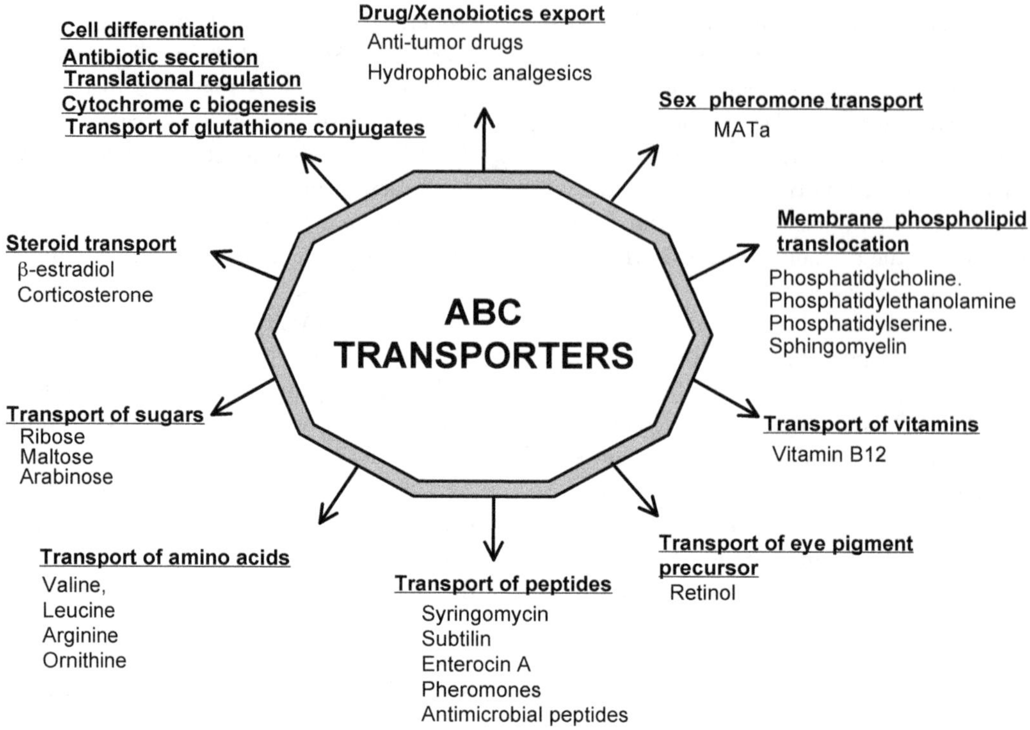

Figure 8.5 Physiological roles of ABC transporters. The figure depicts compilation of some of the physiological roles from different studies.

Transport of human steroids

ScPDR5 and *ScSNQ2* of *S. cerevisiae* and *CaCDR1* of *C. albicans* have been demonstrated to efflux steroid hormones like β-estradiol and corticosterone (Mahe *et al.*, 1996; Kolaczkowski *et al.*, 1996; Krishnamurthy *et al.*, 1998c). Moreover, some of the drugs which are substrates of these multidrug transporters are also capable of competing with β-estradiol and corticosterone thus implying the existence of common binding sites (Krishnamurthy *et al.*, 1998c). Ergosterol, the analogue of mammalian cholesterol is the closest molecule to steroids in yeast cells. It is thus quite possible that these multidrug transporters, like human Mdr1p could be involved in sterol homeostasis in the yeast cells (Kontoyiannis, 2000). Some other ABC transporters like ScAus1p and ScPdr11p mediate intracellular sterol flow from the plasma membrane to the endoplasmic reticulum, in a non-vesicular manner (Wilcox *et al.*, 2002; Li and Prinz, 2004). However, the exact role and mechanism of these proteins in sterol metabolism remains unclear.

Steroids like β-estradiol can not only be effluxed by CaCdr1p and CaCdr2p, they also induce the expression of genes encoding these proteins (Krishnamurthy *et al.*, 1998b; de Micheli *et al.*, 2002). While the presence of steroid binding proteins such as β-estradiol binding protein (EBP) (Madani *et al.*, 1994), corticosteroid binding protein (CBP) (Malloy *et al.*, 1993) and progesterone binding protein (PBP) (Das and Datta, 1985) in yeast is well established, the physiological relevance of these binding proteins remains elusive. Interestingly, it has been shown that steroids can induce a pleiotropic drug resistance (PDR) state in both *C. albicans* and *S. cerevisiae* (Banerjee *et al.*, 2004; Banerjee *et al.*, 2007; Banerjee *et al.*, 2008). This PDR state is mediated by the over-expression of genes encoding PDR determinants such as the ABC transporter encoding genes Ca*CDR1*, Ca*CDR2* and *ScPDR5* which are known to be regulated by the CaTac1p (Transcriptional activator of *CDR* genes protein 1) and the CaPdr1p and ScPdr3p transcription factors, respectively (Moye-Rowley, 2003; Coste *et al.*, 2004). CaTac1p is the first transcription factor identified in *C. albicans*, which positively regulates the expression of *CaCDR1* and *CaCDR2*. Noteworthy, the *CaTAC1* gene itself was found to be responsive to β-estradiol (Cheng *et al.*, 2006) and progesterone (Banerjee *et al.*, 2008).

CDR1 promoter has a stretch spanning −462 to −213 (with respect to ATG), which mediates both drug and steroid response thereby strengthening the transcriptional basis of their response (de Micheli *et al.*, 2002). Besides this the *CDR1* promoter also has two hitherto unknown *cis*-regulatory regions that respond specifically to progesterone and β-estradiol. Other steroid responsive genes in the *C. albicans* genome also contain these sequences, indicating a conserved role for these regions in mediating steroid response in *C. albicans* (Banerjee *et al.*, 2004; Karnani *et al.*, 2004).

Translocation of phospholipids

Phospholipids in the plasma membrane of the eukaryotic cells are asymmetrically distributed across the inner and outer leaflets. The aminophospholipids such as phosphatidylserine (PS) and phosphatidylethanolamine (PE) are predominantly localized in the inner leaflet (Ikeda *et al.*, 2006; Pomorski and Menon, 2006). Maintenance of lipid asymmetry is due to transporters that specifically flip (translocate out-to-in) or flop (translocate in-to-out) phospholipids. P-type ATPases have been identified as phospholipid flippases (Holthuis and Levine, 2005; Ikeda *et al.*, 2006), whereas ABC transporters and members of the family of multidrug resistance proteins such as mouse Mdr1, and human MDR1 and MDR3, act as floppases in the plasma membrane of intact cells (Van Helvoort *et al.*, 1996; Smriti *et al.*, 2002). Direct evidence that ABC transporters can translocate phospholipids was derived from experiments in which purified human MDR1/P-gp, human MRP1 and *Lactococcus lactis* LmrA (Lactococcal multidrug resistance A) (a human P-gp homologue) were reconstituted into proteoliposomes and shown to translocate fluorescent phospholipid analogues in an ATP-dependent fashion (Margolles *et al.*, 1999; Romsicki and Sharom, 2001; Huang *et al.*, 2004; Eckford and Sharom, 2005).

It has been demonstrated that the yeast ABC transporters ScPdr5p and ScYor1p translocate the fluorescent PE analogue, NBD-PE (Decottignies *et al.*, 1998). Similarly, homologues of ScPdr5p, CaCdr1p, CaCdr2p and CaCdr3p in *C. albicans* could translocate phosphoglycerides between the two lipid monolayers of plasma membrane. It is reported that the drug efflux pumps CaCdr1p and CaCdr2p are able to flop fluorescent glycerophospholipids 7-nitro-2–1,3-benzoxadiazol-4-yl phosphatidylcholine (NBD-PC) and NBD- phosphatidylethanolamine (NBD-PE), while CaCdr3p which does not efflux drugs acts as a flippase (Fig. 8.6). It is also shown that the glycerophospholipid translocation ability is restricted only to ABC drug transport proteins since the drug transporter belonging to MFS family are incapable of this property (Smriti *et al.*, 2002).

Recently, purified CaCdr1p has been reconstituted into sealed membrane vesicles (Shukla *et al.*, 2007). A comparison of the ATPase activities of the sealed and permeabilized proteoliposomes indicated that CaCdr1p was asymmetrically reconstituted such that ~70% of the molecules had their ATP binding sites accessible to the extra vesicular space. ATP-dependent transport of the fluorescent lipids into the inner leaflet of the vesicles was observed in such a reconstituted system. The transport of fluorescent glycerophospholipids was reduced in vesicles reconstituted with CaCdr1p variants with impaired ATPase activity and could be competed to different levels by molar excess of drug substrates. These results established a direct link between the ability of CaCdr1p to translocate fluorescent glycrophospholipids and drug efflux. Since only a few members of the ABC superfamily of *Candida* have a well-defined role as drug exporters, lipid translocation mediated by CaCdr1p could reflect its cellular function.

Peptide transport

ABC transporters such as ScMdl1p, ScMdl2p and ScSte6p that are not drug transporters function as peptide transporters in *S. cerevisiae* cells (McGrath and Vershavsky, 1989; Dean *et al.*, 1994). ScMdl1p and ScMdl2p are half transporters, localized in the inner mitochondrial membrane, with a considerable amino acid sequence similarity to

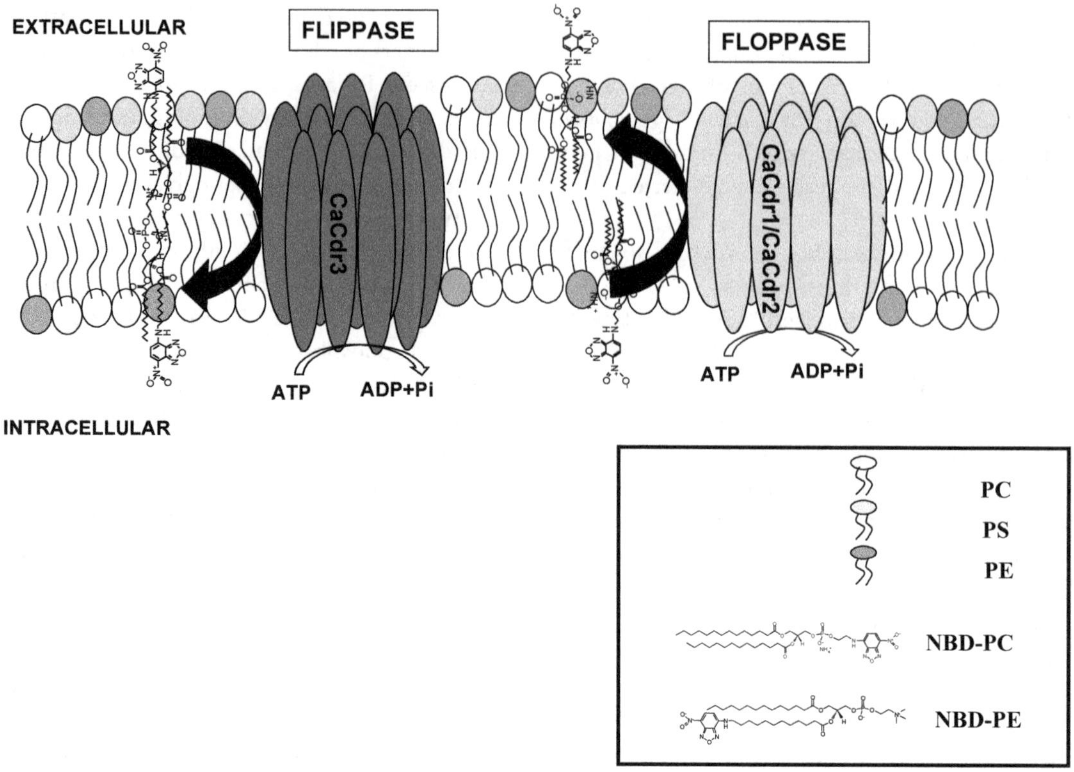

Figure 8.6 ABC proteins function as phospholipid translocaters. Using NBD-PC and NBD-PE which are florescent analogues of PE and PC, it was demonstrated that ABC transporters CaCdr1p and CaCdr2p function as a floppase (in to out) whereas CaCdr3 is a flippase (out to in). PE, phosphatidylethanolamine; PC, phosphatidylcholine, NBD- 7-nitro-2–1,3-benzoxadiazol-4-yl. The model depicts Cdr1p as floppase and Cdr3p as flippase. The box depicts structure of NBD phosphoglycerides (Smriti *et al.*, 2002; Shukla *et al.*, 2007).

the mammalian Tap1p/Tap2p proteins and with the yeast mitochondrial transporter, ScAtm1p. ScAtm1p transports Fe/S clusters from the mitochondria to the cytosol for incorporation into a wide variety of metabolically significant molecules (Leighton and Schatz, 1995). Interestingly, *ScMDL1* can functionally complement defects in iron metabolism of Δ*Scatm1* cells (Chloupkova *et al.*, 2003). The mutant Δ*Scmdl2* displays growth defect in glycerol containing media while the Δ*Scmdl1* null does not show any such defect. These genes are important for peptide transport released by proteases inside the mitochondria. However, the physiological relevance of peptide transport is not clear yet. ScMdl1p is also shown to bind to F_1F_o ATP synthase and is responsible for rescuing cells from oxidant-induced cell death (Chloupkova *et al.*, 2003; Galluhn and Langer, 2004).

ScSte6p of *S. cerevisiae* is a full ABC transporter, which bears a strong similarity to the mammalian drug transporter protein, P-gp. This protein, which is not a drug transporter, is one of the few ABC proteins for which a physiological substrate is known (McGrath and Vershavsky, 1989). Deletion in *ScSTE6* leads to a defect in extracellular secretion of pheromone **a** factor leading to mating defect. Notably, P-gp expression in a Δ*ste6* mutant can partially restore the ability of the cells to mate thus extending their sequence homology to functional similarity (Raymond *et al.*, 1992).

ScSte6p homologues are also known in *S. pombe* and *C. albicans*. SpMam1p gene product of *S. pombe* is known to secrete the **M** pheromone (Christensen *et al.*, 1997a). *CaHST6*, in *C. albicans* was isolated by functional complementation of a Δ*Scste6* mutant strain of *S.*

cerevisiae (Raymond *et al.*, 1998). CaHst6p bears homology to human P-gp (26% identity and 36% similarity), to SpMam1p (29% identity and 41% similarity) and to ScSte6p (35% identity and 48% similarity). Identification of CaHst6p in the asexual *C. albicans* was initially found to be intriguing since at that time existence of a mating cycle was not established. Since then, with the completion of sequencing of the *C. albicans* genome, the Mating-like loci (MTL) has been identified and it is shown that the homozygous diploid *MTL*a and *MTLα* could be generated, which were able to mate both *in vivo* and *in vitro* (Hull and Johnson, 1999; Hull *et al.*, 2000; Magee *et al.*, 2002).

Transport of weak acids

In *S. cerevisiae*, ScPdr12p was identified as an ABC transporter responsible for the spoilage of food (Piper *et al.*, 1998). ScPdr12p is able to transport weak acid metabolites like pentanoic and hexanoic acids, which are routinely used as food preservatives (Piper *et al.*, 1998). The presence of weak acids is perceived as a stress by the yeast cells because of which they induce stress-response pathways. *ScPDR12* expression is induced in the presence of sorbic acid at pH 4.5. ScPdr12p can transport weak acids in an ATP-dependent manner (Holyoak *et al.*, 1999). Δ*Scpdr12* mutant cells are sensitive to sorbate and propionate at pH 4.5. ScWar1p, was identified as a transcription factor that triggers the expression of ScPdr12p in response to weak acids (Kren *et al.*, 2003; Schuller *et al.*, 2004). The induction of *ScPDR12* is independent of the *ScYap1* (Yeast Activator Protein-1) or *ScMsn2*/*ScMsn4* (Multicopy suppressor of Sucrose non-fermenting 1 mutation) transcription factors that are normally involved in stress regulatory pathways (Schuller *et al.*, 2004). Interestingly, *Scpdr1* and *Scpdr3* mutant cells of *S. cerevisiae* display an increased resistance to sorbate indicating a functional cross-talk between the weak acid response and the PDR pathways.

ScPdr12p is also involved in the export of natural products of amino acid catabolism. Fusel acids and alcohols along with their esters normally contribute to the flavours of wine and beer. ScPdr12p is capable of exporting the fusel acids but not fusel alcohols derived from leucine, isoleucine, phenylalanine and tryptophan. This new role of ScPdr12p is directly connected to the metabolic pathway in *S. cerevisiae* demonstrating naturally produced products in yeast as substrate for ScPdr12p (Hazelwood *et al.*, 2006).

Heavy metal detoxification

ABC transporters of *S. cerevisiae* such as ScYcf1p, ScBpt1p and ScYbt1p, are localized in the vacuolar membrane which enable the yeast cells to transport heavy metals as glutathione conjugates into the vacuole (Li *et al.*, 1996; Tommasini *et al.*, 1996; Rebbeor *et al.*, 1998; Sharma *et al.*, 2002; Klein *et al.*, 2002). Based on the ability to confer cadmium resistance, ScYcf1p was isolated in *S. cerevisiae*, which was shown to be localized in plasma membrane. ScYcf1p is homologous to the human MRP and CFTR proteins (Szczypka *et al.*, 1994). Besides conferring cadmium resistance, ScYcf1p is also capable of transporting the glutathione S-conjugate precursor 1-chloro-2,4-dinitrobenzene. Interestingly, mutations of CFTR that generate a defective Cftr1p, when introduced in ScYcf1p, also result in similar phenotypic defects (Wemmie and Moye-Rowley, 1997). This indicates that ScYcf1p can be used as a model protein to study the trafficking and functioning of CFTR. Both human MRP and ScYcf1p, have an extended N-terminal as opposed to CFTR, which is devoid of this extension. The N-terminal extension is required for proper functioning of ScYcf1p but not for its correct localization. The function of Hmt1p in *S. pombe* is analogous to ScYcf1p since it transports phytochelatins and phytochelatin–Cd^{2+} complexes into the vacuole (Ortiz *et al.*, 1992). These ABC transporters therefore play a pivotal role in heavy metal detoxification.

Besides the above functions, ScYcf1p along with another transporter ScBpt1p (bile pigment transporter) transports un-conjugated bilirubin (UCB) into the vacuoles. Thus yeast cells also have dedicated transporters for the uptake of bilirubin (degradation product of haem-containing proteins) that it might see in the environment (Petrovic *et al.*, 2000). ScYbt1p, another vacuolar protein, was isolated from the *S. cerevisiae* vacuoles based on its ability to transport bile acid. It bears homology to the mammalian cMOAT/cMRP1 and MRP transporters and to the ScYcf1p and ScBpt1p transporters (Ortiz *et al.*, 1997).

Mitochondrial iron homeostasis

ScAtm1p was the first ABC protein of the mitochondria to be identified in *S. cerevisiae* (Leighton and Schatz, 1995). It is a half-transporter located in the mitochondrial inner membrane that homo dimerizes in presence of ATP (Chloupkova *et al.*, 2004). Δ*Scatm1* cells are deficient in the holo forms of the haem-containing proteins and does not affect haem biosynthesis. Δ*Scatm1* cells show increased accumulation of iron in the mitochondria suggesting their role in mitochondrial iron homeostasis probably by exporting Fe/S clusters from the mitochondria to the cytosol. The Δ*Scatm1* cells are also hypersensitive for growth in the presence of oxidative reagents, and contain increased levels of the oxidized form of the antioxidant glutathione. Moreover, Δ*Scatm1* cells show a 3-fold higher accumulation of iron in the mitochondria. These levels are much higher than the accumulation observed when the protein frataxin is mutated in neurodegenerative disease such as Friedreich's ataxia (Campuzano *et al.*, 1996). ScYfh1p is the yeast homologue of frataxin and it together with ScAtm1p is presumed to play a role in maintaining normal levels of iron in the mitochondria. The exact mechanism by which ScAtm1p controls the iron levels in the mitochondria remains to be elucidated (Kispal *et al.*, 1997). Recently, though the function of ScAtm1p has been found to be an important determinant of the iron regulon, which includes both the high-affinity cell surface transport system and the Fe-S cluster export from the mitochondria (Rutherford *et al.*, 2005).

ScAtm1p homologue has also been identified in *S. pombe* wherein it plays a role not only in oxidative stress but also in vacuolar functions (Iwaki *et al.*, 2005). Thus, a deletion of *SpATM1* from *S. pombe* not only affects the mitochondrial iron accumulation but also has an effect on vacuolar fusion and endocytosis. This study for the first time established an association of a mitochondrial ABC transporter with vacuolar functions (Iwaki *et al.*, 2005).

ABC proteins are sensitive to membrane lipid environment

The role of the drug efflux pumps as phospholipid translocators are suggestive of the fact that these pump proteins are involved in maintaining lipid homeostasis and thus may be sensitive to fluctuations in membrane lipid environment. Indeed many studies have demonstrated that yeast drug efflux pump proteins are sensitive to fluctuations in membrane lipid milieu (Smriti *et al.*, 1999; Mukhopadhyay *et al.*, 2002; Mukhopadhyay *et al.*, 2004). In this respect, studies done on *erg* mutants of *S. cerevisiae* and *C. albicans* for analysing their drug susceptibilities have shown that ScPdr5p, CaCdr1p and CaCdr2p are mislocalized in these mutant backgrounds. As a result, these pump proteins expressed in *erg* mutant background display decreased resistance to drugs.

Additionally, the same phenomenon could be observed if sphingolipid content of the plasma membrane is varied. Thus, mutants defective in sphingolipid biosynthesis also showed mislocalization of the major drug efflux pumps like ScPdr5p, CaCdr1p and CaCdr2p (Hallstrom *et al.*, 2001; Pasrija *et al.*, 2008). These observations gain importance since membrane sphingolipids and ergosterol are important subset of lipids, which are part of distinct micro domains known as membrane rafts (Martin and Konopka, 2004; Wachtler and Balasubramanian, 2006). These distinct membrane micro domains are known to home various proteins with diverse functions in cellular processes like signal transduction, membrane trafficking, lipid and protein sorting and even receptors for certain pathogens (Dupre and Tsapis, 2003; Fessler *et al.*, 2004; Pike *et al.*, 2005). Recent studies have established that CaCdr1p is exclusively localized within these microdomains on the plasma membrane while a functionally identical drug export protein belonging to MFS family, CaMdr1p show no such preference (Pasrija *et al.*, 2008) (Fig. 8.7).

Coinciding with the co-localization of ergosterol and sphingolipids within membrane rafts and sensitivity of ABC proteins like CaCdr1p to their fluctuations, another interesting observation has emerged which suggest that the genes encoding the drug efflux pump proteins and constituting raft lipid constituents are co-regulated. Microarray analysis has provided an insight into the genes that are co-regulated with the drug resistance genes especially the ones involved in the sphingolipid biosynthetic pathway (Devaux *et al.*, 2002; Coste *et al.*, 2004; Karababa *et al.*, 2004). For example, microarray analyses of *S.*

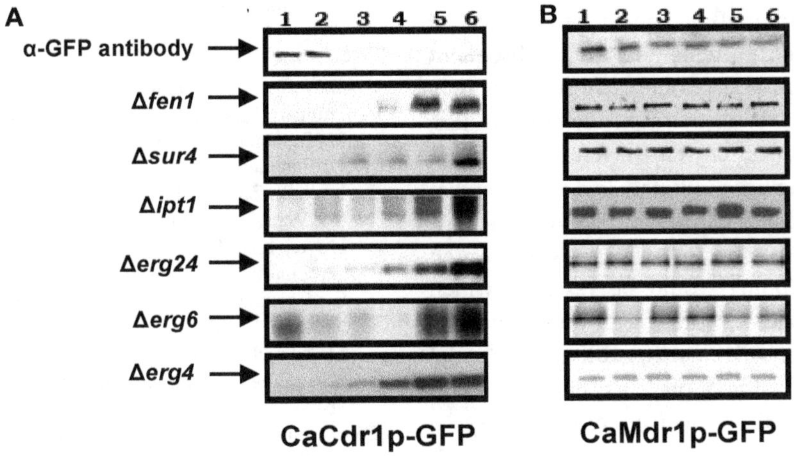

Figure 8.7 CaCdr1p is localized within membrane rafts. **A** Immunoblot analysis with anti-GFP antibody of lipid raft fractions prepared using OPTIPREP gradient shows the presence of CaCDR1p in the two top floating raft fractions while rest of the fractions in the panel show the mislocalization of CaCdr1p in lipid metabolism mutants. **B** The distribution of CaMdr1p on the other hand in the OPTIPREP gradient shows that CaMdr1p is a non raft protein whose localization is not affected in lipid metabolism mutants. *FEN1, SUR4* and *IPT1* are enzymes of the sphingolipid biosynthesis pathway while *ERG24, ERG6* and *ERG4* are enzymes of the ergosterol biosynthesis pathway (Pasrija *et al.*, 2008).

cerevisiae show that some of the genes involved in sphingolipid metabolism (such as *ScIPT1* (Inositol PhosphoTransferase 1) and *ScLAC1* (Longevity-Assurance gene Cognate 1)) and in the homeostasis of its precursors (like *ScRSB1* (Resistance to Sphingoid long-chain Base 1) are up-regulated along with the ABC transporter genes *ScPDR5, ScSNQ2* and *ScYOR1* (DeRisi *et al.*, 2000; Hallstrom *et al.*, 2001; Kihara and Igarashi, 2002; Kolaczkowski *et al.*, 2004). Additionally, genes such as *ScPDR16* (encoding a phosphatidylinositol transfer protein and *ScRTA1* (Resistance to 7-aminocholesterol) which are not a part of the sphingolipid biosynthetic pathway are also up regulated along with ABC transporters genes (Soustre *et al.*, 1996; Saidane *et al.*, 2006). These data signify the importance of the PDR pathways in controlling the lipid milieu of the plasma membrane probably for ensuring the proper localization of the drug efflux pump proteins.

Conclusions and future trends

ABC proteins, which acquired prominence because of their role in MDR have been extensively subjected to structure–function analysis. Such studies hold great promise in elucidating steps involved in drug efflux and thus setting a stage to develop rational pump inhibitors. Additionally, ABC transporters also play a plethora of physiological roles particularly in extruding toxic compounds that the organism encounters during its life cycle. This might be the sole reason for the cells to have these transporters in various organisms throughout the evolutionary scale. However, exact relevance of members of ABC superfamily in cell survival remains an open area of research.

Genome sequencing projects with subsequent genome profiling experiments have facilitated studies related to any set of genes. The proteins of the ABC superfamily also belong to one of these sets of genes. In spite of the availability of information regarding the functions of some of the ABC proteins, it is still not clear what makes a select few of them work as drug transporters leaving the rest to be involved in other unknown cellular functions. The broad substrate specificity of the drug transporters, which includes totally structurally and functionally unrelated compounds, remains a black box to-date. The physiological functions of most of the ABC transporters including those that transport drugs remain elusive. More thrust is being given to areas pertaining to understanding the physiological roles and in dissecting the mechanisms of drug transport for the members of this superfamily. Biochemical purification of these membrane proteins and their subsequent reconstitution in phospholipids vesicles has only begun in yeasts. With these purification studies, attempt to having crystal structures of these proteins has also been initiated. It is expected that with the availability of crystal structures of yeast ABC drug transporters, most of the issues related to their functioning would be resolved which indeed represents a promising future challenge. Yeasts in general will remain a good quality model system to study these proteins since performing genetics in this organism remains easy to handle.

References

Anderson, J.B. (2005). Evolution of antifungal-drug resistance: mechanisms and pathogen fitness. Nat. Rev. Microbiol. *3*, 547–556.

Balan, I., Alarco, A.M., and Raymond, M. (1997). The *Candida albicans CDR3* gene codes for an opaque-phase ABC transporter. J. Bacteriol. *179*, 7210–7218.

Balzi, E., Wang, M., Leterme, S., Van Dyck, L., and Goffeau, A. (1994). *PDR5*, a novel yeast multidrug resistance conferring transporter controlled by the transcription regulator PDR1. J. Biol. Chem. *269*, 2206–2214.

Banerjee, D., Lelandais, G., Shukla, S., Mukhopadhyay, G., Jacq, C., Devaux, F., and Prasad, R. (2008). Responses of pathogenic and nonpathogenic yeast species to steroids reveal the functioning and evolution of multidrug resistance transcriptional networks. Eukaryot. Cell *7*, 68–77.

Banerjee, D., Martin, N., Nandi, S., Shukla, S., Dominguez, A., Mukhopadhyay, G., and Prasad, R. (2007). A genome-wide steroid response study of the major human fungal pathogen *Candida albicans*. Mycopathologia *164*, 1–17.

Banerjee, D., Pillai, B., Karnani, N., Mukhopadhyay, G., and Prasad, R. (2004). Genome-wide expression profile of steroid response in *Saccharomyces cerevisiae*. Biochem. Biophys. Res. Commun. *317*, 406–413.

Bauer, B.E., Wolfger, H., and Kuchler, K. (1999). Inventory and function of yeats ABC proteins: about sex, stress, and pleiotropic drug and heavy metal resistance. Biochim. Biophys. Acta *1461*, 217–236.

Ben-Yaacov, R., Knoller, S., Caldwell, G.A., Becker, J.M., and Koltin, Y. (1994). *Candida albicans* gene encoding resistance to benomyl and methotrexate is a multidrug resistance gene. Antimicrob. Agents Chemother. *38*, 648–652.

Bissinger, P.H. and Kuchler, K. (1994). Molecular cloning and expression of *Saccharomyces cerevisiae STS1* gene product, a yeast ABC transporter conferring mycotoxin resistance. J. Biol. Chem. *269*, 4180–4186.

Blanchard, J.S. (1996). Molecular mechanisms of drug resistance in *Mycobacterium tuberculosis*. Ann. Rev. Biochem. *65*, 215–239.

Borst, P. and Ouelette, M. (1995). New mechanisms of drug resistance in parasitic protozoa. Ann. Rev. Microbiol. *49*, 427–460.

Braun, B.R., van het, H.M., d'Enfert, C., Martchenko, M., Dungan, J., Kuo, A., Inglis, D.O., Uhl, M.A., Hogues, H., Berriman, M., Lorenz, M., Levitin, A., Oberholzer, U., Bachewich, C., Harcus, D., Marcil, A., Dignard, D., Iouk, T., Zito, R., Frangeul, L., Tekaia, F., Rutherford, K., Wang, E., Munro, C.A., Bates, S., Gow, N.A., Hoyer, L.L., Kohler, G., Morschhauser, J., Newport, G., Znaidi, S., Raymond, M., Turcotte, B., Sherlock, G., Costanzo, M., Ihmels, J., Berman, J., Sanglard, D., Agabian, N., Mitchell, A.P., Johnson, A.D., Whiteway, M., and Nantel, A. (2005). A human-curated annotation of the *Candida albicans* genome. PLoS. Genet. *1*, 36–57.

Campuzano, V., Montermini, L., Molto, M.D., Pianese, L., Cossee, M., Cavalcanti, F., Monros, E., Rodius, F., Duclos, F., Monticelli, A., Zara, F., Canizares, J., Koutnikova, H., Bidichandani, S.I., Gellera, C., Brice, A., Trouillas, P., De, M.G., Filla, A., De, F.R., Palau, F., Patel, P.I., Di, D.S., Mandel, J.L., Cocozza, S., Koenig, M., and Pandolfo, M. (1996). Friedreich's ataxia: autosomal recessive disease caused by an intronic GAA triplet repeat expansion. Science *271*, 1423–1427.

Chakraburtty, K. (2001). Translational regulation by ABC systems. Res. Microbiol. *152*, 391–399.

Cheng, G., Yeater, K.M., and Hoyer, L.L. (2006). Cellular and molecular biology of *Candida albicans* estrogen response. Eukaryot. Cell *5*, 180–191.

Chloupkova, M., LeBard, L.S., and Koeller, D.M. (2003). *MDL1* is a high copy suppressor of *ATM1*: evidence for a role in resistance to oxidative stress. J. Mol. Biol. *331*, 155–165.

Chloupkova, M., Reaves, S.K., LeBard, L.M., and Koeller, D.M. (2004). The mitochondrial ABC transporter Atm1p functions as a homodimer. FEBS Lett. *569*, 65–69.

Christensen, P.U., Davey, J., and Nielsen, O. (1997a). The *Schizosaccharomyces pombe* mam1 gene encodes an ABC transporter mediating secretion of M-factor. Mol. Gen. Genet. *255*, 226–236.

Christensen, P.U., Davis, K., Nielsen, O., and Davey, J. (1997b). Abc1: a new ABC transporter from the fission yeast *Schizosaccharomyces pombe*. FEMS Microbiol. Lett. *147*, 97–102.

Coste, A.T., Karababa, M., Ischer, F., Bille, J., and Sanglard, D. (2004). *TAC1*, transcriptional activator of CDR genes, is a new transcription factor involved in the regulation of *Candida albicans* ABC transporters *CDR1* and *CDR2*. Eukaryot. Cell *3*, 1639–1652.

Cui, Z., Shiraki, T., Hirata, D., and Miyakawa, T. (1998). Yeast gene *YRR1*, which is required for resistance to 4-nitroquinoline N-oxide, mediates transcriptional activation of the multidrug resistance transporter gene *SNQ2*. Mol. Microbiol. *29*, 1307–1315.

Das, M. and Datta, A. (1985). Steroid binding protein(s) in yeasts. Biochem. Int. *11*, 171–176.

Daumke, O. and Knittler, M.R. (2001). Functional asymmetry of the ATP-binding-cassettes of the ABC transporter TAP is determined by intrinsic properties of the nucleotide binding domains. Eur. J. Biochem. *268*, 4776–4786.

de Micheli, M., Bille, J., Schueller, C., and Sanglard, D. (2002). A common drug-responsive element mediates the upregulation of the *Candida albicans* ABC transporters *CDR1* and *CDR2*, two genes involved in antifungal drug resistance. Mol. Microbiol. *43*, 1197–1214.

Dean, M., Allikmets, R., Gerrard, B., Stewart, C., Kistler, A., Shafer, B., Michaelis, S., and Strathern, J. (1994). Mapping and sequencing of two yeast genes belonging to the ATP- binding cassete superfamily. Yeast *10*, 377–383.

Decottignies, A. and Goffeau, A. (1997). Complete inventory of the yeast ABC proteins. Nature Genet. *15*, 137–145.

Decottignies, A., Grant, A.M., Nichols, J.W., De Wet, H., McIntosh, D.B., and Goffeau, A. (1998). ATPase and multidrug transport activities of the overexpressed yeast ABC protein Yor1p. J. Biol. Chem. *273*, 12612–12622.

DeRisi, J., Hazel, B.v.d., Marc, P., Balzi, E., Brown, P., Jacq, C., and Goffeau, A. (2000). Genome microarray analysis of transcriptional activation in multidrug resistance yeast mutants. FEBS Lett. *470*, 156–160.

Detmers, F.J.M., Lanfermeijer, F.C., and Poolman, B. (2001). Peptides and ATP binding cassette peptide transporters. Res. Microbiol. *152*, 245–258.

Devaux, F., Carvajal, E., Moye-Rowley, S., and Jacq, C. (2002). Genome-wide studies on the nuclear PDR3-controlled response to mitochondrial dysfunction in yeast. FEBS Lett. *515*, 25–28.

Dogra, S., Krishnamurthy, S., Gupta, V., Dixit, B.L., Gupta, C.M., Sanglard, D., and Prasad, R. (1999). Asymmetric distribution of phosphatidylethanolamine in *C. albicans*: possible mediation by *CDR1*, A multidrug transporter belonging to ATP binding cassette (ABC) superfamily. Yeast *15*, 111–121.

Dupont, B.F., Dromer, F., and Improvisi, L. (1996). The problem of azole resistance in *Candida*. J. Mycol. Med. *6*, 12–18.

Dupre, S. and Tsapis, R.H. (2003). Raft partioning of the yeast Uracil permease during trafficking along the endocytic pathway. Traffic *4*, 83–96.

Eckford, P.D. and Sharom, F.J. (2005). The reconstituted P-glycoprotein multidrug transporter is a flippase for glucosylceramide and other simple glycosphingolipids. Biochem. J. *389*, 517–526.

Egner, R., Bauer, B.E., and Kuchler, K. (2000). The transmembrane domain 10 of the yeast Pdr5p ABC antifungal efflux pump determines the substrate specificity and inhibitor susceptibility. Mol. Microbiol. *35*, 1255–1263.

Egner, R., Rosenthal, F.E., Kralli, A., Sanglard, D., and Kuchler, K. (1998). Genetic separation of FK506 susceptibility and drug transport in the yeast Pdr5 ATP-binding cassette multidrug resistance transporter. Mol. Biol. Cell 9, 523–543.

Ernst, R., Kueppers, P., Klein, C.M., Schwarzmueller, T., Kuchler, K., and Schmitt, L. (2008). A mutation of the H-loop selectively affects rhodamine transport by the yeast multidrug ABC transporter Pdr5. Proc. Natl. Acad. Sci. U. S. A.*105*, 5069–5074.

Fessler, M.B., Arndt, P.G., Frasch, S.C., Lieber, J.G., Johnson, C.A., Murphy, R.C., Nick, J.A., Bratton, D.L., Malcolm, K.C., and Worthen, G.S. (2004). Lipid rafts regulate lipopolysaccharide-induced activation of Cdc42 and inflammatory functions of the human neutrophil. J. Biol. Chem. *279*, 39989–39998.

Fling, M.E., Kopf, J., Tamarkin, A., Gorman, J.A., Smith, H.A., and Koltin, Y. (1991). Analysis of a *Candida albicans* gene that encodes a novel mechanism for resistance to benomyl and methotrexate. Mol. Gen. Genet. *227*, 318–329.

Franz, R., Michel, S., and Morschhauser, J. (1998). A fourth gene from the *Candida albicans CDR* family of ABC transporters. Gene *220*, 91–98.

Galluhn, D. and Langer, T. (2004). Reversible assembly of the ATP-binding cassette transporter Mdl1 with the F1F0-ATP synthase in mitochondria. J. Biol. Chem. *279*, 38338–38345.

Gaur, M., Choudhury, D., and Prasad, R. (2005). Complete inventory of ABC proteins in human pathogenic yeast, *Candida albicans*. J. Mol. Microbiol. Biotechnol. 9, 3–15.

Gauthier, C., Weber, S., Alarco, A.-M., Alqawi, O., Daoud, R., Georges, E., and Raymond, M. (2003). Functional Similarities and Differences between *Candida albicans* Cdr1p and Cdr2p Transporters. Antimicrob. Agents Chemother. *47*, 1543–1554.

Goldman, B.S. and Kranz, R.G. (2001). ABC transporters associated with cytochrome *c* biogenesis. Res. Microbiol. *152*, 323–329.

Gottesman, M.M. and Ambudkar, S.V. (2001). Overview: ABC transporters and human disease. J. Bioenerget. Biomemb. *33*, 453–458.

Gottesman, M.M., Hrycyna, C.A., Schoenlein, P.V., Germann, U.A., and Pastan, I. (1995). Genetic analysis of the multidrug transporter. Ann. Rev. Genet. *29*, 607–649.

Gottesman, M.M. and Pastan, I. (1993). Biochemistry of multidrug resistance mediated by the multidrug transporter. Annu. Rev. Biochem. *62*, 385–427.

Gottesman, M.M., Pastan, I., and Ambudkar, S.V. (1996). P-glycoprotein and multidrug resistance. Curr. Opinion Genet. Dev. *6*, 610–617.

Hallstrom, T.C., Lambert, L., Schorling, S., Balzi, E., Goffeau, A., and Moye-Rowley, W.S. (2001). Coordinate control of sphingolipid biosynthesis and multidrug resistance in *Saccharomyces cerevisiae*. J. Biol. Chem. *276*, 23674–23680.

Hanson, L., May, L., Tuma, P., Keeven, J., Mehl, P., Ferenz, M., Ambudkar, S.V., and Golin, J. (2005). The role of hydrogen bond acceptor groups in the interaction of substrates with Pdr5p, a major yeast drug transporter. Biochemistry *44*, 9703–9713.

Hazelwood, L.A., Tai, S.L., Boer, V.M., De Winde, J.H., Pronk, J.T., and Daran, J.M. (2006). A new physiological role for Pdr12p in *Saccharomyces cerevisiae*: export of aromatic and branched-chain organic acids produced in amino acid catabolism. FEMS Yeast Res. *6*, 937–945.

Higgins, C.F. (1992). ABC Transporters: From Microorganisms to Man. Ann. Rev. Cell Biol. *8*, 67–113.

Higgins, C.F. (2001). ABC transporters: physiology, structure and mechanism-an overview. Res. Microbiol. *152*, 205–210.

Higgins, C.F. and Linton, K.J. (2004). The ATP switch model for ABC transporters. Nat. Struct. Mol. Biol. *11*, 918–926.

Holthuis, J.C. and Levine, T.P. (2005). Lipid traffic: Floppy drives and a Superhighway. Nature *6*, 209–220.

Holyoak, C.D., Bracey, D., Piper, P.W., Kuchler, K., and Coote, P.J. (1999). The *Saccharomyces cerevisiae* weak-acid-inducible ABC transporter Pdr12 transports fluorescein and preservative anions from the cytosol by an energy-dependent mechanism. J. Bacteriol. *181*, 4644–4652.

Huang, Z., Chang, X., Riordan, J.R., and Huang, Y. (2004). Fluorescent modified phosphatidylcholine floppase activity of reconstituted multidrug resistance-associated protein MRP1. Biochim. Biophys. Acta *1660*, 155–163.

Hull, C.M. and Johnson, A.D. (1999). Identification of a mating type-like locus in the asexual pathogenic yeast *Candida albicans*. Science *285*, 1271–1275.

Hull, C.M., Raisner, R.M., and Johnson, A.D. (2000). Evidence for mating of the 'asexual' yeast *Candida albicans* in a mammalian host. Science *289*, 307–310.

Ikeda, M., Kihara, A., and Igarashi, Y. (2006). Lipid asymmetry of the eukaryotic plasma membrane: functions and related enzymes. Biol. Pharm. Bull. *29*, 1542–1546.

Iwaki, T., Fujita, Y., Tanaka, N., Giga-Hama, Y., and Takegawa, K. (2005). Mitochondrial ABC transporter Atm1p is required for protection against oxidative stress and vacuolar functions in *Schizosaccharomyces pombe*. Biosci. Biotechnol. Biochem. *69*, 2109–2116.

Iwaki, T., Giga-Hama, Y., and Takegawa, K. (2006). A survey of all 11 ABC transporters in fission yeast: two novel ABC transporters are required for red pigment accumulation in a *Schizosaccharomyces pombe* adenine biosynthetic mutant. *152*, 2309–2321.

Jha, S., Dabas, N., Karnani, N., Saini, P., and Prasad, R. (2004). ABC multidrug transporter Cdr1p of *Candida albicans* has divergent nucleotide-binding domains which display functional asymmetry. FEMS Yeast Res. *5*, 63–72.

Jha, S., Karnani, N., Dhar, S.K., Mukhopadhayay, K., Shukla, S., Saini, P., Mukhopadhayay, G., and Prasad, R. (2003a). Purification and characterization of the N-terminal nucleotide binding domain of an ABC drug transporter of *Candida albicans*: uncommon cysteine 193 of Walker A is critical for ATP hydrolysis. Biochemistry 42, 10822–10832.

Jha, S., Karnani, N., Lynn, A.M., and Prasad, R. (2003b). Covalent modification of cysteine 193 impairs ATPase function of nucleotide-binding domain of *Candida* drug efflux pump. Biochem. Biophys. Res. Commun. *310*, 869–875.

Jones, T., Federspiel, N.A., Chibana, H., Dungan, J., Kalman, S., Magee, B.B., Newport, G., Thorstenson, Y.R., Agabian, N., Magee, P.T., Davis, R.W., and Scherer, S. (2004). The diploid genome sequence of *Candida albicans*. Proc. Natl. Acad. Sci. U. S. A.*101*, 7329–7334.

Jungwirth, H. and Kuchler, K. (2006). Yeast ABC transporters-- a tale of sex, stress, drugs and aging. FEBS Lett. *580*, 1131–1138.

Karababa, M., Coste, A.T., Rognon, B., Bille, J., and Sanglard, D. (2004). Comparison of gene expression profiles of *Candida albicans* azole-resistant clinical isolates and laboratory strains exposed to drugs inducing multidrug transporters. Antimicrob. Agents Chemother. *48*, 3064–3079.

Karnani, N., Gaur, N.A., Jha, S., Puri, N., Krishnamurthy, S., Goswami, S.K., Mukhopadhyay, G., and Prasad, R. (2004). SRE1 and SRE2 are two specific steroid responsive modules of *Candida* drug resistance gene 1 (*CDR1*) promoter. Yeast, *21*, 219–39.

Katiyar, S.K. and Edlind, T.D. (2001). Identification and expression of multidrug resistance-related ABC transporter gene in *Candida krusei*. Med Mycol *39*, 109–116.

Katzmann, D.J., Hallstrom, T.C., Voet, M., Wysock, W., Golin, J., Volckert, G., and Moye-Rowley, W.S. (1995). Expression of an ATP-binding cassette transporter-encoding gene (*YOR1*) is required for oligomycin resistance in *Saccharomyces cerevisiae*. Mol. Cell. Biol. *15*, 6875–6883.

Kihara, A. and Igarashi, Y. (2002). Identification and characterization of a *Saccharomyces cerevisiae* gene, *RSB1*, involved in sphingoid long-chain base release. J. Biol. Chem. *277*, 30048–30054.

Kispal, G., Csere, P., Guiard, B., and Lill, R. (1997). The ABC transporter Atm1p is required for mitochondrial iron homeostasis. FEBS Lett. *418*, 346–350.

Klein, I., Sarkadi, B., and Varadi, A. (1999). An inventory of the human ABC proteins. Biochim. Biophys. Acta *1461*, 237–262.

Klein, M., Mamnun, Y.M., Eggmann, T., Schuller, C., Wolfger, H., Martinoia, E., and Kuchler, K. (2002). The ATP-binding cassette (ABC) transporter Bpt1p mediates vacuolar sequestration of glutathione conjugates in yeast. FEBS Lett. *520*, 63–67.

Kolaczkowski, M., Kolaczkowska, A., Gaigg, B., Schneiter, R., and Moye-Rowley, W.S. (2004). Differential regulation of ceramide synthase components *LAC1* and *LAG1* in *Saccharomyces cerevisiae*. Eukaryot. Cell *3*, 880–892.

Kolaczkowski, M., van der Rest, M.E., Cybularz-Kolaczkowska, A., Soumillion, J.-P., Konings, W.N., and Goffeau, A. (1996). Anticancer drugs, ionophoric peptides, and steroids as substrates of the yeast multidrug transporter Pdr5p. J. Biol. Chem. *271*, 31543–31548.

Kontoyiannis, D.P. (2000). Efflux mediated resistance to fluconazole could be modulated by sterol homeostasis in *Saccharomyces cerevisiae*. J. Antimicrob. Chemother. *46*, 199–203.

Kren, A., Mamnun, Y.M., Bauer, B.E., Schuller, C., Wolfger, H., Hatzixanthis, K., Mollapour, M., Gregori, C., Piper, P., and Kuchler, K. (2003). War1p, a novel transcription factor controlling weak acid stress response in yeast. Mol. Cell Biol. *23*, 1775–1785.

Krishnamurthy, S., Chatterjee, U., Gupta, V., Prasad, R., Das, P., Snehlata, P., Hasnain, S.E., and Prasad, R. (1998a). Deletion of transmembrane domain 12 of *CDR1*, a multidrug transporter from *Candida albicans*, leads to altered drug specificity: expression of a yeast multidrug transporter in Baculovirus expression system. Yeast *14*, 535–550.

Krishnamurthy, S., Gupta, V., Prasad, R., Panwar, S.L., and Prasad, R. (1998b). Expression of *CDR1*, a multidrug resistance gene of *Candida albicans*: *In vitro* transcriptional activation by heat shock, drugs and human steroid hormones. FEMS Microbiol. Lett. *160*, 191–197.

Krishnamurthy, S., Gupta, V., Snehlata, P., and Prasad, R. (1998c). Characterisation of human steroid hormone transport mediated by Cdr1p, multidrug transporter of *Candida albicans*, belonging to the ATP binding cassette superfamily. FEMS Microbiol. Lett. *158*, 69–74.

Leighton, J. and Schatz, G. (1995). An ABC transporter in the mitochondrial inner membrane is required for normal growth of yeast. EMBO J. *14*, 188–195.

Li, Y. and Prinz, W.A. (2004). ATP-binding cassette (ABC) transporters mediate nonvesicular, raft-modulated sterol movement from the plasma membrane to the endoplasmic reticulum. J. Biol. Chem. *279*, 45226–45234.

Li, Z.S., Szczypka, M., Lu, Y.-P., Thiele, D.J., and Rea, P.A. (1996). The yeast cadmium factor protein (YCF1) is a vacuolar glutathione S-conjugate pump. J. Biol. Chem. *271*, 6509–6517.

Liu, H.Y., Chiang, Y.C., Pan, J., Chen, J., Salvadore, C., Audino, D.C., Badarinarayana, V., Palaniswamy, V., Anderson, B., and Denis, C.L. (2001). Characterization of CAF4 and CAF16 reveals a functional connection between the CCR4–NOT complex and a subset of SRB proteins of the RNA polymerase II holoenzyme. J. Biol. Chem. *276*, 7541–7548.

Madani, N.D., Malloy, P.J., Rodrigues-Pombo, P., Krishnan, A.V., and Feldman, D. (1994). *Candida albicans* estrogen-binding protein gene encodes an oxidoreductase that is inhibited by estradiol. Proc. Natl. Acad. Sci. U.S.A. *91*, 922–926.

Magee, B.B., Legrand, M., Alarco, A.-M., Raymond, M., and Magee, P.T. (2002). Many of the genes required for mating in *Saccharomyces cerevisiae* are also required for mating in *Candida albicans*. Mol. Microbiol. *46*, 1345–1351.

Mahe, Y., Lemoine, Y., and Kuchler, K. (1996). The ATP binding cassette transporters Pdr5 and Snq2 of *Saccharomyces cerevisiae* can mediate transport of steroids *in vivo*. J. Biol. Chem. *271*, 26167–26172.

Malloy, P.J., Zhao, X., Madani, N.D., and Feldman, D. (1993). Cloning and expression of the gene from *Candida albicans* that encodes a high-affinity corticosteroid-binding protein. Proc. Natl. Acad. Sci. U.S.A. *90*, 1902–1906.

Margolles, A., Putman, M., van Veen, H.W., and Konings, W.N. (1999). The purified and functionally reconstituted multidrug transporter Lmr A of *Lactococcus lactis* mediates the transbilayer movement of specific fluorescent phospholipids. Biochemistry *38*, 16298–16306.

Marichal, P., Koymans, L., Willemsens, S., Bellens, D., Verhasselt, P., Luyten, W., Borgers, M., Ramaekers, F.C., Odds, F.C., and Bossche, H.V. (1999). Contribution of mutations in the cytochrome P450 14alpha-demethylase (Erg11p, Cyp51p) to azole resistance in Candida albicans. Microbiology *145*, 2701–2713.

Martin, S.W. and Konopka, J.B. (2004). Lipid rafts Polarization contributes to hyphal growth in *Candida albicans*. Eukaryot. Cell 3, 675–684.

Maurice, T.C., Mazzucco, C.E., Ramanathan, C.S., Ryan, B.M., Warr, G.A., and Puziss, J.W. (1998). A highly conserved intraspecies homolog of the *Saccharomyces cerevisiae* elongation factor-3 encoded by the *HEF3* gene. Yeast *14*, 1105–1113.

McGrath, J.P. and Vershavsky, A. (1989). The yeast *STE6* gene encodes a homologue of the mammalian multidrug resistance P-glycoprotein. Nature *340*, 400–404.

Mendez, C. and Salas, J.A. (2001). The role of ABC transporters in antibiotic-producing organisms: drug secretion and resistance mechanisms. Res. Microbiol. *152*, 341–350.

Miyahara, K., Mizunuma, M., Hirata, D., Tsuchiya, E., and Miyakawa, T. (1996). The involvement of of the *Saccharomyces cerevisiae* multidrug transporters Pdr5p and Snq2p in cation resistance. FEBS Lett. *399*, 317–320.

Miyazaki, H., Miyazaki, Y., Geber, A., Parkinson, T., Hitchcock, C.A., Falconer, D.J., Ward, D.J., Marsden, K., and Bennett, J.E. (1998). Fluconazole resistance associated with drug efflux and increased transcription of a drug transporter gene, *PDH1*, in *Candida glabrata*. Antimicrob. Agents Chemother. *42*, 1695–1701.

Moran, G.P., Sanglard, D., Donnelly, S.M., Shanley, D.B., Sullivan, D.J., and Coleman, D.C. (1998). Identification and expression of multidrug transporters responsible for fluconazole resistance in *Candida dubliniensis*. Antimicrob. Agents Chemother. *42*, 1819–1830.

Morschhauser, J. (2002). The genetic basis of fluconazole resistance development in *Candida albicans*. Biochem. Biophys. Acta. *1587*, 240–248.

Moye-Rowley, W.S. (2003). Transcriptional control of multidrug resistance in the yeast *Saccharomyces*. Prog. Nucleic Acid Res. Mol. Biol. 73, 251–279.

Mukhopadhyay, K., Kohli, A., and Prasad, R. (2002). Drug susceptibilities of yeast cells are affected by membrane lipid composition. Antimicrob. Agents Chemother. *46*, 3695–3705.

Mukhopadhyay, K., Prasad, T., Saini, P., Pucadyil, T.J., Chattopadhyay, A., and Prasad, R. (2004). Membrane Sphingolipid-Ergosterol Interactions Are Important Determinants of Multidrug Resistance in *Candida albicans*. Antimicrob. Agents Chemother. *48*, 1778–1787.

Myers, K.K., Fonzi, W.A., and Sypherd, P.S. (1992). Isolation and sequence analysis of the gene for translation elongation factor 3 from Candida albicans. Nucleic Acids Res. *20*, 1705–1710.

Nagao, K., Taguchi, Y., Arioka, M., Kadokura, H., Takatsuki, A., Yoda, K., and Yamasaki, M. (1995). *bfr1*$^{+}$, a novel gene of *Schizosaccharomyces pombe* which confers brefeldin A resistance, is structurally related to the ATP-binding cassette superfamily. J. Bacteriol. *177*, 1536–1543.

Nakamura, K., Niimi, M., Niimi, K., Holmes, A.R., Yates, J.E., Decottignies, A., Monk, B.C., Goffeau, A., and Cannon, R.D. (2002). Functional expression of *Candida albicans* drug efflux pump Cdr1p in a *Saccharomyces cerevisiae* strain deficient in membrane transporters. Antimicrob. Agents Chemother. *45*, 3366–3374.

Nishi, K., Yoshida, M., Horinouchi, S., and Beppu, T. (1993). Mating of the fission yeast independently of *pmd1+* gene product, a structural homologue of budding yeast *STE6* and mammalian P-glycoproteins. Arch. Microbiol. *160*, 162–165.

Odds, F.C., Brown, A.J., and Gow, N.A. (2003). Antifungal agents: mechanisms of action. Trends Microbiol. *11*, 272–279.

Ogawa, A., Hashida-Okado, T., Endo, M., Tsuruo, T., Takesako, K., and Kato, I. (1998). Role of ABC transporters in aureobasidin A resistance. Antimicrob. Agents Chemother. *42*, 755–761.

Ortiz, D., Kreppel, L., Speiser, D.M., Scheel, G., McDonald, G., and Ow, D.W. (1992). Heavy metal tolerance in the fission yeast requires an ATP-binding cassette-type vacuolar membrane transporter. EMBO J. *11*, 3491–3499.

Ortiz, D.F., Pierre, M.V., Abdulmessih, A., and Arias, I.M. (1997). A yeast ATP-binding cassette-type protein mediating ATP-dependent bile acid transport. J. Biol. Chem. *272*, 15358–15365.

Pagant, S., Brovman, E.Y., Halliday, J.J., and Miller, E.A. (2008). Mapping of inter-domain interfaces required for the functional architecture of Yor1p, a eukaryotic ABC transporter. J. Biol. Chem. In press.

Pagant, S., Kung, L., Dorrington, M., Lee, M.C., and Miller, E.A. (2007). Inhibiting endoplasmic reticulum (ER)-associated degradation of misfolded Yor1p does not permit ER export despite the presence of a diacidic sorting signal. Mol. Biol. Cell *18*, 3398–3413

Pasrija, R., Panwar, S.L., and Prasad, R. (2008). Multidrug transporters CaCdr1p and CaMdr1p of *Candida albicans* display different lipid specificities: both ergosterol and sphingolipids are essential for targeting of CaCdr1p to membrane rafts. Antimicrob. Agents Chemother. *52*, 694–704.

Petrovic, S., Pascolo, L., Gallo, R., Ostrow, J.D., Goffeau, A., Tiribelli, C., and Bruschi, C.V. (2000). The products of *YCF1* and *YLL015w (BPT1)* cooperate for the ATP-dependent vaculoar transport of unconjugated bilirubin in *Saccharomyces cerevisiae*. Yeast *16*, 561–571.

Pike, L.J., Han, X., and Gross, R.W. (2005). Epidermal Growth Factor receptors are localized to lipid rafts that contain a balance of inner and outer leaflet lipids. J. Biol. Chem. *280*, 26796–26804.

Piper, P., Mahe, Y., Thompson, S., Pandjaitan, R., Holyoak, C., Egner, R., Muhlbauer, M., Coote, P., and Kuchler, K. (1998). The Pdr12 ABC transporter is required for the development of weak organic acid resistance in yeast. EMBO J. *17*, 4257–4265.

Pleban, K., Kopp, S., Csaszar, E., Peer, M., Hrebicek, T., Rizzi, A., Ecker, G.F., and Chiba, P. (2005). P-glycoprotein substrate binding domains are located at the transmembrane domain/transmembrane domain interfaces: a combined photoaffinity labeling-protein homology modeling approach. Mol. Pharmacol. *67*, 365–374.

Pomorski, T. and Menon, A.K. (2006). Lipid flippases and their biological functions. Cell Mol. Life Sci. *63*, 2908–2921.

Prasad, R. and Kapoor, K. (2005). Multidrug resistance in yeast *Candida*. Int. Rev. Cytol. *242*, 215–248.

Prasad, R., Snehlata, P., and Smriti (2002). Drug resistance in yeasts- an emerging scenario. In Advances in Microbial Physiology, R.K.Poole, ed. (London: Academic Press), pp. 155–201.

Prasad, R., Worgifosse, P.D., Goffeau, A., and Balzi, E. (1995). Molecular cloning and characterisation of a novel gene of *C.albicans*, *CDR1*, conferring multiple resistance to drugs and antifungals. Curr. Genet. *27*, 320–329.

Purnelle, B., Skala, J., and Goffeau, A. (1991). The product of the YCR105 gene located on the chromosome III from *Saccharomyces cerevisiae* presents homologies to ATP-dependent permeases. Yeast 7, 867–872.

Rai, V., Gaur, M., Kumar, A., Shukla, S., Komath, S.S., and Prasad, R. (2008). A novel catalytic mechanism for ATP hydrolysis employed by the N-terminal nucleotide-binding domain of Cdr1p, a multidrug ABC transporter of *Candida albicans*. Biochim. Biophys. Acta. In press.

Rai, V., Gaur, M., Shukla, S., Shukla, S., Ambudkar, S.V., Komath, S.S., and Prasad, R. (2006). Conserved Asp327 of walker B motif in the N-terminal nucleotide binding domain (NBD-1) of Cdr1p of *Candida albicans* has acquired a new role in ATP hydrolysis. Biochemistry *45*, 14726–14739.

Rai, V., Shukla, S., Jha, S., Komath, S.S., and Prasad, R. (2005). Functional characterization of N-terminal nucleotide binding domain (NBD-1) of a major ABC drug transporter Cdr1p of *Candida albicans*: uncommon but conserved Trp326 of Walker B is important for ATP binding. Biochemistry *44*, 6650–6661.

Raymond, M., Dignard, D., Alarco, A.M., Mainville, N., Magee, B.B., and Thomas, D.Y. (1998). A Ste6p/P-glycoprotein homologue from the asexual yeast *Candida albicans* transports the a-factor mating pheromone in *Saccharomyces cerevisiae*. Mol. Microbiol. *27*, 587–598.

Raymond, M., Gros, P., Whiteway, M., and Thomas, D.Y. (1992). Functional complementation of yeast *ste6* by a mammalian multidrug resistance *mdr* gene. Science *256*, 232–234.

Rebbeor, J.F., Connolly, G.C., Dumont, M.E., and Ballatori, N. (1998). ATP-dependent transport of reduced glutathione on YCF1, the yeast orthologue of mammalian multidrug resistance associated proteins. J. Biol. Chem. *273*, 33449–33454.

Rogers, B., Decottignies, A., Kolaczkowski, M., Carvajal, E., Balzi, E., and Goffeau, A. (2001). The pleiotropic drug ABC transporters from *Saccharomyces cerevisiae*. J. Mol. Microbiol. Biotechnol 3, 207–214.

Romsicki, Y. and Sharom, F.J. (2001). Phospholipid flippase activity of the reconstituted P-glycoprotein multidrug transporter. Biochemistry *40*, 6937–6947.

Ruetz, S. and Gros, P. (1994). A mechanism of P-glycoprotein action in multidrug resistance: are we there yet? Trends Pharmacol. Sci. *15*, 260–263.

Rutherford, J.C., Ojeda, L., Balk, J., Muhlenhoff, U., Lill, R., and Winge, D.R. (2005). Activation of the iron regulon by the yeast Aft1/Aft2 transcription factors depends on mitochondrial but not cytosolic iron-sulfur protein biogenesis. J. Biol. Chem. *280*, 10135–10140.

Saidane, S., Weber, S., De, D., X, St-Germain, G., and Raymond, M. (2006). *PDR16*-mediated azole resistance in *Candida albicans*. Mol. Microbiol. *60*, 1546–1562.

Saier, M.H.Jr. (1994). Computer-aided analyses of transport protein sequences: gleaning evidence concerning function, structure, biogenesis, and evolution. Microbiol. Rev. *58*, 71–93.

Saini, P., Gaur, N.A., and Prasad, R. (2006). Chimeras of the ABC drug transporter Cdr1p reveal functional indispensability of transmembrane domains and nucleotide-binding domains, but transmembrane segment 12 is replaceable with the corresponding homologous region of the non-drug transporter Cdr3p. Microbiology *152*, 1559–1573.

Saini, P., Prasad, T., Gaur, N.A., Shukla, S., Jha, S., Komath, S.S., Khan, L.A., Haq, Q.M., and Prasad, R. (2005). Alanine scanning of transmembrane helix 11 of Cdr1p ABC antifungal efflux pump of *Candida albicans*: identification of amino acid residues critical for drug efflux. J. Antimicrob. Chemother. *56*, 77–86.

Sandbaken, M.G., Lupisella, J.A., DiDomenico, B., and Chakraburtty, K. (1990). Protein synthesis in yeast. Structural and functional analysis of the gene encoding elongation factor 3. J. Biol. Chem. *265*, 15838–15844.

Sanglard, D., Ischer, F., and Bille, J. (2001). Role of ATP-binding-cassette transporter genes in high-frequency acquisition of resistance to azole antifungals in *Candida glabrata*. Antimicrob. Agents Chemother. *45*, 1174–1183.

Sanglard, D., Ischer, F., Calabrese, D., Majcherczyk, P.A., and Bille, J. (1999). The ATP binding cassette transporter gene *CgCDR1* from *Candida glabrata* is involved in the resistance of clinical isolates to azole antifungal agents. Antimicrob. Agents Chemother. *43*, 2753–2765.

Sanglard, D., Ischer, F., Monod, M., and Bille, J. (1997). Cloning of *Candida albicans* genes conferring resistance to azole antifungal agents: Characterisation of *CDR2*, a new multidrug ABC transporter gene. Microbiology *143*, 405–416.

Sanglard, D., Kuchler, K., Ischer, F., Pagani, J.-L., Monod, M., and Bille, J. (1995). Mechanisms of resistance to azole antifungal agents in *Candida albicans* isolates from AIDS patients involve specific multidrug transporters. Antimicrob. Agents Chemother. *39*, 2378–2386.

Sarthy, A.V., McGonigal, T., Capobianco, J.O., Schmidt, M., Green, S.R., Moehle, C.M., and Goldman, R.C. (1998). Identification and kinetic analysis of a functional homolog of elongation factor 3, *YEF3* in *Saccharomyces cerevisiae*. Yeast *14*, 239–253.

Saurin, W., Hofnung, M., and Dassa, E. (1999). Getting in or out: early segregation between importers and exporters in the evolution of ATP-binding cassette (ABC) transporters. J. Mol. Evol. *48*, 22–41.

Schneider, E. (2001). ABC transporters catalyzing carbohydrate uptake. Res. Microbiol. *152*, 303–310.

Schneider, E. and Hunke, S. (1998). ATP-binding-cassette (ABC) transport systems: Functional and structural aspects of the ATP-hydrolyzing subunit/domain. FEMS Microbiol. Rev. *22*, 1–20.

Schuller, C., Mamnun, Y.M., Mollapour, M., Krapf, G., Schuster, M., Bauer, B.E., Piper, P.W., and Kuchler, K. (2004). Global phenotypic analysis and transcriptional profiling defines the weak acid stress response regulon in *Saccharomyces cerevisiae*. Mol. Biol. Cell *15*, 706–720.

Servos, J., Haase, E., and Brendel, M. (1993). Gene *SNQ2* of *Saccharomyces cerevisiae*, which confers resistance to 4-nitroquinoline-N-oxide and other chemicals, encodes a 169 kDa protein homologous to ATP-dependent permeases. Mol. Gen. Genet. *236*, 214–218.

Shani, N. and Valle, D. (1996). A *Saccharomyces cerevisiae* homolog of the human adrenoleukodystrophy transporter is a heterodimer of two half ATP-binding cassette transporters. Proc. Natl. Acad. Sci. U. S. A.*93*, 11901–11906.

Shani, N., Watkins, P.A., and Valle, D. (1995). *PXA1*, a possible *Saccharomyces cerevisiae* ortholog of the human adrenoleukodystrophy gene. Proc. Natl. Acad. Sci. U. S. A.*92*, 6012–6016.

Sharma, K.G., Mason, D.L., Liu, G., Rea, P.A., Bachhawat, A.K., and Michaelis, S. (2002). Localization, regulation, and substrate transport properties of Bpt1p, a *Saccharomyces cerevisiae* MRP-type ABC transporter. Eukaryotic Cell *1*, 391–400.

Shukla, S., Ambudkar, S.V., and Prasad, R. (2004). Substitution of threonine-1351 in the multidrug transporter Cdr1p of *Candida albicans* results in hypersusceptibility to antifungal agents and threonine-1351 is essential for synergic effects of calcineurin inhibitor FK520. J. Antimicrob. Chemother. *54*, 38–45.

Shukla, S., Rai, V., Saini, P., Banerjee, D., Menon, A.K., and Prasad, R. (2007). Candida Drug Resistance Protein 1, a Major Multidrug ATP Binding Cassette Transporter of *Candida albicans*, Translocates Fluorescent Phospholipids in a Reconstituted System. Biochemistry *46*, 12081–12090.

Shukla, S., Saini, P., Smriti, Jha, S., Ambudkar, S.V., and Prasad, R. (2003). Functional characterization of *Candida albicans* ABC transporter Cdr1p. Eukaryot. Cell 2, 1361–1375.

Simon, S., Roy, D., and Schindler, M. (1994). Intracellular pH and the control of multidrug resistance. Proc. Natl. Acad. Sci. U.S.A. *91*, 1128–1132.

Skovsgaard, T., Nielsen, D., Maare, C., and Wasserman, K. (1994). Cellular resistance to cancer chemotherapy. Int. Rev. Cytol. *156*, 77–157.

Smriti, Krishnamurthy, S., and Prasad, R. (1999). Membrane fluidity affects functions of Cdr1p, a multidrug ABC transporter of *Candida albicans*. FEMSMicrobiol. Lett. *173*, 475–481.

Smriti, Krishnamurthy, S., Dixit, B.L., Gupta, C.M., Milewski, S., and Prasad, R. (2002). ABC transporters Cdr1p, Cdr2p and Cdr3p of a human pathogen *Candida albicans* are general phospholipid translocators. Yeast *19*, 303–318.

Soustre, I., Letourneux, Y., and Karst, F. (1996). Characterization of the *Saccharomyces cerevisiae RTA*1 gene involved in 7-aminocholesterol resistance. Curr. Genet. *30*, 121–125.

Sturtevant, J., Cihlar, R., and Calderone, R. (1998). Disruption studies of a *Candida albicans* gene, *ELF1*: a member of the ATP-binding cassette family. Microbiology *144*, 2311–2321.

Suzuki, Y., Shimozawa, N., Imamura, A., Fukuda, S., Zhang, Z., Orii, T., and Kondo, N. (2001). Clinical, biochemical and genetic aspects and neuronal migration in peroxisome biogenesis disorders. J. Inherit. Metab Dis. *24*, 151–165.

Szczypka, M.S., Wemmie, J.A., Moye-Rowley, W.S., and Thiele, D.J. (1994). A yeast metal resistance protein similar to human cystic fibrosis transmembrane conductance regulator (CFTR) and multidrug resistance-associated protein. J. Biol. Chem. *269*, 22853–22857.

Taglicht, D. and Michaelis, S. (1998). *Saccharomyces cerevisiae* ABC proteins and their relevance to human health and disease. Meth. Enzymol. *292*, 130–162.

Theiss, S., Kretschmar, M., Nichterlein, T., Hof, H., Agabian, N., Hacker, J., and Kohler, G.A. (2002). Functional analysis of a vacuolar ABC transporter in wild-type *Candida albicans* reveals its involvement in virulence. Mol. Microbiol. *43*, 571–584.

Tommasini, A., Evers, R., Vogt, E., Mornet, C., Zaman, G.J.R., Schinkel, A.H., Borst, P., and Martinoia, E. (1996). The human multidrug resistance-associated protein functionally complements the yeast cadmium resistance factor 1. Proc. Natl. Acad. Sci. U.S.A. *93*, 6743–6748.

Van Helvoort, A., Smith, A.J., Sprong, H., Fritzche, I., Schinkel, A.H., Borst, P., and Van Meer, G. (1996). MDR1 P-glycoprotein is a lipid translocase of broad specifity, while MDR3 P-glycoprotein specifically translocates phosphatidylcholine. Cell *87*, 507–517.

Vazquez de Aldana, C.R., Marton, M.J., and Hinnebusch, A.G. (1995). GCN20, a novel ATP binding cassette protein, and GCN1 reside in a complex that mediates activation of the eIF-2 alpha kinase GCN2 in amino acid-starved cells. EMBO J. *14*, 3184–3199.

Verrier, P.J., Bird, D., Burla, B., Dassa, E., Forestier, C., Geisler, M., Klein, M., Kolukisaoglu, U., Lee, Y., Martinoia, E., Murphy, A., Rea, P.A., Samuels, L., Schulz, B., Spalding, E.J., Yazaki, K., and Theodoulou, F.L. (2008). Plant ABC proteins--a unified nomenclature and updated inventory. Trends Plant Sci. *13*, 151–159.

Wachtler, V. and Balasubramanian, M.K. (2006). Yeast lipid rafts?--an emerging view. Trends Cell Biol. *16*, 1–4.

Walker, J.E., Sarsate, M., Runswick, M., and Gay, N.J. (1982). Distantly related sequences in the a- and b-subunits of ATP synthase, myosin, kinases and other ATP- requiring enzymes and a common nucleotide binding fold. EMBO J. *1*, 945–951.

Wemmie, J.A. and Moye-Rowley, W.S. (1997). Mutational analysis of the *Saccharomyces cerevisiae* ATP-binding cassette transporter protein Ycf1p. Mol. Microbiol. *25*, 683–694.

Wilcox, L.J., Balderes, D.A., Wharton, B., Tinkelenberg, A.H., Rao, G., and Sturley, S.L. (2002). Transcriptional profiling identifies two members of the ATP-binding cassette transporter superfamily required for sterol uptake in yeast. J. Biol. Chem. *277*, 32466–32472.

Wolfger, H., Mahe, Y., Parle-McDermott, A., Delahodde, A., and Kuchler, K. (1997). The yeast ATP binding cassette (ABC) protein genes *PDR10* and *PDR15* are novel targets for the Pdr1 and Pdr3 transcriptional regulators. FEBS Lett. *418*, 269–274.

Wolfger, H., Mamnun, Y.M., and Kuchler, K. (2001). Fungal ABC proteins: pleiotropic drug resistance, stress response and cellular detoxification. Res. Microbiol. *152*, 375–389.

ABC Transporter Blockers and Reversal of Drug Resistance in Microorganisms

9

Alicia Ponte-Sucre, Maritza Padrón-Nieves and Emilia Díaz

Abstract

One of the major causes of drug resistance and chemotherapeutic failure in both cancer and anti-infective therapies is the decrease of effective drug concentrations at their target place, due to the increased action of ATP-binding cassette (ABC) transporters. ABC transporters encompass membrane proteins that couple the energy derived from ATP hydrolysis to the translocation of solutes across biological membranes. Their functions include ancient and conserved mechanisms related to physiological mechanisms in prokaryotes and eukaryotes, emphasizing the biological relevance of ABC transporters. Since ABC transporter blockers can be used in combination with current drugs to increase their effective intracellular concentration, the possible impact of ABC transporter inhibitors is of great clinical interest. Special concern exists since scarce ABC transporter blockers that may be useful to increase the efficacy of current drugs have entered clinical trials and are available to be used in therapeutic regimes. Herein we review the progress made in recent years in the identification, design, availability and applicability of compounds that may work as ABC transporter blockers, as well as the roles of these compounds not only by directly blocking ABC transporters but also in the pharmacokinetics and pharmacodynamics of therapeutic drugs used against infectious diseases. These data may be helpful in the design of strategies to circumvent drug resistance in microorganisms including parasites in clinical circumstances.

Introduction

Chemotherapy has been one of the most potent tools of contemporary medicine, enabling successful treatment of many microbial infections and cancer diseases. However, the great achievements of chemotherapy have been endangered by the development of mutational or phenotypic resistance to the drug effect (Borowski *et al.*, 2005).

Drug resistance is the capacity of disease-causing microorganisms or cancer cells to withstand exposure to drugs previously toxic to them. That is, the ability (or efficacy) of a compound to kill or to decrease the growth velocity or the virulence of a microorganism is dramatically reduced. Efficacy involves many functional steps, from signal transduction and generation of second messengers to regulation of gene expression and modulation of the biological response (Galandrin *et al.*, 2007). Comprehensive understanding of cell metabolism, differentiation checkpoints and intracellular cascades is therefore fundamental to assist in drug research, especially in the field of drug resistance. Fortunately, molecular and biochemical data regarding genome sequences of many bacteria and fungi and of many unicellular parasites are now available at http://www.sanger.ac.uk/Projects/Microbes.

Development of drug resistance is a relatively new concept; it originated in the twentieth century. Resistance to a particular compound was immediately associated with physiological changes that could be overcome if other drugs were used. Indeed, the use of different drugs addressing different targets could circumvent changes in their intracellular concentrations, due to changes in the equilibrium between influx vs. efflux; decreased efficacy due to structural changes in the drug-target; and increased levels of detoxification systems or activation of DNA repair systems (Fig. 9.1) (Cohen, 1992).

About 60 years ago a more complex scenario appeared for both cancer and infectious diseases, and the concept of multi-drug resistance (MDR) had to be incorporated into the lexicon. It referred to the loss of a drug's efficacy associated with the selection of cells, or strains of bacteria, fungi or parasites, that were simultaneously resistant to more than one type of drug (Borowski *et al.*, 2005).

MDR was immediately associated with the over-expression of a specific group of transporters (Juliano and Ling, 1976), the ATP binding cassette (ABC) transporters. Movement of molecules through ABC transporters is coupled to ATP hydrolysis (Holland and Blight, 1999; Saurin *et al.*, 1999). More than 100 members of this family have been described (Pedersen, 2005; Sheps and Ling, 2006); however, the role of many of these conserved proteins still waits to be characterized.

As described in Chapter 5, ABC transporters in bacteria may export substrates, including drugs and antibiotics, or mediate the uptake of essential nutrients (Sheps and Ling, 2006). In fact, ABC transporters from prokaryote systems mediate the uptake of small solutes like histidine, maltose, peptides or ribose (Ehrmann *et al.*, 1998; Holland and Blight, 1999). On the other hand, in fungi and parasites most ABC transporters function as exporters, mediating the translocation of substrates from the ATP-rich cytosol out of the cell through the plasma membrane or into intracellular organelles, where they collaborate to sequester the drug (Saurin *et al.*, 1999; Sheps and Ling, 2006; Teodori *et al.*, 2006). Most cases of drug resistance are due to the increased expression of ABC transporters that actively decrease the concentration of the drug in the cytosol.

It is clear then that, at least theoretically, inhibition of ABC transporters could increase the drug sensitivity of a drug-resistant cell and therefore restore the 'original' activity and efficacy of the given drug. Consequently, among other strategies the development of molecules that might inhibit the function of a given ABC transporter has been the subject of intensive research.

One of the main goals of this research has been to identify the minimal characteristics that a compound must possess in order to function as an ABC transporter blocker and hence therapeutically help to restore the original activity of a given drug. The present chapter summarizes the progress made in recent years in the identification, design, feasibility and value of these so-called ABC transporter blockers. Since the main work has been done in the field of cancer, these data will also be mentioned and incorporated

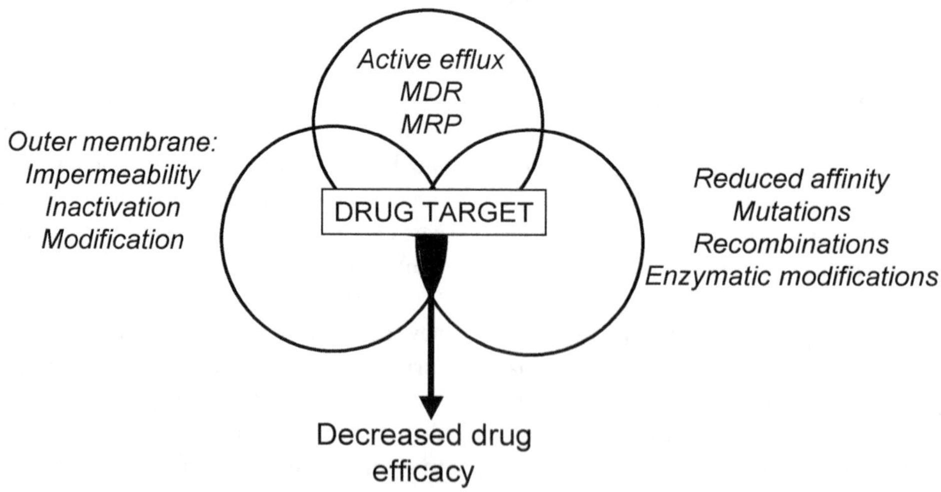

Figure 9.1 Main resistance mechanisms to drugs observed in cellular systems.

into the scenario of infectious diseases caused by microorganisms.

Drug-resistant microorganisms: physiological implications for infectivity and virulence

It has been suggested that continued therapy upon emergence of drug resistance may be justified as a way to maintain a less healthy microorganism and to reduce the load of microorganisms without drug-mediated cytostatic or cytotoxic effects. This proposal is based on the assumption that, when microorganisms acquire drug resistance, they often have a reduced ability to replicate and cause disease (Lenski, 1998; Geretti, 2005; Quinones-Mateu and Arts, 2006; Gagneux *et al.*, 2006). However, the extent of impairment in microorganism replication varies widely in resistant strains; additionally, a broad variability in the capacity of microorganisms to replicate also exists in wild-type strains due to the natural genetic polymorphism. Moreover, the 'cost' paid to achieve resistance, for example in bacteria, depends on the specific mutation that confers drug resistance and on the strain's genetic background. This 'cost' can be reversed by compensatory mutations (McAdams *et al.*, 2004; Geretti, 2005; Gagneux *et al.*, 2006) and, therefore, the dynamic value of the regulatory adaptations involved is difficult to estimate (McAdams *et al..*, 2004). Specifically, each system within an organism makes a contribution that is linked to its overall fitness strategy to survive in a particular environment. We still do not know which implications these processes could have for drug resistance, but understanding the plasticity of microorganisms, as well as the pressures that modulate their metabolic and functional states, could certainly help to establish general guidelines for clinical decisions in the case of patients suspected of harboring drug-resistant microorganisms.

Decrease in fitness (Natera *et al.*, 2007) has also been a cause of debate regarding parasites (Williams and Day, 2001; Peyron *et al.*, 2004; Hastings and Donelly, 2005). In general, it has been suggested that drug-resistant parasites have a lower reproductive efficiency and a compromised ability to compete with other isolates. However, although drug-resistant mutant parasites have been found to grow more slowly in the presence of drug pressure than its drug-sensitive progenitor, they grow faster when seeded in the absence of drug (Hastings and Watkins, 2005). This finding suggests that compensatory mutations allow the parasite to lead a successful life. Another concept that has been used to define fitness in parasites is virulence, a function that is essential for the survival and pathogenesis of the parasite (Uliana *et al.*, 1999; Beverley, 2001; Williams and Day, 2001; Hastings and Watkins 2005) and changes upon drug resistance development. In fact, continued therapy upon emergence of drug resistance may be viewed as maintenance of a less healthy (virulent) variant and as means to reduce the parasitic load in the absence of actual drug-mediated inhibition. However, the evolution of new levels of virulence in response to medical interventions such as drugs is an issue not yet solved in parasite populations.

Due to the importance that decreased fitness may clinically have, a proposal has been made to create novel organisms with decreased fitness and a minimal probability of developing into resistant microorganisms (McAdams *et al.*, 2004). These organisms should then compete with the original strains. Removal of the best-fit (resistant) strains could allow the less-fit strains to expand into the newly vacated niche space within a host but with less virulence. However, a less competent drug-resistant microorganism could eventually evolve into a competent isolate, i.e., one that is stable and that upon removal of drug pressure can be transmitted to new human recipients. These drawbacks therefore impose a word of caution and indicate that the maintenance of less efficient resistant strains is still certainly not a tool to be used in the clinic in order to circumvent drug resistance.

Types of ABC transporters

As defined in chapters 1, 2 and 3, an ABC transporter is a transport ATPase. ABC means for 'ATP binding cassette', a phrase used by Hyde (1990) and Higgins (1992) to indicate the distinguishing attribute of this transport ATPase class (Pedersen, 2005). ABC transporters contain transmembrane domains, which anchor the protein to the membrane and form a pore through which the transport of substrates occurs. Additionally they have the so called 'Walker A and B consensus motifs' (see Chapter 1) twice in the same polypeptide chain, denoting the presence of two nucleotide binding sites; however, depending on the studied organisms, these two nucleotide binding sites can also be found in different polypeptides (McKeegan *et al.*, 2004; Pedersen, 2005). The nomenclature of ABC transporters is provided and frequently updated at the following URL: http://www.gene.ucl.ac.uk/nomenclature/genefamily/abc.html.

The best studied ABC transporter is the organic cation pump P-gp (mammalian P-glycoprotein) that is the product of the *ABCB1* gene (Leonard *et al.*, 2003; Higgins, 2007) and is associated with MDR. This is a membrane protein whose hydropathy profile contains an extensive transmembrane region similar to the one present in bacterial transporters belonging to a multi-functional import-export protein family. Both types of proteins (from bacteria and humans) are pore-forming plasma membrane molecules that contain C-terminal sequences homologous to the nucleotide-binding domains of ATP-binding proteins (Gerlach *et al.*, 1986).

This homology, as well as the close similarity of the P-gp with the haemolysin B transport protein, prompted the recognition of P-gp as a eukaryotic homologue of this multifunctional family of bacterial transporters, later known as the ABC super-family (Sheps and Ling, 2006). Additionally, this data confirmed that ABC transporters, including P-gp, the multidrug resistance associated protein (MRP) (also known as ABCC1) (Cole and Deeley., 1998; McKeegan *et al.*, 2004), and the breast cancer resistance proteins (BCRP, ABCG2; mitoxantrone resistance proteins, MXR) (Doyle *et al.*, 1998; Maliepaard *et al.*, 1999) are pumps, which transport an immense variety of substrates (Gerlach *et al.*, 1986; Hyde *et al.*, 1990).

ABC transporters belong to a superfamily that involves evolutionarily conserved molecules spanning from bacteria to humans (Saurin, 1999; Dassa and Boiuge, 2001; Lage, 2003) and perform a large variety of functions. It is therefore fundamental to determine the structure of many of these proteins, with and without substrates, for a comprehensive understanding of their interaction (McKeegan *et al.*, 2004; Higgins 2007). However, the three-dimensional structures that have been resolved up to now provide only a coarse and static view of the molecule (McKeegan *et al.*, 2004), especially since high-resolution structural data are available for only few of the ABC transporters (Holland and Blight, 1999; McKeegan *et al.*, 2004; Higgins, 2007).

Due to limited writing space and since this chapter will cover only a specific area of the literature related to ABC transporters and their blockers, readers are referred to the following comprehensive reviews: Hyde *et al.* (1990); Higgins (1992); van Veen and Konings (1997); Ambdukar *et al.* (1999); Holland and Blight (1999); Dassa and Bouige (2001); Paulsen *et al.* (2001); Paulsen (2003); McKeegan *et al.* (2004); Jones and George (2005); Dean *et al.* (2006); Shilling *et al.* (2006); Higgins (2007); Ponte-Sucre (2007). Readers are also referred to Chapters 1, 2 and 3, dedicated to the structure and function of ABC transporters including details on their evolution, distribution, biology and physiology.

Strategies to circumvent drug resistance

There are several ways to circumvent drug resistance: to apply higher concentrations of the drugs subjected to resistance, to search for new antibiotics and anti-parasitic drugs that are not ABC transporter substrates, or to block the extrusion of compounds by decreasing the activity of ABC transporters. This chapter deals especially with the third approach; however, the advantages and drawbacks of each strategy will be briefly discussed herein.

Higher concentrations of compounds may partly overcome resistance, but they may increase side effects, and the toxicity of the compound might exceed its benefits. On the other hand, since the substrate specificity of ABC transporters is so broad, the probability of finding and identifying compounds that are not substrates of ABC transporters is rather low (Fojo and Bates, 2003). The search for new antibiotics and anti-parasitic drugs that are not ABC transporter substrates has been going on since the drug efflux mediated by these proteins was identified as a mechanism of drug resistance. In fact, new ketolides, macrolides, and tetracyclines as well as fluoroquines are designed against both Gram-positive and Gram-negative bacteria (Lomovskaya and Watkins, 2001, Lomovskaya *et al.*, 2001). For example, the new tetracyclines (glycylcyclines) do not generate drug resistance because they are not recognized by ABC transporters, and therefore are not transported out of the cell. But approaches to modify existing classes of drugs to enable them to avoid efflux pumps have been less successful for Gram-negative than for Gram-positive bacteria. One of the reasons for this is that drugs have to be amphiphilic in order to cross both membranes of the Gram-negative bacteria, a property that makes them good substrates for ABC transporters (Lomovoskaya and Watkins, 2001).

The most trustworthy avenue to overcoming drug resistance, then, is the development of agents that increase the pharmacodynamics of the existing drugs by inhibiting ABC transporters. The aim of this latter approach is to interfere with either the expression or the function of the transporters. These goals could be attained using genetic approaches (to decrease the ABC transporter gene expression) or through more pharmacological approaches such as the design of compounds that modulate the activity and the function of the ABC transporter proteins (Borowski *et al.*, 2005).

Although ABC transporters have a strongly conserved primary sequence, few orthologous pairs of transporters are shared between phyla (Sheps *et al.*, 2004). This characteristic, together with the broad substrate recognition of ABC transporters, suggests that redundancy in ABC transporters for the same substrate renders gene losses tolerable (Sheps *et al.*, 2004) and therefore that the regulation of ABC transporter gene expression as a tool to revert drug resistance would not necessarily be reflected in a better therapy for the patient.

On the other hand, in the field of cancer research much attention has been paid to the control of MDR expression by interfering mRNA translation with antisense oligonucleotides (Borowski *et al.*, 2005). This methodology is dealt with in Chapter 10 and will therefore not be discussed in the present chapter.

As already mentioned, pharmacological approaches to modulate the activity and the function of ABC transporters could result in a better therapy for the patient. During the remaining part of this chapter we will concentrate on the use of this strategy to circumvent drug resistance in infectious diseases. Initially we will discuss issues that are fundamental to understand how difficult it has been to use this approach and which advantages and drawbacks have resulted from the knowledge we have so far.

ABC transporters contribute to the pharmacokinetics and pharmacodynamics of drugs

ABC transporters are promiscuous

Despite the structural variations found among ABC transporters, most of these proteins recognize similar dyes, toxic ions, and clinically relevant antibiotics and anticancer drugs (van Veen, 2001). Regrettably, the broad spectrum of substrates that are accepted by ABC transporters and the poor specificity and low affinity that the substrates hold for the binding site, have impaired a successful development of effective blockers that might enter the clinical pipeline and be approved for therapeutic use (Teodori *et al.*, 2006).

In fact, questions that stimulate scientists regarding ABC transporters include: how can these proteins recognize the so many different substrates they should accommodate? Do they discriminate the substrates at separated recognition sites? Are they specific, are they promiscuous; is it necessary to inhibit all the transporters in a cell to obtain a therapeutic response? Or is it enough to inhibit a small percentage of them? (Teodori *et al.*, 2006). The information obtained up to now suggests that one large polymorphous recognition domain accommodates a great variety of substrates (Loo and Clarke, 2000; Loo and Clarke, 2001; Kaur, 2002; Pajeva and Wiese, 2002; Avendaño and Menendez, 2002, 2004; Yu *et al.*, 2003; Murray *et al.*, 2004; Pajeva *et al.*, 2004; Pleban *et al.*, 2005; Teodori *et al.*, 2005, 2006).

ABC transporters are ubiquitous

The wide distribution and multiple functions of ABC transporters complicate more than facilitate the strategies to overcome drug resistance. In fact, their ubiquity affects the absorption, distribution and clearance of many families of drugs both at the cellular and at the systemic level. Moreover, compounds that inhibit ABC transporters can affect their own distribution.

ABC transporters are extensively localized in normal tissues. Their participation in many functions may control the tissue levels of exogenous compounds, medicaments and xenobiotics (van Veen and Konings, 1997). This means that although a key issue in designing ABC transporter blockers is the inhibition of drug efflux to enhance its effect, additional (side) effects related to the physiological functions of ABC transporters should be expected (Holland and Blight, 1999).

More challenging is the fact that any foodstuff or metabolite that modulates the activity of ABC transporters affects not only the movement of substrates but also the pharmacokinetics of any drug present. This all means that the physiological role of P-gp and sister proteins requires modulators that are more potent and with proper selectivity and pharmacokinetics in order to avoid unwanted side effects (Teodori *et al.*, 2006). The resolution of the structures of strictly related transporters allows the improvement of models that would give useful information on the drug recognition site of P-gp and MRP1. Consequently, medicinal chemists are changing their random screening to rational design of molecules, tailored to interact with high affinity with the recognition site of extruding pumps and lacking the unwanted properties (Teodori *et al.*, 2006).

It is known that ABC transporters located in the gastrointestinal tract may limit the absorption of drugs and an area currently not explored regarding the availability of antibiotics and anti-parasitic drugs is the use of ABC transporter blockers to enhance the systemic levels of drugs that have been orally administered (Sun *et al.*, 2004). The use of ABC transporter blockers may therefore increase the systemic availability of drugs whose transport back to the lumen of the gastrointestinal track depends on ABC transporters. The impact of this could be an immediate change in drug pharmacokinetics (Leonard *et al.*, 2003; Higgins, 2007).

Plasma proteins mimic the binding of compounds to ABC transporters

The binding properties of ABC transporters associated with chemo-resistance are very similar to those of the plasma protein α1-acid glycoprotein (AGP). AGP is a component of plasma and is normally present at concentrations between 0.5 to 1.0 $mg\,ml^{-1}$. Together with albumin, AGP belongs to the most important proteins in plasma that determine the free drug concentrations available for the pharmacological effect. In general it is said that AGP and ABC transporters associated with drug resistance might act synergistically in protecting cells from xenobiotics. This is especially true since both proteins have a considerable overlap in the type of compounds they bind and the molecules that can induce their expression (Zsila and Iwao, 2007) (Box 9.1). This means that

Box 9.1 Inducers of AGP and P-gp

Antibiotics
- Azithromycin
- Elindamycin
- Erythromycin
- Lincomycin
- Rifampicin
- Roxithromycin
- Vancomycin

Antimalarials
- Artemisin
- Chloroquine
- Mefloquine
- Primaquine
- Quinidine
- Quinine

HIV protease inhibitors
- Amprenavir
- Stazenavir
- Indinavir
- lopinaveir
- Nelfinavir
- Ritonavir
- Sequinavir

Calcium antagonists
- Bepridil
- Diltiazem
- Gallopamil
- Felodipine
- Mibefradil
- Nicardipine
- Nifedipine
- Nimodipine
- Verapamil

Modified from Zsila and Iwao, 2007

understanding the binding characteristics of AGP could help in the prediction of drug interaction and disposition in chemo-sensitive and chemo-resistant scenarios. Moreover, an analysis of the behaviour of AGP should be considered in situations where optimization of pharmacotherapy is desired, especially since its manipulation can pragmatically lead to changes in the systemic levels of drugs.

AGP belongs to the so-called group of 'acute phase proteins' and its levels correlate with the severity of a disease, the stage of a disease, and the success of the therapy (Zsila and Iwao, 2007). In fact, at least in the case of antimalarial drugs, AGP has been identified as the source of failure in the reversal of chloroquine resistance in *Plasmodium falciparum* (Zsila *et al.*, 2008). Furthermore, the development of a promising anti-HIV protease inhibitor was discontinued because its extensive binding to AGP markedly decreased its anti-HIV properties (Zsila and Iwao, 2007). Also, it has been demonstrated that the effectiveness of verapamil-type compounds is greatly reduced because they extensively bind to AGP rather than to ABC transporters (Gbotosho *et al.*, 2006).

Biology of ABC transporter substrates and modulators

It has been recognized that substances that permeate the membrane slowly can be transported out of the cell by P-gp, whereas those that permeate rapidly act as P-gp inhibitors (Sheps and Ling, 2006). This means, that at least theoretically, P-gp inhibitors can be cycled repeatedly through the cell membrane and in this way prevent the export of the more slowly cycled accompanying drug, increasing its concentration in the cell; in this case, the modulator acts as a competitive inhibitor (Varma *et al.*, 2003). Therefore, the speed at which P-gp and ABC transporters in general move substrates across membranes is crucial, and the analysis of resistance of a cell to a given compound must thus consider the membrane and the transporter as a functionally integrated unit (Sheps and Ling, 2006). For example, this is fundamental in the case of hydrophobic compounds and their interaction with P-gp, (Etian *et al.*, 1996). Furthermore, changes in membrane dynamics might allow drugs to enter the cell more rapidly than P-gp can export them, thereby reducing resistance (Kamau *et al.*, 2005). Taken together, this means that the critical parameter determining whether a cell resists a drug is not the extracellular concentration of drug, but rather the intracellular concentration set by the equilibrium between drug inflow through the membrane and drug export by the cell's competent transporters (Sheps and Ling, 2006).

Classification of ABC transporter blockers

The importance of transporter blockers to circumvent drug resistance was originally recognized by Tsuruo (1983). His group was the first one to describe that the calcium channel inhibitor verapamil and the phenotiazine derivative trifluoperazin potentiate the activity of vincristine (Tsuruo, 1983). Since then, the discovery and/or design of transporter blockers has become a clinically relevant endeavor.

P-gp modulation occurs by diverse means for different classes of compounds. P-gp inhibitors may interact directly with the binding site of the ABC transporter through one or more of the P-gp binding sites, thereby blocking (competitively or non-competitively) the transport of substrates. Interestingly, the data suggest that P-gp has multiple, interacting drug-binding sites, but can also possess one single, large, flexible pocket (Higgins, 2007). Alternatively, an inhibitor may prevent ATP binding, ATP hydrolysis, or the coupling of ATP hydrolysis to the translocation of substrates (Ambudkar *et al.*, 1999).

Presently ABC transporter blockers are classified in three generations. First generation inhibitors are pharmaceuticals already in use for other treatments but also able to block P-gp, such as calcium channel inhibitors like verapamil, immunosupressants like cyclosporin A, anti-arrhythmics and neuroleptics like quinidine, reserpine, and yohimbine, and antiestrogens like tamoxifen and toremifene (Varma *et al.*, 2003). The clinical efficacy of these compounds is limited by their toxicity (Khrishna and Mayer, 2000; McKeegan *et al.*, 2004). For example, verapamil produces cardiac cytotoxicity at the concentration needed to inhibit drug resistance. This fundamental drawback has stimulated the development of modulators lacking the toxic side effects of first generation compounds.

Second generation P-gp modulators like R-verapamil, GF120918, MS-209, PSC-833, VX-710, or VX-853 were derived from first generation P-gp modulators with the specific purpose of decreasing drug resistance. For example, the compound PSC-833 (valspodar) was developed in an effort to find a non-immune-suppressant analogue of the natural chemosensitizer cyclosporine A (Boesch *et al.*, 1991). This compound is 10-fold more active than cyclosporine A in blocking the ABC transporters associated to drug resistance, and has already undergone clinical trials (Fracasso, 2001). Unfortunately, although second generation modulators are much less toxic than first generation modulators, they can still produce extreme side effects (Krishna and Mayer, 2001; Fracasso, 2001; McKeegan *et al.*, 2004), especially due to an inhibition of multiple cell ABC transporters and to drug–drug interactions (Balayssac *et al.*, 2005; Varma *et al.*, 2003).

Third generation modulators, like biricodar, laniquidar, zosuquidar, LY335979, OC144093, R101933 and XR9576, are highly selective inhibitors of P-gp still under development (Dantzig *et al.*, 2003). Clinical trials with third generation modulators specifically developed for MDR reversal are in progress. The results however are not encouraging and it may be that the perfect reverser does not exist (Nobili *et al.*, 2006).

Structure–activity relationship in ABC transporter blockers

An ABC transporter blocker should selectively block the ABC transporters associated with drug resistance without affecting its own pharmacokinetics or the pharmacokinetics of the targeted compound (Seelig, 1998) (Box 9.2).

In the 1990s, a minimum of structural features needed for compounds to be substrates of ABC transporters was proposed. These features included a basic nitrogen atom and two planar aromatic domains. Furthermore, at that time it was stressed that the aromatic domains should adopt a well-defined conformation, although at that same time it was demonstrated that compounds lacking these domains could also interact with ABC transporters. Until now, being hydrophobic and amphiphilic, as well as possessing a basic amine have emerged as the main structural characteristic of compounds that are substrates of ABC transporters (Seelig, 1998). Following these guidelines, numerous compounds have been either synthesized or isolated from plants, and tested as ABC transporter blockers especially against P-gp and MRP-1 (Table 9.1) (Avendaño and Menendez, 2004).

The literature involving the structure–activity relationship (SAR) of ABC transporters in general and especially the proteins involved in drug resistance, is large and complex and is extensively discussed in Chapter 3. Herein we will briefly summarize the results that are fundamental for our discussion.

Although the various approaches focus mainly on the development of ABC transporter blockers or modulators, in many cases addressing the substrates of these transporters cannot be avoided (Stouch and Gudmundsson, 2002). Furthermore, it is often difficult to differentiate between substrates, modulators and inhibitors. In

Box 9.2 Definition of ABC transporter modulators or chemosensitizers

- Pharmacological agents that reverse drug resistance by occupying the ABC transporter active or allosteric site
- Pharmacological agents that alter the physicochemical properties of biomembranes
- Pharmacological agents that
  - interfere with drug sequestration by organelles
  - decrease the expression of ABC transporters

Most chemosensitizers bind with transmembrane domains of ABC transporters; however, steroids and flavonoids are recently introduced chemosensitizers which inhibit transporters by binding with nucleotide binding domains.

Modified from Seelig, 1998

Table 9.1 Examples of ABC transporter modulators

Arylalkylamines	verapamil and analogues, lomerizine, haloperidol
Aryloxypropanolamines	propafenone and benzofenone analogues, carvedilol, zosuquidar
Anthranylamides, silicylamides	tariquidar
Nitrogen heterocycles	lamellarin I, ningalin B hexamethylether, staurosporine analogues, bromocriptine analogues, imidazol derivatives, cinchonine, sulfonamide derivatives of chloroquine, camptothecin, pyronacridine, iodoazido arylprazosin, methylen blue
Oxygen heterocycles	natural and semisynthetic pyran derivatives, chalcones, galangin, ONO-1078
Carbohydrates and glycosides	atractylsucrose I, O-2-fluoropyranosylalizarine blue derivatives
Cyclic peptides and depsipeptides	cyclosporin analogues, aureobasidins, haplosins
Macrolactones	macrolactones
Steroids	progesterone analogues, toremifene
Terpenes	paclitaxel analogues
Miscellaneous	isothiocianates

fact, as is described in Chapter 3, information is still not complete regarding the structural design of drug and modulator binding sites, the way in which the energy released from the ABC domains is transferred to the transport module, and the details of the transport mechanism (Pedersen, 2005; Sheps and Ling, 2006; Higgins, 2007). This information is essential for designing drugs to modulate the activity of ABC transporters (van der Heide and Poolman, 2002; Shilling *et al.*, 2006).

On the other hand, as already mentioned, ABC transporters translocate many compounds and it is not easy to find compounds that do not behave as substrates. Therefore, the analysis and definition of the available compounds is normally biased towards the characteristics substrates should have. Furthermore, many of the compounds are pesticides, and many are structurally related. These characteristics prevent the determination of a truly general SAR especially if methods employing structural features or fragments as descriptors are used (Stouch and Gudmundsson, 2002).

An additional issue is the variability in the methods, cell lines, assays and record systems employed to collect the data. In many cases, the data cannot be compared, making the task to develop a realistic and valid SAR for ABC transporters even more difficult. In fact, several factors, including the assays employed, the binding sites, the cell lines, the inducer and the mechanism used to stimulate the expression of the ABC transporters, the membrane partition of the compounds, and the permeation rates of them could be so varied that the data obtained cannot be definitively compared (Stouch and Gudmundsson, 2002).

On top of this, in the conditional case that all factors considered are equivalent, an intrinsic dilemma arises: as described in Chapter 3, a minimum of two binding sites have been described, further complicating the situation (Wiese and Pajeva, 2001). Additional problems are the cytotoxicity of the compounds or the alteration of membrane properties promoted by ABC transporter blockers, or re-entry of the compounds into the cell, thus producing an unspecific component in the function of ABC transporters (Stouch and Gudmundsson, 2002).

Troubles usually include the fact that binding or affinities have been used as terms of functionality instead of using substrate efflux or ATP degradation. Binding does not necessarily confer activity and therefore the use of this concept may induce an additional complexity and should be avoided. Additionally, the use of descriptors such as *log P* (logarithm of the ratio of the concentrations of the un-ionized solute in the solvents), van der Waals forces or additional molecular properties has not been consistent, and imposes additional complexities that complicate the real meaning of the SAR studies. (Stouch and Gudmundsson, 2002).

As a result of this complicated panorama many studies have abandoned the quantitative analysis of the SAR and have simply classified the compounds as substrates and inhibitors. Although with this approach the effect of weakly active molecules is lost, at least a sensitive and sensible result is obtained. There is consensus in that the following properties should be fulfilled by the pharmacophore: (i) the molecule should have a *log P* value of at least 2.92 or higher to allow a hydrophobic van der Waals interaction with the ABC transporter (which is lipophilic); (ii) the molecule should have a long chain axis higher than 18 carbon atoms, to cover more than one unit of protein to increase the strength of the binding; (iii) the molecule should promote a good nucleophillic interaction of the molecule with the ABC transporter (that is, the energy should favor a better nucleophillic attack); and (iv) in order to form a cation at physiological pH the molecule should have at least one tertiary amine, so that the ionic/hydrogen bond is strengthened (Wang *et al.*, 2002) (Box 9.3).

Of the many SAR studies found in the literature for ABC transporters (Ecker and Chiba, 1995; Seelig, 1998; Krishna and Mayer, 2001; Geney *et al.*, 2002; Stouch and Gudmundsson, 2002; Wang *et al.*, 2002, Dantzig *et al.*, 2003, McKeegan *et al.*, 2004; Tsakovska and Pajeva, 2006), the one by Seelig (1998) is especially elegant. Although already somewhat old, the description made by this author is very sophisticated, relates the compounds to their activity and classifies them as pharmacophores, i.e., substrates; non-substrates; and inducers.

In her description of the features that should be fulfilled by compounds, Seelig (1998) describes two recognition elements based on electron donor characteristics: one (named type I), is constituted by two electron donor groups separated 2.5 ± 0.3 Å; a second one (named type II), is constituted by two electron donor groups separated 4.6 ± 0.6 Å; in some cases a third electron donating group may be located between the outer other two. Seelig (1998) analysed 100 compounds from the literature and could identify the following characteristics: compounds classi-

Box 9.3 Pharmacophore model for substrate recognition by ABC transporters

Minimal requirements:

- Two hydrophobic sub-structures
- Three hydrogen-bond acceptors
- One hydrogen-bond donor

Physicochemical properties:

- *Log P** higher than 2.92
- Molecular axis of at least 18 atoms
- High energy value for nucleophilic attack
- One tertiary basic nitrogen
- Electric charge allowing the compounds to induce membrane permeation

*Octanol-water partition coefficient (P): measure of the propensity of neutral compounds to differentially dissolve in immiscible phases. It is usually referred to as the logarithmic ratio, logP, and is a quantitative descriptor of lipophilicity.

LogP is commonly used in quantitative structure activity relationship (QSAR) studies and drug design since it relates to drug absorption, bioavailability, metabolism, and toxicity .

With ACD/LogP DB software, accurate values of logP (in octanol/water, calculated under standard conditions, at 25°C) are evaluated and extensive libraries of experimental logP data are reviewed.

Source: http://www.acdlabs.com/products/phys_chem_lab/logp/

fied as inducers contained one type II recognition element and many type I recognition elements, and their capacity to form hydrogen bonding correlated with their capacity to induce the expression of ABC transporters. Unfortunately, inhibitors were hard to find and only seven could be analysed; additionally, although somewhat similar to the substrates, they did not present the recognition elements.

In the model of Seelig (1998), partitioning into the membrane is the rate-limiting step in the interaction between ABC transporters and compounds; additionally, dissociation of the compound from the transporter is related to the strength of the hydrogen bonds. However, Seelig did not take in consideration molecular weight and hydrophobicity; a compound can bind and still have structural modifications that render it inactive, originating an inhibitor or blocker.

The good correlation between the number of recognition elements and the strength of the binding suggests that ABC transporters may have more than one binding site, all of them similar or equivalent. Additionally, the structural variety of substrates means that different drugs can occupy different receptor sites in different binding modes, the most active compounds being those that occupy simultaneously the highest number of receptor sites (Avendaño and Menendez, 2004).

A further attempt to make SAR analysis for ABC transporters was done by Stouch and Gudmundsson (2002). They performed a study where descriptors such as molecular weight, shape, lipophilicity, complexity, and electronic conformation were analysed to evaluate most of the compounds used by Seelig (1998). Unfortunately their analysis could only modestly differentiate between substrates and blockers; the best descriptors were molecular shape and electrostatic charge across the surface. In an additional attempt to make a SAR analysis for ABC transporters, flavonoids with 3-OH group have been described to play a role as antioxidants, while flavonoids with 5-OH and/or 7-OH groups show a higher cytotoxicity and flavonoids with 3′-OMe and/or 5′-OMe groups are good as P-gp inhibitors. Of note, flavonoids with 6-OMe groups are not good P-gp inhibitors (Jeong *et al.*, 2007). It is believed that these SAR results can be taken into account for the development of flavonoids with high therapeutic index (Jeong *et al.*, 2007).

From the previous discussion it is clear that additional data are urgently needed in order to have a better knowledge of the SAR of ABC transporters. However, although many problems need to be solved, the information obtained to date has been helpful for understanding the architecture of the binding sites for drug and modulator and the link between energy generation and drug translocation functions, as well as for the rational design of drugs to block the activity of ABC transporters.

Examples of the real use of ABC transporter blockers

As we already mentioned, the design and isolation of specific blockers for ABC transporters is one of the most promising approaches to overcome the limited success of antimicrobial and anti-parasitic chemotherapy. For example, quinolone derivatives and other alkoxyquinoline derivatives as well as semi-synthetic derivatives of tetracyclines inhibit the efflux of antibiotics via ABC transporters and restore, at least partially, the susceptibility of bacteria to different antibiotics (Nelson and Levy, 1999; Mallea *et al.*, 2003; McKeegan *et al.*, 2004). However, the success of this approach has so far been limited. Further, the activity of classical ketolides and macrolides against Gram-positive macrolide-resistant bacteria is enhanced by the presence of an ABC transporter blocker, but this is not the case for newly designed macrolides or fluoroquinolines (Lomovskaya and Watkins, 2001). The reader is referred to Chapter 4 for a deeper discussion on this theme.

The analysis of compounds isolated from natural sources is still in its infancy and the number of groups searching for these compounds is very small. Plants should have evolved compounds to evade (or promote) mechanisms of accumulation of xenobiotics in different ecological scenarios and this information should be very useful in the design of ABC transporter blockers (Stavri *et al.*, 2007).

Natural inhibitors include flavonoids extracted from plants. They have a very low toxicity and increase the cellular accumulation of substances like the fluorescent substrate rhodamine 123 in KB-C2 cells overexpressing P-gp, due to a selective inhibition of ABC transporters (Kitagawa,

2006). These flavonoids are many times more active than verapamil, their inhibitory activity was found to correlate with their hydrophobicity as measured through their capacity to dissolve in octanol, and are much safer than previously designed chemosensitizers (Choi *et al.,* 2002). Of note, flavonoids do not bind to the transmembrane domains but rather to the nucleotide-binding domain of ABC transporters, although in a site different from that occupied by ATP (Dayan *et al.,* 1997).

Although Chapters 6 and 7 discuss which compounds are inhibitors of ABC transporters in *Leishmania* and *Plasmodium* we want to mention very briefly the pharmacological value of dihydriethanoanthracene derivatives (Pradines *et al.,* 2002) and dihydro-β-agarofuran sesquiterpenes isolated from the roots of *Mayetna magallanica* and *M. chubutensis* (Kennedy *et al.,* 2001; Pradines *et al.,* 2002) against parasitic microorganisms. Dihydriethanoanthracene derivatives restore the therapeutic efficacy of chloroquine against resistant parasites when simultaneously administered. On the other hand, dihydro-β-agarofuran sesquiterpenes revert the resistant phenotype and modulate intracellular drug accumulation in *Leishmania* parasites, probably affecting ABC transporters. These results suggest that the administration of an additional compound available at relatively low cost may be an effective strategy to enhance the activity of the common antiparasitic drugs.

It is also worth to mention that Hayeshi *et al.* (2006) described that out of 21 classical anti-parasitic drugs most of them flavonoids, 14 inhibit ABC transporters expressed in human intestinal epithelial CaCo-2 cells. Of these 21, only quinine, which also classified as the most active of all 21 compounds, is simultaneously an inhibitor and a substrate of the ABC transporters expressed in these cells; the rest of the anti-parasitic drugs act only as inhibitors. Finally, ABC transporter blockers have been used efficiently *in vivo* to increase the potency of classical drugs and reduce the size of lesions in BALB/c mice infected with drug-resistant *Leishmania* (Serrano-Martin *et al.,* 2006). Interestingly, the drug-resistant parasites used in this assay were selected and characterized using previously described protocols (Ponte-Sucre *et al.,* 1997; Silva *et al.,* 2004; Uzcátegui *et al.,* 2005; Machuca *et al.,* 2006; Padrón-Neives *et al.,* 2009). The herein described results suggest that, at least in experimental settings, the simultaneous administration of ABC transporter blockers increases the efficiency of the primarily administered drug.

Recently designed compounds include inhibitors having the 17-azapentacyclo[6,6,5,$0^{2,7}$,$0^{9,14}$,$0^{15,19}$]-nonadeca-2,4,6,9(14),10,12-esene-16,18 dione scaffold. These compounds inhibited the translocation of fluorescent rhodamine 123 through ABC transporters expressed in L5178 mouse T-cell lymphoma cells infected with the pHq MDR1/A retrovirus (Bisi *et al.,* 2006). Importantly, compounds with the dione scaffold inhibit drug translocation 5-fold more effectively than verapamil.

Finally, improving the bioavailability of orally administered drugs is a fundamental challenge that if resolved would mean the real introduction of ABC transporter modulators in the combination therapy with the concomitant result of a decreased development of drug resistance (Edafiogho *et al.,* 2007). In an attempt to reach this goal, Edafiogho *et al.* (2007) proposed to develop enaminone derivatives that retain their physicochemical properties pertinent to P-gp recognition. This proposal is based on the fact that enaminones are weak substrates of P-gp, and therefore could act as potent inhibitors of the pump. Interestingly, the lack of specificity for action on P-gp is considered to be an inherent disadvantage for most if not all ABC transporters modulators. Pharmacokinetic complications could occur from the interaction of the ABC transporter inhibitors with cytochromes, involved in drug metabolism. Fortunately, in the case of the enaminones like LY335979 or OC144–093, an increased specificity and consequently a decreased occurrence of pharmacokinetic alterations has been demonstrated.

Many P-gp inhibitors have also been tested against MRP-1 or BCRP but unfortunately, most of them had no inhibitory effect (Allen *et al.,* 2002; Boumendjel *et al.,* 2005). This was somehow predictable, as P-gp recognizes hydrophobic compounds while MRP-1 recognizes hydrophilic substrates (Boumendjel *et al.,* 2005). However, this result was difficult to understand since BCRP transports structurally and functionally diverse organic substrates, including hydrophobic compounds, weak bases, organic anions, and glucuronide-, sulfate-, glutamylate- and glutathione-conjugates (van Herwaarden and Schinkel, 2006). Additionally, P-gp inhibitors should prevent drug extrusion from cells by ABC transporters and thus increase the cytotoxic action of drugs transported by ABC transporters, but since they act on multiple ABC transporters, they could also induce an enhanced level of toxicity at different non-target sites (Sheps and Ling, 2006).

Conclusions and future trends

The increasing development of drug resistance is one of the major reasons why cancer chemotherapy fails, and is predicted to constitute one of the main impediments to successful chemotherapy against microorganisms in the near future. Although the design of new drugs not transported by ABC transporters is in progress, the isolation or design of blockers to ABC transporters constitutes one of the most promising approaches to overcoming the lack of success in the aforementioned chemotherapeutic approaches. Several ABC transporter blockers have been successfully tried against bacteria and parasites; however, ABC transporter inhibitors are not yet in standard use for the treatment of bacterial or parasitic infections or for cancer diseases. Reasons for this are numerous and include the low substrate specificity and the multiple functions subserved by ABC transporters.

The central paradigm of SAR is that compounds with similar structures act at the same site and with the same mechanism. Unfortunately, as already discussed, inhibition and/or binding at P-gp, MRP-1, or ABC transporters in general can occur via different mechanisms as well as at a number of different sites, and a successful design of ABC transporter substrates, modulators, and blockers has thus been hindered by an incomplete understanding of the mechanisms and biology of these transporters.

In this regard, various organizations are devoted to encourage the development of anti-infective drugs in general and ABC transporter modulators in particular. In this regard, two Web sites, http://www.TDRtargets.org and http://www.dndi.org/, focus on many diseases that have not been high on the pharmaceutical industry's priority list because most of them largely afflict people from developing countries.

Finally, as already mentioned by Higgins (2007), probably the time has come not to fight against drug resistance but to find alternatives to avoid and avert it. In this regard, a promising avenue is the use of ABC transporter blockers to actively prevent the emergence of drug resistance. For example, it would be interesting to explore whether the use of ABC transporter blockers prevents the selection of resistant strains with increased expression of ABC transporters like MDR or MRP. Prevention of chemo-resistance through the combination of ABC transporter blockers with the selected drug has, however, the potential drawback that resistance to the inhibitor itself could appear. Hence, the risk/benefit profile of the joint use of such combinations should be evaluated on a case-by-case basis.

Acknowledgements

The authors are grateful for the financing support received from the Coordination for Research, Faculty of Medicine, UCV; and the Council for Scientific and Humanistic Research (CDCH), Universidad Central de Venezuela. Likewise they are grateful for the support conferred by FONACIT to Maritza Padrón Nieves.

References

Allen, J.D., van Loevezijn, A., Lakhai, J.M., van der Valk, M., van Tellingen, O., Reid, G., Schellens, J.H., Koomen, G.J., and Schinkel, A.H. (2002). Potent and specific inhibition of the breast cancer resistance protein multidrug transporter *in vitro* and in mouse intestine by a novel analogue of fumitremorgin. C. Mol. Cancer Ther. *1*, 417–425.

Ambudkar, S. V., Dey, S., Hrycyna, C.A., Ramachandra, M., Pastan, I., and Gottesman, M.M. (1999). Biochemical and pharmacological aspects of the multidrug transporter. Annu. Rev. Pharmacol. Toxicol. *39*, 361–398.

Avendaño, C., and Menéndez, J.C. (2002). Inhibitors of multidrug resistance to antitumor agents (MDR). Curr. Med. Chem. *9*, 159–193.

Avendaño, C., and Menéndez, J.C. (2004). Recent advances in multidrug resistance modulators Medicinal Chemistry On line *1*, 419–444 (http://www.ingentaconnect.com/content/ben/mcro/2004/00000001/00000004/art00006).

Balayssac, D., Authier, N., Cayre, A., and Coudore, F. (2005). Does inhibition of P-glycoprotein lead to drug–drug interactions? Toxicol. Lett. *156*, 319–329.

Beverley, S.M. (2001). Genetic and genomic approaches to the analysis of *Leishmania* virulence. In: Marr J, Nilsen T, Komunecki R, eds. Molecular and Medical Parasitology. Amsterdam: Elsevier.

Bisi, A., Gobbi, S., Rampa, A., Belluti, F., Piazzi, L., Valenti, P., Gyemant, N., and Molnár, J. (2006). New potent P-glycoprotein inhibitors carrying a polycyclic scaffold. J. Med. Chem. *49*, 3049–3051

Boesch, D., Gaveriaux, C., Jachez, B., Pourtier-Manzanedo, A., Bollinger, P., and Loor, F. (1991). *In vivo* circumvention of P-glycoprotein-mediated multidrug resistance of tumor cells with SDZ PSC 833. Cancer Res. *51*, 4226–4233.

Borowski, E., Bontemps-Gracz, M.M., and Piwkowska A. (2005). Strategies for overcoming ABC transporters-mediated multidrug resistance (MDR) of tumor cells. Acta Biochim. Pol. *52*, 609–627.

Boumendjel, A., Baubichon-Cortay, H., Trompier, D., Perrotton, T., and Di Pietro, A. (2005). Anticancer multidrug resistance mediated by MRP1: Recent advances in the discovery of reversal agents. Med. Res. Rev. *25*, 453–472.

Cohen, M.L. (1992). Epidemiology of drug resistance: implications for a post-antimicrobial era. Science. *257*, 1050–1082.

Cole, S., and Deeley, R.G. (1998). Multidrug resistance mediated by the ATP-binding cassette transporter protein MRP. Bio Essays. *20*, 931–940.

Choi, C.H., Sun, K.H., An, C.S., Yoo, J.C., Hahm, K.S., Lee, I.H., Sohng, J.K., and Kim, Y.C. (2002). Reversal of P-glycoprotein-mediated multidrug resistance by 5, 6, 7, 3′, 4′-pentamethoxyflavone (Sinensetin). Biochem.Biophys.Res.Commun.295, 832–840.

Dantzig, A.H., Alwis, D.P., and Burgess, M. (2003). Considerations in the design and development of transport inhibitors as adjuncts to drug therapy. Adv. Drug Deliv. Rev. *56*, 133–150.

Dassa, E., and Bouige, P. (2001). The ABC of ABCs: A phylogenetic and functional classification of ABC systems in living organisms. Res. Microbiol. *152*, 211–229.

Dayan, G., Jault, J.M., Baubichon-Cortay, H., Baggetto, L.G., Renoir, J.M., Baulieu, E.E., Gros, P., and Di Pietro, A. (1997). Binding of steroid modulators to recombinant cytosolic domain from mouse P-glycoprotein in close proximity to the ATP site. Biochemistry 36, 15208–15215.

Dean, M., Rzhetsky, A., and Allikmets, R. (2006). The human ATP-binding cassette (ABC) transporter superfamily. Genome Res. *11*, 1156–1166.

Doyle, L.A., Yang, W., Abruzzo, L.V., Krogmann, T., Gao, Y., Rishi, A.K., and Ross, D.D. (1998). A multidrug resistance transporter from human MCF-7 breast cancer cells. Proc. Natl. Acad. Sci. U.S.A. 95, 15665–15670.

Ecker, G., and Chiba, P. (1995). Structure-activity-relationship studies on modulators of the multidrug transporter P-glycoprotein-an overview. Wien. Klin. Wochenschr. *107*, 681–686.

Edafiogho, I.O., Kombian, S.B., Ananthalakshmi, K.V., Salama, N.N., Eddington, N.D., Wilson, T.L., Alexander, M.S., Jackson, P.L., Hanson, C.D., and Scott, K.R. (2007). Enaminones: Exploring additional therapeutic activities. J. Pharm. Sci. *96*, 2509–2531.

Ehrmann, M., Ehrle, R., Hofmann, E., Boos, W., and Schlosser, A. (1998). The ABC maltose transporter. Mol. Microbiol. 29, 685–694.

Etian, G.D., Regev, R., Oren, G., and Asaraf, Y.G. (1996). The role of passive trans bilayer drug movement in multidrug resistance and its modulation. J. Biol. Chem. *271*, 12897–12902.

Fojo, T., and Bates, S. (2003). Strategies for reversing drug resistance. Oncogene *22*, 7512–7523.

Fracasso, P.M. (2001). Overcoming drug resistance in ovarian carcinoma. Curr. Oncol. Rep. *3*, 19–26.

Gagneux, S., Long, C.D., Small, P.M., Van, T., Schoolnik, G.K., and Bohannan, B.J. (2006). The competitive cost of antibiotic resistance in *Mycobacterium tuberculosis*. Science. *312*, 1944–1946.

Galandrin, S., Oligny-Longpré, G., and Bouvier, M. (2007). The evasive nature of drug efficacy: implications for drug discovery. Trends. Pharmacol. Sci. *28*, 423–430.

Gbotosho, G.O., Ogundahunsi, O.A., Happi, C.T., Kyle, D.E., Gerena, L., Milhous, W.K., Sowunmi, A., Oduola, A.M., and Salako, L.A. (2006). The effects of alpha1-acid glycoprotein on the reversal of chloroquine resistance in *Plasmodium falciparum*. Ann. Trop. Med. Parasitol. *100*, 571–578.

Geney, R., Ungureanu, M., Li, D., and Ojima, I. (2002). Overcoming multidrug resistance in taxane chemotherapy. Clin. Chem. Lab. Med. *40*, 918–925.

Geretti, A.M. (2005). The clinical significance of viral fitness. J. HIV. Ther. *10*, 6–10.

Gerlach, J., Endicott, J., Juranka, P., Henderson, G., Sarangi, F., Deuchars, K., and Ling, V. (1986). Homology between P-glycoprotein and a bacterial hemolysin transport protein suggests a model for multidrug resistance. Nature. *324*, 485–489.

Hastings, I.M., and Donnelly, M.J. (2005). The impact of antimalarial drug resistance mutations on parasite fitness, and its implications for the evolution of resistance. Drug Resist. Updat. *8*, 43–50.

Hastings, I.M., and Watkins, W.M. (2005). Intensity of malaria transmission and the evolution of drug resistance. Acta Trop. *94*, 218–29.

Hayeshi, R., Masimirembwa, C., Mukanganyama, S., and Ungell, A.L.B. (2006). The potential inhibitory effect of antiparasitic drugs and natural products on p-glycoprotein mediated efflux. Eur. J. Pharm. Sci. *29*, 70–81.

Higgins, C.F. (1992). ABC transporters: from microorganisms to man. Annu. Rev. Cell. Biol. *8*, 67–113.

Higgins, C.F. (2007). Multiple molecular mechanisms for multidrug resistance transporters. Nature. *446*, 749–757.

Holland, I.B., and Blight, M.A. (1999). ABC-ATPases, adaptable energy generators fueling transmembrane movement of a variety of molecules in organisms from bacteria to humans. J. Mol. Biol. 293, 381–399.

Hyde, S.C., Emsley, P., Harthshorn, M.J., Mimmack, M.M., Gileadi, U., Pearce, S. R., Gallager, M.P., Gill, D. R., Hubbard, R.E., and Higgins, C.F. (1990). Structural model of ATP-binding proteins associated with cystic fibrosis, multidrug resistance and bacterial transport. Nature. *346*, 362–365.

Jeong, J.M., Choi, C.H., Kang, S.K., Lee, I.H., Lee, J.Y., and Jung, H. (2007). Antioxidant and chemosensitizing effects of flavonoids with hydroxy and/or methoxy groups and structure–activity relationship. J. Pharm. Pharm. Sci. *10*, 537–546.

Jones, P.M., and George, A.M. (2005). Multidrug resistance in parasites: ABC transporters, p-glycoproteins and molecular modeling. Int. J. Parasitol. *35*, 555–566.

Juliano, R.L., and Ling, V. (1976). A surface glycoprotein modulating drug permeability in Chinese hamster ovary cell mutants. Biochim. Biophys. Acta. *455*, 152–162.

Kamau, S.W., Kramer, S.D., Matja, G., and Wunderli-Allenspach, H. (2005). Effect of the modulation of the membrane lipid composition on the localization and function of P-glycoprotein in mdr1-mdck cells. In Vitro Cell Dev. Biol. Anim. *41*, 207–216.

Karászi, E., Jakab, K., Homolya, L., Szakács, G., Holló, Z., Telek, B., Kiss, A., Rejtǫ, L., Nahajevsky, S., Sarkadi, B., and Kappelmayer, J. (2001). Calcein assay for multidrug resistance reliably predicts therapy response and survival rate in acute myeloid leukaemia. Brit. J. Haematol. *112*, 308–314.

Kaur, P. (2002). Multidrug resistance: can different keys open the same lock? Drug Resist. Updat. *5*, 61–64.

Kennedy, M., Cortés-Selva, F., Pérez-Victoria, J.M., Jiménez, I.A., González, A G, Muñoz, O.M., Gamarro, F., Castanys, S., and Ravelo, A.G. (2001). Chemosensitization of a multidrug resistant *Leishmania tropica* line by new sesquiterpenes from *Maytenus magellanica* and *Maytenus chubutensis*. J. Med. Chem. *44*, 4668–4676.

Krishna, R., and Mayer, L.D. (2000). Multidrug resistance (MDR) in cancer. Mechanisms, reversal using modulators of MDR and the roles of MDR modulators in influencing the pharmacokinetics of anticancer drugs. Eur. J. Pharm. Sci. *11*, 265–283.

Krishna, R., and Mayer, L.D. (2001). Modulation of P-glycoprotein (PGP) mediated multidrug resistance (MDR) using chemosensitizers: recent advances in the design of selective MDR modulators. Curr. Med. Chem. Anticancer Agents *1*, 163–174.

Kitagawa, S. (2006). Inhibitory effects of polyphenols on P-glycoprotein-mediated transport. Biol. Pharm. Bull. *29*, 1–6.

Lage, H. (2003). ABC transporters: implications on drug resistance from microorganisms to human cancers. Int. J. Antimicrob. Agents. 22, 188–199.

Lenski, R.E. (1998). Bacterial evolution and the cost of antibiotic resistance. Int. Microbiol. *1*, 265–270.

Leonard, G.D., Fojo, T., and Bates, S.E. (2003). The role of ABC transporters in clinical practice. Oncologist. *8*, 411–424.

Lomovskaya, O. and Watkins, W.J. (2001). Efflux pumps: their role in antibacterial drug discovery. Curr. Med. Chem. *8*, 1699–1771.

Lomovskaya, O., Warren, M.S., Lee, A., Galazzo, J., Fronko, R., Lee, M., Blais, J., Cho, D., Chamberland, S., and Renau, T. (2001). Identification and characterization of inhibitors of multidrug resistance efflux pumps in *P. aeruginosa*: novel agents for combination. Antimicrob. Agents. Chemother. *45*, 105–116.

Loo, T.W., and Clarke, D.M. (2000). Identification of residues within the drug-binding domain of the human multidrug resistance P-glycoprotein by cysteine-scanning mutagenesis and reaction with dibromobimane. J. Biol. Chem. *275*, 39272–39278

Loo, T.W., and Clarke, D.M. (2001). Defining the drug-binding site in the human multidrug resistance P-glycoprotein using a methanethiosulfonate analog of verapamil, MTS-verapamil. J. Biol. Chem. *276*, 14972–14979.

Machuca, C., Rodríguez, A., Herrera, M., Silva, S., and Ponte-Sucre, A. (2006). Metabolic adaptations induced by resistance to glibenclamide in *Leishmania amazonensis*. Exp. Parasitol.114, 1–9.

Maliepaard, M., van Gastelen, M.A., de Jong, L.A., Pluim, D., van Waardenburg, R.C., Ruevekamp-Helmers, M.C., Floot, B.G., and Schellens, J.H. (1999). Overexpression of the BCRP/MXR/ABCP gene in a topotecan-selected ovarian tumor cell line. Cancer Res. *59*, 4559–4563.

Mallea, M., Mahamoud, A., Chevalier, J., Albert-Franco, S., Brounant, P., Barbe, J., and Pages, J.M. (2003). Alkylaminoquinolones inhibit the antibiotic efflux pump in multidrug-resistant clinical isolates. Biochem. J. *376*, 801–805.

McAdams, H.H., Srinivasan, B., and Arkin, A.P. (2004). The evolution of genetic regulatory systems in bacteria. Nat. Rev. Genet.5, 169–178.

McKeegan, K.S., Borges-Walmsley, M.I., and Walmsley, A.R. (2004). Structural understanding of efflux-mediated drug resistance: potential routes to efflux inhibition. Curr. Opin. Pharmacol. *4*, 479–486.

Murray, D.S., Schumacher, M.A., and Brennan, R.G. (2004). Crystal structuRes. of QacR–diamidine complexes reveal additional multidrug-binding modes and a novel mechanism of drug charge neutralization. J. Biol. Chem. *279*, 14365–14371.

Natera, S., Machuca, C., Padrón-Nieves, M., Romero, A., Díaz, E., and Ponte-Sucre, A. (2007). *Leishmania* spp.: proficiency of drug-resistant parasites. Int. J. Antimicrob. Agents. *29*, 637–642.

Nelson, M.L., and Levy, S.B. (1999). Reversal of tetracycline resistance mediated by different bacterial tetracycline determinants by an inhibitor of the Tet(B) antiport protein. Antimicrob. Agents Chemother. *43*, 1719–1724.

Padrón-Nieves, M., Díaz, E., Machuca, C., Romero, A., and Ponte-Sucre, A. (2009). Glibenclamide modulates glucantime activity and disposition in *Leishmania major*. Exp. Parasitol. *121*, 331–337.

Nobili, S., Landini, I., Giglioni, B., and Mini, E. (2006). Pharmacological strategies for overcoming multidrug resistance. Curr. Drug Targets. *7*, 861–879.

Pajeva, I.K., Globisch, C., and Wiese, M. (2004). Structure–function relationships of multidrug resistance P-glycoprotein. J. Med. Chem. *47*, 2523–2533.

Pajeva, I.K., and Wiese, M. (2002). Pharmacophore model of drugs involved in P-glycoprotein multidrug resistance: explanation of structural variety (hypothesis). J. Med. Chem. *45*, 5671–5686.

Paulsen, I.T. (2003). Multidrug efflux pumps and resistance: regulation and evolution. Curr. Opin. Microbiol. *6*, 446–451.

Paulsen, I.T., Chen, J., Nelson, K.E., and Saier, M.H. Jr. (2001) Comparative genomics of microbial drug efflux systems. J. Mol. Microbiol. Biotechnol. *3*, 145–150.

Pedersen, P.L. (2005). Transport ATPases: structure, motors, mechanism and medicine: a brief overview. J. Bioenerg. Biomembr. *37*, 349–357.

Peyron, F., Eudes, N., de Monbrison, F., Wallon, M., and Picot, S. (2004). Fitness of *Toxoplasma gondii* is not related to DHFR single-nucleotide polymorphism during congenital toxoplasmosis. Int. J. Parasitol. *34*, 1169–1175.

Pleban, K., Kaiser, D., Kopp, S., Peer, M., Chiba, P., and Ecker, G.F. (2005). Targeting drug-efflux pumps a pharmacoinformatic approach. Acta Biochim. Pol. *52*, 737–740.

Ponte-Sucre A. (2007). Availability and applications of ATP-binding cassette (ABC) transporter blockers. Appl. Microbiol. Biotechnol. *76*, 279–286.

Ponte-Sucre, A., Campos, Y., Vasquez, J., Moll, H., and Mendoza-León, A. (1997)Sensitivity of *Leishmania* spp. to glibenclamide and 4-aminopyridine: a tool for the study of drug resistance development. Mem. Inst. Oswaldo Cruz *92*, 601–606.

Pradines, B., Albert-Franco, S., Houdoin, C., Mosnier, J., Santelli-Rouvier, C., Papa, V., Rogier, C., Fusai, T., Barbe, J., and Parzy, D. (2002). *In vitro* reversal of chloroquine resistance in *Plasmodium falciparum* with dihydroethanoanthracene derivatives. Am. J. Trop. Med. Hyg. *66*, 661–666.

Quinones-Mateu, M.E., and Arts, E.J. (2006). Virus fitness: concept, quantification, and application to HIV population dynamics. Curr. Top. Microbiol. Immunol. *299*, 83–140.

Saurin, W., Hofnung, M., and Dassa, E. (1999). Getting in or out: early segregation between importers and exporters in the evolution of ATP-binding cassette (ABC) transporters. J. Mol. Evol. *48*, 22–41.

Seelig, A. (1998). A general pattern for substrate recognition by P-glycoprotein. Eur. J. Biochem. *251*, 252–261.

Serrano-Martin, X., Payares, G., and Mendoza-León, A. (2006). Glibenclamide, a blocker of K^{+}(ATP) channels, shows antileishmanial activity in experimental murine cutaneous leishmaniasis. Antimicrob. Agents Chemother. *50*, 4214–4216. (URL: http://genomebiology.com/2004/5/3/R15)

Sheps, J.A., and Ling, V. (2006). Preface: the concept and consequences of multidrug resistance. Pflug. Arch. Eur. J. Physiol. *453*, 545–553.

Sheps J, A., Ralph, S., Zhao, Z-Y., Baillie, D.L., and Ling, V. (2004). The ABC transporter gene family of *Caenorhabditis elegans* has implications for the evolutionary dynamics of multidrug resistance in eukaryotes. Genome Biol. 5, R15.

Shilling, R.A., Venter, H., Velamakanni, S., Bapna, A., Woebking, B., Shahi, S., and van Veen, H.W. (2006). New light on multidrug binding by and ATP-binding-cassette transporter. Trends. Pharmacol. Sci. *27*, 195–203.

Silva, N., Camacho, N., Figarella, K., and Ponte-Sucre, A. (2004). Cell differentiation and infectivity of *Leishmania mexicana* are inhibited in an ABC-transporter blocker resistant strain. Parasitology. *128*, 9.629–634.

Stavri, M., Piddock, L.J., and Gibbons, S. (2007). Bacterial efflux pump inhibitors from natural sources. J. Antimicrob. Chemother. *59*, 1247–1260.

Stouch, T.R., and Gudmundsson, O. (2002). Progress in understanding the structure–activity relationships of P-glycoprotein. Adv. Drug Deliv. Rev. *54*, 315–328.

Sun, J., He, Z.G., Cheng, G., Wang, S.J., Hao, X.H., and Zou, M.J. (2004). Multidrug resistance P-glycoprotein: crucial significance in drug disposition and interaction. Med. Sci. Monit. 10, RA5-RA14.

Teodori, E., Dei, S., Garnier-Suillerot, A., Gualtieri, F., Manetti, D., Martelli, C., Romanelli, M.N., Scapecchi, S., Sudwan, P., and Salerno, M. (2005). Exploratory chemistry toward the identification of a new class of multidrug resistance reverters inspired by pervilleine and verapamil models. J. Med. Chem. *48*, 7426–7436.

Teodori, E., Dei, S., Martelli, C., Scapecchi, S., and Gualtieri, F. (2006). The functions and structure of ABC transporters: implications for the design of new inhibitors of Pgp and MRP1 to control multidrug resistance (MDR). Curr. Drug. Targets. *7*, 893–909.

Tsakovska, I., and Pajeva, I. (2006). Phenothiazines and structurally related compounds as modulators of cancer multidrug resistance. Curr. Drug Targets. *7*, 1123–1134.

Tsuruo, T. (1983). Reversal of acquired resistance to vinca alkaloids and anthracycline antibiotics. Cancer Treat. Rep. *67*, 889–894.

Uliana, S.R., Goyal, N., Freymüller, E., and Smith, D.F. (1999). Overexpression and comparative structural analysis of the stage-regulated meta-1 gene. Exp. Parasitol. (Jena) 92, 183–191.

Uzcategui, N.L., Figarella, K., Camacho, N., and Ponte-Sucre A. (2005). Substrate preferences and glucose uptake in glibenclamide-resistant *Leishmania* parasites. Compar. Biochem. Physiol. *140*, 9.395–402.

van der Heide, T., and Poolman, B. (2002). ABC transporters: one, two or four extracytoplasmic substrate binding sites? EMBO . Rep. 3, 938–943.

van Herwaarden, A.E., and Schinkel, A.H. (2006). The function of breast cancer resistance protein in epithelial barriers, stem cells and milk secretion of drugs and xenotoxins. Trends. Pharmacol. Sci. 27, 10–16.

van Veen, H.W., and Konings, W.N. (1997). Multidrug transporters from bacteria to man: similarities in structure and function. Cancer Biol. *8*, 183–191.

van Veen, H.W. (2001). Towards the molecular mechanism of prokaryotic and eukaryotic multidrug transporters. Cell. Devel. Biol. *12*, 239–245.

Varma, M.V., Ashokraj, Y., Dey, C.S., and Panchagnula, R. (2003). P-glycoprotein inhibitors and their screening: a perspective from bioavailability enhancement. Pharmacol. Res. *48*, 347–359.

Wang, S., Ryder, H., Pretswell, I., Depledge, P., Milton, J., Hancox, T.C., Dale, I., Dangerfield, W., Charlton, P., Faint, R., Dodd, R., and Hassan, S. (2002). Studies on quinazolinones as dual inhibitors of Pgp and MRP1 in multidrug resistance. Bioorg. Med. Chem. Lett. *12*, 571–574.

Wiese, M., and Pajeva, I.K. (2001). Structure-activity of multidrug resistance reversers. Curr. Med. Chem. *8*, 685–713.

Williams, P.D., and Day, T. (2001). Interactions between sources of mortality and the evolution of parasite virulence. Proc. Biol. Sci. *268*, 2331–2337.

Yu, E.W., McDermott, G., Zgurskaya, H.I., Nikaido, H. and Koshland, D.E.Jr. (2003). Structural basis of multiple drug-binding capacity of the AcrB multidrug efflux pump. Science. *300*, 976–980.

Zsila, F. and Iwao, Y. (2007). The drug binding site of human alpha1-acid glycoprotein: insight from induced circular dichroism and electronic absorption spectra. Biochim. Biophys. Acta. *1770*, 797–809.

Zsila, F., Visy, J., Mády, G. and Fitos, I. (2008). Selective plasma protein binding of antimalarial drugs to alpha1-acid glycoprotein. Bioorg. Med. Chem. *16*, 3759–3772.

ABC Transporters as Targets for RNA Interference-mediated Reversal of Multidrug Resistance: Implications in Microorganisms

10

Hermann Lage

Abstract

Drug resistance is a common clinical problem that occurs in patients suffering from infectious diseases, and in patients suffering from cancer. Prokaryotic and eukaryotic microorganisms as well as neoplastic cells are often found to be refractory to multiple structurally unrelated compounds used during chemotherapy. This phenomenon has been termed multidrug resistance (MDR). In both, prokaryotic and eukaryotic organisms, MDR is frequently associated with overexpression of trans-membrane xenobiotic transport molecules belonging to the superfamily of ATP-binding cassette (ABC) transporters. Inhibition of ABC transporters by low-molecular weight compounds has been extensively investigated in cancer patients; however, the clinical results have been disappointing. Consequently, innovative experimental therapeutic strategies for overcoming MDR are urgently needed and are under investigation. These strategies include the application of the novel RNA interference (RNAi) technology. Various RNAi strategies have been applied to reverse MDR in different tumour models *in vitro* and *in vivo* indicating that this technology is tremendously effective in reversing ABC transporter-mediated MDR in cancer and is therefore a promising strategy for overcoming MDR by gene therapeutic applications. The RNAi pathway is restricted to eukaryotic cells. This observation suggests that the RNAi technology appears to be only useful for reversal of ABC transporter-mediated MDR in eukaryotic infection disease-causing microorganisms, such as parasites. However, besides the data that has been obtained using this technology in pathogenic parasites, efforts have been made to trigger gene silencing by RNAi in prokaryotic cells. Should this approach be successful in the future, RNAi technology could also be considered for overcoming MDR in infection diseases mediated by prokaryotic pathogens.

Introduction

Resistance to chemotherapeutic agents, i.e. anti-microbial or antineoplastic drugs, is one of the major reasons why treatment of infectious diseases or malignant tumours fails. Pathogenic microorganisms as well as human cancer cells have developed various biological mechanisms that provide a defence against the cytotoxic attack by those drugs. Ordinarily, drug resistance is specific to a single class of antibiotics or anticancer drugs. One of the best-investigated examples of specific microbial resistance to β-lactam antibiotics such as penicillins or cephalosporins is that resistance is promoted by plasmid-encoded β-lactamases (Rossolini and Docquier, 2006). Microorganisms, as well as human cancer cells, may also exhibit a cross-resistant phenotype against several unrelated drugs that differ widely with respect to their molecular structure and target specificity. This phenomenon has been termed multidrug resistance (MDR). Various biological mechanisms have been described to be involved in MDR of pathogenic microorganisms or human cancer cells (Lage, 2008).

The first mechanism that was identified to mediate a multidrug-resistant phenotype in cancer cells was the activity related to the well-known membrane-embedded drug extrusion pump MDR1/P-glycoprotein (MDR1/P-gp or ABCB1) member of the family of ATP-binding cassette (ABC) transporters (Lage, 2003). This transporter was originally isolated from the plasma membranes of Chinese hamster ovary cells displaying a 'classical' MDR phenotype (Juliano and Ling, 1976). Evolution and details of the molecular structures of MDR1/P-gp and other ABC transporters are deeply discussed in chapters 1–3. The 'classical' MDR phenotype is characterized by a typical cross-resistance pattern against natural product-related anticancer agents, such as vinca alkaloids, epipodophyllotoxins, anthracyclines, or taxanes (Table 10.1). Furthermore, the 'classical' MDR phenotype can by reversed by specific agents – termed chemosensitizers or MDR modulators – such as verapamil and cyclosporin A derivatives (Table 10.1; see Chapter 9). Alternative MDR phenotypes with different cross-resistance pattern and/or different molecular mechanisms are often designated as 'MDR1/P-gp-independent' or 'atypical' MDR. Following the identification of MDR1/P-gp, a large number of drug-transporting ABC transporters have been identified in cancer cells (Lage, 2008) as well as in pathogenic microorganisms (Van Bambeke *et al.*, 2000).

Besides drug extrusion pumps, many alternative mechanisms have been described that cause a multidrug-resistant phenotype in cancer cells. These mechanisms include the modulation of cellular death pathways, or alteration and repair of target molecules. Together these mechanisms can build a complex network of cellular pathways and molecular mechanisms mediating an individual MDR phenotype to a given cancer (Lage, 2008).

Efforts to develop chemical compounds which target drug resistance-associated cellular pathways have been made over many years. These so called 'chemosensitizer' or 'MDR modulators' were predominantly designed as ABC transporter inhibitors, in particular MDR1/P-gp-inhibiting agents (Szakács *et al.*, 2006). Unfortunately, until now these ABC transporter inhibitors failed in clinical trials. Thus, additionally to these classical attempts to overcome MDR with the help of low molecular weight pharmacological active compounds, preclinical studies have been performed with experimental therapeutics including molecules triggering the RNA interference (RNAi) pathway for specific knockdown of ABC transporter-encoding mRNAs (Lage, 2006).

Table 10.1 Examples of drugs that belong to the 'classical' MDR spectrum and 'classical' MDR modulators (Lage, 2008)

Drugs transported by MDR1/P-gp	Modulators of MDR1/P-gp activity
Anthracyclines: daunorubicin, doxorubicin	Calcium channel blockers: verapamil, nifedipine, azidopine, dihydropyridines
Vinca alkaloids: vinblastine, vincristine, vindesine	Immunosuppressants and derivatives: cyclosporin A, valspodar (PSC833), tacrolimus
Epipodophyllotoxines: etoposide, teniposide	Antiarrhythmics: quinine, quinidine, amiodarone
Antibiotics: actinomycin D, dactinomycin, mitomycin C	Antihypertensives: reserpine, yohimbine
Taxanes: paclitaxel	Antibiotics: hydrophobic cephalosporins
Others: colchicine, topotecan, valinomyycin, puromycin, emetine, digoxin, imatinib	Steroid hormones and derivatives: progesterone, tamoxifen
Many other hydrophobic amphipatic drugs and derivatives	HIV protease inhibitors: sequinavir, indinavir, retanavir
	Herbal constituents: curcumin
	Others: elacridar (GF120918), zosuquidar (LY335979), tariquidar (XR9576), laniquidar (R101933)

Multidrug transporters

One of the most important MDR mechanisms is the activity of transmembrane drug transporters that extrude anti-microbial compounds or antineoplastic agents from the cells. Although some drug transporters are more or less specific for a given drug or class of drugs, other transport molecules, the multidrug transporters, mediate the unidirectional transport of a broad range of compounds across biological membranes. These transporters promote different MDR phenotypes in pathogenic microorganisms (Van Bambeke *et al.*, 2000) and human cancer cells (Lage, 2008) by extrusion of drugs. Multidrug transporters belong to a limited number of protein families (Blackmore *et al.*, 2001) and the result of their pump activity is a decreased accumulation of cytotoxic drugs in the intracellular milieu.

On the basis of energy source and structural relationships, these transport systems have been shown to exist in all organisms investigated up to now, and can be divided into two major classes of transport molecules. The first class is composed of primary active transporters that are energized by the hydrolysis of ATP. These primary transporters are represented by the superfamily of ABC transporters (Higgins, 1992) and by some recently described Ral-binding proteins, that can act as multispecific, ATP-utilizing xenobiotic transporters that have no sequence homology with the ABC transporter family, and differ from those in several other important aspects (Awasthi *et al.*, 2007).

The second class of transporters consists of secondary active transporters that mediate the drug efflux reaction in a coupled exchange with protons or sodium ions along a concentration gradient as symport or antiport (Paulsen *et al.*, 1996). Members of this class are the small MDR (SMR) superfamily, the multidrug and toxic compound extrusion (MATE) family, the resistance-nodulation-cell division (RND) family, and the major facilitator (MFS) superfamily (Putman *et al.*, 2000). Primary transporters or ABC transporters play the most important role in drug efflux in eukaryotic cells, including human cancer cells, whereas in prokaryotic organisms the secondary active multidrug transporters are predominantly active in drug extrusion (Van Bambeke *et al.*, 2000).

ABC transporters

ABC transporters are not identical with ABC proteins. Both types of proteins are defined by the presence of a highly conserved approximately 215 amino acids consensus sequence, the nucleotide-binding domain (NBD). The NBD contains two short peptide motifs: a glycine-rich Walker A and a hydrophobic Walker B motif (Walker *et al.*, 1982), both involved in ATP binding and commonly present in all nucleotide-binding proteins. A third consensus sequence is named the ABC signature (Hyde *et al.*, 1990) and is unique in ABC domains. ABC signature-containing proteins couple the phosphate bond energy of ATP hydrolysis to many cellular processes that are not necessarily restricted to transport functions. However, the proper meaning of the term ABC transporter is satisfied only when the ABC-protein is associated with a hydrophobic, membrane-embedded transmembrane domain (TMD), usually composed of at least, six transmembrane (TM) α-helices. The TMDs are believed to determine the specificity of recognition for the substrate molecules transported by the ABC transporter protein. The minimal structural requirement for a biological active ABC transporter seems to be two TMDs and two NBDs $[TMD\text{-}NBD]_2$. In 'full-transporters', this structural arrangement may be formed by a single polypeptide sequence, and in multiprotein complexes, by more than one polypeptide chain. Most of the prokaryotic genes encoding ABC transporters are organized in operons that contain NBDs and TMDs as separate subunits requiring assembly as a biologically active transporter. In some ABC transporter encoding genes, the different domains are already fused into higher structural units, e.g. in 'half-size transporters' [TMD-NBD] or 'full-size transporters' $[TMD\text{-}NBD]_2$. The organization of ABC transporter encoding genes in eukaryotic organisms also shows wide variation. Nevertheless, the different domain combinations are commonly distributed in one or two genes encoding 'half-' or 'full-size transporters'.

ABC transporters in bacterial drug resistance

Most of the prokaryotic drug transporters belong to the class of secondary active transport molecules, in particular drug-proton exchange systems

(Paulsen *et al.*, 1996). Nevertheless, various drug transporting systems utilize the energy of ATP hydrolysis to drive drug extrusion and belong to the ABC transporter family. Table 10.2 lists some of the best studied representatives, a summary of which will be briefly discussed here. A detailed overview about prokaryotic ABC transporters is given in Chapter 5.

Most bacterial drug transport molecules mediate single drug resistance (SDR) in Gram-positive bacteria. SDR transporters are specific for a single pharmacological compound or a group of closely related drugs, such as Ard1 from *Streptomyces capreolus* that confers resistance to the aminonucleoside antibiotic A201A (Barrasa *et al.*, 1995), the *Streptomyces logisporoflavus* transporter TnrB (Linton *et al.*, 1994) conferring SDR to polyether-ionophore antibiotic tetronasin, or the plasmid-encoded *Staphylococcus epidermidis* MsrA protein. Plasmid encoded MsrA mediates active efflux of the macrolide antibiotic erythromycin from *Staphylococcus aureus* (Ross *et al.*, 1990). Of great clinical importance is the observation that antibiotic resistance of some important human pathogens, such as *Staphylococcus aureus* or *Staphylococcus epidermidis* is mediated by ABC transporters; in particular, the dramatic increase of infections with methicillin-resistant *Staphylococcus aureus* (MRSA) has become a widespread clinical problem with nearly 20,000 deaths per year only in the United States (Cosgrove and Fowler, 2008).

Examples of bacteria that must have a dual behavior are also observed. For example, bacteria which synthesize antibiotic active compounds must protect themselves from the cytotoxic effects of these secondary metabolites. In fact, *Streptomyces peucetius* the producer of the clinical commonly used anticancer drugs, daunorubicin and doxorubicin, expresses the SDR transporter DrrAB that mediates resistance to both toxic substances (Guifoile *et al.*, 1991).

LmrA, the first ATP-dependent prokaryotic MDR transporter identified in *Lactococcus lactis* (Bolhuis *et al.*, 1994) is a very interesting example that should be mentioned. The pharmacological characteristics of LmrA are very similar to the human MDR1/P-gp and include the spectrum of its substrates (Table 10.1). Furthermore, LmrA is involved in resistance to 17 out of 21 clinically most used antibiotics (Poelarends *et al.*, 2000) including aminoglycosides, lincosamides, macrolides, quinolones, streptogramins, tetracyclines, and the chloramphenicol broad-spectrum family of antibiotics. A slightly enhanced resistance to certain β-lactam antibiotics has also been observed.

The ABC transporter HorA of *Lactobacillus brevis* confers resistance to iso-α-acids from hop (Simpson and Smith, 1992). These acids give beer a bitter taste and exert bacteriostatic effects on most Gram-positive bacteria due to their ability to dissipate the proton motive force. Furthermore, HorA mediates resistance to the structurally unrelated compounds novobiocin, ethidium bromide, and Hoechst 33342 (Sakamoto *et al.*, 2001). Of major importance is that HorA is plasmid-encoded. Thus, HorA may be exchanged between pathogenic microorganisms and may be responsible for acquired clinical MDR of bacteria.

Interestingly, so far no ABC transporters involved in SDR or MDR in Gram-negative bacteria have been identified. Although sequence homology searches identified putative drug transporters in Gram-negative bacteria, no experimental evidence demonstrated any drug transport by these molecules (Young and Holland, 1999). Although the genome of *Escherichia coli* (Blattner *et al.*, 1997) encodes more than 100 ABC transporters which genes occupy 5% of its genome (Linton and Higgins, 1998), and represent the largest family of proteins in *Escherichia coli,* none of them show any functions associated with drug transport. So far, the reasons why ABC transporter-basing drug-transporting systems are absent in Gram-negative bacteria in view of the relatively frequent occurrence of such ABC transporters in these organisms are not sufficiently satisfying elucidated.

Last but not least, Archaebacteria also contain ATP-dependent MDR-transporter systems. The halophilic archaeon *Haloferax volcanii* expresses a drug efflux system conferring the transport of anthracyclines, vinca alkaloids, ethidium bromide, monensin, and rhodamine 123 (Miyauchi *et al.*, 1992). Moreover, this ATP-dependent drug efflux system can be inhibited by inhibitors of human MDR1/P-gp such as the calcium channel antagonists verapamil or nifedipine. However, the gene encoding this ABC-like MDR transporter remains to be cloned.

ABC transporters in drug resistance of fungi

Two major classes of xenobiotics transport molecules are involved in drug resistance of fungi, the family of primary active ABC transporters and the family of proton-motive force driven MFS-transporters. In contrast to prokaryotic microorganisms, in fungi the family of ABC transporters comprises the largest number of membrane-spanning efflux pumps (Higgins, 1992) (Table 10.2). A detailed description of ABC transporters from fungi is expounded in Chapter 6. From at least 29 putative ABC transporter encoding genes in *Saccharomyces cerevisiae* (Decottignies and Goffeau, 1997), 5 genes encode ABC transporters, Pdr5p, Pdr12p, Snq2p, Ycf1p, and Yor1p, involved in MDR. Pdr5p confers resistance to multiple chemically unrelated compounds, including different classes of clinically important antimycotics, antibiotics, anticancer drugs, mycotoxins, agricultural fungicides, and herbicides (Kolaczkowski *et al.*, 1998). Pdr12p mediates resistance to food preservatives such as sorbic, benzoic, and propionic acid (Holyoak *et al.*, 1999). Snq2p confers resistance to the mutagens 4-nitroquinoline-N-oxide and triaziquone, as well as to the chemicals sulphomethuron methyl and phenanthroline (Servos *et al.*, 1993). Ycf1p, and Yor1p exhibit broad substrate specificity with partial overlap to that of Pdr5p (Katzmann *et al.*, 1999).

In other yeast species, such as the fission yeast *Schizosaccharomyces pombe,* additional ABC transporters are found. Sequencing of the *Schizosaccharomyces pombe* genome revealed at least six ABC transporter encoding genes (Wood *et al.*, 1992). Two of these ABC transporters mediate MDR. The first protein, Pmd1, confers resistance to the antimycotic agent leptomycin B, as well as to various other cytotoxic compounds (Nishi *et al.*, 1992). The other ABC transporter, Bfr1 is involved in resistance to various compounds, including actinimycin, cerulenin, cytochalasin B and brefeldin A (Nagao *et al.*, 1995).

In the past decades, *Candida albicans,* isolated in more than 50% of fungal infection cases (Nguyen *et al.*, 1996), has become an increasing clinical problem. The increase in *Candida albicans* infections is considered to be a consequence of the growing number of immuno-compromised patients, e.g. organ transplant, cancer, and HIV patients. As a consequence of the frequent application of antimycotic agents, drug resistance has become problematic for the treatment of candidiasis (Sanglard and Odds, 2002). In *Candida albicans* five ABC transporters have been identified. Two of them, Cdr1p and Cdr2p, have been demonstrated to play a major role in drug resistance. Cdr1p confers resistance to the clinically important azole antimycotics fluconazole and miconazole, as well as to allylamines, morpholines, and several other drugs (Prasad *et al.*, 1995). Cdr2p is involved in resistance to azole antifungal agents, other antimycotic agents, such as terbinafine and amorolfine, and to a variety of metabolic inhibitors (Sanglard *et al.*, 1997).

ABC transporters in parasites

For the chemotherapeutic treatment of diseases caused by protozoan parasites, drug resistance has become a major clinical problem. More than 2 billion people are at risk of malaria, caused by four different species of the parasite *Plasmodium.* The parasites primarily affect poor populations in tropical and subtropical areas whereby approximately 500 million cases and 1 million deaths from malaria are occurring annually (Greenwood *et al.*, 2008). Although a detailed description of ABC transporters expressed in parasites is presented in chapters 6 and 7 a brief description is useful for this discussion. The non-ABC transporter PfCRT is associated with resistance of *Plasmodium falciparum* to the primary antimalaria agent chloroqine (Fidock *et al.*, 2000). Additionally, at least five ABC transporter encoding genes have been identified in *Plasmodium falciparum,* MDR1/Phg1, MAL3P1, MDR2/Phg2, MAL1P3 and PfGCN20 (Ouellette and Légaré, 2003). MDR1/Phg1 and MAL1P3 contribute to quinoline resistance in *Plasmodium falciparum* (Henry *et al.*, 2008) (Table 10.2). So far MDR2/Phg2 has not been associated with drug transport but this protein functions as an efflux pump of heavy metals (Rosenberg *et al.*, 2006). However, it appears most likely that the cloroquine-resistant phenotype in *Plasmodium falciparum* is a complex event mediated by multiple genes with important contribution from ABC transporters.

Table 10.2 ABC transporters associated with drug resistance in microorganisms

Organism	ABC transporter	Transported drugs	References
Archaebacteria			
Haloferax volcanii	not identified	antrhracyclines, vinca alkaloids, monensin	Miyauchi *et al.*, 1992; Kaidoh *et al.*, 1996
Eubacteria			
Gram-positive			
Streptomyces capreolus	Ard1	A201A	Barrasa *et al.*, 1995
Streptomyces logisporoflavus	TnrB	tetronasin	Linton *et al.*, 1994
Streptomyces peucetius	DrrAB	daunorubicin, doxorubicin	Guilfoile and Hutchinson, 1991
Staphylococcus epidermidis	MsrA	erythromycin	Ross *et al.*, 1990
Lactococcus lactis	LmrA	Drugs from the 'classical' MDR spectrum (Table 10.1), anthracyclines, vinca alkaloids, aminoglycosides, lincosamides, macrolides, quinolones, streptogramins, tetracyclins, chloramphenicol	Bolhuis *et al.*, 1994; Poelarends *et al.*, 2000
Lactobacillus brevis	HorA	novobiocin	Sakamoto *et al.*, 2001
Gram-negative		ABC transporters are not involved in drug transport	Young and Holland, 1999
Fungi			
Saccharomyces cerevisiae	Pdr5p	Drugs from the 'classical' MDR spectrum (Table 10.1), antimycotics, antibiotics, anticancer drugs, fungicides, herbicides	Bissinger and Kuchler, 1994; Kolaczkowski *et al.*, 1996; Kolaczkowski *et al.*, 1998
	Pdr12p	food preservatives	Piper *et al.*, 1998; Holyoak *et al.*, 1999
	Snq2p	4-nitroquinoline-N-oxide, triaziquone, phenanthroline	Servos *et al.*, 1993
	Ycf1p	Drugs from the 'classical' MDR spectrum (Table 10.1), similar to Pdr5	Rebbeor *et al.*, 1998
	Yor1p	Drugs from the 'classical' MDR spectrum (Table 10.1), similar to Pdr5	Katzmann *et al.*, 1995; Decottignies *et al.*, 1998; Katzmann *et al.*, 1999
Schizosaccharomyces pombe	Pmd1	antimycotics	Nishi *et al.*, 1992
	Bfr1	antimycotics	Nagao *et al.*, 1995; Turi and Rose, 1995

Organism	ABC transporter	Transported drugs	References
Candida albicans	Cdr1p	antimycotics	Prasad *et al.*, 1995; Sanglard *et al.*, 1997
	Cdr2p	antimycotics	Sanglard *et al.*, 1997
Parasites			
Plasmodium falciparum	MDR1/ Phg1	chloroquine, mefloquine, quinine, halofantrine, artemisinin	Foote *et al.*, 1989; Reed *et al.*, 2000
	MAL1P3	chloroquine	Henry *et al.*, 2008
Leishmania tarentolae	PGPA	antimony drugs, arsenic drugs	Callahan and Beverley, 1991; Papadopoulou *et al.*, 1997; Légaré *et al.*, 1997
Leishmania species	MDR1	Drugs from the 'classical' MDR spectrum (Table 10.1), similar to LmrA	Henderson *et al.*, 1992; Chow *et al.*, 1993; Gueiros-Filho *et al.*, 1995; Chiquero *et al.*, 1998; Katakura *et al.*, 1999
Entamoeba histolytica	EhP-gp1	amoebicides	Cedeno and Krogstad, 1983; Samarawickrema *et al.*, 1997; Samuelson *et al.*, 1990; Borst and Ouellette, 1995; Gomez *et al.*, 1996
	EhP-gp2	amoebicides	Cedeno and Krogstad, 1983; Samarawickrema *et al.*, 1997; Samuelson *et al.*, 1990; Borst and Ouellette, 1995; Gomez *et al.*, 1996
Trypanosoma species	MDR1/P-gp-like		Kerboeuf *et al.*, 2003
Haemonchus contortus	MDR1/P-gp-like	anthelmintics	Kerboeuf *et al.*, 2003

Among parasitic infections, leishmaniasis has increased dramatically and has become the second leading cause of death worldwide (Hirst and Stapley, 2000). Chemotherapy remains the only effective way to control infections caused by the intracellular *Leishmania* species. First line drugs for the treatment of leishmaniasis are pentavalent antimonials in the form of glucantime and pentostam (Marsella and Ruiz de Gopegui, 1998), but these antimony-based and alternative metal-based drugs are not very efficient due to their toxicity and the increased appearance of drug resistance in some areas of the world (Faraut-Gambarelli *et al.*, 1997). In different *Leishmania* species, at least 12 ABC transporters have been identified that could be involved in drug resistance (Ouellette and Légaré, 2003) and the ABC transporter PGPA contributes to resistance against antimony- and arsenic-based antiprotozoan drugs in *Leishmania*; however, high level drug resistance requires additional factors (Légaré *et al.*, 1997). The availability of these so far unknown factors may differ in various *Leishmania* species. In several *Leishmania* species the ABC transporter MDR1, a homologue to the human MDR1/P-gp has been also characterized (Henderson *et al.*, 1992). This ABC transporter causes a multidrug-resistant phenotype directed against anticancer agents of the human classical MDR spectrum. Although, these drugs are not used in the treatment of leishmaniasis, *Leishmania* MDR1 may extrude cytotoxic compounds from the cell suggests that it is required for detoxification.

Amoebiasis is an infectious disease caused by the protozoan *Entamoeba histolytica*. This protozoan infects 500 million people, provokes 50 million cases of dysentery or liver abscesses, and kills 40,000–100,000 humans each year worldwide (Ackers and Mirelman, 2006). Treatment of amoebiasis is primarily by amoebicides, mainly metronidazole and emetine. Since differences in drug susceptibility have been reported from *Entamoeba histolytica* isolates as well as laboratory strains, it was suggested that the parasite can develop a MDR phenotype in its human host (Samarawickrema *et al.*, 1997). Experiments demonstrated that verapamil-sensitive ABC transporters may be involved in resistance against amoebicides in drug-resistant parasites (Borst and Ouellette, 1995). So far, six homologues to the human MDR1/P-gp, EhP-gp1–6, have been characterized (Gomez *et al.*, 1996). Apparently, four of these, EhP-gp1, EhP-gp2, EhP-gp5, and EhP-gp6, are expressed in drug-resistant mutants of *Entamoeba histolytica* whereby only the expression level of EhP-gp1 and EhP-gp2 correlates with drug resistance in the trophozoites.

Drug resistance in worm infections is not a common problem in human patients but remains as an important problem in veterinary medicine (Geerts and Gryseels, 2001). Different ABC transporters have been identified from both phyla of parasitic helminths, Platyhelmintha and Nematoda. Alterations in the expression levels and allele frequencies of ABC transporters in anthelmintics-resistant populations have been observed (Kerboeuf *et al.*, 2003). ABC transporter molecules that are homologous to the human MDR1/P-gp appear to contribute to anthelmintics resistance, e.g. in the sheep parasite *Haemonchus contortus* (Table 10.2).

ABC transporters in human cancers

Worldwide, over 22 million people are cancer patients. Approximately 10 million people are newly diagnosed with cancer each year and more than 6 million cancer patients die of the disease annually (Stewart and Kleihues, 2003). MDR of neoplastic cells is as old as the chemotherapeutic treatment of human cancers. Besides other mechanisms, the enhanced activity of different ABC transporters is of major impact for cancer MDR. Since completion of the human genome sequence, 48 different ABC transporters have been identified in humans and were divided by their phylogenetic characteristics into 7 subfamilies, ABCA, ABCB, ABCC, ABCD, ABCE, ABCF, and ABCG (Dean *et al.*, 2001). At least 18 human ABC transporters have been described to be associated with drug transport *in vitro* (Table 10.3). Detailed characteristics of these ABC transporters are discussed in Chapters 1–3. Thus, it seems likely that different ABC transporter expression patterns play an important role in inter-individual differences in the drug sensitivity of different tumours.

RNA interference (RNAi) technology

Ten years ago the RNA interference (RNAi) phenomenon was first described in the model organism worm *Caenorhabditis elegans* (Fire *et al.*, 1998). RNAi is an ancient physiological defence mechanism protecting eukaryotic cells including protists, fungi, plants, invertebrates and mammals from viruses and transposable genetic elements. Furthermore, RNAi plays an important role in post-transcriptional gene silencing (PTSG), a common biological mechanism of gene regulation. In plants RNAi is mediated by short 25 nucleotides double-stranded RNA (dsRNA) molecules, so-called short interfering RNAs (siRNAs) (Hamilton and Baulcombe, 1999).

Moreover, multiple microRNAs (miRNAs) have been identified in almost all metazoan genomes including mammals. They are generally 21–25nucleotide, non-coding RNAs that are derived from larger precursors that form imperfect stem-loop structures. MiRNAs have diverse expression patterns and regulate gene expression in a sequence-specific manner (He and Hannon, 2004). Fundamentally, siRNAs and miRNAs are similar in terms of their molecular characteristics, biogenesis and effector functions. The distinctions between these two types of RNAs might be arbitrary, and might simply reflect the different history through which they were originally discovered. Meanwhile it is clear that miRNAs play important roles in various diseases including cancer. In human malignancies, miRNAs are probably involved in all aspects of cancer biology,

Table 10.3 Human ABC transporters associated with drug resistance (Lage, 2008)

HUGO nomenclature	Common name(s)	Drugs	Physiological substrates	References
ABCA2	ABC2	Estramustine, mitoxantrone	Steroids	Laing *et al.*, 1998; Vulevic *et al.*, 2001; Boonstra *et al.*, 2004; Mack *et al.*, 2007
ABCA3	ABC3	Doxorubicin	Surfactant production	Steinbach *et al.*, 2006
ABCB1	MDR1/P-gp, P-170, P-gp, MDR1, PGY1	'Classical' MDR spectrum (table 10.1)	Phospholipids, neutral and cationic organic compounds	Riordan and Ling, 1979; Kartner *et al.*, 1983; Lage, 2003; Szakács *et al.*, 2006
ABCB2	TAP1	Mitoxantrone, epipodophyllotoxins	Peptides	Izquierdo *et al.*, 1996; Lage *et al.*, 2001; Lage, 2003
ABCB3	TAP2	Mitoxantrone, epipodophyllotoxins	Peptides	Izquierdo *et al.*, 1996; Lage *et al.*, 2001; Lage, 2003
ABCB4	MDR3/P-gp, MDR3, PGY3	Paclitaxel, vinca alkaloids	Phosphatidylcholine	Ruetz and Gros, 1994; Gottesman *et al.*, 2002
ABCB5		Doxorubicin, camptothecin, 5-fluorouracil	Pigment transport?	Huang *et al.*, 2004; Chen *et al.*, 2005; Frank *et al.*, 2005
ABCB6	MTABC3	Cisplatin, camptothecin	Mitochondrial porphyrin uptake	Yasuí *et al.*, 2004; Paterson *et al.*, 2007
ABCB11	BSEP, SPGP, ABC16, PGY4	Paclitaxel	Bile salts	Childs *et al.*, 1998; Gerloff *et al.*, 1998; Szakács *et al.*, 2006
ABCC1	MRP, MRP1	Anthracyclines, vinca alkaloids, epipodophyllotoxins, methotrexate	Glutathione and other conjugates, organic anions, leukotrienes	Cole *et al.*, 1992; Jedlitschky *et al.*, 1996; Borst *et al.*, 2000; Haimeur *et al.*, 2004
ABCC2	MRP2, cMOAT	Platin-drugs, anthracyclines, vinca alkaloids, epipodophyllotoxins, camptothecins, methotrexate	Glutathione and other conjugates, organic anions, leukotriene C_4	Taniguchi *et al.*, 1996; Cui *et al.*, 1999; König *et al.*, 1999; Materna *et al.*, 2005; Haimeur *et al.*, 2004
ABCC3	MRP3, MOAT-D, MLP2	Vinca alkaloids, epipodophyllotoxins, methotrexate, cisplatin	Glucuronides, bile salts, peptides	Kool *et al.*, 1999; Zeng *et al.*, 1999; de Jong *et al.*, 2001; Borst *et al.*, 2007
ABCC4	MRP4, MOAT-B	Nucleotide analogues, methotrexate	Organic anions	Schuetz *et al.*, 1999; Chen *et al.*, 2001; Chen *et al.*, 2002; Borst *et al.*, 2007
ABCC5	MRP5, MOAT-C	Nucleotide analogues	Organic anions, cyclic nucleotides	Jedlitschky *et al.*, 2000; Wijnholds *et al.*, 2000; Borst *et al.*, 2007

Table 10.3 Continued

HUGO nomenclature	Common name(s)	Drugs	Physiological substrates	References
ABCC6	MRP6	Anthracyclines, epipodophyllotoxins, cisplatin	Glutathione conjugates	Belinsky *et al.*, 2002; Szakács *et al.*, 2006
ABCC10	MRP7	Taxanes, vinca alkaloids	Organic anions, cyclic nucleotides, bile salts, leukotriene C_4	Hopper-Borge *et al.*, 2004; Kruh *et al.*, 2007
ABCC11	MRP8	5-Fluorouracil	Organic anions, cyclic nucleotides, leukotriene C_4	Kruh *et al.*, 2007; Oguri *et al.*, 2007
ABCG2	BCRP, MXR, ABCP	Mitoxantrone, anthracyclines, camptothecins, topotecan	Prazosin	Allikmets *et al.*, 1998; Doyle *et al.*, 1998; Miyake *et al.*, 1999; Lage and Dietel, 2000; Doyle and Ross, 2003

such as carcinogenesis, proliferation, apoptosis, invasion/metastasis, angiogenesis and drug resistance (Lee and Dutta, 2008).

The major breakthrough for the RNAi technology in biomedical research was the observation that delivery of siRNAs to mammalian cells also caused sequence-specific gene silencing by triggering the RNAi pathway (Elbashir *et al.*, 2001a). Therewith, it was demonstrated that RNAi could be potentially applied as a specific means to inhibit the expression of any gene of interest in mammalian – including human – cells in a cell-type- or pathway-specific manner. Biological active siRNAs can be obtained relatively easy and inexpensive by chemically synthesizing or expression from DNA matrices. For that reason, the siRNA-directed gene-silencing methodology has sparked a revolution in somatic cell genetics, allowing the rapid analysis of gene function in eukaryotic cell systems. Consequently, RNAi technology has been successfully introduced in many biomedical laboratories both in basics research as well as in applied sciences. On top of that, the RNAi technology has been and is considered as a potential new therapeutic strategy against various disorders caused by the over expression of one or more defined 'pathogenic' genes, including ABC transporters associated with drug resistance in cancer and infectious diseases; therefore, primarily eukaryotic pathogens appear to be suited therapeutic targets of RNAi-utilizing gene therapeutic approaches. Additional information can be found in the following RNAi-related web links: www.nature.com/focus/rnai/animations/animation/animation.htm, www.nature.com/mt/webfocus/rnai/index.html#2008 and www.silencinggenomes.org/.

The gene expression-regulating RNAi pathway is restricted to eukaryotic organisms, but homologous proteins of this pathway were also identified in prokaryote (Hutvagne and Simard, 2008). Furthermore, small noncoding RNAs have been also identified in various bacteria, whereby most of the small prokaryotic RNAs of known function regulate gene expression by binding to specific mRNAs or proteins (Vogel and Wagner, 2007). Thus, in the future it should not be principally impossible to develop RNAi-related applications in prokaryotic organisms.

The 'RNAi hype' is remarkable since the specific inhibition of the expression of a gene of interest by different nucleic acids-based technologies is not novel. In the 'pre-RNAi-era', various RNA- and DNA-based molecules were used for the construction of antisense oligonucleotides, hammerhead ribozymes, DNAzymes or aptamers which were characterized in detail for many gene expression knock out applications. Although such gene-silencing molecules were successfully applied against various targets, these technologies have been difficult to apply universally (Sullenger and Gilboa, 2002). In general, RNAi technology is assessed to be more potent than those alternative gene-silencing methodologies. Nevertheless, controversial data are available whether RNAi technology is indeed always the methodology of choice with the most pronounced efficiency and specificity. An experimental study directly analysed and compared the biological efficacies of siRNA molecules, antisense oligonucleotides and ribozymes against the same RNA target molecule (Miyagishi *et al.*, 2003). These analyses demonstrated that the siRNAs showed most prominent biological activity. In contrast, the application of antisense oligonucleotides directed against the FasL-specific mRNA (Ji *et al.*, 2003) showed a slightly more pronounced gene-silencing activity than the utilization of anti-FasL siRNAs (Ji *et al.*, 2005). In the case of stable down regulation of the MDR1/P-gp encoding mRNA in multidrug-resistant human cancer cells, hammerhead ribozyme constructs (Kowalski 2005) as well as a RNAi-based approach (Stege *et al.*, 2004) could completely reverse the drug-resistant phenotype.

Advantages and disadvantages of different nucleic acids technology-based gene silencing approaches are summarized in Table 10.4. The enormous interest in RNAi technology probably raised by the simultaneous completion of the human and other genome sequence projects at the time of introduction of RNAi technology in laboratory practice. Since then the target-specific modulation of the expression of each of the human and other species genes appears to be a real possibility, including the silencing of genes that are pathogenic to the host organism. Last but not least, the RNAi technology is enormously supported by biotech and pharmaceutical companies. Therefore, in 2006 the US pharmaceutical giant Merck & Co paid 1.1 billion US dollars for Sirna Therapeutics Inc., a tiny biotechnology firm focused on drugs based on RNA interference. In 2007, another billion-dollar deal was made by the Swiss pharmaceutical company Roche and the small biopharmaceutical company Alnylam Pharmaceuticals, Inc. As a start, Roche transferred 331 million dollars to Alnylam. Following the achievement of milestones, additional money will be paid.

The RNAi technology utilizes the physiological gene knock-out mechanism by entering the RNAi pathway on the level of siRNA duplexes (Fig. 10.1). Naturally, cells express long dsRNA molecules required for PTSG (He and Hannon, 2004). Physiologically, the first step of RNAi is initiated by the processing of gene regulatory dsRNAs by the ribonuclease III (RNase III)-like enzyme Dicer to smaller siRNAs (Hammond *et al.*, 2000). These siRNAs have two- to three-nucleotide 3′-overhanging ends and contain 5′-phosphate and free 3′-hydroxyl termini, which are characteristic for all RNase III-processing

Table 10.4 Advantages and disadvantages of gene- and gene product-silencing strategies (Lage, 2005)

Strategy	Advantages	Disadvantages
Low molecular weight agents	• Easy to administrate • Often inexpensive	• Often non-specific • May not be available • Off-target effects
Antisense oligonucleotides	• Easy to synthesize • Inexpensive • Modification to improve selectivity and efficacy	• Only exogenous delivery • Protein binding • Off-target effects • Induction of interferon response
Ribozymes	• Simple catalytic domain • Delivery as free molecules and by expression vectors	• Requirement of specific cleavage triplets • Protein binding
DNAzymes	• Inexpensive • Good catalytic properties • Modification for systemic delivery	• Only exogenous delivery • Off-target effects
RNAi	• High effective at low concentrations • Delivery as free siRNA molecules and by shRNA expression vectors	• Off-target effects • Induction of interferon response • Expensiveness of siRNAs

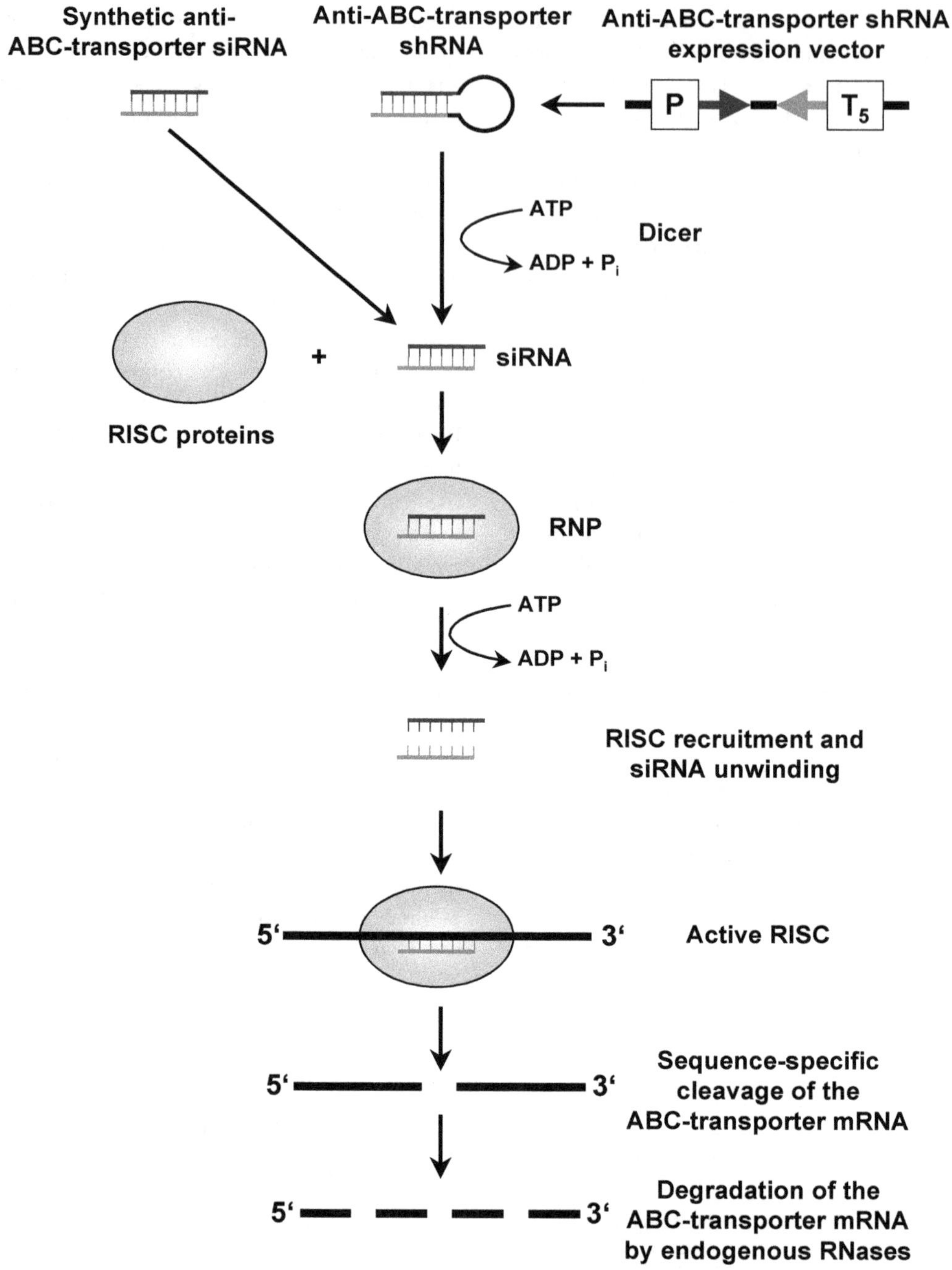

Figure 10.1 RNAi pathway utilized for suppression of ABC transporter-encoding mRNA expression in eukaryotic organisms. RNAi-mediating RNA molecules directed against ABC transporter can be chemically synthesized as siRNAs, or they can be synthesized from expression vectors such as plasmids, transposons or viruses as short hairpin RNAs (shRNAs) which will be intracellular cleaved into the corresponding siRNAs by the endogenous RNase III enzyme Dicer. DNA encoded expression cassettes or siRNAs are delivered into target cells by different methodologies. The intracellular siRNAs are recruited by members of the Argonaute family via their PAZ domain to form a ribonucleotide protein complex (RNP). The siRNAs are unwound before activation of RISC (RNA-induced silencing complex) and the strand complementary to the ABC transporter target mRNA is incorporated into RISC. Beside other proteins, RISC contains an endonuclease that cleaves only the target mRNA within the hybridized region. **P**, RNA-polymerase III promoter; T_5, RNA-polymerase III termination signal.

reactions (Bernstein *et al.*, 2001). In experimental as well as potential therapeutical approaches, the endoribonucleolytic activity of Dicer is merely used for the generation of siRNAs from short hairpin RNAs (shRNAs) encoded by plasmid-, transposon- or virus-based expression vectors (see paragraph 'Delivery of shRNAs'). In contrast to naturally produced siRNAs, chemically synthesized siRNAs commonly used for experimental analysis of gene function and 'therapeutically' siRNAs for specific knock-down of disease-associated gene products are dsRNA duplexes composed of a 21-nt sense and a 21-nt antisense strand. Both strands are hybridized in a manner to have a 2-nt 3′-overhang whereby duplexes with 3′-TT overhangs can be used without effect on the specificity of the mRNA target recognition or gene silencing efficiency (Elbashir *et al.*, 2001c). By interaction of these siRNAs with the protein interaction domain PAZ (PIWI-Argonaute-Zwille/Pinhead) (Cerutti *et al.*, 2000) of members of the Argonaute or PPD (PAZ/PIWI domain) protein family, the siRNAs are incorporated into a multiprotein RNA-inducing silencing complex (RISC). For incorporation of siRNAs into RISC, it is necessary that siRNAs are phosphorylated at the 5′-terminus (Nykanen *et al.*, 2001). RISC becomes activated by unwinding the siRNA duplex. The antisense RNA strand guides RISC to the homologous sequence on the target mRNA for endoribonucleolytic digestion. The cleavage of the target mRNA occurs at a single site, 10 nt from the 5′-phosphate residue of the antisense strand of the unwound siRNA molecule (Elbashir *et al.*, 2001b). Due to the loss of the 7-methylguanine cap structure of the 3′-cleavage product and the loss of the poly(A)-tail of the 5′-cleavage product, the digested target mRNA is no longer protected against endogenous RNases and can be degraded.

Delivery of siRNAs or shRNAs to human cells

The major obstacle in gene therapy is the unsolved problem of delivery of therapeutic genetic elements (Sanguino *et al.*, 2008). Accordingly, the delivery of therapeutic RNAi-mediating molecules is the most important problem that has to be solved before RNAi technology can become a realistic therapeutical option in clinical practice. In experimental studies, RNAi-mediated gene-silencing in human cancer cells can be achieved by transfection, e.g. using liposomes or electroporation, of synthetic siRNA molecules, or by gene transfer using expression cassettes consisting of DNA encoding RNA polymease III promoter-regulated dsRNA molecules. All these delivery strategies were already used for reversal of multidrug-resistant phenotypes in different tumour cell models (Lage, 2006).

Delivery of siRNAs

The most straightforward technique for intracellular activation of the RNAi pathway is the use of chemically synthesized siRNAs. However, besides siRNAs produced by chemical synthesis, siRNAs may be generated from long dsRNAs *in vitro* via recombinant Dicer (Myers *et al.*, 2003), by *in vitro* transcription using T7 RNA polymerase (Donze and Picard, 2002). Alternatively, siRNAs can be isolated from Drosophila embryo extracts (Holen *et al.*, 2002). Classic transfection of siRNA molecules using different physical methods such as liposome-mediated transfection (Gitlin *et al.*, 2002), electroporation (Siegmund *et al.*, 2002), or single-cell microinjection (Usui *et al.*, 2003) has been successfully applied. Commonly, treatment of mammalian cells with siRNAs typically results in a transient down regulation of the specific target mRNA after 1–2 days for 3–5 days.

Following the impressive demonstrations which show that several siRNAs delivery approaches can mediate specific gene-silencing in various cell culture models, different strategies were tested for *in vivo* delivery. Successful *in vivo* delivery of siRNAs to mice was reported by direct injection of therapeutic RNAs into the tail veins. By this approach, a down regulation of 90% of endogenous hepatitis C virus mRNA transcripts in the majority of liver cells after a single siRNA injection could be achieved (McCaffrey *et al.*, 2002). In the liver, the RNAi effect persists only for several days. Another study demonstrated that delivery of anti-Fas siRNAs into mice by pulse injection into the tail vein effectively reduced Fas protein expression in hepatocytes for 10 days and protected mice from Fas-induced fulminant hepatitis (Song *et al.*, 2003). Although, these studies demonstrated that *in vivo* delivery of siRNAs is possible, effective delivery to tissues *in vivo* is difficult. Since siRNAs tolerate only a few chemical modifications they are not readily taken up into tissues. Furthermore, dsRNA molecules including siRNAs are susceptible to degradation by nucleases in the bloodstream. A relatively simple solution of the problem of siRNA delivery was reported by Soutschek *et al.* (Soutschek *et al.*, 2004). In that study, siRNAs targeting the apolipoprotein B encoding mRNA were applied. These molecules contained selective stabilizing modifications and were joined to a cholesterol group chemically linked to the 3′-hydroxyl group of the sense strand. Intravenous injections of these siRNA–cholesterol conjugates in mice resulted in uptake into several tissues, whereby the levels of apolipoprotein B encoding mRNA was decreased by more than 50% in the liver and by 70% in the jejunum. This reduction resulted in a lowering of the levels of blood cholesterol comparable to that observed in apolipoprotein B knock out mice.

Delivery of shRNAs

A major limitation in the application of siRNAs is the transient nature of the gene-silencing effect. According to conventional low molecular weight pharmacological active compounds, a metabolic decomposition of the therapeutic siRNAs occurs in the target cells, here by RNases. Thus, more effective and stable long-termed RNAi-mediating systems have been developed. These strategies utilize vectors containing an expression cassette for the production of shRNAs (Brummelkamp *et al.*, 2002) that afterwards is intracellularly cleaved into the corresponding siRNAs by Dicer; alternatively, the expression cassettes could drive the production of sense and antisense strands separately, before the two strands hybridize inside the cell to form functional active siRNA molecules (Miyagishi and Taira, 2002; Lee *et al.*, 2002).

Although it would be possible to utilize RNA-polymerase II-driven promoters for expression of such short transcripts, in these gene-silencing expression vectors, in general RNA polymerase III-specific promoters are used, i.e. the H1-RNA promoter (Brummelkamp *et al.*, 2002), or the U6-RNA promoter (Paul *et al.*, 2002). These RNA-polymerase III-depending promoters have a defined transcription starting point, and a termination signal consisting of five consecutive thymidines (T5). Therewith, RNA polymerase III-depending promoters can be used to direct the synthesis of small RNA molecules of interest lacking a poly-adenosine tail. Cleavage of the RNA transcript at the termination site occurs after the second of the five uridines which are encoded by the T5 signal on the DNA matrix. Thus, the result of RNA-polymerase III promoter-depending RNA synthesis is the production of small RNA molecules that are similar to the ends of chemically synthesized siRNAs. Furthermore, they contain two 3′-overhanging uridines instead of two thymidines that are present in chemically synthesized siRNAs, which do not negatively influence the biological activity and correspond to the typical 3′-overhanging nucleotide residues of Dicer-processed RNAs. In shRNAs, the sequence of interest consists of a 19-nt sequence homologous to the target mRNA, linked with a 3–11-nt spacer sequence to the reverse complement of the same 19-nt target-specific sequence. The synthesized RNA transcript folds back to its complementary strand to form a 19-base pair shRNA molecule, which is then processed by Dicer to a corresponding siRNA molecule and passed into the RNAi pathway.

One of the most promising strategies to deliver shRNA expression cassettes to humans is the use of viral vectors. Already various studies are available reporting successful gene suppression by RNAi using lentiviral expression systems *in vitro* (An *et al.*, 2003; Van den Haute *et al.*, 2003) or applying adenovirus-based systems (Hosono *et al.*, 2004). Additionally, bacteria such as non-pathogenic *Escherichia coli* have been successfully utilized to deliver therapeutic shRNA molecules to human cancer cells (Xiang *et al.*, 2006). The *E. coli* were engineered to transcribe anti-catenin β-1 shRNAs from a plasmid containing the invasin gene Inv and the listeriolysin O gene HlyA, which encode two bacterial factors needed for transfer of therapeutic shRNAs into mammalian cells. Invasin permits non-invasive *E. coli* to enter β-1-integrin-expressing mammalian cells and listeriolysin O mediates the lysis of entry vesicles with the result that genetic material encoding therapeutic shRNAs escape from the membrane-surrounded vesicles. Upon oral or intravenous administration of these genetically modified *E. coli* a significant gene silencing in the intestinal epithelium and in human colon cancer xenografts in mice was induced. This study demonstrates the potential of bacteria-mediated RNAi for development of clinically compatible RNAi-based therapies.

Moreover, alternative delivery strategies may inject gene-silencing RNAi-mediating expression cassettes directly into the cancerous tissue via jet-injection (Walther *et al.*, 2004). Finally, alternative delivery strategies, based on the utilization of shRNA-encoding transposons (Heggestad *et al.*, 2004) are also used; however, new methods to deliver siRNAs or shRNAs into cells are continuously being developed and will continue to incorporate innovative features.

Clinical trials with RNAi-based therapeutics

The RNAi technology has been rapidly advanced from basic research to clinical trials. Currently, two different targets are under clinical investigation with RNAi-triggering therapeutics, the vascular endothelial growth factor (VEGF) pathway which contributes to growth of excess blood vessels for the wet form of age-related macular degeneration (AMD) and on the respiratory syncytial virus (RSV) genome, for the treatment of RSV infection (de Fougerolles *et al.*, 2007). For targeting the VEGF pathway, Cand5, an unmodified siRNA that targets VEGF-A has completed a phase II trial in patients with serious progressive wet AMD and has been reported to provide dose-related benefits including near vision and lesion size. Likewise, Cand5 is also being tested for efficacy against diabetic macular oedema in a phase II trial. On the other hand, Sirna-027, a chemically modified siRNA that targeting VEGF receptor 1 completed phase I trials in patients with wet AMD. The well-tolerated therapeutics Sirna-027 was demonstrated to stabilize or improve visual acuity in a subset of patients. Furthermore, two phase I intranasal trials with ALN-RSV01, an siRNA directed against the nucleocapsid-encoding mRNA of RSV, have been completed. This siRNA was found to be safe and well tolerated. These targets have been selected because they are easily accessible for siRNA therapeutics. In the case of AMD, direct injection of the RNAi-mediating molecules into the vitreal cavity efficiently targets siRNAs to the retina. Another advantage of the ocular compartment is that it is relatively free of nucleases. In the case of RSV infections, intranasal delivery to the lung by inhalation is non-invasive and directly targets the lung epithelium since lung epithelial cells are capable of naked siRNA uptake.

However, recent experiments in mice suggest that the siRNA treatment of the VEGF pathway in AMD does not work through the RNAi mechanism; instead it works through activation of Toll-like receptor 3 (TLR3) and induction of components of the innate immune response (Kleinman *et al.*, 2008). Thus, the clinical RNAi trials will not be the anticipated proof-of-concept for this technology in humans.

So far, no clinical RNAi trials have been started with cancer patients or patients suffering from infection diseases. Accordingly, ABC transporters as targets in clinical trials applying the RNAi platform are far away from being realistic.

RNAi-mediated inhibition of ABC transporters in human cancer cells

Various *in vitro* studies using siRNAs or shRNAs have demonstrated the potential use of the RNAi phenomenon to treat cancers (Lage, 2005). These strategies aim to particular features of neoplastic cells including the MDR phenotype. The first studies showing the proof of principle of RNAi-mediated reversal of cancer MDR by MDR1/P-gp inhibition were published in 2003 simultaneously (Nieth *et al.*, 2003; Wu *et al.*, 2003). In these studies, a transient down regulation of MDR1/P-gp expression by means of siRNAs reversed the multidrug-resistant phenotype in different cancer cell lines. Meanwhile, various additional studies were published demonstrating siRNA- or shRNA-dependent RNAi-mediated down-regulation of MDR1/P-gp in various cell models from different cancer entities; Table 10.5 lists a summary of some of these studies. Approaches with transient siRNA-mediated down regulation of the MDR1/P-gp expression resulted in relatively moderate inhibition in different cell systems (Table 10.5). Unfortunately these studies used dissimilar experimental techniques and different proliferation assay protocols or clonogenic assays, to determine the extent of drug resistance reversal; for that reason, the efficiencies of the different transient RNAi approaches

Table 10.5 Gene therapeutic approaches to inhibit ABC transporters in human cancer cells by RNA interference[1]

Cell lines	Tumour origin	Duration of RNAi	RNAi-mediating RNAs	Vector	References
MDR1/P-gp (ABCB1)					
A2780Dx5	Ovarian carcinoma	Transient	siRNA	–	Wu *et al.*, 2003; Arora *et al.*, 2004
Bel7402/5-Fu	Hepatocellular carcinoma	Transient	siRNA	–	Ren, 2006
BT325	Glioma	Stable	shRNA	Retrovirus	Zhao *et al.*, 2008
Caco-2	Colon carcinoma	Transient; stable	siRNA; shRNA	Plasmid	Celius *et al.*, 2004; Watanabe *et al.*, 2005
EPG85–257RDB	Gastric carcinoma	Transient	siRNA	–	Nieth et al, 2003
EPG85–257RDB	Gastric carcinoma	Stable	shRNA	Plasmid	Stege *et al.*, 2004
EPG85–257RDB	Gastric carcinoma	Transient	shRNA	Adenovirus	Kaszubiak *et al.*, 2007
EPP85–181RDB	Pancreatic carcinoma	Transient	siRNA	–	Nieth et al, 2003
EPP85–181RDB	Pancreatic carcinoma	Transient	shRNA	Adenovirus	Kaszubiak *et al.*, 2007
HCT-8	Colon cancer	Stable	shRNA	Retrovirus	Pichler *et al.*, 2005
HCT-15	Colon cancer	Stable	shRNA	Plasmid	Park *et al.*, 2008
HepG2/ADM	Hepatocellular carcinoma	Stable	shRNA	Plasmid	Chen *et al.*, 2006
K562/A02	CML	Transient; stable	siRNA; shRNA	Plasmid	Peng *et al.*, 2004; Gu *et al.*, 2006; Xiao *et al.*, 2006
K562/ADR	CML	Transient	siRNA	–	Lim *et al.*, 2007; Wei *et al.*, 2007
K562/Dox	CML	Stable	shRNA	Transposon	Widmer *et al.*, 2007
KB-8–5[2]	Cervix carcinoma, HeLa-derived cell line[2]	Transient	siRNA	–	Logashenko *et al.*, 2004
KB-C1[2]	Cervix carcinoma, HeLa-derived cell line[2]	Stable	shRNA	scAAV[6]	Xu *et al.*, 2005
KBv200[2]	Cervix carcinoma, HeLa-derived cell line[2]	Stable	shRNA	Plasmid	Shi *et al.*, 2006b
KD30	CML	Stable	shRNA	Plasmid	Yagüe *et al.*, 2004
MaTu/ADR[3]	Cervix carcinoma, HeLa-derived cell line[3]	Stable *in vitro*, transient *in vivo*	shRNA	Plasmid	Stein *et al.*, 2008
MCF-7[4]	Breast cancer	Transient	siRNA	–	Su *et al.*, 2005
MCF-7/ADR[5]	Breast cancer/ ovarian carcinoma[5]	Stable	shRNA	Plasmid	Li *et al.*, 2004

Cell lines	Tumour origin	Duration of RNAi	RNAi-mediating RNAs	Vector	References
MCF-7/AdrR[5]	Breast cancer/ ovarian carcinoma[5]	Transient, stable	siRNA; shRNA	Plasmid	Wu *et al.*, 2003; Gan *et al.*, 2005
MCF-7/BC-19	Breast cancer	Transient	siRNA	–	Wu *et al.*, 2003; Arora *et al.*, 2004
MCF-7-R	Breast cancer	Transient	siRNA	–	Stierle *et al.*, 2004; Sierle *et al.*, 2005; Li *et al.*, 2006; Stierle *et al.*, 2007
NCI/ADR-RES[5]	Breast cancer/ ovarian carcinoma[5]	Transient; stable	siRNA; shRNA	Plasmid	Xu *et al.*, 2004
NCI/ADR-RES[5]	Breast cancer/ ovarian carcinoma[5]	Stable	shRNA	scAAV[6]	Xu *et al.*, 2005
NCI-H460	Lung carcinoma	Transient	siRNA	–	Xiao *et al.*, 2008
NIH3T3.MDR	Mouse fibroblast NIH3T3-derived cell line	Transient	Altriol-modified siRNA	–	Fisher *et al.*, 2007
NIH-MDR-G185	Mouse fibroblast NIH3T3-derived cell line	Transient	siRNA	–	Stierle *et al.*, 2004
OVCAR8TR	Ovarian carcinoma	Transient; stable	siRNA; shRNA	Plasmid	Duan *et al.*, 2004; Lou *et al.*, 2006
SGC7901/VCR	Gastric carcinoma	Transient	siRNA	–	Gao *et al.*, 2006
SKOV-3TR	Ovarian carcinoma	Transient; stable	siRNA; shRNA	Plasmid	Duan *et al.*, 2004
MRP1 (ABCC1)					
K562/ADM	CML	transient	siRNA	–	Song *et al.*, 2008
MRP2 (ABCC2)					
A2780/cp70	Ovarian carcinoma	transient; stable	shRNA	Plasmid	Ma *et al.*, 2008
A2780RCIS	Ovarian carcinoma	transient; stable	siRNA; shRNA	Plasmid	Materna *et al.*, 2006
CNE2	Nasopharyngeal carcinoma	stable	shRNA	Lentivirus	Xie *et al.*, 2008
BCRP (ABCG2)					
BeWo	Choriocarcinoma	transient	siRNA	–	Ee *et al.*, 2004
EPG85–257RNOV	Gastric carcinoma	transient; stable	siRNA; shRNA	Plasmid	Priebsch *et al.*, 2006

Table 10.5 Continued

Cell lines	Tumour origin	Duration of RNAi	RNAi-mediating RNAs	Vector	References
MCF-7/AdrVp	Breast cancer	transient	siRNA	Plasmid	Priebsch *et al.*, 2006
MCF-7/BCRP	Breast cancer	transient, stable	shRNA	Plasmid	Lv *et al.*, 2007
MCF-7/MX	Breast cancer	transient	siRNA	Plasmid	Priebsch *et al.*, 2006
MCF-7/MX100	Breast cancer	stable	shRNA	Plasmid	Li *et al.*, 2005

1 Studies using RNAi only for functional genomics applications are not considered.

2 KB cells were originally thought to be derived from an epidermal carcinoma of the mouth, but were subsequently found to have been established via contamination with the cervix carcinoma cell line HeLa (Lacroix, 2008).

3 The MaTu cell line was originally described to be derived from a human mammary carcinoma, but was recently found to may be a mutated derivative of HeLa cells (Stein *et al.*, 2008).

4 MDR1/P-gp expression following adrenaline stimulation.

5 Since the exact designation of cell lines is not unambiguously in all studies, the multidrug-resistant cell line MCF7/ADR appear to be identical with the cell line MCF7/AdrR. By karyotyp analyses it could be demonstrated that MCF7/AdrR cells are not derived from MCF7 breast cancer cells. These cells – later re-designated as NCI/ADR-RES – are derived from OVCAR-8 ovarian adenocarcinoma cells (Liscovitch and Ravid, 2007).

6 Self-complementary adeno-associated viral vector.

are not directly comparable. However, the most pronounced siRNA-mediated transient MDR reversal of nearly 90% was achieved in the pancreatic carcinoma-derived cell line EPP85–181RDB (Nieth *et al.*, 2003). These strong gene-silencing effects might be explained by the well-assessable secondary structure of the MDR1-P-gp encoding mRNA in this region that was experimental demonstrated by hammerhead ribozyme cleavage experiments previously (Kowalski *et al.*, 2005). Similar observations, i.e. a siRNA directed against a well-characterized hammerhead ribozyme cleavage site showed more biological activity than a siRNA designed according the usual recommendations, were also made by the experimental characterization of siRNAs directed against the ABC transporters MRP2 (Materna *et al.*, 2006) and BCRP (Priebsch *et al.*, 2006). Altogether these data indicate that the efficacy of the RNAi effect may be siRNA sequence-dependent as well as cell line-dependent.

Moreover, stable anti-MDR1/P-gp shRNA expression vectors have been used to modulate the MDR phenotype (Table 10.5). In one of these studies, the stable RNAi effect was similar or less effective than a transient approach in the same cell model, the ovarian carcinoma-derived cell line SKOV-3TR (Duan *et al.*, 2004). A more pronounced efficacy of a stable RNAi approach than a transient RNAi strategy was achieved in the breast carcinoma cell line NCI/ADR-RES (Xu *et al.*, 2004); however, also in this system the reversal of the multidrug-resistant phenotype was not complete. In these studies U6-RNA promoter-driven shRNA expression vectors were used. In contrast, two other investigations used H1-RNA promoter-depending shRNA expression cassettes (Stege *et al.*, 2004; Yagüe *et al.*, 2004). In these studies, the multidrug-resistant phenotype of the gastric carcinoma cell line EPG85–257RDB or the chronic myelogenous leukaemia (CML) K562-derived cell line KD30 was completely reversed by shRNA-mediated RNAi. However, a H1 promoter-driven RNAi phenomenon in the multidrug-resistant cancer cell line MCF-7/ADR could not decrease the MDR1/P-gp mRNA expression completely (Li *et al.*, 2004). These data indicate that the optimal anti-MDR1/P-gp shRNA expression cassette is cell system-depended and should be chosen empirically.

Besides these plasmid vector-based strategies for delivery of anti-MDR1/P-gp shRNAs also viral vectors containing an anti-MDR1/P-gp shRNA expression cassette were constructed. By using a self-complementary recombinant adeno-associated virus (AAV) shRNAs were efficiently delivered into multidrug-resistant human breast cancer and oral cancer cells (Xu *et al.* 2005). Using this viral delivery strategy, the level of reduction in cell surface MDR1/P-gp expression was up to 90%. By application of an adenovirus expressing shRNAs directed against the MDR1/P-gp-specific transcript, a complete knock down of the MDR1/P-gp expression could be achieved in the multidrug-resistant gastric carcinoma cell line EPG85–257RDB as well as in the MDR pancreatic carcinoma cell model EPP85–181RDB (Kaszubiak *et al.*, 2007).

In vivo application of RNAi technology for ABC transporter inhibition

Different reports are available in the literature that illustrates the *in vivo* application of the RNAi technology to modulate the MDR1/P-gp activity in human cancer cells. For example, multidrug-resistant HeLa cervix carcinoma-derived KBv200 cells (Lacroix, 2008) were stably transfected with an anti-MDR1 shRNA expression plasmid, and grown as xenografts in nude mice (Shi *et al.*, 2006b). In these cells, vincristine treatment inhibited the growth of the shRNA clone-derived solid tumour by 42%. A similar approach was used to treat human HCT-8 colon carcinoma cells; these cells were transduced with an anti-MDR1 shRNA expressing retrovirus (Pichler *et al.*, 2005). In these studies, a shRNA-mediated down-regulation of MDR1/P-gp transport activity could be measured. Therewith, both studies demonstrated RNAi-mediated *in vivo* modulation of cancer MDR; however, the drawback from these studies was the lack of a strategy for efficient delivery of shRNAs to cancer patients in the clinic. In a third study, mice were injected intravenously with an shRNA-expressing plasmid DNA or with synthetic siRNAs (Matsui *et al.*, 2005). The endogenous murine mdr1a mRNA expression level in the liver was then determined and was found to decreased around 50% demonstrating that *in vivo* RNAi directed against endogenous genes including ABC transporters is a promising strategy to reverse drug resistance.

A successful *in vivo* reversal of cancer MDR by the RNAi technology using an efficient shRNA delivery system which already demonstrated its clinical applicability in a Phase I clinical trial (Walther *et al.*, 2008) was reported by Stein *et al.* (2008). In this study, a nonviral jet-injection technology was applied to deliver naked shRNA-vector constructs for direct intratumoural *in vivo* transfer. The highly efficient anti-MDR1 shRNA expression vectors were applied in the human MDR1/P-gp overexpressing MaTu/ADR cancer xenograft bearing mice. The expression level of the MDR1/P-gp encoding mRNA decreased to more than 90% and the corresponding ABC transporter protein was no longer detectable in the tumours after anti-MDR1/P-gp shRNA vector injection. Two jet-injections of shRNA vectors into the tumours combined with two intravenous administrations of doxorubicin were sufficient for complete reversal of the drug-resistant phenotype. The study demonstrates that *in vivo* delivery of therapeutic shRNA-encoding DNA by the jet-injection technique is extreme effective, here in reversing MDR1/P-gp-mediated MDR *in vivo*, and is therefore a promising strategy for the development of novel therapeutic RNAi-basing strategies for the potential treatment of cancer patients.

RNAi pathway homologues in prokaryota

Although the RNAi pathway is restricted to eukaryotic organisms, the expression of homologues of this pathway is also found in prokaryote. In this kingdom the RNAi pathway is well represented by members of the family of the multidomain Argonaute proteins. Argonautes are organized in four distinct domains: the N-terminal, PAZ, Mid and PIWI domains (Hutvagne and Simard, 2008). In eukaryote, the PAZ domain consists of a modified oligonucleotide/oligosaccharide-binding fold and may be responsible for the binding to the 3′-ends of siRNA molecules. PIWI domains are found only in Argonaute proteins; their exact function is not yet known. In prokaryote, the functions of Argonaute proteins are still unknown but based on their structural similarity to eukaryotic Argonaute proteins they could be involved in nucleic acid-directed regulatory pathways related to RNA silencing (Hall, 2005). Unfortunately, so far this has not been demonstrated, although efforts have been done to trigger gene silencing by siRNAs in prokaryotic cells. Preliminary data suggest the successful silencing of prokaryotic genes in Gram-positive and Gram-negative bacteria as well as in Mycobacteria (Patzel *et al.* 2006). In these investigations the observed phenotypes ranged from transient gene knockdown to persistent gene knockout. However, the observed gene-silencing effects could not be unambiguously separated from potential antisense effects of the antisense strands of the double–stranded siRNA molecules. Thus, this all means that research on this field should continue to demonstrate whether the RNAi strategy would be useful for prokaryotic organisms. Before RNAi can be considered for therapeutic applications, e.g. targeting ABC transporter encoding transcripts in prokaryota, functional genomics will demonstrate whether RNAi may be really an option in those organisms.

RNAi in protozoa

The phenomenon of RNAi in protozoan parasites was first reported in *Trypanosoma brucei* in 1998 (Ngo *et al.*, 1998). In this organism, the RNAi pathway shares many of the characteristic features of known RNAi pathways from model organisms, including the generation of siRNAs (Djikeng *et al.*, 2003), as well as the requirement of Argonaute and Dicer proteins (Shi *et al.*, 2006a). Various studies followed and analysed whether other protozoan parasites express biological active RNAi pathways and demonstrated that some, but not all, protozoan parasites are endowed with components of the RNAi pathway, although the presence or absence of RNAi machinery cannot be explained by parasite phylogeny (Table 10.6). It rather appears that the phenomenon of RNAi has been repeatedly lost during evolution, due presumably to an absence of selective pressure (Militello *et al.*, 2008). Taxonomic analyses of the distribution and the phylogenetic relationships of key components of the RNAi pathway suggested that relatively complex RNAi machinery was already present in the last common ancestor of eukaryotes. It consisted, at a minimum, of one Argonaute-like polypeptide, one PIWI-like protein, one Dicer, and one RNA-dependent RNA polymerase (Cerutti and Casas-Mollano, 2006). The ancestral biological role of these components may have been fundamental in defence responses against genomic parasites such as transposable elements and viruses. The RNAi pathways in eukaryotic ancestors may have been capable of both small RNA-mediated transcript degradation as well as transcriptional repression. Both roles appear to be widespread among living eukaryotes and this diversification of function could account for the evolutionary conservation of duplicated Argonaute-PIWI proteins. In contrast, additional RNAi-mediated pathways such as RNA-directed DNA methylation, programmed genome rearrangements, meiotic silencing by unpaired DNA and miRNA-mediated gene regulation may have evolved independently in specific lineages.

Despite the expression of the RNAi pathways in protozoan parasites, the current utility of this technology in parasite basic and clinical research is questionable. First, problems have arisen with the specificity of RNAi. Treatment of the parasites with dsRNA resulted, in many cases, in non-specific effects. Second, the current RNAi methods have a limited efficiency and effects are sometimes difficult to reproduce (Geldhof *et al.*, 2007). The future application of RNAi in parasite functional genomics will greatly depend on how these difficulties can be solved. Optimization of RNAi-mediating delivery methods and *in vitro* culture conditions will be the major challenges in functional studies. Accordingly, potential therapeutic RNAi-based strategies in protozoa including the silencing of ABC transporters are far away from clinical development and application. Currently, merely the use of RNAi technology for the functional characterization and for the analysis of the potential role of ABC transporters in drug resistance of parasites appears to be a realistic option.

Conclusions and future trends

In conclusion, RNAi technology has an enormous potential for the development of gene therapeutic strategies against human cancer cells and other disorders including infectious diseases.

Table 10.6 Analyses of the RNAi pathway in protozoan parasites (Militello et al., 2008)

Organism	Methods for RNAi analysis	Dicer and Argonaute homologues	RNAi reports	siRNA generation	Endogenous siRNAs
Trypanosoma brucei	Bioinformatic and functional RNAi support – RNAi use is routine	+ (Shi *et al.*, 2006a)	+ (Ngo *et al.*, 1998)	+ (Djikeng *et al.*, 2001)	+ (Djikeng *et al.*, 2001)
Trypanosoma congolense	Bioinformatic and functional RNAi support – RNAi use is routine	+ (Shi *et al.*, 2006a)	+ (Inoue *et al.*, 2002)		
Trypanosoma cruzi	No bioinformatic and functional RNAi support	– (Ullu *et al.*, 2004)	– (DaRocha *et al.*, 2004)		
Leishmania major	No bioinformatic and functional RNAi support	– (Ullu *et al.*, 2004)	– (Robinson and Beverley, 2003)		
Leishmania infantum	No bioinformatic and functional RNAi support	– (Peacock *et al.*, 2007)			
Cryptosporidium spp.	No bioinformatic and functional RNAi support	– (Militello *et al.*, 2008)			
Theileria spp.	No bioinformatic and functional RNAi support	– (Militello *et al.*, 2008)			
Babesia bovis	No bioinformatic and functional RNAi support	– (Militello *et al.*, 2008)			
Eimeria tenella	No bioinformatic and functional RNAi support	– (Militello *et al.*, 2008)			
Leishmania braziliensis	Bioinformatic or functional RNAi support – RNAi use is not routine	+ (Peacock *et al.*, 2007)			

Table 10.6 Continued

Organism	Methods for RNAi analysis	Dicer and Argonaute homologues	RNAi reports	siRNA generation	Endogenous siRNAs
Giardia intestinalis	Bioinformatic or functional RNAi support – RNAi use is not routine	+ (Cerutti and Casas–Mollano, 2006)			+ (Ullu *et al.*, 2005)
Entamoeba histolytica	Bioinformatic or functional RNAi support – RNAi use is not routine	+ (Abed and Ankri, 2005)	+ (Kaur and Lohia, 2004)		
Trichomonas vaginalis	Bioinformatic or functional RNAi support – RNAi use is not routine	+ (Carlton *et al.*, 2007)			
Toxoplasma gondii	Bioinformatic or functional RNAi support – RNAi use is not routine	+ (Ullu *et al.*, 2004)	+ (Al–Anouti *et al.*, 2003)		
Plasmodium spp.	Bioinformatic or functional RNAi support – RNAi use is not routine	– (Cerutti and Casas–Mollano, 2006)	+ (McRobert and McConkey, 2002)	+ (Mohmmed *et al.*, 2003)	– (Rathjen *et al.*, 2006)

In respect of the development of therapeutic strategies, RNAi technology may complete the available chemotherapeutic anticancer strategies including conventional cytotoxic compounds and targeted agents such as therapeutic antibodies or small molecule kinase inhibitors. The utilization of the RNAi platform for reversal of cancer MDR by inhibition of ABC transporters may be one of the most interesting applications. Further animal experiments will demonstrate the therapeutic efficacy of the RNAi technology for silencing various targets *in vivo* including the targeting of pathogens involved in infectious diseases. Furthermore, the on-going clinical trials have to demonstrate the proof-of-concept for treatment of humans with the RNAi approach.

However, the principal challenge of all gene therapy strategies is the delivery of the therapeutic agents to the target tissue at an efficacious dose. The main obstacles to successful RNAi delivery *in vivo* involve achieving proper pharmacokinetics to ensure drug exposure to the relevant target cell type and harnessing mechanisms to promote cellular uptake and the release of the drug into the cytoplasm. There are some promising delivery strategies on the horizon for viral therapies, employing replication competent viruses. On the other hand, there are some non-viral delivery technologies, such as chemically synthesized nanoparticles, the use of covalent conjugation and non-covalent complexes or direct jet-injection of the therapeutic nucleic acids into the target tissue. These strategies offer new possible opportunities. But targeted delivery has yet to be optimized for most cell types and tissues. Solutions of these problems will probably need to be individually tailored for each target cell type and disease indication. In optimizing treatment protocols, it is vital that the key parameters of efficacy and safety are simultaneously assessed to rapidly arrive at the formulations which have the greatest therapeutic window. Thus, for treatment of human cancers, the improvement of therapeutical RNAi applications will be coupled with the progress in the development of these delivery technologies. If progress in delivery of RNAi-mediating agents will be made, it will be possible to advance RNAi therapeutics rapidly into clinical trials for many diseases, including those which remain untreatable or poorly treated by conventional small molecular weight drugs.

Currently, the therapeutic application of the RNAi technology in eukaryotic microorganism is far away from being realistic, but the use of RNAi strategies in somatic cell genetics of these organisms has become laboratory practice. Future developments may help to design experimental concepts to improve this technology for potential therapeutic options. Likewise, ABC transporters in eukaryotic microorganisms appear as potential therapeutic targets applying RNAi technology.

At this time, it is not clear whether the RNAi technology is suited in the future for functional investigation of prokaryotic molecules. Although efforts were made to trigger gene silencing by siRNAs in bacteria the major breakthrough is missing. However, if this approach will be successful in the future, the RNAi technology could also be considered for reversal of ABC transporter-mediated drug resistance in prokaryotic pathogens.

References

Abed, M., and Ankri, S. (2005). Molecular characterization of *Entamoeba histolytica* RNase III and AGO2, two RNA interference hallmark proteins. Exp. Parasitol. *110*, 265–269.

Ackers, J.P., and Mirelman, D. (2006). Progress in research on *Entamoeba histolytica* pathogenesis. Curr. Opin. Microbiol. *9*, 367–373.

Al-Anouti, F., Quach, T., and Ananvoranich, S. (2003). Double-stranded RNA can mediate the suppression of uracil phosphoribosyltransferase expression in *Toxoplasma gondii*. Biochem. Biophys. Res. Commun. *302*, 316–323.

Allikmets, R., Schriml, L.M., Hutchinson, A., Romano-Spica, V., and Dean, M. (1998). A human placenta-specific ATP-binding cassette gene (ABCP) on chromosome 4q22 that is involved in multidrug resistance. Cancer Res. *58*, 5337–5339.

An, D.S., Xie, Y., Mao, S.H., Morizono, K., Kung, S.K., and Chen, I.S. (2003). Efficient lentiviral vectors for short hairpin RNA delivery into human cells. Hum. Gene Ther. *14*, 1207–1212.

Arora, S., Yang, J.M., Utsumi, R., Okamoto, T., Kitayama, T., and Hait, W.N. (2004). P-glycoprotein mediates resistance to histidine kinase inhibitors. Mol. Pharmacol. *66*, 460–467.

Awasthi, Y.C., Sharma, R., Yadav, S., Dwivedi, S., Sharma, A. and Awasthi, S. (2007). The non-ABC drug transporter RLIP76 (RALBP-1) plays a major role in the mechanisms of drug resistance. Curr. Drug Metab. *8*, 315–323.

Barrasa, M.I., Tercero, J.A., Lacalle, R.A., and Jimenez, A. (1995). The ard1 gene from *Streptomyces capreolus* encodes a polypeptide of the ABC transporters superfamily which confers resistance to the aminonucleoside antibiotic A201A. Eur. J. Biochem. *228*, 562–569.

Belinsky, M.G., Chen, Z.S., Shchaveleva, I., Zeng, H., and Kruh, G.D. (2002). Characterization of the drug resistance and transport properties of multidrug resistance protein 6 (MRP6, ABCC6). Cancer Res. *62*, 6172–6177.

Bernstein, E., Caudy, A.A., Hammond, S.M., and Hannon, G.J. (2001). Role for a bidentate ribonuclease in the initiation step of RNA interference. Nature *409*, 363–366.

Bissinger, P.H., and Kuchler, K. (1994). Molecular cloning and expression of the *Saccharomyces cerevisiae* STS1 gene product. A yeast ABC transporter conferring mycotoxin resistance. J. Biol. Chem. *269*, 4180–4186.

Blackmore, C.G., McNaughton, P.A., and van Veen, H.W. (2001). Multidrug transporters in prokaryotic and eukaryotic cells: physiological functions and transport mechanisms. Mol. Membr. Biol. *18*, 97–103.

Blattner, F.R., Plunkett, G. 3rd, Bloch, C.A., Perna, N.T., Burland, V., Riley, M., Collado-Vides, J., Glasner, J.D., Rode, C.K., Mayhew, G.F., Gregor, J., Davis, N.W., Kirkpatrick, H.A., Goeden, M.A., Rose, D.J., Mau, B., and Shao, Y. (1997). The complete genome sequence of *Escherichia coli* K-12. Science *277*, 1453–1474.

Bolhuis, H., Molenaar, D., Poelarends, G., van Veen, H.W., Poolman, B., Driessen, A.J., and Konings, W.N. (1994). Proton motive force-driven and ATP-dependent drug extrusion systems in multidrug-resistant *Lactococcus lactis*. J. Bacteriol. *176*, 6957–6964.

Boonstra, R., Timmer-Bosscha, H., van Echten-Arends, J., van der Kolk, D.M., van den Berg, A., de Jong, B., Tew, K.D., Poppema, S., and de Vries, E.G. (2004). Mitoxantrone resistance in a small cell lung cancer cell line is associated with ABCA2 upregulation. Br. J. Cancer *90*, 2411–2417.

Borst, P., de Wolf, C., and van de Wetering, K. (2007). Multidrug resistance-associated proteins 3, 4, and 5. Pflugers Arch. *453*, 661–673.

Borst, P., Evers, R., Kool, M., and Wijnholds, J. (2000). A family of drug transporters: the multidrug resistance-associated proteins. J. Natl. Cancer Inst. *92*, 1295–1302.

Borst, P., and Ouellette, M. (1995). New mechanisms of drug resistance in parasitic protozoa. Annu. Rev. Microbiol. *49*, 427–460.

Brummelkamp, T.R., Bernards, R., and Agami, R. (2002). A system for stable expression of short interfering RNAs in mammalian cells. Science *296*, 550–553.

Callahan, H.L., and Beverley, S.M. (1991). Heavy metal resistance: a new role for P-glycoproteins in *Leishmania*. J. Biol. Chem. *266*, 18427–18430.

Carlton, J.M., Hirt, R.P., Silva, J.C., Delcher, A.L., Schatz, M., Zhao, Q., Wortman, J.R., Bidwell, S.L., Alsmark, U.C., Besteiro, S., Sicheritz-Ponten, T., Noel, C.J., Dacks, J.B., Foster, P.G., Simillion, C., Van de Peer, Y., Miranda-Saavedra, D., Barton, G.J., Westrop, G.D., Müller, S., Dessi, D., Fiori, P.L., Ren, Q., Paulsen, I., Zhang, H., Bastida-Corcuera, F.D., Simoes-Barbosa, A., Brown, M.T., Hayes, R.D., Mukherjee, M., Okumura, C.Y., Schneider, R., Smith, A.J., Vanacova, S., Villalvazo, M., Haas, B.J., Pertea, M., Feldblyum, T.V., Utterback, T.R., Shu, C.L., Osoegawa, K., de Jong, P.J., Hrdy, I., Horvathova, L., Zubacova, Z., Dolezal, P., Malik, S.B., Logsdon, J.M. Jr., Henze, K., Gupta, A., Wang, C.C., Dunne, R.L., Upcroft, J.A., Upcroft, P., White, O., Salzberg, S.L., Tang, P., Chiu, C.H., Lee, Y.S., Embley, T.M., Coombs, G.H., Mottram, J.C., Tachezy, J., Fraser-Liggett, C.M., and Johnson, P.J. (2007). Draft genome sequence of the sexually transmitted pathogen *Trichomonas vaginalis*. Science *315*, 207–212.

Cedeno, J.R., and Krogstad, D.J. (1983). Susceptibility testing of *Entamoeba histolytica*. J. Infect. Dis. *148*, 1090–1095.

Celius, T., Garberg, P., and Lundgren, B. (2004). Stable suppression of MDR1 gene expression and function by RNAi in Caco-2 cells. Biochem. Biophys. Res. Commun. *324*, 365–371.

Cerutti, H., and Casas-Mollano, J.A. (2006). On the origin and functions of RNA-mediated silencing: from protists to man. Curr. Genet. *50*, 81–99.

Cerutti, L., Mian, N., and Bateman, A. (2000). Domains in gene silencing and cell differentiation proteins: the novel PAZ domain and redefinition of the Piwi domain. Trends Biochem. Sci. *25*, 481–482.

Chen, Z.S., Lee, K., and Kruh, G.D. (2001). Transport of cyclic nucleotides and estradiol 17-beta-D-glucuronide by multidrug resistance protein 4. Resistance to 6-mercaptopurine and 6-thioguanine. J. Biol. Chem. *276*, 33747–33754.

Chen, Z.S., Lee, K., Walther, S., Raftogianis, R.B., Kuwano, M., Zeng, H., and Kruh, G.D. (2002). Analysis of methotrexate and folate transport by multidrug resistance protein 4 (ABCC4): MRP4 is a component of the methotrexate efflux system. Cancer Res. *62*, 3144–3150.

Chen, K.G., Szakács, G., Annereau, J.P., Rouzaud, F., Liang, X.J., Valencia, J.C., Nagineni, C.N., Hooks, J.J., Hearing, V.J., and Gottesman, M.M. (2005). Principal expression of two mRNA isoforms (ABCB 5alpha and ABCB 5beta) of the ATP-binding cassette transporter gene ABCB 5 in melanoma cells and melanocytes. Pigment Cell Res. *18*, 102–112.

Chen, X.P., Wang, Q., Guan, J., Huang, Z.Y., Zhang, W.G., and Zhang, B.X. (2006). Reversing multidrug resistance by RNA interference through the suppression of MDR1 gene in human hepatoma cells. World J. Gastroenterol. *12*, 3332–3337.

Childs, S., Yeh, R.L., Hui, D., and Ling, V. (1998). Taxol resistance mediated by transfection of the liver-specific sister gene of P-glycoprotein. Cancer Res. *58*, 4160–4167.

Chiquero, M.J., Perez-Victoria, J.M., O'Valle, F., González-Ros, J.M., del Moral, R.G., Ferragut, J.A., Castanys, S., and Gamarro, F. (1998). Altered drug membrane permeability in a multidrug-resistant *Leishmania tropica* line. Biochem. Pharmacol. *55*, 131–139.

Chow, L.M., Wong, A.K., Ullman, B., and Wirth, D.F. (1993). Cloning and functional analysis of an extrachromosomally amplified multidrug resistance-like gene in *Leishmania enriettii*. Mol. Biochem. Parasitol. *60*, 195–208.

Cole, S.P., Bhardwaj, G., Gerlach, J.H., Mackie, J.E., Grant, C.E., Almquist, K.C., Stewart, A.J., Kurz, E.U., Duncan,

A.M. and Deeley, R.G. (1992). Overexpression of a transporter gene in a multidrug-resistant human lung cancer cell line. Science *258*, 1650–1654.

Cosgrove, S.E., and Fowler, V.G. Jr. (2008). Management of methicillin-resistant *Staphylococcus aureus* bacteremia. Clin. Infect. Dis. *46 Suppl. 5*, S386-S393.

Cui, Y., König, J., Buchholz, J.K., Spring, H., Leier, I., and Keppler, D. (1999). Drug resistance and ATP-dependent conjugate transport mediated by the apical multidrug resistance protein, MRP2, permanently expressed in human and canine cells. Mol. Pharmacol. *55*, 929–937.

DaRocha, W.D., Otsu, K., Teixeira, S.M., and Donelson, J.E. (2004). Tests of cytoplasmic RNA interference (RNAi) and construction of a tetracycline-inducible T7 promoter system in *Trypanosoma cruzi*. Mol. Biochem. Parasitol. *133*, 175–186.

de Fougerolles, A., Vornlocher, H.P., Maraganore, J., and Lieberman, J. (2007). Interfering with disease: a progress report on siRNA-based therapeutics. Nat. Rev. Drug Discov. *6*, 443–453.

de Jong, M.C., Slootstra, J.W., Scheffer, G.L., Schroeijers, A.B., Puijk, W.C., Dinkelberg, R., Kool, M., Broxterman, H.J., Meloen, R.H., and Scheper, R.J. (2001). Peptide transport by the multidrug resistance protein MRP1. Cancer Res. *61*, 2552–2557.

Dean, M., Rzhetsky, A., and Allikmets, R. (2001). The human ATP-binding cassette (ABC) transporter superfamily. Genome Res. *11*, 1156–1166.

Decottignies, A., and Goffeau, A. (1997). Complete inventory of the yeast ABC proteins. Nat. Genet. *15*, 137–145.

Decottignies, A., Grant, A.M., Nichols, J.W., de Wet, H., McIntosh, D.B., Goffeau, A. (1998). ATPase and multidrug transport activities of the overexpressed yeast ABC protein Yor1p. J. Biol. Chem. *273*, 12612–12622.

Djikeng, A., Shi, H., Tschudi, C., and Ullu, E. (2001). RNA interference in *Tripanosoma brucei*: cloning of small interfering RNAs provides evidence for retroposon-derived 24–26-nucleotide RNAs. RNA *7*, 1522–1530.

Djikeng, A., Shi, H., Tschudi, C., Shen, S., and Ullu, E. (2003). An siRNA ribonucleoprotein is found associated with polyribosomes in *Trypanosoma brucei*. RNA *9*, 802–808.

Donze, O., and Picard, D. (2002). RNA interference in mammalian cells using siRNAs synthesized with T7 RNA polymerase. Nucleic Acids Res. *30*, e46.

Doyle, L.A., and Ross, D.D. (2003). Multidrug resistance mediated by the breast cancer resistance protein BCRP (ABCG2). Oncogene *22*, 7340–7358.

Doyle, L.A., Yang, W., Abruzzo, L.V., Krogmann, T., Gao, Y., Rishi, A.K., and Ross, D.D. (1998). A multidrug resistance transporter from human MCF-7 breast cancer cells. Proc. Natl. Acad. Sci. U.S.A. *95*, 15665–15670.

Duan, Z., Brakora, K.A., and Seiden, M.V. (2004). Inhibition of ABCB1 (MDR1) and ABCB4 (MDR3) expression by small interfering RNA and reversal of paclitaxel resistance in human ovarian cancer cells. Mol. Cancer Ther. *3*, 833–838.

Ee, P.L., He, X., Ross, D.D., and Beck, W.T. (2004). Modulation of breast cancer resistance protein (BCRP/ABCG2) gene expression using RNA interference. Mol. Cancer Ther. *3*, 1577–1583.

Elbashir, S.M., Harborth, J., Lendeckel, W., Yalcin, A., Weber, K., and Tuschl, T. (2001a). Duplexes of 21-nucleotide RNAs mediate RNA interference in cultured mammalian cells. Nature *411*, 494–498.

Elbashir, S.M., Lendeckel, W., and Tuschl, T. (2001b). RNA interference is mediated by 21- and 22-nucleotide RNAs. Genes Dev. *15*, 188–200.

Elbashir, S.M., Martinez, J., Patkaniowska, A., Lendeckel, W., and Tuschl, T. (2001c). Functional anatomy of siRNAs for mediating efficient RNAi in *Drosophila melanogaster* embryo lysate. EMBO J. *20*, 6877–6888.

Faraut-Gambarelli, F., Piarroux, R., Deniau, M., Giusiano, B., Marty, P., Michel, G., Faugère, B., and Dumon, H. (1997). In vitro and *in vivo* resistance of *Leishmania infantum* to meglumine antimoniate: a study of 37 strains collected from patients with visceral leishmaniasis. Antimicrob. Agents Chemother. *41*, 827–830.

Fidock, D.A., Nomura, T., Talley, A.K., Cooper, R.A., Dzekunov, S.M., Ferdig, M.T., Ursos, L.M., Sidhu, A.B., Naudé, B., Deitsch, K.W., Su, X.Z., Wootton, J.C., Roepe, P.D., and Wellems, T.E. (2000). Mutations in the *P. falciparum* digestive vacuole transmembrane protein PfCRT and evidence for their role in chloroquine resistance. Mol. Cell *6*, 861–871.

Fire, A., Xu, S., Montgomery, M.K., Kostas, S.A., Driver, S.E., and Mello, C.C. (1998). Potent and specific genetic interference by double-stranded RNA in *Caenorhabditis elegans*. Nature *391*, 806–811.

Fisher, M., Abramov, M., Van Aerschot, A., Xu, D., Juliano, R.L., and Herdewijn, P. (2007). Inhibition of MDR1 expression with altritol-modified siRNAs. Nucleic Acids Res. *35*, 1064–1074.

Foote, S.J., Thompson, J.K., Cowman, A.F., and Kemp, D.J. (1989). Amplification of the multidrug resistance gene in some chloroquine-resistant isolates of *P. falciparum*. Cell *57*, 921–930.

Frank, N.Y., Margaryan, A., Huang, Y., Schatton, T., Waaga-Gasser, A.M., Gasser, M., Sayegh, M.H., Sadee, W., and Frank, M.H. (2005). ABCB5-mediated doxorubicin transport and chemoresistance in human malignant melanoma. Cancer Res. *65*, 4320–4333.

Gan, H.Z., Zhang, G.Z., Zhao, J.S., Zhang, F.C., Bu, L.S., Yang, S.J., Piao, S.L., Du, Z.W., Gao, S., and Zheng, D.M. (2005). Reversal of MDR1 gene-dependent multidrug resistance using short hairpin RNA expression vectors. Chin. Med. J. (Engl). *118*, 893–902.

Gao, F.L., Wang, F., Wu, J.L., Le, X.P., and Zhang, Q.X. (2006). [Screening effective sequences of small interfering RNAs targeting MDR1 gene in human gastric cancer SGC7901/VCR cells – article in Chinese] Zhonghua Zhong Liu Za Zhi *28*, 178–182.

Geerts, S., and Gryseels, B. (2001). Anthelmintic resistance in human helminths: a review. Trop. Med. Int. Health *6*, 915–921.

Geldhof, P., Visser, A., Clark, D., Saunders, G., Britton, C., Gilleard, J., Berriman, M., and Knox, D. (2007). RNA interference in parasitic helminths: current situation, potential pitfalls and future prospects. Parasitology *134*, 609–619.

Gerloff, T., Stieger, B., Hagenbuch, B., Madon, J., Landmann, L., Roth, J., Hofmann, A.F., and Meier,

P.J. (1998). The sister of P-glycoprotein represents the canalicular bile salt export pump of mammalian liver. J. Biol. Chem. *273*, 10046–10050.

Gitlin, L., Karelsky, S., and Andino, R. (2002). Short interfering RNA confers intracellular antiviral immunity in human cells. Nature *418*, 430–434.

Gomez, M.D., Perez, D.G., Ayala, P., Samuelson, J., and Orozco, E. (1996). Physiology and molecular biology of multidrug resistance in *Entamoeba histolytica*. Arch. Med. Res. *27*, 421–425.

Gottesman, M.M., Fojo, T., and Bates, S.E. (2002). Multidrug resistance in cancer: role of ATP-dependent transporters. Nat. Rev. Cancer *2*, 48–58.

Greenwood, B.M., Fidock, D.A., Kyle, D.E., Kappe, S.H., Alonso, P.L., Collins, F.H., and Duffy, P.E. (2008). Malaria: progress, perils, and prospects for eradication. J. Clin. Invest. *118*, 1266–1276.

Gu, J.W., Zhang, T., Chen, B.B., Lu, Y., and Lin, G.W. (2006). [Functional study of mdr1 and GSTpi expression reversed by hairpin siRNA in K562/A02 cell line – article in Chinese] Zhonghua Xue Ye Xue Za Zhi *27*, 17–20.

Gueiros-Filho, F.J., Viola, J.P., Gomes, F.C., Gomes, F.C., Farina, M., Lins, U., Bertho, A.L., Wirth, D.F., and Lopes, U.G. (1995). *Leishmania amazonensis*: multidrug resistance in vinblastine-resistant promastigotes is associated with rhodamine 123 efflux, DNA amplification, and RNA overexpression of a *Leishmania* mdr1 gene. Exp. Parasitol. *81*, 480–490.

Guilfoile, P.G., and Hutchinson, C.R. (1991). A bacterial analog of the mdr gene of mammalian tumor cells is present in *Streptomyces peucetius*, the producer of daunorubicin and doxorubicin. Proc. Natl. Acad. Sci. U.S.A. *88*, 8553–8557.

Haimeur, A., Conseil, G., Deeley, R.G., and Cole S.P. (2004). The MRP-related and BCRP/ABCG2 multidrug resistance proteins: biology, substrate specificity and regulation. Curr. Drug Metab. *5*, 21–53.

Hall, T.M. (2005). Structure and function of argonaute proteins. Structure *13*, 1403–1408.

Hamilton, A.J., and Baulcombe, D.C. (1999). A species of small antisense RNA in posttranscriptional gene silencing in plants. Science *286*, 950–952.

Hammond, S.M., Bernstein, E., Beach, D., and Hannon, G.J. (2000). An RNA-directed nuclease mediates post-transcriptional gene silencing in *Drosophila* cells. Nature *404*, 293–296.

He, L., and Hannon, G.J. (2004). MicroRNAs: small RNAs with a big role in gene regulation. Nat. Rev. Genet. *5*, 522–531.

Heggestad, A.D., Notterpek, L., and Fletcher, B.S. (2004). Transposon-based RNAi delivery system for generating knockdown cell lines. Biochem. Biophys. Res. Commun. *316*, 643–650.

Henderson, D.M., Sifri, C.D., Rodgers, M., Wirth, D.F., Hendrickson, N., and Ullman, B. (1992). Multidrug resistance in *Leishmania donovani* is conferred by amplification of a gene homologous to the mammalian mdr1 gene. Mol. Cell Biol. *12*, 2855–2865.

Henry, M., Alibert, S., Rogier, C., Barbe, J., and Pradines, B. (2008). Inhibition of efflux of quinolines as new therapeutic strategy in malaria. Curr. Top. Med. Chem. *8*, 563–578.

Higgins, C.F. (1992). ABC transporters: from microorganisms to man. Annu. Rev. Cell Biol. *8*, 67–113.

Hirst, S.I., and Stapley, L.A. (2000). Parasitology: the dawn of a new millennium. Parasitol. Today *16*, 1–3.

Holen, T., Amarzguioui, M., Wiiger, M.T., Babaie, E., and Prydz, H. (2002). Positional effects of short interfering RNAs targeting the human coagulation trigger tissue factor. Nucleic Acids Res. *30*, 1757–1766.

Holyoak, C.D., Bracey, D., Piper, P.W., Kuchler, K., and Coote, P.J. (1999). The *Saccharomyces cerevisiae* weak-acid-inducible ABC transporter Pdr12 transports fluorescein and preservative anions from the cytosol by an energy-dependent mechanism. J. Bacteriol. *181*, 4644–4652.

Hopper-Borge, E., Chen, Z.S., Shchaveleva, I., Belinsky, M.G., and Kruh, G.D. (2004). Analysis of the drug resistance profile of multidrug resistance protein 7 (ABCC10): resistance to docetaxel. Cancer Res. *64*, 4927–4930.

Hosono, T., Mizuguchi, H., Katayama, K., Xu, Z.L., Sakurai, F., Ishii-Watabe, A., Kawabata, K., Yamaguchi, T., Nakagawa, S., Mayumi, T., and Hayakawa, T. (2004). Adenovirus vector-mediated doxycycline-inducible RNA interference. Hum. Gene Ther. *15*, 813–819.

Huang, Y., Anderle, P., Bussey, K.J., Barbacioru, C., Shankavaram, U., Dai, Z., Reinhold, W.C., Papp, A., Weinstein, J.N., and Sadée, W. (2004). Membrane transporters and channels: role of the transportome in cancer chemosensitivity and chemoresistance. Cancer Res. *64*, 4294–4301.

Hutvagne, G., and Simard, M.J. (2008). Argonaute proteins: key players in RNA silencing. Nat. Rev. Mol. Cell Biol. *9*, 22–32.

Hyde, S.C., Emsley, P., Hartshorn, M.J., Mimmack, M.M., Gileadi, U., Pearce, S.R., Gallagher, M.P., Gill, D.R., Hubbard, R.E., and Higgins, C.F. (1990). Structural model of ATP-binding proteins associated with cystic fibrosis, multidrug resistance and bacterial transport. Nature *346*, 362–365.

Inoue, N., Otsu, K., Ferraro, D.M., and Donelson, J.E. (2002). Tetracycline-regulated RNA interference in *Trypanosoma congolense*. Mol. Biochem. Parasitol. *120*, 309–313.

Izquierdo, M.A., Neefjes, J.J., Mathari, A.E., Flens, M.J., Scheffer, G.L., and Scheper, R.J. (1996). Overexpression of the ABC transporter TAP in multidrug-resistant human cancer cell lines. Br. J. Cancer *74*, 1961–1967.

Jedlitschky, G., Leier, I., Buchholz, U., Barnouin, K., Kurz, G., and Keppler, D. (1996). Transport of glutathione, glucuronate, and sulfate conjugates by the MRP gene-encoded conjugate export pump. Cancer Res. *56*, 988–994.

Jedlitschky, G., Burchell, B., and Keppler, D. (2000). The multidrug resistance protein 5 functions as an ATP-dependent export pump for cyclic nucleotides. J. Biol. Chem. *275*, 30069–30074.

Ji, J., Wernli, M., Buechner, S., and Erb, P. (2003). Fas ligand downregulation with antisense oligonucleotides in cells and in cultured tissues of normal skin epidermis and basal cell carcinoma. J. Invest. Dermatol. *120*, 1094–1099.

Ji, J., Wernli, M., Mielgo, A., Buechner, S.A., and Erb, P. (2005). Fas-ligand gene silencing in basal cell carcinoma tissue with small interfering RNA. Gene Ther. *12*, 678–684.

Juliano, R.L., and Ling, V. (1976). A surface glycoprotein modulating drug permeability in Chinese hamster ovary cell mutants. Biochim. Biophys. Acta *455*, 152–162.

Kaidoh, K., Miyauchi, S., Abe, A., Tanabu, S., Nara, T., and Kamo, N. (1996). Rhodamine 123 efflux transporter in *Haloferax volcanii* is induced when cultured under 'metabolic stress' by amino acids: the efflux system resembles that in a doxorubicin-resistant mutant. Biochem. J. *314*, 355–359.

Kartner, N., Riordan, J.R., and Ling, V. (1983). Cell surface P-glycoprotein associated with multidrug resistance in mammalian cell lines. Science *221*, 1285–1288.

Kaszubiak, A., Holm, P.S., and Lage, H. (2007). Overcoming the classical multidrug resistance phenotype by adenoviral delivery of anti-MDR1 short hairpin RNAs and ribozymes. Int. J. Oncol. *31*, 419–430.

Katakura, K., Iwanami, M., Ohtomo, H., Fujise, H., and Hashiguchi, Y. (1999). Structural and functional analysis of the LaMDR1 multidrug resistance gene in *Leishmania amazonensis*. Biochem. Biophys. Res. Commun. *255*, 289–294.

Katzmann, D.J., Hallstrom, T.C., Voet, M., Wysock, W., Golin, J., Volckaert, G., and Moye-Rowley, W.S. (1995). Expression of an ATP-binding cassette transporter-encoding gene (YOR1) is required for oligomycin resistance in *Saccharomyces cerevisiae*. Mol. Cell Biol. *15*, 6875–6883.

Katzmann, D.J., Epping, E.A., and Moye-Rowley, W.S. (1999). Mutational disruption of plasma membrane trafficking of *Saccharomyces cerevisiae* Yor1p, a homologue of mammalian multidrug resistance protein. Mol. Cell Biol. *19*, 2998–3009.

Kaur, G., and Lohia, A. (2004). Inhibition of gene expression with double strand RNA interference in *Entamoeba histolytica*. Biochem. Biophys. Res. Commun. *320*, 1118–1122.

Kerboeuf, D., Blackhall, W., Kaminsky, R., and von Samson-Himmelstjerna, G. (2003). P-glycoprotein in helminths: function and perspectives for anthelmintic treatment and reversal of resistance. Int. J. Antimicrob. Agents *22*, 332–346.

Kleinman, M.E., Yamada, K., Takeda, A., Chandrasekaran, V., Nozaki, M., Baffi, J.Z., Albuquerque, R.J., Yamasaki, S., Itaya, M., Pan, Y., Appukuttan, B., Gibbs, D., Yang, Z., Karikó, K., Ambati, B.K., Wilgus, T.A., DiPietro, L.A., Sakurai, E., Zhang, K., Smith, J.R., Taylor, E.W., and Ambati, J. (2008). Sequence- and target-independent angiogenesis suppression by siRNA via TLR3. Nature *452*, 591–597.

Kolaczkowski, M., van der Rest, M., Cybularz-Kolaczkowska, A., Soumillion, J.P., Konings, W.N., and Goffeau, A. (1996). Anticancer drugs, ionophoric peptides, and steroids as substrates of the yeast multidrug transporter Pdr5p. J. Biol. Chem. *271*, 31543–31548.

Kolaczkowski, M., Kolaczowska, A., Luczynski, J., Witek, S., and Goffeau, A. (1998). In vivo characterization of the drug resistance profile of the major ABC transporters and other components of the yeast pleiotropic drug resistance network. Microb. Drug Resist. *4*, 143–158.

König, J., Nies, A.T., Cui, Y., Leier, I., and Keppler, D. (1999). Conjugate export pumps of the multidrug resistance protein (MRP) family: localization, substrate specificity, and MRP2-mediated drug resistance. Biochim. Biophys. Acta *1461*, 377–394.

Kool, M., van der Linden, M., de Haas, M., Scheffer, G.L., de Vree, J.M., Smith, A.J., Jansen, G., Peters, G.J., Ponne, N., Scheper, R.J., Elferink, R.P., Baas, F., and Borst, P. (1999). MRP3, an organic anion transporter able to transport anti-cancer drugs. Proc. Natl. Acad. Sci. U.S.A. *96*, 6914–6919.

Kowalski, P., Surowiak, P., and Lage, H. (2005). Reversal of different drug-resistant phenotypes by an autocatalytic multitarget multiribozyme directed against the transcripts of the ABC transporters MDR1/P-gp, MRP2, and BCRP. Mol. Ther. *11*, 508–522.

Kruh, G.D., Guo, Y., Hopper-Borge, E., Belinsky, M.G., and Chen, Z.S. (2007). ABCC10, ABCC11, and ABCC12. Pflugers Arch. *453*, 675–684.

Lacroix, M. (2008). Persistent use of 'false' cell lines. Int. J. Cancer *122*, 1–4.

Lage, H. (2003). ABC transporters: implications on drug resistance from microorganisms to human cancers. Int. J. Antimicrob. Agents *22*, 188–199.

Lage, H. (2005). Potential applications of RNA interference technology in the treatment of cancer. Fut. Oncol. *1*, 103–113.

Lage, H. (2006). MDR1/P-glycoprotein (ABCB1) as target for RNA interference-mediated reversal of multidrug resistance. Curr. Drug Targets *7*, 813–821.

Lage, H. (2008). An overview of cancer multidrug resistance: a still unsolved problem. Cell. Mol. Life Sci. *65*, 3145–3167.

Lage, H., and Dietel, M. (2000). Effect of the breast-cancer resistance protein on atypical multidrug resistance. Lancet Oncol. *1*, 169–175.

Lage, H., Perlitz, C., Abele, R., Tampé, R., Dietel, M., Schadendorf, D., and Sinha P. (2001). Enhanced expression of human ABC-transporter tap is associated with cellular resistance to mitoxantrone. FEBS Lett. *503*, 179–184.

Laing, N.M., Belinsky, M.G., Kruh, G.D., Bell, D.W., Boyd, J.T., Barone, L., Testa, J.R., and Tew K.D. (1998). Amplification of the ATP-binding cassette 2 transporter gene is functionally linked with enhanced efflux of estramustine in ovarian carcinoma cells. Cancer Res. *58*, 1332–1337.

Lee, N.S., Dohjima, T., Bauer, G., Li, H., Li, M.J., Ehsani, A., Salvaterra, P., and Rossi, J. (2002). Expression of small interfering RNAs targeted against HIV-1 rev transcripts in human cells. Nat. Biotechnol. *20*, 500–505.

Lee, Y.S., and Dutta, A. (2008). MicroRNAs in cancer. Annu. Rev. Pathol. *4*, 199–227.

Légaré, D., Papadopoulou, B., Roy, G., Mukhopadhyay, R., Haimeur, A., Dey, S., Grondin, K., Brochu, C., Rosen, B.P., and Ouellette, M. (1997). Efflux systems and increased trypanothione levels in arsenite-resistant *Leishmania*. Exp. Parasitol. *87*, 275–282.

Li, C.B., Zhang, F., Shi, Y.R., Wei, X.Y., Yang, Y., and Niu, R.F. (2004). [Reversing Multidrug Resistance in Breast Cancer Cell Line MCF-7/ADR by Small Interfering RNA – article in Chinese] Ai Zheng *23*, 1605–1610.

Li, L., Xu, J., Min, T., Huang, W. (2006). Reversal of MDR1 gene-dependent multidrug resistance using low concentration of endonuclease-prepared small interference RNA. Eur. J. Pharmacol. *536*, 93–97.

Li, W.T., Zhou, G.Y., Song, X.R., Chi, W.L., Ren, R.M., and Wang, X.W. (2005). Modulation of BCRP mediated atypical multidrug resistance phenotype by RNA interference. Neoplasma *52*, 219–224.

Lim, M.N., Lau, N.S., Chang, K.M., Leong, C.F., and Zakaria, Z. (2007). Modulating multidrug resistance gene in leukaemia cells by short interfering RNA. Singapore Med. J. *48*, 932–938.

Linton, K.J., Cooper, H.N., Hunter, I.S., and Leadlay, P.F. (1994). An ABC-transporter from *Streptomyces longisporoflavus* confers resistance to the polyether-ionophore antibiotic tetronasin. Mol. Microbiol. *11*, 777–785.

Linton, K.J., and Higgins, C.F. (1998). The *Escherichia coli* ATP-binding cassette (ABC) proteins. Mol. Microbiol. *28*, 5–13.

Liscovitch, M., and Ravid, D. (2007). A case study in misidentification of cancer cell lines: MCF-7/AdrR cells (re-designated NCI/ADR-RES) are derived from OVCAR-8 human ovarian carcinoma cells. Cancer Lett. *245*, 350–352.

Logashenko, E.B., Vladimirova, A.V., Repkova, M.N., Venyaminova, A.G., Chernolovskaya, E.L., and Vlassov, V.V. (2004). Silencing of MDR 1 gene in cancer cells by siRNA. Nucleosides Nucleotides Nucleic Acids *23*, 861–866.

Lou, J.Y., Peng, Z.L., Zheng, Y., Wang, H., He, B., and Wang, H.J. (2006). [Reversal of multi-drug resistance in ovarian cancer cell by RNA interference – article in Chinese] Zhonghua Fu Chan Ke Za Zhi *41*, 413–416.

Lv, H., He, Z., Liu, X., Yuan, J., Yu, Y., and Chen, Z. (2007). Reversal of BCRP-mediated multidrug resistance by stable expression of small interfering RNAs. J. Cell. Biochem. *102*, 75–81.

Ma, J.J., Chen, B.L., and Xin, X.Y. (2009). Inhibition of multi-drug resistance of ovarian carcinoma by small interfering RNA targeting to MRP2 gene. Arch. Gynecol. Obstet. *279*, 144–157.

Mack, J.T., Townsend, D.M., Beljanski, V., and Tew, K.D. (2007). The ABCA2 transporter: intracellular roles in trafficking and metabolism of LDL-derived cholesterol and sterol-related compounds. Curr. Drug Metab. *8*, 47–57.

Marsella, R., and Ruiz de Gopegui, R. (1998). Leishmaniasis: a re-emerging zoonosis. Int. J. Dermatol. *37*, 801–814.

Materna, V., Liedert, B., Thomale, J., and Lage, H. (2005). Protection of platinum-DNA adduct formation and reversal of cisplatin resistance by anti-MRP2 hammerhead ribozymes in human cancer cells. Int. J. Cancer *115*, 393–402.

Materna, V., Stege, A., Surowiak, P., Priebsch, A., and Lage, H. (2006). RNA interference-triggered reversal of ABCC2-dependent cisplatin resistance in human cancer cells. Biochem. Biophys. Res. Commun. *348*, 153–157.

Matsui, Y., Kobayashi, N., Nishikawa, M., and Takakura, Y. (2005). Sequence-specific suppression of mdr1a/1b expression in mice via RNA interference. Pharm. Res. *22*, 2091–2098.

McCaffrey, A.P., Meuse, L., Pham, T.T., Conklin, D.S., Hannon, G.J., and Kay, M.A. (2002). RNA interference in adult mice. Nature *418*, 38–39.

McRobert, L., and McConkey, G.A. (2002). RNA interference (RNAi) inhibits growth of *Plasmodium falciparum*. Mol. Biochem. Parasitol. *119*, 273–278.

Militello, K.T., Refour, P., Comeaux, C.A., and Duraisingh, M.T. (2008). Antisense RNA and RNAi in protozoan parasites: working hard or hardly working? Mol. Biochem. Parasitol. *157*, 117–126.

Miyagishi, M., Hayashi, M., and Taira, K. (2003). Comparison of the suppressive effects of antisense oligonucleotides and siRNAs directed against the same targets in mammalian cells. Antisense Nucleic Acid Drug Dev. *13*, 1–7.

Miyagishi, M., and Taira, K. (2002). U6 promoter-driven siRNAs with four uridine 3′ overhangs efficiently suppress targeted gene expression in mammalian cells. Nat. Biotechnol. *20*, 497–500.

Miyake, K., Mickley, L., Litman, T., Zhan, Z., Robey, R., Cristensen, B., Brangi, M., Greenberger, L., Dean, M., Fojo, T., and Bates, S.E. (1999). Molecular cloning of cDNAs which are highly overexpressed in mitoxantrone-resistant cells: demonstration of homology to ABC transport genes. Cancer Res. *59*, 8–13.

Miyauchi, S., Komatsubara, M., and Kamo, N. (1992). In archaebacteria, there is a doxorubicin efflux pump similar to mammalian P-glycoprotein. Biochim. Biophys. Acta *1110*, 144–50.

Mohmmed, A., Dasaradhi, P.V., Bhatnagar, R.K., Chauhan, V.S., and Malhotra, P. (2003). In vivo gene silencing in *Plasmodium berghei*—a mouse malaria model. Biochem. Biophys. Res. Commun. *309*, 506–511.

Myers, J.W., Jones, J.T., Meyer, T., and Ferrell, J.E. Jr. (2003). Recombinant Dicer efficiently converts large dsRNAs into siRNAs suitable for gene silencing. Nat. Biotechnol. *21*, 324–328.

Nagao, K., Taguchi, Y., Arioka, M., Kadokura, H., Takatsuki, A., Yoda, K., and Yamasaki, M. (1995). bfr1+, a novel gene of *Schizosaccharomyces pombe* which confers brefeldin A resistance, is structurally related to the ATP-binding cassette superfamily. J. Bacteriol. *177*, 1536–1543.

Ngô, H., Tschudi, C., Gull, K., and Ullu, E. (1998). Double-stranded RNA induces mRNA degradation in *Trypanosoma brucei*. Proc. Natl. Acad. Sci. U.S.A. *95*, 14687–14692.

Nguyen, M.H., Peacock, J.E., and Morris, A.J. (1996). The changing face of candidemia – emergence of non-*Candida albicans* species and antifungal resistance. Am. J. Med. *100*, 617–623.

Nieth, C., Priebsch, A., Stege, A., and Lage, H. (2003). Modulation of the classical multidrug resistance (MDR) phenotype by RNA interference (RNAi). FEBS Lett. *545*, 144–150.

Nishi, K., Yoshida, M., Nishimura, M., Nishikawa, M., Nishiyama, M., Horinouchi, S., and Beppu, T. (1992).

A leptomycin B resistance gene of *Schizosaccharomyces pombe* encodes a protein similar to the mammalian P-glycoproteins. Mol. Microbiol. *6*, 761–769.

Nykanen, A., Haley, B., and Zamore, P.D. (2001). ATP requirements and small interfering RNA structure in the RNA interference pathway. Cell *107*, 309–321.

Oguri, T., Bessho, Y., Achiwa, H., Ozasa, H., Maeno, K., Maeda, H., Sato, S., and Ueda R. (2007). MRP8/ABCC11 directly confers resistance to 5-fluorouracil. Mol. Cancer Ther. *6*, 122–127.

Ouellette, M., and Légaré, D. (2003). Drug resistance mediated vby ABC transporters in parasites of humans. In: Holland, I.B., Cole, S.P.C., Kuchler, K., and Higgins, C.F. (eds.) ABC proteins from bacteria to man. Academic Press, SanDiego; CA, USA. pp317–334.

Papadopoulou, B., Roy, G., Dey, S., Rosen, B.P., and Ouellette, M. (1994). Contribution of the *Leishmania* P-glycoprotein-related gene ltpgpA to oxyanion resistance. J. Biol. Chem. *269*, 11980–11986.

Park, S.Y., Lee, W., Lee, J., and Kim, I.S. (2008). Combination gene therapy using multidrug resistance (MDR1) gene shRNA and herpes simplex virus-thymidine kinase. Cancer Lett. *261*, 205–214.

Paterson, J.K., Shukla, S., Black, C.M., Tachiwada, T., Garfield, S., Wincovitch, S., Ernst, D.N., Agadir, A., Li, X., Ambudkar, S.V., Szakacs, G., Akiyama, S., and Gottesman, M.M. (2007). Human ABCB6 localizes to both the outer mitochondrial membrane and the plasma membrane. Biochemistry *46*, 9443–9452.

Patzel, V., Dietrich, I., and Kaufmann, S.H. (2006). RNA Silencing in the struggle against disease. Ann. N. Y. Acad. Sci. *1082*, 44–46.

Paul, C.P., Good, P.D., Winer, I., and Engelke, D.R. (2002). Effective expression of small interfering RNA in human cells. Nat. Biotechnol. *20*, 505–508.

Paulsen, I.T., Brown, M.H., and Skurray, R.A. (1996). Proton-dependent multidrug efflux systems. Microbiol. Rev. *60*, 575–608.

Peacock, C.S., Seeger, K., Harris, D., Murphy, L., Ruiz, J.C., Quail, M.A., Peters, N., Adlem, E., Tivey, A., Aslett, M., Kerhornou, A., Ivens, A., Fraser, A., Rajandream, M.A., Carver, T., Norbertczak, H., Chillingworth, T., Hance, Z., Jagels, K., Moule, S., Ormond, D., Rutter, S., Squares, R., Whitehead, S., Rabbinowitsch, E., Arrowsmith, C., White, B., Thurston, S., Bringaud, F., Baldauf, S.L., Faulconbridge, A., Jeffares, D., Depledge, D.P., Oyola, S.O., Hilley, J.D., Brito, L.O., Tosi, L.R., Barrell, B., Cruz, A.K., Mottram, J.C., Smith, D.F., and Berriman, M. (2007). Comparative genomic analysis of three *Leishmania* species that cause diverse human disease. Nat. Genet. *39*, 839–847.

Peng, Z., Xiao, Z., Wang, Y., Liu, P., Cai, Y., Lu, S., Feng, W., and Han, Z.C. (2004). Reversal of P-glycoprotein-mediated multidrug resistance with small interference RNA (siRNA) in leukemia cells. Cancer Gene Ther., *11*, 707–712. – Identical data were also published in: Peng, Z., Xiao, Z.J., Wang, Y., Liu, P., Cai, Y.L., Feng, W.L., and Han, Z.C. (2004). [Reversal of multi-drug resistance in K562/A02 cells by small interference RNA of mdr1 gene – article in Chinese] Zhonghua Xue Ye Xue Za Zhi *25*, 5–7.

Pichler, A., Zelcer, N., Prior, J.L., Kuil, A.J., and Piwnica-Worms, D. (2005). In vivo RNA interference-mediated ablation of MDR1 P-glycoprotein. Clin Cancer Res. *11*, 4487–4494.

Piper, P., Mahe, Y., Thompson, S., Pandjaitan, R., Holyoak, C., Egner, R., Mühlbauer, M., Coote, P., and Kuchler, K. (1998). The pdr12 ABC transporter is required for the development of weak organic acid resistance in yeast. EMBO J. *17*, 4257–4265.

Poelarends, G.J., Mazurkiewicz, P., Putman, M., Cool, R.H., Veen, H.W., and Konings, W.N. (2000). An ABC-type multidrug transporter of *Lactococcus lactis* possesses an exceptionally broad substrate specificity. Drug Resist. Updat. *3*, 330–334.

Prasad, R., De Wergifosse, P., Goffeau, A., and Balzi, E. (1995). Molecular cloning and characterization of a novel gene of Candida albicans, CDR1, conferring multiple resistance to drugs and antifungals. Curr. Genet. *27*, 320–329.

Priebsch, A., Rompe, F., Tönnies, H., Kowalski, P., Surowiak, P., Stege, A., Materna, V., and Lage, H. (2006). Complete Reversal of ABCG2-depending atypical multidrug resistance (MDR) by RNA interference in human carcinoma cells. Oligonucleotides *16*, 263–274.

Putman, M., van Veen, H.W., and Konings, W.N. (2000). Molecular properties of bacterial multidrug transporters. Microbiol. Mol. Biol. Rev. *64*, 672–693.

Rathjen, T., Nicol, C., McConkey, G., and Dalmay, T. (2006). Analysis of short RNAs in the malaria parasite and its red blood cell host. FEBS Lett. *580*, 5185–5188.

Rebbeor, J.F., Connolly, G.C., Dumont, M.E., and Ballatori, N. (1998). ATP-dependent transport of reduced glutathione on YCF1, the yeast orthologue of mammalian multidrug resistance associated proteins. J. Biol. Chem. *273*, 33449–33454.

Reed, M.B., Saliba, K.J., Caruana, S.R., Kirk, K., and Cowman, A.F. (2000). Pgh1 modulates sensitivity and resistance to multiple antimalarials in *Plasmodium falciparum*. Nature *403*, 906–909.

Ren, Y.Y. (2006). [Suppression of MDR1 expression and restoration of sensitivity to chemotherapy in multidrug-resistant hepatocellular carcinoma cell line Bel7402/5-Fu by RNA interference – article in Chinese] Zhong Nan Da Xue Xue Bao Yi Xue Ban *31*, 872–876.

Riordan, J.R., and Ling, V. (1979). Purification of P-glycoprotein from plasma membrane vesicles of Chinese hamster ovary cell mutants with reduced colchicine permeability. J. Biol. Chem. *254*, 12701–12705.

Robinson, K.A., and Beverley, S.M. (2003). Improvements in transfection efficiency and tests of RNA interference (RNAi) approaches in the protozoan parasite *Leishmania*. Mol. Biochem. Parasitol. *128*, 217–128.

Rosenberg, E., Litus, I., Schwarzfuchs, N., Sinay, R., Schlesinger, P., Golenser, J., Baumeister, S., Lingelbach, K., and Pollack, Y. (2006). pfmdr2 confers heavy metal resistance to *Plasmodium falciparum*. J. Biol. Chem. *281*, 27039–27045.

Ross, J.I., Eady, E.A., Cove, J.H., Cunliffe, W.J., Baumberg, S., and Wootton, J.C. (1990). Inducible erythromycin resistance in staphylococci is encoded by a member of the ATP-binding transport super-gene family. Mol. Microbiol. *4*, 1207–1214.

Rossolini, G.M., and Docquier, J.D. (2006). New beta-lactamases: a paradigm for the rapid response of bacterial evolution in the clinical setting. Future Microbiol. *1*, 295–308.

Ruetz, S., and Gros, P. (1994). Phosphatidylcholine translocase: a physiological role for the mdr2 gene. Cell *77*, 1071–1081.

Sakamoto, K., Margolles, A., van Veen, H.W., and Konings, W.N. (2001). Hop resistance in the beer spoilage bacterium *Lactobacillus brevis* is mediated by the ATP-binding cassette multidrug transporter HorA. J. Bacteriol. *183*, 5371–5375.

Samarawickrema, N.A., Brown, D.M., Upcroft, J.A., Thammapalerd, N., and Upcroft, P. (1997). Involvement of superoxide dismutase and pyruvate:ferredoxin oxidoreductase in mechanisms of metronidazole resistance in *Entamoeba histolytica*. J. Antimicrob. Chemother. *40*, 833–840.

Samuelson, J., Ayala, P., Orozco, E., Wirth, D. (1990). Emetine-resistant mutants of *Entamoeba histolytica* overexpress mRNAs for multidrug resistance. Mol. Biochem. Parasitol. *38*, 281–290.

Sanglard, D., Ischer, F., Monod, M., and Bille, J. (1997). Cloning of *Candida albicans* genes conferring resistance to azole antifungal agents: characterization of CDR2, a new multidrug ABC transporter gene. Microbiology *143*, 405–416.

Sanglard, D., and Odds, F.C. (2002). Resistance of *Candida* species to antifungal agents: molecular mechanisms and clinical consequences. Lancet Infect. Dis. *2*, 73–85.

Sanguino, A., Lopez-Berestein, G., and Sood, A.K. (2008). Strategies for *in vivo* siRNA delivery in cancer. Mini Rev. Med. Chem. *8*, 248–255.

Schuetz, J.D., Connelly, M.C., Sun, D., Paibir, S.G., Flynn, P.M., Srinivas, R.V., Kumar, A., and Fridland, A. (1999). MRP4: A previously unidentified factor in resistance to nucleoside-based antiviral drugs. Nat. Med. *5*, 1048–1051.

Servos, J., Haase, E., and Brendel, M. (1993). Gene SNQ2 of *Saccharomyces cerevisiae*, which confers resistance to4-nitroquinoline-N-oxide and other chemicals, encodes a 169 kDa protein homologous to ATP-dependent permeases. Mol. Gen. Genet. *236*, 214–218.

Shi, H., Tschudi, C., and Ullu, E. (2006a). An unusual Dicer-like1 protein fuels the RNA interference pathway in *Trypanosoma brucei*. RNA *12*, 2063–2072.

Shi, Z., Liang, Y.J., Chen, Z.S., Wang, X.W., Wang, X.H., Ding, Y., Chen, L.M., Yang, X.P., and Fu, L.W. (2006b). Reversal of MDR1/P-glycoprotein-mediated multidrug resistance by vector-based RNA interference *in vitro* and *in vivo*. Cancer Biol. Ther. *5*, 39–47.

Siegmund, D., Hadwiger, P., Pfizenmaier, K., Vornlocher, H.P., and Wajant, H. (2002). Selective inhibition of FLICE-like inhibitory protein expression with small interfering RNA oligonucleotides is sufficient to sensitize tumor cells for TRAIL-induced apoptosis. Mol. Med. *8*, 725–732.

Simpson, W.J., and Smith, A.R. (1992). Factors affecting antibacterial activity of hop compounds and their derivatives. J. Appl. Bacteriol. *72*, 327–334.

Song, E., Lee, S.K., Wang, J., Ince, N., Ouyang, N., Min, J., Chen, J., Shankar, P., and Lieberman, J. (2003). RNA interference targeting Fas protects mice from fulminant hepatitis. Nat. Med. *9*, 347–351.

Song, Z.Y., Hu, H.Y., Deng, L., Wu, B.Y., Guo, K.Y., and Zhang, M.X. (2008). [Effect of small interfering RNA targeting multidrug resistance-related protein and bcl-2 on drug resistance and apoptosis of K562 and K562/ADM cells – article in Chinese] Nan Fang Yi Ke Da Xue Xue Bao. *28*, 1306–1308.

Soutschek, J., Akinc, A., Bramlage, B., Charisse, K., Constien, R., Donoghue, M., Elbashir, S., Geick, A., Hadwiger, P., Harborth, J., John, M., Kesavan, V., Lavine, G., Pandey, R.K., Racie, T., Rajeev, K.G., Rohl, I., Toudjarska, I., Wang, G., Wuschko, S., Bumcrot, D., Koteliansky, V., Limmer, S., Manoharan, M., and Vornlocher, H.P. (2004). Therapeutic silencing of an endogenous gene by systemic administration of modified siRNAs. Nature *432*, 173–178.

Stege, A., Priebsch, A., Nieth, C., and Lage, H. (2004). Stable and complete overcoming of MDR1/P-glycoprotein-mediated multidrug resistance in human gastric carcinoma cells by RNA interference. Cancer Gene Ther. *11*, 699–706.

Stein, U., Walther, W., Stege, A., Kaszubiak, A., Fichtner, I. and Lage, H. (2008). Complete *in vivo* reversal of the multidrug resistance phenotype by jet-injection of anti-MDR1 short hairpin RNA-encoding plasmid DNA. Mol. Ther. *16*, 178–186.

Steinbach, D., Gillet, J.P., Sauerbrey, A., Gruhn, B., Dawczynski, K., Bertholet, V., de Longueville, F., Zintl, F., Remacle, J., and Efferth, T. (2006). ABCA3 as a possible cause of drug resistance in childhood acute myeloid leukemia. Clin. Cancer Res. *12*, 4357–4363.

Stewart, B.W., and Kleihues, P. (eds.) (2003). World cancer report. IARC Press, Lyon. pp 11–19.

Stierle, V., Laigle, A., and Jolles, B. (2004). The reduction of P-glycoprotein expression by small interfering RNAs is improved in exponentially growing cells. Oligonucleotides *14*, 191–198.

Stierlé, V., Laigle, A., and Jollès, B. (2005). Modulation of MDR1 gene expression in multidrug resistant MCF7 cells by low concentrations of small interfering RNAs. Biochem. Pharmacol. *70*, 1424–1430.

Stierlé, V., Laigle, A., and Jollès, B. (2007). Modulation of the efficiency of a siRNA directed against MDR1 expression in MCF7-R cells when combined with a second siRNA. Biochimie *89*, 1033–1036.

Su, F., Ouyang, N., Zhu, P., Ouyang, N., Jia, W., Gong, C., Ma, X., Xu, H., and Song, E. (2005). Psychological stress induces chemoresistance in breast cancer by upregulating mdr1. Biochem. Biophys. Res. Commun. *329*, 888–897.

Sullenger, B.A., and Gilboa, E. (2002). Emerging clinical applications of RNA. Nature *418*, 252–258.

Szakács, G., Paterson, J.K., Ludwig, J.A., Booth-Genthe, C., and Gottesman, M.M. (2006). Targeting multidrug resistance in cancer. Nat. Rev. Drug Discov. *5*, 219–234.

Taniguchi, K., Wada, M., Kohno, K., Nakamura, T., Kawabe, T., Kawakami, M., Kagotani, K., Okumura, K., Akiyama, S., and Kuwano, M. (1996). A human canalicular multispecific organic anion transporter (cMOAT) gene is overexpressed in cisplatin-resistant human cancer cell lines with decreased drug accumulation. Cancer Res. *56*, 4124–4129.

Turi, T.G., and Rose, J.K. (1995). Characterization of a novel *Schizosaccharomyces pombe* multidrug resistance transporter conferring brefeldin A resistance. Biochem. Biophys. Res. Commun. *213*, 410–418.

Ueda, K., Yoshida, A., and Amachi, T. (1999). Recent progress in P-glycoprotein research. Anticancer Drug Des. *14*, 115–121.

Ullu, E., Lujan, H.D., and Tschudi, C. (2005). Small sense and antisense RNAs derived from a telomeric retroposon family in *Giardia intestinalis*. Eukaryot. Cell *4*, 1155–1157.

Ullu, E., Tschudi, C., and Chakraborty, T. (2004). RNA interference in protozoan parasites. Cell. Microbiol. *6*, 509–519.

Usui, I., Imamura, T., Huang, J., Satoh, H., and Olefsky, J.M. (2003). Cdc42 is a Rho GTPase family member which can mediate insulin signaling to glucose transport in 3T3-L1 adipocytes. J. Biol. Chem. *278*, 13765–13774.

Van Bambeke, F., Balzi, E., and Tulkens, P.M. (2000). Antibiotic efflux pumps. Biochem. Pharmacol. *60*, 457–470.

Van den Haute, C., Eggermont, K., Nuttin, B., Debyser, Z., and Baekelandt, V. (2003). Lentiviral vector-mediated delivery of short hairpin RNA results in persistent knockdown of gene expression in mouse brain. Hum. Gene Ther. *14*, 1799–1807.

Vogel, J., and Wagner, E.G. (2007). Target identification of small noncoding RNAs in bacteria. Curr. Opin. Microbiol. *10*, 262–270.

Vulevic, B., Chen, Z., Boyd, J.T., Davis, W.Jr., Walsh, E.S., Belinsky, M.G., and Tew, K.D. (2001). Cloning and characterization of human adenosine 5′-triphosphate-binding cassette, sub-family A, transporter 2 (ABCA2). Cancer Res. *61*, 3339–3347.

Walker, J.E., Saraste, M., Runswick, M.J., and Gay, N.J. (1982). Distantly related sequences in the alpha- and beta-subunits of ATP synthase, myosin, kinases and other ATP-requiring enzymes and a common nucleotide binding fold. EMBO J. *1*, 945–951.

Walther, W., Stein, U., Fichtner, I., and Schlag, P.M. (2004). Low-volume jet injection for efficient nonviral *in vivo* gene transfer. Mol. Biotechnol. 28, 121–128.

Walther, W., Siegel, R., Kobelt, D., Knösel, T., Dietel, M., Bembenek, A., Aumann, J., Schleef, M., Baier, R., Stein, U., and Schlag, P.M. (2008).Novel jet-injection technology for nonviral intratumoral gene transfer in patients with melanoma and breast cancer. Clin. Cancer Res. *14*, 7545–7553.

Watanabe, T., Onuki, R., Yamashita, S., Taira, K., and Sugiyama, Y. (2005). Construction of a functional transporter analysis system using MDR1 knockdown Caco-2 cells. Pharm. Res. *22*, 1287–1293.

Wei, H.L., Gao, L.P., Jing, T., Zhao, H.S., Yi, J., Sun, J., and Han, J. (2007). [siRNA silences mdr1 gene expression and reverses apoptosis resistance of K562/ADM cells line – article in Chinese] Zhonghua Xue Ye Xue Za Zhi *28*, 388–390.

Widmer, N., Rumpold, H., Untergasser, G., Fayet, A., Buclin, T., and Decosterd, L.A. (2007). Resistance reversal by RNAi silencing of MDR1 in CML cells associated with increase in imatinib intracellular levels. Leukemia *21*, 1561–1562.

Wijnholds, J., Mol, C.A., van Deemter, L., de Haas, M., Scheffer, G.L., Baas, F., Beijnen, J.H., Scheper, R.J., Hatse, S., De Clercq, E., Balzarini, J., and Borst, P. (2000). Multidrug-resistance protein 5 is a multispecific organic anion transporter able to transport nucleotide analogs. Proc. Natl. Acad. Sci. U.S.A. *97*, 7476–7481.

Wood, V., Gwilliam, R., Rajandream, M.A., et al. (2002). The genome sequence of *Schizosaccharomyces pombe*. Nature *415*, 871–880.

Wu, H., Hait, W.N., and Yang, J.M. (2003). Small interfering RNA-induced suppression of MDR1 (P-glycoprotein) restoRes. sensitivity to multidrug-resistant cancer cells. Cancer Res. *63*, 1515–1519.

Xiang, S., Fruehauf, J., and Li, C.J. (2006). Short hairpin RNA–expressing bacteria elicit RNA interference in mammals. Nature Biotechnol. *24*, 697–702.

Xiao, X.B., Xie, Z.X., and Qin, Q. (2006). [Reversal of multidrug resistance by MDR1 shRNA expression vector in human leukemia K562/A02 cells – article in Chinese] Zhonghua Zhong Liu Za Zhi *28*, 422–425.

Xiao, H., Wu, Z., Shen, H., Luo, A.L., Yang, Y.F., Li, X.B., and Zhu, D.Y. (2008). In vivo reversal of P-glycoprotein-mediated multidrug resistance by efficient delivery of stealth RNAi. Basic Clin. Pharmacol. Toxicol. *103*, 342–348.

Xie, S.M., Fang, W.Y., Liu, Z., Wang, S.X., Li, X., Liu, T.F., Xie, W.B., and Yao, K.T. (2008). Lentivirus-mediated RNAi silencing targeting ABCC2 increasing the sensitivity of a human nasopharyngeal carcinoma cell line against cisplatin. J. Transl. Med. *6*, 55.

Xu, D., Kang, H., Fisher, M., and Juliano, R.L. (2004). Strategies for inhibition of MDR1 gene expression. Mol. Pharmacol. *66*, 268–275.

Xu, D., McCarty, D., Fernandes, A., Fisher, M., Samulski, R.J., and Juliano, R.L. (2005). Delivery of MDR1 small interfering RNA by self-complementary recombinant adeno-associated virus vector. Mol. Ther. *11*, 523–530.

Yagüe, E., Higgins, C.F., and Raguz, S. (2004). Complete reversal of multidrug resistance by stable expression of small interfering RNAs targeting MDR1. Gene Ther. *11*, 1170–1174.

Yasui, K., Mihara, S., Zhao, C., Okamoto, H., Saito-Ohara, F., Tomida, A., Funato, T., Yokomizo, A., Naito, S., Imoto, I., Tsuruo, T., and Inazawa, J. (2004). Alteration in copy numbers of genes as a mechanism for acquired drug resistance. Cancer Res. *64*, 1403–1410.

Young, J., and Holland, I.B. (1999). ABC transporters: bacterial exporters – revisited five years on. Biochim. Biophys. Acta *1461*, 177–200.

Zeng, H., Bain, L.J., Belinsky, M.G., and Kruh, G.D. (1999). Expression of multidrug resistance protein-3 (multispecific organic anion transporter-D) in human embryonic kidney 293 cells confers resistance to anticancer agents. Cancer Res. *59*, 5964–5967.

Zhao, P., Zhang, Y., Sun, M., and He, Y. (2008). Reversion of multidrug resistance in human glioma by RNA interference. Neurol. Res. *30*, 562–566. – Identical data were also published in: (1) Zhao, P., Zhang, Y.Z., and Sun, M.Z. (2005). [Regulatory effect of small interfering RNA targeting multidrug resistant protein 1 on chemosensitivity of human multiforme glioblastoma cell line BT325 – article in Chinese] Ai Zheng *24*, 1436–1441; and (2) Zhao, P., Hu, W., Zhang, Y.Z., Sun, M.Z., and He, Y. (2006). [Reversion of multidrug resistance (MDR) in human glioma cells by RNA interference (RNAi) – article in Chinese]. Zhonghua Zhong Liu Za Zhi *28*, 183–187.

Index

A

B

C

D

F

M

N

S

T

U

V

W

X

Colour Plates

```
MsbA_S.typhimurium        1    EKDEGKRVIDRATGDLEFRNVTFTYPGREVPALRNINLKIPAGKTVALVGRSGSGKSTIASLITRFY..DIDEGHILMDGHD.........LREYT
MsbA_E.coli               1    EKDEGKRVIERATGDVEFRNVTFTYPGRDVPALRNINLKIPAGKTVALVGRSGSGKSTIASLITRFY..DIDEGEILMDGHD.........LREYT
Sav1866_S.aureus          1 DYDIKNGVGAQPIEIKQGRIDIDHVSFQYNDNEAPILKDINLSIEKGETVAFVGMSGGGKSTLINLIPRFY..DVTSGQILIDGHN.........IKDFL
LmrA_L.lactis             1                  MLSARHVDFAYDDSEQ.ILRDISFEAQPNSIIAFAGPSGGGKSTIFSLLERFY..QPTAGEITIDGQP.........IDNIS
HlyB_E.coli               1                   DITFRNIRFRYKPDSPVILDNINLSIKQGEVIGIVGRSGSGKSTLTKLIQRFY..IPENGQVLIDGHD.........LALAD
MetN_E.coli               1                 MIKLSNITKVFHQGTRTIQALNNVSLHVPAGQIYGVIGASGAGKSTLIRCVNLLE..RPTEGSVLVDGQ..........ELTTL
HisP_S.typhimurium        1                MMSENKLHVIDLHKRYGGHEVLKGVSLQARAGDVISIIGSSGSGKSTFLRCINFLE..KPSEGAIIVNGQNINLVRDKDGQLKVA
MalK_P.horikoshii         1          GNNIEVIKMVEVKLENLTKRFG.NFTAVNKLNLTIKDGEFLVLLGPSGCGKTTTLRMIAGLE..EPTEGRIYFGDRD.........VTYLP
1VCI_P.horikoshii         1         MGNNIEVIKMVEVKLENLTKRFG.NFTAVNKLNLTIKDGEFLVLLGPSGCGKTTTLRMIAGLE..EPTEGRIYFGDRD.........VTYLP
MalK_E.coli               1                 MASVQLQNVTKAWG.EVVVSKDINLDIHEGEFVVFVGPSGCGKSTLLRMIAGLE..TITSGDLFIGEKR.........MNDTP
ModC_A.fulgidus           1                   MFLKVRAEKRLG.NFRLN...VDFEMGRDYCVLLGPTGAGKSVFLELIAGIV..KPDRGEVRLNGAD.........ITPLP
ModC_M.acetivorans        1                   MIEIESLSRKWK.NFSLDNLSLKVESG.EYFVILGPTGAGKTLFLELIAGFH..VPDSGRILLDGKD.........VTDLS
CysA_A.acidocaldarius     1       MRGHHHHHHGSMTIEFVGVEKIYPGGARSVRGVSFQIREGEMVGLLGPSGSGKTTILRLIAGLE..RPTKGDVWIGGKR.........VTDLP
MJ1267_E.coli             1                 MIKLKNVTKTYKMGEEIIYALKNVNLNIKEGEFVSIMGPSGSGKSTMLNIIGCLD..KPTEGEVYIDNIK.........TNDLD
MJ0796_M.janaschii        1                 MIKLKNVTKTYKMGEEIIYALKNVNLNIKEGEFVSIMGPSGSGKSTMLNIIGCLD..KPTEGEVYIDNIK.........TNDLD
GlcV_S.solfataricus       1                MVRIIVKNVSKVFKKGK..VVALDNVNINIENGERFGILGPSGAGKTTFMRIIAGLD..VPSTGELYFDDRL.........VASNG
SufC_E.coli               1                      MLSIKDLHVSVEDKAILRGLSLDVHPGEVHAIMGPNGSGKSTLSATLAGREDYEVTGGTVEFKGKD.........LLALS
SufC_T.thermophilus       1                     MSQLEIRDLWASIDGETILKGVNLVVPKGEVHALMGPNGAGKSTLGKILAGDPEYTVERGEILLDGEN.........ILELS
TM0544_T.maritima         1       MGSDKIHHHHHHMGAVVVKDLRKRIGKKEILKGISFEIEEGEIFGLIGPNGAGKTTTLRIISTLIK..PSSGIVTVFGKN.........VVEEP
MJ1267_M.janaschii        1              MRDTMEILRTENIVKYFGEFKALDGVSISVNKGDVTLIIGPNGSGKSTLINVITGFL..KADEGRVYFENKD.........ITNKE
BtuD_E.coli               1                        MSIVMQLQDVAESTRLGPLSGEVRAGEILHLVGPNGAGKSTLLARMAGMT...SGKGSIQFAGQP.............
HI1470_H.influenzae       1                 MNKALSVENLGFYYQAENFLFQQLNFDLNKGDILAVLGQNGCGKSTLLDLLLGIH..RPIQGKIEVYQSIG...........
                                                                                   Walker A

MsbA_S.typhimurium       86 LAS....LRNQVALVSQNVHLFN...DTVANNIAYARTEEYSREQIEEAARMAYAMDFINKMDNGLDT.IIGENGVL.LSGGQRQRIAIARALLRDS...
MsbA_E.coli              86 LAS....LRNQVALVSQNVHLFN...DTVANNIAYARTEQYSREQIEEAARMAYAMDFINKMDNGLDT.VIGENGVL.LSGGQRQRIAIARALLRDS...
Sav1866_S.aureus         90 TGS....LRNQIGLVQQDNILFS...DTVKENILLGR.PTATDEEVVEAAKMANAHDFIMNLPQGYDT.EVGERGVK.LSGGQKQRLSIARIFLNNP...
LmrA_L.lactis            71 LEN....WRSQIGFVSQDSAIMA...GTIRENLTYGLEGDYTDEDLWQVLDLAFARSFVENMPDQLNT.EVGERGVK.ISGGQRQRLAIARAFLRNP...
HlyB_E.coli              72 PNW....LRRQVGVVLQDNVLLN...RSIIDNISLAN.PGMSVEKVIYAAKLAGAHDFISELREGYNT.IVGEQGAG.LSGGQRQRIAIARALVNNP...
MetN_E.coli              73 SESELTKARRQIGMIFQHFNLLS...SRTVFG.NVALPLELDN...TPKDEVKRRVTELLSLVGLGDK.HD.SYPSN.LSGGQKQRVAIARALASNP...
HisP_S.typhimurium       84 DKNQLRLLRTRLTMVFQHFNLWS...HMTVLENVMEAPIQVLG...LSKHDARERALKYLAKVGIDER.AQGKYPVH.LSGGQQQRVSIARALAMEP...
MalK_P.horikoshii        80 P......KDRNISMVFQSYAVWP...HMTVYE.NIAFPLKIKK...FPKDEIDKRVRWAAELLQIEE..LLNRYPAQ.LSGGQRQRVAVARAIVVEP...
1VCI_P.horikoshii        81 P......KDRNISMVFQSYAVWP...HMTVYE.NIAFPLKIKK...FPKDEIDKRVRWAAELLQIEE..LLNRYPAQ.LSGGQRQRVAVARAIVVEP...
MalK_E.coli              72 P......AERGVGMVFQSYALYP...HLSVAE.NMSFGLKLAG...AKKEVINQRVNQVAEVLQLAH..LLDRKPKA.LSGGQRQRVAIGRTLVAEP...
ModC_A.fulgidus          67 P......ERRGIGFVPQDYALFP...HLSVYR.NIAYGLRNVE...RVER..DRRVREMAEKLGIAH..LLDRKPAR.LSGGERQRVALARALVIQP...
ModC_M.acetivorans       69 P......EKHDIAFVYQNYSLFP...HMNVKK.NLEFGMR.MK...KIKD..PKRVLDTARDLKIEH..LLDRNPLT.LSGGEQQRVALARALVTNP...
CysA_A.acidocaldarius    84 P......QKRNVGLVFQNYALFQ...HMTVYD.NVSFGLREKR...VPKDEMDARVRELLRFMRLES..YANRFPHE.LSGGQQQRVALARALAPRP...
MJ1267_E.coli            74 DDELTKIRRDKIGFVFQQFNLIP...LLTALE.NVELPLIFKYRGAMSGEERRKRALECLKMAELEER.FANHKPNQ.LSGGQQQRVAIARALANNP...
MJ0796_M.janaschii       74 DDELTKIRRDKIGFVFQQFNLIP...LLTALE.NVELPLIFKYRGAMSGEERRKRALECLKMAELEER.FANHKPNQ.LSGGQQQRVAIARALANNP...
GlcV_S.solfataricus      74 KLIVPPEDR.KIGMVFQTWALYP...NLTAFE.NIAFPLTNMK...MSKEEIRKRVEEVAKILDIHH..VLNHFPRE.LSGGQQQRVALARALVKDP...
SufC_E.coli              72 PED...RAGEGIFMAFQYPVEIPGVSNQFFLQTALNAVRSYRGQETLDRFDFQDLMEEKIALLKMPED.LLTRSVNVGFSGGEKKRNDILQMAVLEP...
SufC_T.thermophilus      74 PDE...RARKGLFLAFQYPVEVPGVTIANFLRLALQAKLGR....EVGVAEFWTKVKKALELLDWDES.YLSRYLNEGFSGGEKKRNEILQLLVLEP...
TM0544_T.maritima        84 HEV.....RKLISYLPEEAGAYR.......NMQGIEYLRFVAGFYASSSSEIEEMVERATEIAGLGEK.IKDRVSTY.SKGMVRKLLIARALMVNP...
MJ1267_M.janaschii       76 PAELYHYGIVRTFQTPQPLKEMTVLENLLIGEINPGESPLNSLFYKKWIPKEEEMVEKAFKILEFLKLSHLYDRKAGELSGGQMKLVEIGRALMTNP...
BtuD_E.coli              63 ......LEAWSATKLALHRAYLS.....QQQTPPFATPVWHYLTLHQHDKTRTELLNDVAGALALDDK..LGRSTNQ.LSGGEWQRVRLAAVVLQITPQA
HI1470_H.influenzae      70 ..........FVPQFFSSPFAYS......VLDIVLMGRSTHIN...TFAKPKSHDYQVAMQALDYLNLTHLAKREFTSLSGGQRQLILIARAIASEC...
                                Q-loop                                   Structural diverse region   C-loop         Pro-loop
```

```
MsbA_S.typhimurium      174 ....PILILDEATSALDTESERAIQAALDELQK..NRTSLVIAHRLS.TIEQADEIVVVEDGIIVERGTHSELLAQ.HGVYAQLHKMQFGQ
MsbA_E.coli             174 ....PILILDEATSALDTESERAIQAALDELQK..NRTSLVIAHRLS.TIEKADEIVVVEDGVIVERGTHNDLLEH.RGVYAQLHKMQFGQ
Sav1866_S.aureus        177 ....PILILDEATSALDLESESIIQEALDVLSK..DRTTLIVAHRLS.TITHADKIVVIENGHIVETGTHRELIAK.QGAYEHLYSIQNL
LmrA_L.lactis           159 ....KILMLDEATASLDSESESMVQKALDSLMK..GRTTLVIAHRLS.TIVDADKIYFIEKGQITGSGKHNELVAT.HPLYAKYVSEQLTVGQ
HlyB_E.coli             159 ....KILIFDEATSALDYESEHVIMRNMHKICK..GRTVIIIAHRLS.TVKNADRIIVMEKGKIVEQGKHKELLSEPESLYSYLYQLQSD
MetN_E.coli             160 ....KVLLCDEATSALDPATTRSILELLKDINRRLGLTILLITHEMDVVKRICDCVAVISNGELIEQDTVSEVFSHPKTPLAQKFIQSTLHLDIPEDYQE
HisP_S.typhimurium      173 ....DVLLFDEPTSALDPELVGEVLRIMQQLAE.EGKTMVVVTHEMGFARHVSSHVIFLHQGKIEEEGDPEQVFGNPQSPRLQQFLKGSLKKLEH
MalK_P.horikoshii       161 ....DVLLMDEPLSNLDAKLRVAMRAEIKKLQQKLKVTTIYVTHDQVEAMTMGDRIAVMNRGQLLQIGSPTEVYLRPNSVFVATFIGAPEMNILEVSVGD
1VCI_P.horikoshii       162 ....DVLLMDEPLSNLDAKLRVAMRAEIKKLQQKLKVTTIYVTHDQVEAMTMGDRIAVMNRGQLLQIGSPTEVYLRPNSVFVATFIGAPEMNILEVSVGD
MalK_E.coli             153 ....SVFLLDQPLSNLDAALRVQMRIEISRLHKRLGRTMIYVTHDQVEAMTLADKIVVLDAGRVAQVGKPLELYHYPADRFVAGFIGSPKMNFLPVKVTA
ModC_A.fulgidus         146 ....RLLLLDEPLSAVDLKTKGVLMEELRFVQREFDVPILHVTHDLIEAAMLADEVAVMLNGRIVEKGKLKELFS.AKNGEVAEFL...........SAR
ModC_M.acetivorans      147 ....KILLLDEPLSALDPRTQENAREMLSVLHKKNKLTVLHITHDQTEARIMADRIAVVMDGKLIQVGKPEEIFEKPVEGRVASFVGFENVLKGRVISAE
CysA_A.acidocaldarius   165 ....QVLLFDEPFAAIDTQIRRELRTFVRQVHDEMGVTSVFVTHDQEEALEVADRVLVLHEGNVEQFGTPEEVYEKPGTLFVASFIGESNVWTRAVQNGR
MJ1267_E.coli           165 ....PIILADEPTGALDSKTGEKIMQLLKKLNEEDGKTVVVVTHD..................................INVARFG..............
MJ0796_M.janaschii      165 ....PIILADEPTGALDSKTGEKIMQLLKKLNEEDGKTVVVVTHD..................................INVARFG..............
GlcV_S.solfataricus     160 ....SLLLLDEPFSNLDARMRDSARALVKEVQSRLGVTLLVVSHDPADIFAIADRVGVLVKGKLVQVGKPEDLYDNPVSIQVASLIGEINELEGKVTNEG
SufC_E.coli             165 ....ELCILDESDSGLDIDALKVVADGVNSLRDGKRSFIIVT.HYQRILDYIKPDYVHVLYQGRIVKSGDFTLVKQLEEQGYGWLTEQQ
SufC_T.thermophilus     163 ....TYAVLDETDSGLDIDALKVVARGVNAMRGPNFGALVIT.HYQRILNYIQPDKVHVMMDGRVVATGGPELALELEAKGYEWLKEKVKEGA
TM0544_T.maritima       166 ....RLAILDEPTSGLDVLNAREVRKILKQASQE.GLTILVSSHNMLEVEFLCDRIALIHNGTIVETGTVEELKERYKAQNIEEVFEEVVKCSENF
MJ1267_M.janaschii      173 ....KMIVMDQPIAGVAPGLAHDIFNHVLELKAK.GITFLIIEHRLDIVLNYIDHLYVMFNGQIIAEGRGEEEIKNVLSDPKVVEIYIGE
BtuD_E.coli             149 NPAGQLLLLDEPMNSLDVAQQSALDKILSALCQQ.GLAIVMSSHDLNHTLRHAHRAWLLKGGKMLASGRREEVLTPPNLAQAYGMNFRRLDIEGHRMLIS
HI1470_H.inluenzae      148 ....KLILLDEPTSALDLANQDIVLSLLIDLAQSQNMTVVFTTHQPNQVVAIANKTLLLNKQNFKFGETRNILTSENLTALFHLPMFEQQAQYKESFFTH
                                Walker B D-loop                         H-loop
```

Plate 1.1 Amino acid sequence alignment of the nucleotide binding domains of prokaryotic ABC transporters whose structures have been solved. The conserved motifs are colour-coded: Walker A (red), Q-loop (magenta), structural diverse region (SDR; black), C-loop (pale green), Pro-loop (orange), Walker B (blue), D-loop (cyan), and H-loop (green). The function of the distinct residues is explained in detail in the text.

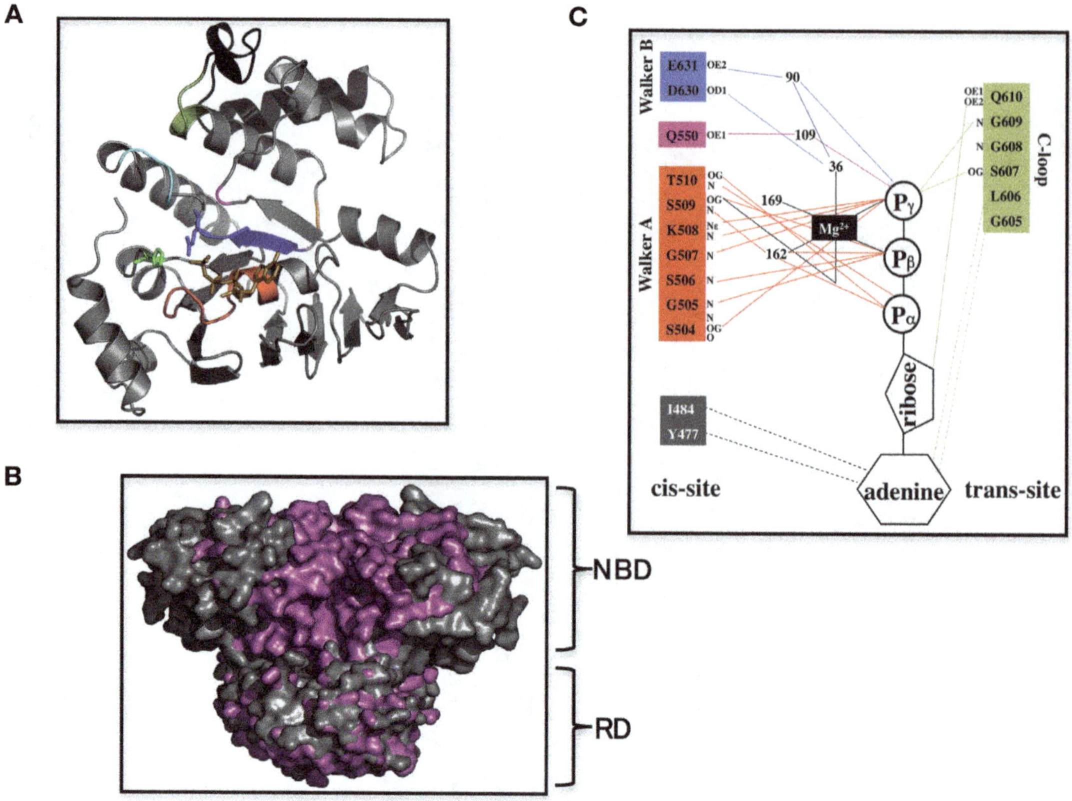

Plate 1.2 **A** Crystal structure of HisP, the nucleotide binding domain of the histidine permease from *S.* Typhimurium with bound ATP (pdb entry 1B0U). The helical domain contains the C-loop (pale green) and the structurally diverse region (SDR; black) and is linked via the Q-loop (magenta) and the Pro-loop (orange) to the catalytic domain. The catalytic domain contains the Walker A (red) and Walker B (blue) motifs and the D-loop (cyan) as well as the H-loop (green). The glutamate at the end of Walker B and the histidine of the H-loop are depicted in stick-representation as they are crucial in ATP hydrolysis. **B** Structural superposition of the nucleotide free (gray) and ATP-bound (magenta) structure of the nucleotide binding domain MalK (pdb entries: 1Q1E, 1Q12). The NBDs come into close contact upon ATP binding whereas only minor changes can be detected in the regulatory domains (RD). **C** Detailed network of ATP coordination in the binding pocket of the HlyB-NBD. Solid lines indicate hydrogen bonds and dashed lines show van-der-Waals interactions. Colour-coding is equivalent to Plate 1.2A. The black numbers indicate water molecules that are part of the extensive network of interactions. Letters next to the amino acid labels indicate involvement via main-chain or side-chain interactions.

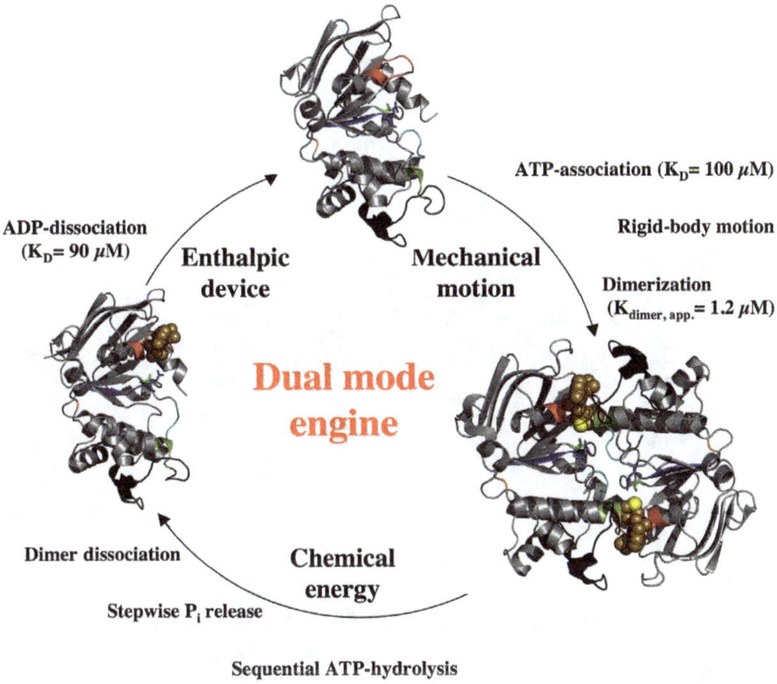

Plate 1.3 Schematic representation of the catalytic cycle of the HlyB-NBD. Colour-coding is identical to Plate 1.2A. The cycle starts with the nucleotide free state of the monomer (pdb entry: 1MT0). Binding of ATP to both monomers induces dimer formation (ATP in brown and Mg^{2+} in yellow; pdb entry: 1XEF). After hydrolysis of ATP the dimer dissociates and ADP is subsequently released (pdb entry: 2FF7). A detailed description of each step can be found in the corresponding section of the text.

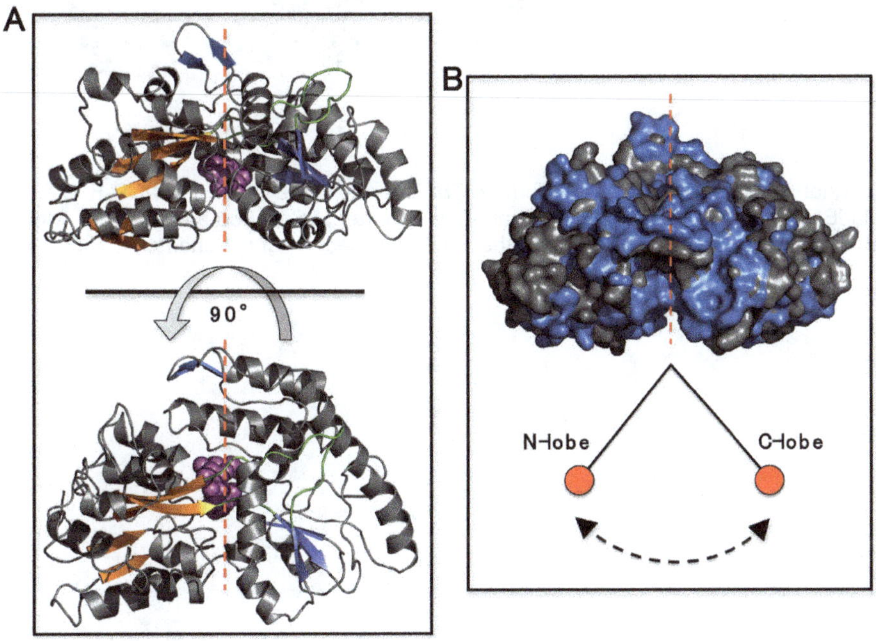

Plate 1.4 **A** In the upper part the side view of the periplasmic binding protein BtuF with bound vitamin B_{12} (magenta) is shown (pdb entry 1N2Z). The five-stranded antiparallel β-sheets are depicted in orange and blue, respectively. N- and C-lobe are connected by loop regions; here shown in green. In the lower part BtuF is rotated about 90° to look at the protein from above. **B** Structural superposition of the ligand-free (grey) and ligand-bound (blue) structure of the maltose binding protein. Upon binding of maltose a domain displacement leading to closure of the lobes of roughly 30° is observed.

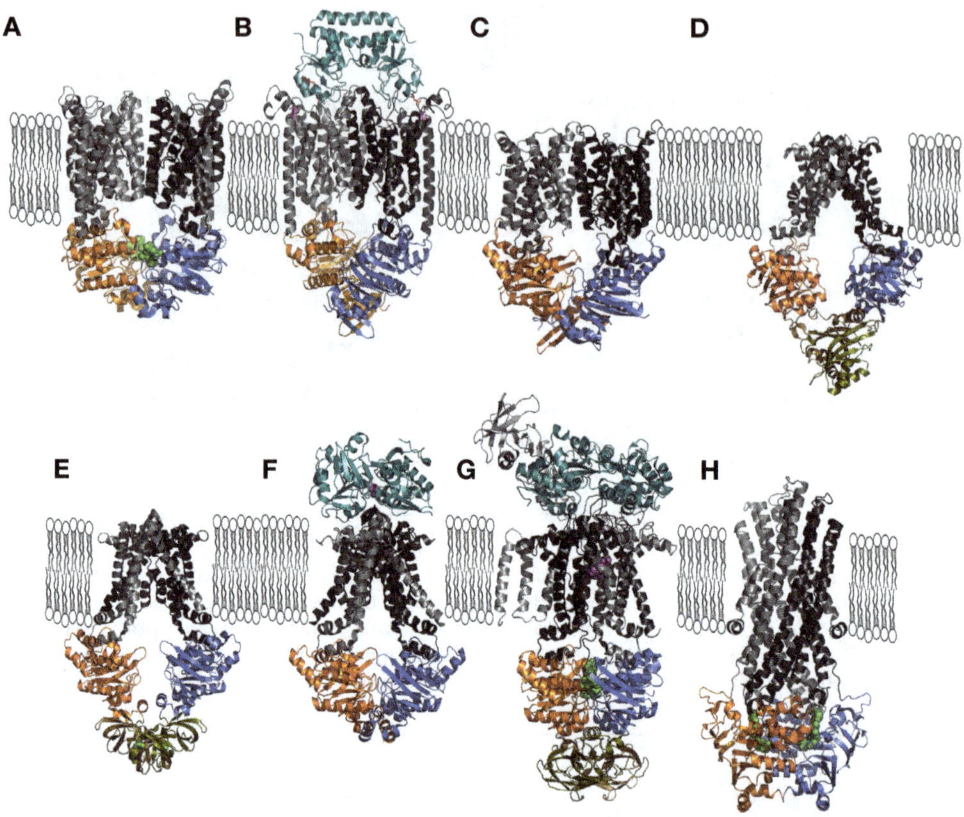

Plate 1.5 Crystal structures of ABC transporters. The importers **A** BtuCD, **B** BtuCD-F, **C** HI1470/1471, **D** MetNI, **E** ModBC, **F** ModBC-A, **G** MalFGK-E, and the exporter **H** Sav1866 are depicted. More information about a particular transporter can be found in the respective section of the chapter.

A C B D

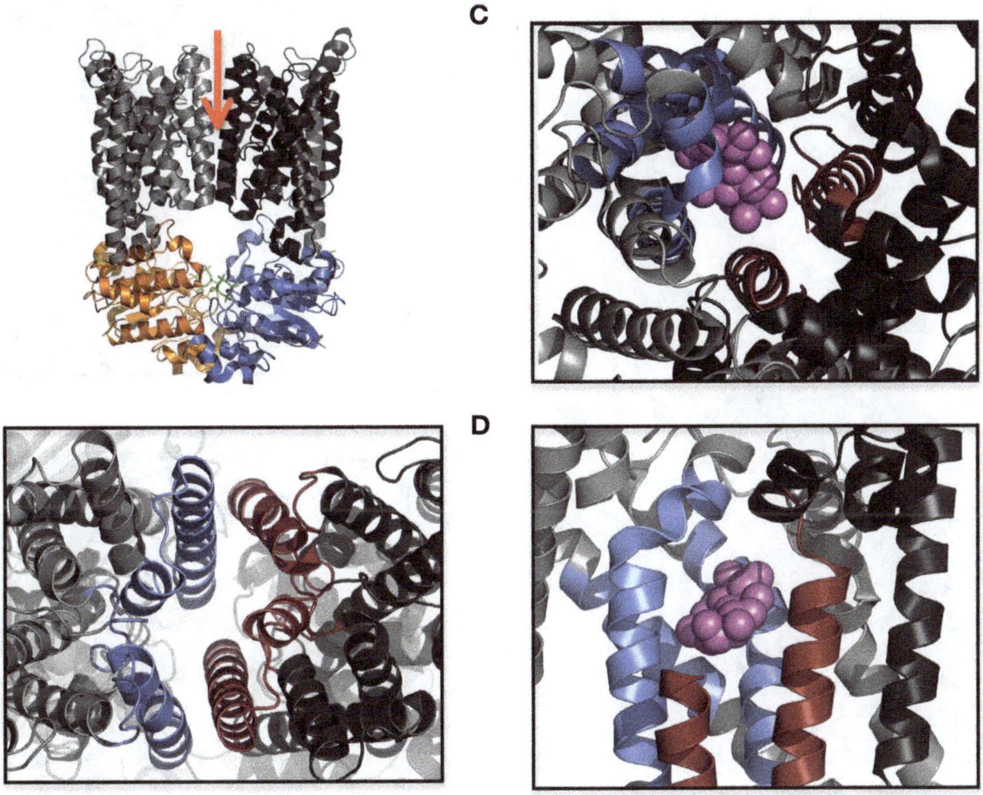

Plate 1.7 A Quaternary structure of the vitamin B_{12} importer BtuCD. **B** BtuCD is viewed from above through the postulated substrate translocation pore. Helices from each TMD subunit lining the translocation pore are coloured red and blue, respectively. **C** View from above into the translocation pore of the MalEFGK$_2$ transporter with bound maltose (magenta). **D** Side view of the maltose binding site in the transmembrane domains of the MalEFGK$_2$ importer complex. For clarity some helices have been omitted.

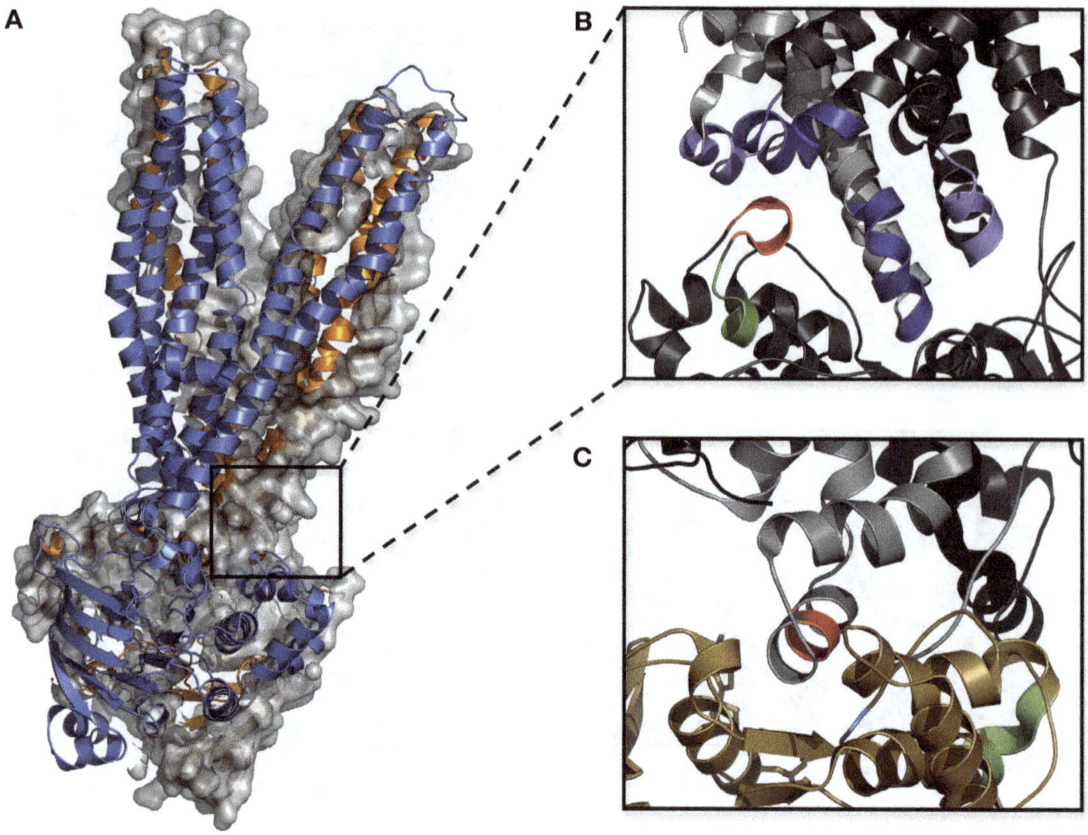

Plate 1.8 A Cartoon representation of the ABC exporter Sav1866. **B** The inset is a magnification of the postulated interaction site between TMDs and NBDs. In blue and slate intracellular loops from both membrane subunits are shown. Residues in red represent the TEVGERG sequence found to interact extensively with the intracellular loops. The C-loop is depicted in pale green. **C** Interaction site between TMD and NBD of the $MalEFGK_2$ complex. The transmembrane part is shown in grey and the NBD in brown. In the L-loop from the transmembrane domain the EAA motif characteristic for ABC importers (red) is highlighted. The C-loop is depicted in pale green and the D-loop in cyan.

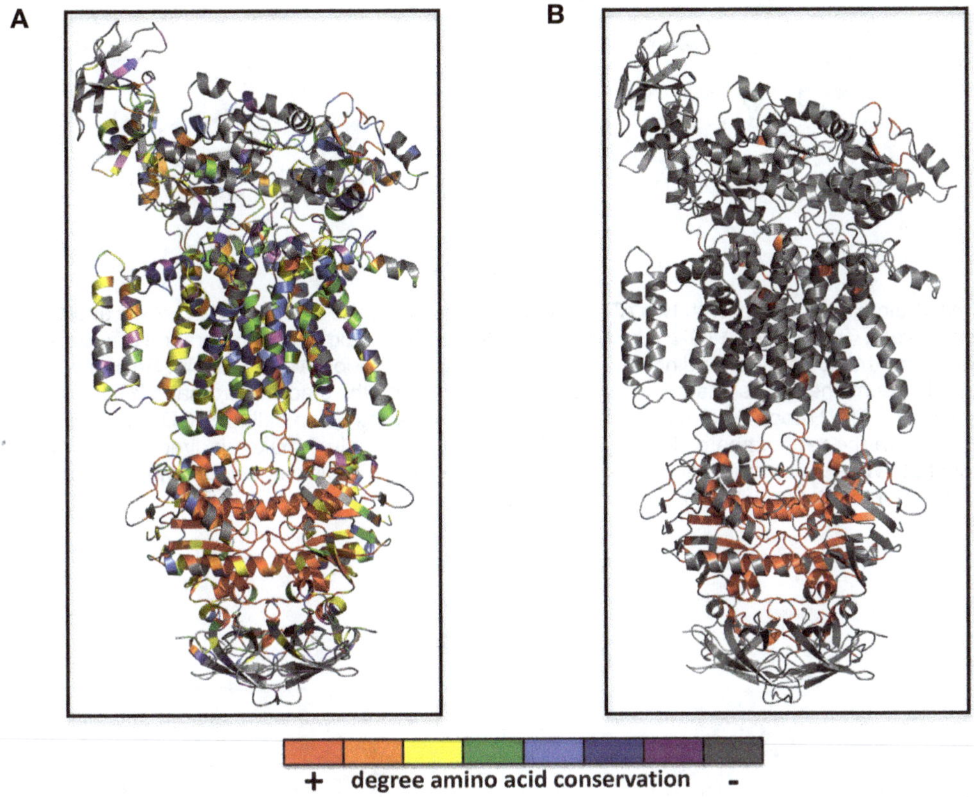

Plate 1.9 A Analysis of conserved amino acids in the MalEFGK$_2$ complex. The amino acid sequence of this importer was compared with 50 other ABC transporters and conserved amino acids are highlighted with the ConSurf Server (http://consurf.tau.ac.il/). **B** Only the highest conserved amino acids are shown in red. As explained in the text, the highest conservation is found in the nucleotide binding domains.

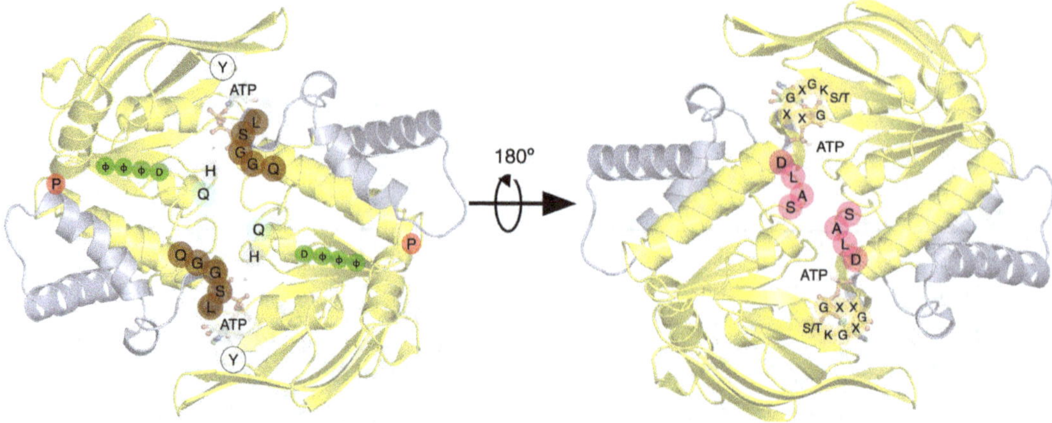

Plate 3.1 NBD dimer in complex with two ATP molecules. The core lobe is rendered in yellow and the small helical lobe in slate. The Walker A (GXXGXGKS/T) and Walker B (ϕϕϕϕD) motifs (orange and dark green, respectively) form the nucleotide-binding fold also observed in the P-loop ATPase family. Residues from the Walker B (dark green), the ABC signature motif (LSGGQ; brown), and the stacking aromatic (Y), form the bulk of the binding site. The H-loop (H grey) histidine and D-loop aspartate (also called SALD-motif; pink) make direct contact with the γ-phosphate. The Q-loop glutamine (Q; light blue) makes indirect contact with the ATP γ-phosphate via a water molecule. The almost invariable P-loop proline is rendered in red. Plates were generated with PyMol (DeLano, 2002).

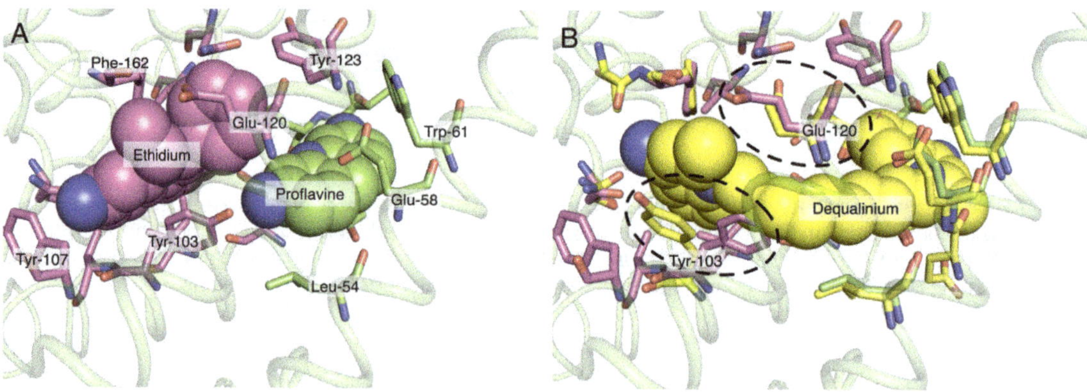

Plate 3.2 A Plasticity in the QacR multidrug binding site. Ethidium and proflavine are bound to QacR simultaneously (pdb code 1QVU). The drugs are stabilised by hydrophobic stacking and polar interactions. **B** Dequalinium occupies a site equivalent to the volume of ethidium and proflavine combined (pdb code 1JT6). Side chains are superimposed for improved clarity and visualization of the structural rearrangements to accommodate the dequalinium moiety. Broken circles indicate residues undergoing rearrangement to accommodate different drugs in the binding site. Plates were generated with PyMol (DeLano, 2002).

www.ingramcontent.com/pod-product-compliance
Lightning Source LLC
LaVergne TN
LVHW062226200926
848484LV00006B/1487
* 9 7 8 1 9 0 4 4 5 5 4 9 3 *